Bacteriocins, Microcins and Lantibiotics

NATO ASI Series

Advanced Science Institutes Series

A series presenting the results of activities sponsored by the NATO Science Committee, which aims at the dissemination of advanced scientific and technological knowledge, with a view to strengthening links between scientific communities.

The Series is published by an international board of publishers in conjunction with the NATO Scientific Affairs Division

A Life Sciences B Physics	Plenum Publishing Corporation London and New York
C Mathematical and Physical Sciences D Behavioural and Social Sciences E Applied Sciences	Kluwer Academic Publishers Dordrecht, Boston and London
F Computer and Systems Sciences G Ecological Sciences H Cell Biology I Global Environmental Change	Springer-Verlag Berlin Heidelberg New York London Paris Tokyo Hong Kong Barcelona Budapest

NATO-PCO DATABASE

The electronic index to the NATO ASI Series provides full bibliographical references (with keywords and/or abstracts) to more than 30000 contributions from international scientists published in all sections of the NATO ASI Series. Access to the NATO-PCO DATABASE compiled by the NATO Publication Coordination Office is possible in two ways:

- via online FILE 128 (NATO-PCO DATABASE) hosted by ESRIN, Via Galileo Galilei, I-00044 Frascati, Italy.

- via CD-ROM "NATO-PCO DATABASE" with user-friendly retrieval software in English, French and German (© WTV GmbH and DATAWARE Technologies Inc. 1989).

The CD-ROM can be ordered through any member of the Board of Publishers or through NATO-PCO, Overijse, Belgium.

Series H: Cell Biology, Vol. 65

Bacteriocins, Microcins and Lantibiotics

Edited by

Richard James

School of Biological Sciences
University of East Anglia
Norwich NR4 7TJ
Norfolk, United Kingdom

Claude Lazdunski

Centre National de la Recherche Scientifique
Centre de Biochimie et de Biologie Moléculaire
31, chemin Joseph Aiguier, B. P. 71
13402 Marseille Cédex 9, France

Franc Pattus

European Molecular Biology Laboratory
Meyerhofstrasse 1, Postfach 10 22 09
6900 Heidelberg, Germany

Springer-Verlag
Berlin Heidelberg New York London Paris Tokyo
Hong Kong Barcelona Budapest
Published in cooperation with NATO Scientific Affairs Division

Proceedings of the NATO Advanced Research Workshop on Bacterial Plasmid-Coded Toxins: Bacteriocins, Microcins and Lantibiotics, held at île de Bendor, France, September 22-26, 1991

ISBN-13: 978-3-642-76976-4 e-ISBN-13: 978-3-642-76974-0
DOI: 10.1007/978-3-642-76974-0

Library of Congress Cataloging-in-Publication Data
Bacteriocins, microcins, and lantibiotics / edited by Richard James, Claude Lazdunski, and Franc Pattus.
(NATO ASI series. Series H, Cell biology ; vol. 65)
"Proceedings of the NATO Advanced Research Workshop on Bacterial Plasmid-Coded Toxins: Bacteriocins, microcins, and lantibiotics, held at île de Bendor, France, September 22-26, 1991" -- Copr.
Includes bibliographical references and index.

1. Bacteriocins--Congresses. I. James, Richard. II. Lazdunski, Claude. III. Pattus, Franc. IV. NATO Advanced Research Workshop on lantibiotics (1991 : île de Bendor, France) V. Series.
QR92.B3B33 1992 589.9'019246--dc20

Softcover reprint of the hardcover 1st edition 1992

Typesetting: Camera ready by authors
31/3145 - 5 4 3 2 1 0 - Printed on acid-free paper

Preface

The EMBO-FEMS-NATO Advanced Research Workshop on Bacteriocins, Microcins and Lantibiotics was held on the Isle de Bendor, France from September 22nd-26th, 1991.

Bacteriocins are protein antibiotics produced by bacteria. They differ from traditional antibiotics because they only kill bacteria which are closely related to the producing strain. The producing strain usually shows immunity towards the bacteriocin which it produces. Bacteriocins are of high molecular weight and are typically plasmid-encoded proteins. The most intensively studied group of bacteriocins, the colicins, are antibiotic proteins produced by some strains of *Escherichia coli* and closely related enterobacteria such as *Shigella sonnei*. Colicins are subdivided into groups based largely upon the receptors to which they bind on the surface of sensitive *E.coli* cells; ie. the E colicins bind to the product of the *btuB* gene of *E.coli*.

Bacteriocins have been subjected to intensive study in the last ten years and it has now become clear that they have a contribution to make in addressing some fundamental problems in modern biology.

1. Being large proteins, how are bacteriocins secreted across both the cytoplasmic and outer membrane of a Gram-ve host cell, and how do they get into and then kill sensitive cells ? A considerable amount is known about the uptake mechanism into sensitive cells but there are considerable unresolved questions about how colicins are secreted from the producing cell.
2. The mechanism of insertion of the pore-forming colicins E1 and A into *E.coli* membranes has generated considerable interest amongst membrane physiologists as well as those who are more interested in these molecules as colicins.
3. What is the mechanism of immunity to a bacteriocin ? The specific nature of the interaction between a colicin and its cognate immunity protein provide a model system for studying protein-protein interactions.
4. Families of E colicins which show close homology have been identified. How have these proteins, and their cognate immunity proteins, evolved ?

These are some of the interesting questions addressed by the later sessions of the workshop in Bendor.

I have worked with colicins for 12 years since being introduced to the subject by Pearl Cooper here in Norwich. In that time I have not been to a conference at which more than a handful of other bacteriocin workers had been present, usually as a peripheral part of the main theme of the conference. I thought for some time as to why there had been so few bacteriocin meetings and concluded that, although there were a large number of interesting problems being studied which involved bacteriocins, no single problem had the critical mass of scientists required to organise and attract funding for a meeting. A related problem to this was that researchers working with colicins did not have much contact with those working with the bacteriocins produced by Gram +ve bacteria. I therefore decided to try and organize a workshop with the aim to encompass the breadth of research problems being addressed using bacteriocins. I wrote to as many people as I knew working in the field and was gratified to receive an enthusiastic response to the idea of a workshop. I was lucky that Claude Lazdunski and Franc Pattus agreed to be co-organisers. They had considerable input into the structure of the workshop programme, the selection of the invited speakers and the general planning of the workshop. Without them the workshop would have been impossible to organise. Not least in importance was their inspired choice of the Isle de Bendor as the workshop location. Everyone who attended the workshop will I am sure have happy memories of Bendor and Provence. Like me, I am sure that many will hope to return.

In order to broaden the scope of the workshop, we also included sessions on microcins and lantibiotics. Microcins are a family of low molecular weight antibiotics (< 10,000 daltons) produced by diverse strains of *Enterobacteriaceae*. Microcins, like other traditional antibiotics, are produced as the cultures enter the stationary phase of growth. They can therefore serve as excellent model systems in which to study stationary phase phenomena. As a bonus, microcins are secreted into the medium by dedicated export mechanisms and are therefore model systems for studying signal-sequence independent protein secretion.

Lantibiotics are distinguished from other bacteriocins of Gram-positive bacteria because of their unique structural features. The term lantibiotic is an abbreviation for **lanthionine-containing peptide antibiotic**. Besides lanthionine and its analogue 3-methyllanthionine, all lantibiotics contain didehydroalanine and/or didehydrobutyrine. There is considerable interest in determining the solution structures of these unusual molecules and the mechanisms by which the unusual amino-acid residues are introduced into lantibiotics. It is of particular interest that one member of the lantibiotics, nisin, is now approved for use as a food preservative. This is perhaps the first example of an **important** commercial use for a member of the bacteriocins, microcins and lantibiotics. Epidermin, another lantibiotic, has potential as a highly specific therapeutic drug against acne. If we consider that the primary role of bacteriocins in nature is as "agents of bacteriological warfare" against other bacteria, then it is possible to envisage bacteriocins, other than nisin, being used as biological control agents. In the event this would dramatically alter the funding prosepects for work with these molecules, and would also ensure that there are many more workshops like Bendor in the future.

The workshop co-organisers invited some 40 participants and then selected 36 others from those who applied in response to advertizements. The participants came from a total of 17 countries, with most of the major research groups in the world being represented. A list of participants is included at the back of this book. I am sure that all participants will agree with me that the quality of the invited lectures was very high, with considerable discussion following each one. The workshop timetable did not try to pack in too many lectures. I believe that this proved to be a suitable format for the exchange of views, techniques and data betwen the participants. The workshop was also enhanced by the large number of very high quality posters on display. It is unfortunate that the rival attractions of the French cuisine, especially at dinner, restricted the time for participants to view the posters. I am therefore glad that some of the poster presentations have been written up as manuscripts and are included in the second half of this book.

The reader will hopefully notice that the contributions to this book have been type-set. I personally do not like to see the array of typefaces and formats which appear in the typical Proceedings of a workshop. I therefore decided to use an Apple Mac desk top publishing system, available in Norwich, to impose a common format and typeface on the contributions. With the benefit of hindsight, considering the amount of work involved, I would probably not do this again unless the authors provided their contributions in a suitable electronic fomat for direct input into the desk top publishing system. However, I hope that both contributors and readers appreciate the efort involved and agree that it improves the appearance and readibility of the book. I hope that the publication of this volume will remind those who were there what a marvellous workshop it was, and will perhaps persuade someone that it would be useful to organize a follow up meeting in two years time. In addition I hope that it will attract new workers to this wonderful research field.

Lastly I should like on behalf of my co-organisers to thank Deborah Clemitshaw the ARW secretary. Without her considerable organisational and secretarial skills, both in Norwich and in Bendor, the planning and smooth running of this Workshop, as well as the production of this book, would have been made impossible.

Richard James
Norwich

Acknowledgements

In addition to the major financial support from EMBO, FEMS and NATO for this ARW, we are also grateful to the CNRS, France for providing financial support for the travel expenses of French-based speakers.

Contents

INTRODUCTION TO THE MICROCIN SESSION

Roberto Kolter[1] and Felipe Moreno[2]

The microcins constitute a family of low-molecular-weight antibiotic substances produced by diverse strains of *Enterobacteriaceae*. Clinical isolates of *Enterobacteriaceae* have been screened for the production substances that can pass through cellophane membranes with a molecular weight cutoff of about 10,000 daltons and inhibit the growth of lawns of standard laboratory strains of *Escherichia coli*. Using this as an operational definition, almost twenty different microcins have been identified to date.

Like most antibiotics, microcins are synthesized as cultures enter stationary phase. Since microcins are produced by *Escherichia coli* their synthesis can be studied using the powerful approaches afforded by that organism's well developed genetics. Therefore the microcins can serve as excellent model systems in which to study broad questions regarding stationary phase phenomena as well as the specifics of antibiotic production. In addition, since microcins are secreted into the medium by dedicated export mechanisms they can also serve as model systems in which to study signal sequence-independent secretion.

The best studied microcins are known as B17 and C7. Microcin B17 is a 43 amino acid peptide which inhibits DNA gyrase while microcin C7 contains seven amino acids and kills cells by inhibiting protein synthesis. From genetic and biochemical studies of these molecules much of the overall biology of these substances is known. The mechanisms underlying the production and the mode of action of these microcins is now understood in great detail.

The synthesis of microcin B17 requires the action of seven plasmid borne genes whose function is clear. Aside from the structural, three genes are required for post-translational modification of the precursorpeptide. Two genes are required for the export of microcin B17 from the cell and one gene is required for immunity against the action of the peptide. In addition, at least five chromosomal genes are required for the normal synthesis of B17: *ompR*, *himA*, *himB* and *mprA* which code for transcriptional regulators and *pmbA*, whose product appears to cleave the N-terminal peptide to generate mature microcin B17.

The structure of microcin B17 is quite unusual. Sixty-percent of the amino acid residues are glycine. Of the remaining residues, four of six serines and four of four cysteines are post-translationally modified. Although this combination of modified serines and cysteines could suggest the presence of lanthionine in

[1] Department of Microbiology and Molecular Genetics, Harvard Medical School, 200 Longwood Avenue, Boston, Ma 02115, U.S.A. [2] Unidad de Genética Molecular, Hospital Ramón y Cajal, Carretera de Colmenar, km 9,100, 28034 Madrid, Spain

NATO ASI Series, Vol. H 65
Bacteriocins, Microcins and Lantibiotics
Edited by R. James, C. Lazdunski and F. Pattus

microcin B17, the evidence presented indicates that the post-translational modifications do not result in lanthionine. While the exact structure of the modifications has not been elucidated, indications are that four heteroatomic chromophores are present in mature microcin B17.

ESCHERICHIA COLI GENES REGULATING THE PRODUCTION OF MICROCINS MCCB17 AND MCCC7

F. Moreno, J. L. San Millán, I. del Castillo, J. M. Gómez, M. C. Rodríguez-Sáinz, J. E. González-Pastor and L. Díaz-Guerra.
Unidad de Genética Molecular
Hospital Ramón y Cajal
Carretera de Colmenar, km 9,100
28034 Madrid
Spain

Microcins are a family of low-molecular-weight antibiotics produced by diverse strains of enteric bacteria. These antibiotics are active on many different species of the *Enterobacteriaceae* family (Baquero & Moreno, 1984). Microcin B17 (MccB17) is a 3,250 Da peptide that inhibits DNA replication by blocking DNA gyrase (Davagnino *et al.*, 1986; Herrero & Moreno, 1986; Vizán *et al.*, 1991). As a consequence, SOS repair functions are induced, and finally DNA is completely degraded (Herrero & Moreno, 1986; Mayo *et al.*, 1988).

Microcin C7 (MccC7) is a 1,000 Da peptide that inhibits protein synthesis *in vivo* (García-Bustos *et al.*, 1984; García-Bustos *et al.*, 1985) and *in vitro* (J.L. San Millán, unpublished results). Wild-type strains producing microcins are specifically resistant/immune to the microcin they produce. (Baquero & Moreno, 1984).

The MccB17 and MccC7 genetic determinants for production and immunity were cloned from plasmids pMccB17 and pMccC7, respectively (San Millán *et al.*, 1985a; Novoa *et al.*, 1986). Figure 1 summarizes our current knowledge of the plasmid genetic systems encoding these antibiotics. The seven *mcb* (MccB17) genes have been sequenced, and their products (except McbE) have been visualized in SDS-polyacrylamide gels (San Millán *et al.*, 1985b; Davagnino *et al.*, 1986; Garrido *et al.*, 1988; Genilloud *et al.*, 1989). Five regions, each containing one or more *mcc* genes (MccC7), have been identified within the MccC7 plasmid system (Novoa *et al.*, 1986; our unpublished results). The sequencing of α and the upstream region has revealed the existence of a twenty-one nucleotide stretch of DNA which probably codes for the seven amino acids present in the mature microcin (Díaz-Guerra *et al.*, submitted). The other MccC7 regions are being sequenced.

The transcription of microcin genes depends on the growth phase

Both microcins, MccB17 and MccC7, are produced when cells cease exponential growth and enter

NATO ASI Series, Vol. H 65
Bacteriocins, Microcins and Lantibiotics
Edited by R. James, C. Lazdunski and F. Pattus

the stationary phase. To understand the molecular basis of this growth-phase dependent regulation, the *lac* operon and the *lacZ* gene were fused to different microcin genes and the β-galactosidase activity synthesized from these fusions was determined at different times of growth, in different genetic constructs, and in different chromosomal backgrounds. By Sl nuclease protection and primer extension experiments transcription start sites were determined and putative promoters identified. (Connell *et al.*, 1987; Genilloud *et al.*, 1989). From these studies emerged the transcriptional patterns shown in Figure 1. All *mcb* genes are transcribed in the same direction from the promoter P_{mcb}, that is located to the left of the *mcbA* gene. *mcbD*, and presumably the genes downstream, are also transcribed from the promoter P2 that is located within *mcbC*. Yet most of the transcription (80-90%) of these distal genes is driven from P_{mcb} (Hernández-Chico *et al.*, 1986; Genilloud *et al.*, 1989).

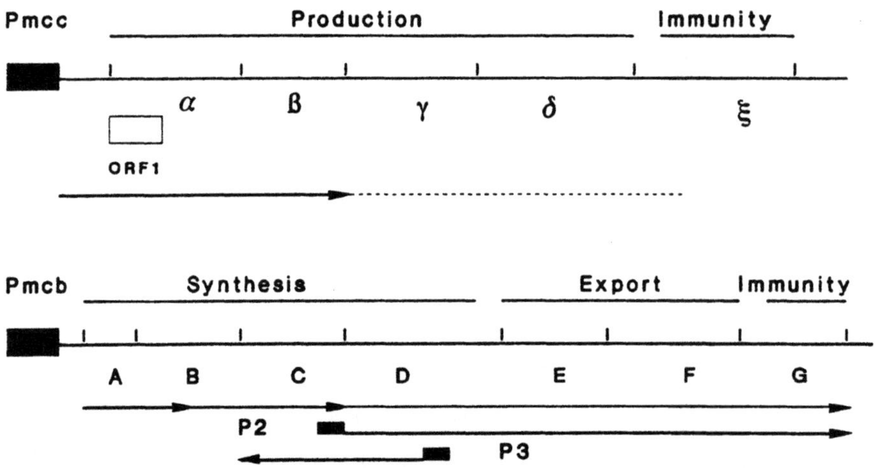

Figure 1. Structure of the genetic systems MccC7 and MccB17. Upper part: MccC7. ORFl indicates the open-reading-frame that probably encodes the primary structure of MccC7. The β and δ regions, when cloned separately, confer partial resistance to exogenous MccC7 (Novoa *et al.*, 1986). The ξ region codes for immunity (J. L. San Millán, unpublished results). Lower part: MccB17. P3 promoter directs transcription of an untranslated mRNA (Hernández-Chico *et al.*, 1986; Genilloud *et al.* 1989). P_{mcc} and P_{mcb} are the growth-phase regulated promoters. Arrows indicate the direction of transcription. Each system includes about 5 kilobases. For further information, see the text.

Transcription of the proximal genes *(mcbA-mcbC)* greatly increases at the approach of the stationary phase, whereas the transcription of the other *mcb* genes is kept at the same level as in the exponential phase (Hernández-Chico *et al.*, 1986; Genilloud *et al.*, 1989). Other data concerning *mcb* transcription are shown in Figure 1.

Similarly, all *mcc* genes are transcribed in the same direction (Novoa *et al.*, 1986). The promoter P_{mcc} directs transcription of the proximal genes (those included in the α region, and presumably those in the β region). This transcription, as that originated in P_{mcb}, is activated by the cessation of cell growth (Díaz-Guerra *et al.*, 1989). Whether the downstream genes are transcribed from P_{mcc} is unknown.

Thus the production of microcins seems to be dependent on the transcription of microcin-specific genes from promoters that become activated when cells cease exponential growth due to nutrient deprivation. In fact, transcripts from P_{mcb} and P_{mcc} were only detected in stationary cells.

To further study the regulation of the P_{mcb} promoter, the *mcbA-lacZ* gene fusion described by Connell *et al.* (1987) was used and, occasionally, the *mcbB-lac* operon fusion described by Hernández-Chico *et al.* (1986). The regulation of P_{mcc} was studied using the *mcc α-lacZ* gene fusion described by Díaz-Guerra *et al.* (1989). The three fusions were present at a single copy per cell.

As judged from the activity of these fusions, P_{mcb} and P_{mcc} are regulated not only by the growth-phase, but also by the growth rate and by the nature of the growth media. An inverse correlation between the growth rate and the activity of both promoters was observed when cells were grown in minimal media supplemented with different carbon sources (Connell *et al.*, 1987; Díaz-Guerra *et al.*, submitted). The influence of the growth media on the activity of these promoters is illustrated by the following. P_{mcc} is very poorly active in both LB and M63 media (5-10 β-galactosidase units)[1] during exponential growth, but it directs 10 fold more transcription in LB than in M63 (2,000 units versus 200 units) during the stationary phase (Díaz-Guerra *et al.*, submitted). On the contrary, P_{mcb} is similarly active in these media during stationary growth (around 800-1,000 units), but its activity is much lower in LB (10-20 units) than in M63 (150-200 units) during exponential growth (Connell *et al.*, 1987; our unpublished data). These results indicate that these promoters are fine-tuned by different physiological factors or conditions that merit further exploration.

E. coli chromosome and microcin production

By different experimental approaches we have identified several genes of *E. coli* affecting the production of one or more microcins. The characterization of mutants impaired in MccB17 production led to the identification of three of these genes. Two of them (*hisT* and *ompR*) were previously known, but not the third one (*pmbA*). The *hisT*::Tn*5* insertion we isolated provoked a moderate reduction in the production of microcins B17, C7, H47, and ColV. It is known that the absence of HisT produces a reduction of the

[1] Units are as defined by Miller (1972)

translation elongation rate (Palmer *et al.*, 1983). Most probably this is the cause of the impairment in microcin production (Rodríguez-Sáinz *et al.*, 1991).

The *pmbA* gene, which maps at min 96 of the *E. coli* map, was cloned and sequenced. It encodes a polypeptide of 50 kDa. Cell-fractionation experiments indicated that this protein is cytoplasmic, but it may be loosely associated with the inner membrane. From the study of the behaviour of $PmbA^-$ cells harbouring microcin-producing plasmids, and from other studies, it was proposed that these mutant cells synthesize a MccB17-like molecule (pro-MccB17) able to inhibit DNA replication, but unable to be exported from the cytoplasm. PmbA should facilitate the antibiotic export by completing its maturation (Rodríguez-Sáinz *et al.*, 1990). This maturation could consist of the processing of the 26 amino acid sequence that is present in the amino end of the McbA precursor, but absent in the mature microcin (Rodríguez-Sáinz *et al.*, 1990).

Genes regulating P_{mcb} and P_{mcc} promoters

P_{mcb} activators

Mutations drastically reducing MccB17 production were mapped to the *ompR* gene (Hernández-Chico *et al.*, 1982). This gene encodes a 27 kDa transcriptional activator of genes coding for the major outer membrane porins OmpC and OmpF (Hall & Silhavy, 1981). OmpR also regulates transcription from the major promotor P_{mcb} (Hernández-Chico *et al.*, 1986).

As judged from the β-galactosidase activity synthesized from *mcbA-lac* and *mcb-lac* operon fusions, the absence of OmpR produced a decrease of 8-10 times in the P_{mcb} activity in stationary cells (Hernández-Chico *et al.*, 1986). In these experiments, which were performed in minimal M63 medium, the typical growth-phase-dependent induction could not be detected in mutant cells, and we concluded that OmpR was required for the stationary phase activation. However, when the experiments were performed in LB medium, a clear growth-phase induction was also observed in mutant cells, indicating that the growth-phase dependent activation of P_{mcb} is OmpR-independent (Connell *et al.*, 1987). It is now clear that the "excessive" basal expression of P_{mcb} in M63 medium masked the actual induction and led us to a wrong conclusion. The OmpR-independence of the growth-phase-dependent induction was further demonstrated by deleting the DNA sequence located upstream of the nucleotide at position -54 in the promoter region (the start transcription site being at position +l) (Bohannon *et al.*, 1991). This deletion caused an important decrease in the activity of the *mcbA-lacZ* fusion in exponential and stationary cells, but the promoter retained the ability to be induced at the beginning of the stationary phase with a similar efficiency in mutant and wild-type cells. A corollary of these results is that the region upstream of -54 is essential for the OmpR activation, and that the downstream region is sufficient for the promoter to display stationary-phase induction (Bohannon *et al.*, 1991).

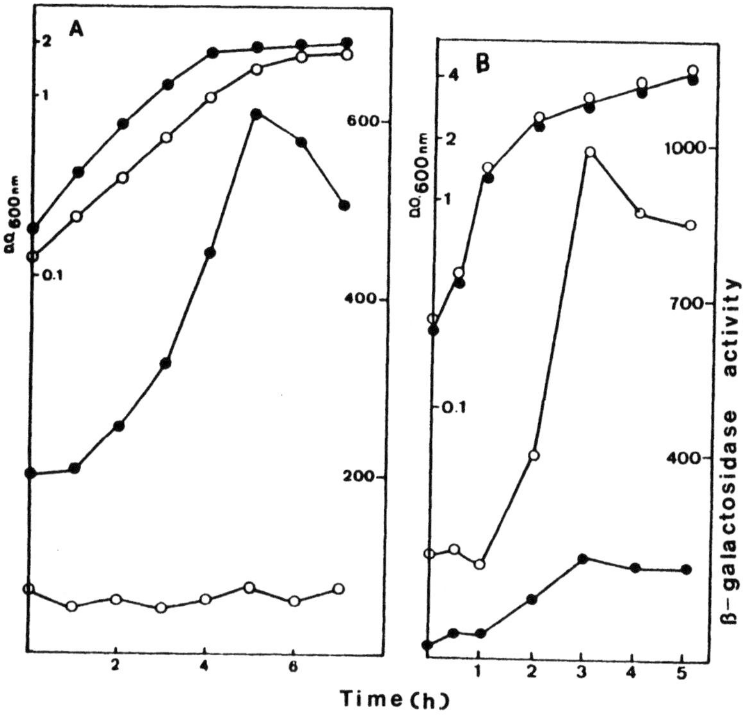

Figure 2. A: Effect of the *himA*Δ*82* mutation on the activity of the *mcbA-lacZ* fusion. Cells were grown in a M63 glucose medium at 30°C.
B: Effect of the *mprA560* mutation on the activity of the *mcbB-lacZ* operon fusion. Cells were grown in a LB medium at 37°C. *mprA560 is* a null insertion/deletion mutation (del Castillo *et al.*, 1991). Symbols: ● wild-type strain; O mutant strain.

Another activator of transcription from P_{mcb} is the integration host factor (IHF). IHF is a heterodimer made up of two subunits, A and B, which are the products of the genes *himA* and *himD*, respectively (Friedman, 1988). IHF binds to specific sites on DNA, producing a characteristic DNA bending. In this way, IHF may participate in site-specific recombination, transposition, and transcription. Indeed, this histone-like protein exerts a direct positive or negative role in the expression of a number of genes (Drlica & Rouviere-Yaniv, 1987; Friedman, 1988). As shown in Figure 2A the activity of the *mcbA-lacZ* gene fusion was decreased 10-fold in stationary cells when a null *himA* mutant was introduced into the wild-type cells harbouring the fusion. Furthermore, no microcin activity was detected when *himA* and *himD* mutants harbouring pMccB17 were checked for microcin production on plates. It should be noted that the expression of the porin *ompC* gene, which shares some regulatory features with P_{mcb} is inhibited by IHF (Hernández-Chico *et al.*, 1986; Huang *et al.*, 1990).

P_{mcc} activators

Transcription from P_{mcc} does not appear to be activated by OmpR and IHF but by two other chromosomal gene products, those encoded by the genes *crp* and *appR*. The cAMP receptor protein (CRP) is known to regulate the transcription of a number of genes. The major role of CRP is to activate transcription of catabolite sensitive promoters, but in several cases this protein represses transcription (Magasanik & Neidhardt, 1987). The expression of the *mcc* α*-lacZ* fusion was 200-fold lower in stationary cells carrying *crp* or *cyaA* deletions, and both mutant strains were unable to produce MccC7. The fusion activity and microcin production were recovered when 5mM cAMP was added to the *cya* strain, but not to the *crp* strain (Díaz-Guerra *et al.*, submitted). Together these results indicated that P_{mcc} is activated by CRP and subjected to catabolite repression.

The AppR product, also called KatF, activates, directly or indirectly, the expression of many genes and functions known to be induced in the transition to the stationary phase (Lange & Hengge-Aronis, 1991a). Specifically, it activates the transcription of the genes *appA* (Touati *et al.*, 1986; Díaz-Guerra *et al.*, 1989), *katE* (Mulvey *et al.*, 1990), *xthA* (Sak *et al.*, 1989), and *bolA* (Bohannon *et al.* 1991; Lange & Hengge-Aronis, 1991b). On the basis of the close sequence relationship between KatF and known σ factors (Mulvey & Lowen, 1989), KatF has been suggested to be a RNA polymerase σ subunit required to express genes belonging to a large, starvation/stationary phase *E. coli* regulon (Lange & Hengge-Aronis, 1991a). As indicated above, the *mcc* α*-lacZ* gene fusion, which is poorly expressed during exponential phase, is strikingly induced when cells enter the stationary phase (induction rate: 20-40 in M63 glucose, and 200-400 in LB). In *appR/katF* mutants the expression in the stationary phase was decreased 10-fold but the activation induced by the phase-transition was maintained (Díaz-Guerra *et al.*, 1989; see also Figure 3). This indicates that the growth-phase dependent induction of P_{mcc} does not depend on *appR*. Rather AppR ensures full expression of P_{mcc} as OmpR does for P_{mcb} expression.

P_{mcb} and P_{mcc} inhibitors

Two chromosomal genes regulate P_{mcb} negatively. The first one, *mprA*, was identified by its ability to suppress, when in high copy number, the deletereous effects exhibited by microcin-producing cells lacking the immunity gene *mcbG*. The molecular basis of the suppression was shown to be due to a decrease in MccB17 production. The gene was sequenced and shown to encode a polypeptide of 176 residues, with an apparent molecular weight of 19 kDa (del Castillo *et al.*, 1991). Null mutants were constructed in vitro and then substituted for *mprA*$^+$ on the bacterial chromosome to examine their effect on microcin production and *mcb* gene expression. As expected these mutations derepressed the expression of the *mcbA-lacZ* fusion (del Castillo *et al.*, 1991) and that of the *mcbB-lac* operon fusion (Figure 2B), and mutant cells produced more microcin. It should be added that *mprA*, when in high copy number, caused a strong repression of MccC7 production and transcription of the *mccα-lacZ* gene fusion, and blocked the osmoinduction of the *proU* locus in high osmolarity media (del Castillo *et al.*, 1990). Surprisingly, neither the osmoinduction nor the *mccα* expression was significantly affected in MprA⁻ mutants (del Castillo *et al.*, 1991).

The second gene exerting a negative control on the P_{mcb} system was found by mutagenizing the strain carrying the *mcbA-lacZ* fusion with Tn*10*. A dark blue colony was isolated from LB plates supplemented with the X-gal indicator. The Tn*10* insert was mapped to locus *bglY*, and shown by DNA sequencing to interrupt the gene (J. M. Gómez *et al.*, in preparation). *bglY*, also called *osmZ*, encodes the histone-like protein HI, which is a major component of the bacterial nucleoid. HI regulates the expression of bacterial genes involved in many different functions *(bgl* operon, osmoregulation, thermoregulation, virulence, frequency of chromosomal deletions, etc., reviewed by Higgins *et al.*, 1990).

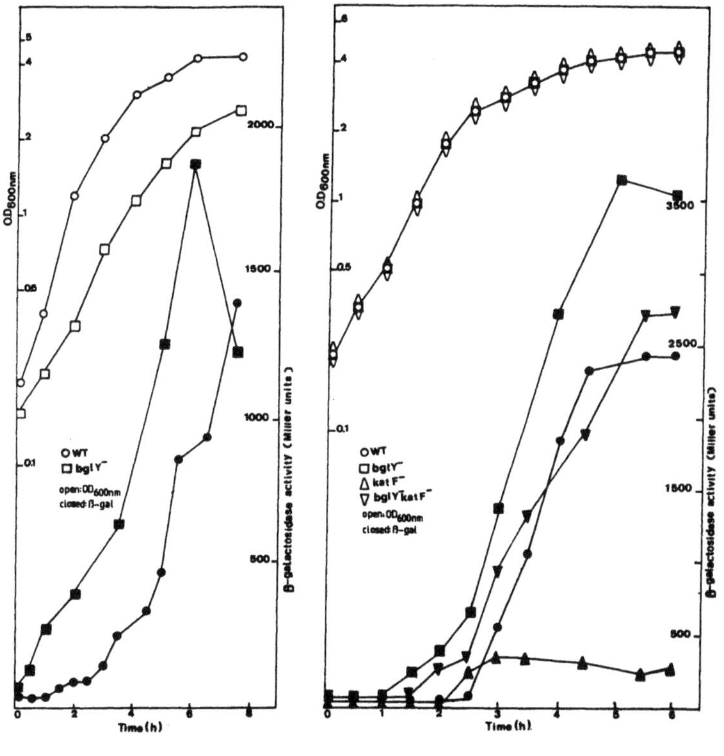

Figure 3. Effect of the *bglY*::Tn*10* mutation on the activity of the *mcbA-lacZ* and the *mccα-lacZ* gene fusions. Left panel: *mcb-lacZ*. Cells were grown in LB medium at 30°C. Right panel: *mcc-lacZ*. Cells were grown in LB at 37°C. The *katF* mutation is the deletional insertion mutation *katF*::Km^r described by Bohannon *et al.* (1991).

The *bglY*::Tn*10* mutation provoked a high constitutive expression of the osmoregulated *proU* locus, and allowed mutant cells to grow in minimal medium containing the β-glucoside salicin as a unique carbon source. As shown in Figure 3, the expression of the *mcbA-lacZ* and *mccα-lacZ* fusions was highly derepressed in the mutant. In both cases the growth phase induction was premature, as if HI were blocking the activity of the promoters without affecting their intrinsic ability to be induced by the transition to the stationary phase. Similarly, the inhibitor effect mediated by MprA on these promoters was independent of the growth-phase (del Castillo *et al.*, 1990).

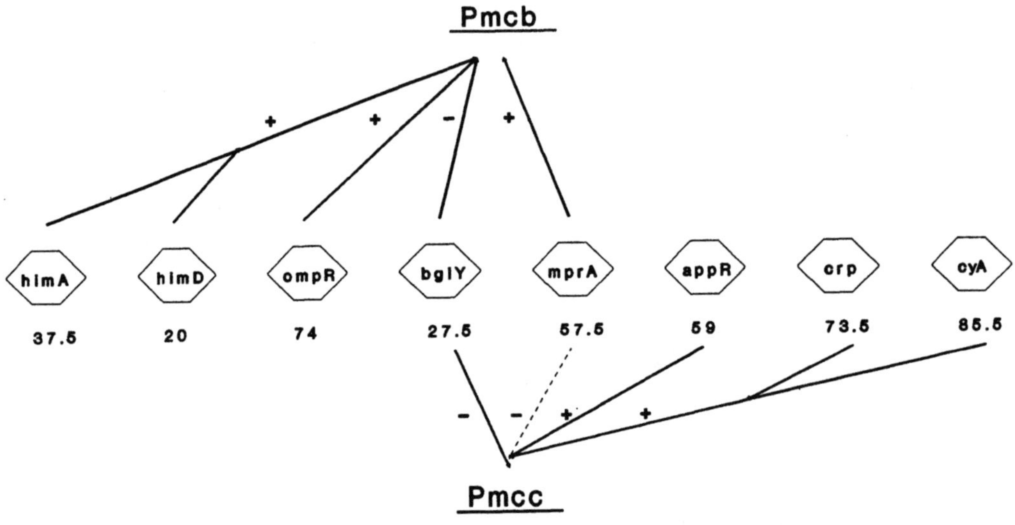

Figure 4. *Escherichia coli* chromosomal genes regulating P_{mcb} and P_{mcc}. (+) indicates activation of promoter; (-) inhibition of promoter. The dashed line indicates that the P_{mcc} is clearly inhibited only in the presence of many copies of *mprA*. Numbers under the genes indicate their location on the genetic map.

Concluding remarks

Figure 4 summarizes our knowledge about the regulation of microcin promoters by chromosomal genes. Each promoter is activated by a different pair of proteins, and both are inhibited to a different degree by another pair of proteins. These activators and inhibitors affect the expression of many other genes and functions that are induced under stress conditions. Expression of microcin operons has been formerly conceived as a mere activation induced by the transition to the stationary phase. This view seems to be incomplete at present. In fact this kind of regulation may also be formulated as repression of microcin operons during exponential growth and derepression at the beginning of stationary growth. A finding illustrating this alternative view is in Figure 3, where it is shown that Kat F is not required to fully express P_{mcc} during the stationary phase when HI is absent. On the other hand, this result raises the question whether or not Kat F is a stationary-phase-specific σ factor as suggested by others (Lange & Hengge-Aronis, 1991a).

We believe that the interaction of "inhibitor" proteins with the region upstream of each microcin promoter ensures inactivity of these promoters during vegetative exponential growth. During the transition to the stationary phase, the activators would occupy these interaction sites displacing the inhibitors. This could create local changes in DNA conformation favouring the binding of RNA polymerase to the promoters. Indeed short DNA sequences matching the consensus for OmpR and IHF binding-sites are present upstream of P_{mcb} and DNA sequences homologous to CRP-binding sites are present upstream of P_{mcc} The extent of DNA supercoiling, which increases when cells enter in the stationary phase (Dorman *et al.*, 1988), may modulate the proposed removal of the inhibitors by the activators.

Acknowledgements

This research has been supported by the Fondo de Investigaciones Sanitarias and the Comisión Interministerial de Ciencia y Technologia of Spain.

References

Baquero F, Moreno F (1984) The microcins. FEMS Microbiol Lett 23:117-124

Bohannon DE, Connell N, Keener J, Tormo A, Espinosa-Urgel M, Zambrano MM, Kolter R (1991) Stationary-phase-inducible "gearbox" promoters: differential effects of the *katF* mutations and role of σ^{70}. J Bacteriol 173:4482-4492

Connell N, Han Z, Moreno F, Kolter R (1987) An *E. coli* promoter induced by the cessation of growth. Mol Microbiol 1:195-201

Davagnino J, Herrero M, Furlong D, Moreno F, Kolter R (1986) The DNA replication inhibitor microcin B17 is a forty-three-amino acid protein containing sixty percent glycine. Proteins 1:230-238

del Castillo I, Gómez JM, Moreno F (1990) *mprA*, an *Escherichia coli* gene that reduces growth-phase-dependent synthesis of microcins B17 and C7 and blocks osmoinduction of *proU* when cloned on a high-copy number plasmid. J Bacteriol 172:437-445

del Castillo I, González-Pastor JE, San Millán JL, Moreno F (1991) Nucleotide sequence of the *Escherichia coli* regulatory gene *mprA*-deficient mutants. J Bacteriol 173:3942-3929

Díaz-Guerra L,Moreno F, San Millán JL (1989) *app*R gene product activates transcription of microcin C7 plasmid genes. J Bacteriol 171:2906-2908

Dorman CJ, Barr GC, Nibhriain N, Higgins CF (1988) DNA supercoiling and the anaerobic and growth phase regulation of *tonB* gene expression. J Bacteriol 170:2816-2826

Drlica K, Rouviere-Yaniv J (1987) Histone-like proteins of bacteria. Microbiol Rev 51:301-319

Friedman DI (1988) Integration host factor: a protein for all reasons. Cell 51:545-554

García-Bustos JF, Pezzi N, Asensio C (1984) Microcin 7: purification and properties. Biochem Biophys Res Commun 119:779-785

García-Bustos JF, Pezzi N, Méndez E (1985) Structure and mode of action of microcin 7, an antibacterial peptide produced by *Escherichia coli.* Antimicrob Agents Chemother 27:791-797

Garrido MC, Herrero M, Kolter R, Moreno F (1988) The export of the DNA replication inhibitor microcin B17 provides immunity for the host cell. EMBO J 7:1853-1862

Genilloud O, Moreno F, Kolter R (1989) DNA sequence products, and transcriptional pattern of the genes involved in production of the DNA replication inhibitor microcin B17. J Bacteriol 171:1126-1135

Hall MN, Silhavy TJ (1981) Genetic analysis of the *ompB* locus in *Escherichia* K12. J Mol Biol 151:1-15

Hernández-Chico C, Herrero M, Rejas M, San Millán JL, Moreno F (1982) Gene *ompR* and regulation of microcin B17 and colicin E2 synthesis. J Bacteriol 152:897-900

Hernández-Chico C, San Millán JL, Kolter R, Moreno F (1986) Growth phase and OmpR regulation of transcription of microcin B17 genes. J Bacteriol 167:1058-1065

Herrero M, Moreno F (1986) Microcin B17 blocks DNA replication and induces the SOS system in *Escherichia coli.* J Gen Microbiol 132:393-402

Higgins CF, Hinton JCD, Hulton CSF, Owen-Hughes T, Pavitt GD, Seirafi A (1990) Protein HI: a role for chromatine structure in the regulation of bacterial gene expresion and virulence? Mol Microbiol 4:2007-2012

Huang L, Tsui P, Freundlich M (1990) Integration host factor is a negative effector of *in vivo* and *in vitro* expression of *ompC* in *Escherichia coli.* J Bacteriol 172:5293-5298

Lange R, Hengge-Aronis R (1991a) Identification of a central regulator of stationary phase gene expression in *Escherichia coli.* Mol Microbiol 5:49-59

Lange R, Hengge-Aronis R (1991b) Growth phase-regulated expression of *bolA* and morphology of stationary-phase *Escherichia coli* cells are controlled by the novel sigma factor σ^s. J Bacteriol 173:4474-4481

Magasanik B, Neidhardt FC (1987) Regulation of carbon and nitrogen utilization In: J. L. Ingrahan, K. B. Low, B. Magasanik, M. Schaechter, H. E. Umbarger, and F. C. Neidhardt (eds) *Escherichia coli* and *Salmonella typhimurium:* cellular and molecular biology. American Society for Microbiology, Washington, D, p1318-1324

Mayo O, Hernández-Chico C, Moreno F (1988) Microcin B17, a novel tool for preparation of maxicells: identification of polypeptides encoded by the IncFII minireplicon pMccB17. J.Bacteriol 170:2414-2417

Miller JH (1972) Experiments in molecular genetics. Cold Spring Harbour Laboratory, Cold Spring Harbour, New York

Mulvey MR, Loewen PC (1989) Nucleotide sequence of *katF* of *Escherchia coli* suggests KatF protein is a novel σ transcription factor. Nucleic Acids Res 17:9979-9991

Mulvey MR, Switala J, Borys A, Loewen PC (1990) Regulation of transcription of KatE and KatF in *Escherichia coli.* J Bacteriol 153:357-363

Novoa MA, Díaz-Guerra L, San Millán JL, Moreno F 1986) Cloning and mapping of the genetic determinants for microcin C7 production and immunity. J Bacteriol 168:1384-1391

Palmer DT, Blum PH, Artz SW (1983) Effects of the *hisT* mutation of *Salmonella thyphimurium* on translation elongation rate. J Bacteriol 153:357-363

Rodríguez-Sáinz MC, Hernández-Chico C, Moreno F (1990) Molecular characterization of *pmbA*, an *Escherichia coli* chromosomal gene required for the production of the antibiotic peptide MccB17. Mol Microbiol 4:1921-1932

Rodríguez-Sáinz MC, Hernández-Chico C, Moreno F (1991) A *hisT*::Tn*5* insertion affects production of microcins B17, C7, and H47 and colicin V. J Bacteriol 173:7018-7020

Sak BD, Eisenstark A, Touati D (1989) Exonuclease III and the catalase hydroperoxidase II in *Escherichia coli* are both regulated by the *katF* gene product. Proc Nat Acad Sci USA 86:3271-3275

San Millán JL, Hernández-Chico C, Pereda P, Moreno F (1985a) Cloning and mapping of the genetic determinants for microcin B17 production and immunity. J Bacteriol 163:275-281

San Millán JL, Kolter R, Moreno F (1985b) Plasmid genes required for microcin B17 production. J Bacteriol 163:1016-1020

Touati E, Dassa E, Boquet PL (1986) Pleiotropic mutations in *appR* reduce pH 2.5 acid phophatase expresion and restore succinate utilization in CRP-deficient strains of *Escherichia coli*. Molec Gen Genet 202:257-264

Vizan JL, Hernandez-Chico C, del Castillo I, Moreno F (1991) The peptide antibiotic microcin B17 induces double strand cleavage of DNA mediated by gyrase. EMBO J 10:467-476

UPTAKE AND MODE OF ACTION OF THE PEPTIDE ANTIBIOTIC MICROCIN B17

C. Hernández-Chico, O. Mayo, J. L. Vizán, M. Laviña[1], and F. Moreno.
Unidad de Genética Molecular
Hospital Ramón y Cajal
Carretera de Colmenar, km 9,100
28034 Madrid
SPAIN

Microcin B17 is an antibiotic peptide produced by wild-type *Escherichia coli* strains carrying pMccB17 or other related plasmids (Baquero *et al.*, 1978; San Millán *et al.*, 1987). The genes involved in the biosynthesis and secretion of this antibiotic as well as in the regulation of its production are described in an accompanying paper (Moreno *et al.*, 1991). MccB17 is a glycine-rich peptide of about 3,200 daltons (Davagnino *et al.*, 1986). It inhibits semiconservative DNA replication, induces the SOS repair system, and finally causes DNA degradation (Herrero & Moreno, 1986; Mayo *et al.*, 1988).

To study the uptake and mode of action of MccB17, a genetical approach was undertaken. Microcin resistant *E. coli* mutants were isolated and characterized. Spontaneous mutants were obtained at a frequency of about 10^{-6} by plating sensitive cells on selective agar plates containing microcin. The mutations were mapped to three genes: *ompF*, *ompR*, and *sbmA*. *ompF* is the structural gene for the outer membrane OmpF porin, and OmpR is an activator of the *ompF* gene. Thus mutations in these genes may produce defective OmpF or no protein. In conclusion, MccB17 requires a functional OmpF protein to penetrate the cells. This protein is also involved in the uptake of some colicins (Pugsley, 1984).

A specific MccB17 receptor on the inner membrane

The previously unidentified *sbmA* gene (for sensivity to B microcin) was mapped at 8.7 min of the *E. coli* genetic map, closely linked *to phoA* (Laviña *et al.*, 1986). Mutations in this gene confer a very high resistance to exogenous microcin but none to endogenous microcin (Laviña *et al.*, 1986; Herrero *et al.*, 1986). This indicates that *sbmA* could be the specific microcin receptor on the bacterial envelope. This hypothesis was further supported by showing that MccB17 is able to inhibit DNA replication in *sbmA* mutant cells treated with toluene (unpublished results).

The nucleotide sequence of *sbmA* was determined. The gene is 1,218 nucleotides long and encodes a polypeptide of 406 amino acids. The deduced amino acid sequence indicates that it is quite hydrophobic, 66% of residues being nonpolar. The Kyte and Doolittle hydropathy plot predicts seven putative transmembrane domains (Mayo & Moreno, manuscript in preparation). Fractionation experiments have

[1] Present address: Instituto de Investigaciones Biológicas "Clemente Estable". Montevideo.

NATO ASI Series, Vol. H 65
Bacteriocins, Microcins and Lantibiotics
Edited by R. James, C. Lazdunski and F. Pattus

shown that SbmA is located in the inner membrane. Thus *sbmA* specifically transports MccB17 into the cell cytoplasm.

MccB17 blocks DNA gyrase

Selection of mutants affected in the intracellular target of the antibiotic is hampered by the relatively high frequency of cell-envelope-resistant mutants. However, two independent mutants were isolated by following two different convenient procedures (Vizán *et al.*, 1991). These mutants were 8 to 10-fold less susceptible to microcin than the wild-type isogenic strain. After showing that both mutations were located very close to a coumermycin A_1 resistant mutation, one of these mutations was cloned together with the Cou^R mutation, using one of the mini-Mu defective cloning vectors prepared by Groisman *et al.* (1984) and selecting for coumermycin resistance. Additional experiments showed that the microcin-resistant mutation affected the *gyrB* gene (Vizán *et al.*, 1991).

High-copy plasmids carrying *gyrB*$^+$ did not modify the phenotype of Mcc^S hosts but significantly reduced the level of resistance of Mcc^R hosts. High-copy plasmids carrying the mutant allele did not modify the level of resistance of Mcc^R hosts, but conferred a slight resistance to wild-type (Mcc^S) cells. By exchanging homologous DNA fragments between the mutant and the wild-type genes and characterizing the phenotypes conferred by the hybrid genes to wild-type (Mcc^S) and mutant (Mcc^R) strains, the microcin-resistant mutation was located within a 3' end fragment of *gyrB*. The sequencing of the mutant fragment showed that the Mcc^R mutation consisted of a transition AT→GC at position 2,251, which results in a substitution of the tryptophan 751 by arginine. The other Mcc^R mutation was also shown to contain the same amino-acid change. The above results strongly suggest that the DNA gyrase is, or is part of, the intracellular target of MccB17. The fact that the two independent spontaneous mutations produced the same amino acid replacement suggests that Trp-751, and/ or the region around it, may be essential for MccB17 interaction with gyrase.

DNA gyrase is a bacterial type II topoisomerase required for DNA replication. The *E. coli* enzyme is a tetramer made up of two A subunits and two B subunits. It negatively supercoils DNA in a reaction driven by ATP hydrolysis (Cozzarelli, 1980; Gellert, 1981). This reaction implies site-specific cleavage of both DNA strands, formation of a transient covalent bond between the 5' phosphoryl groups of DNA and tyrosines at position 122 of the A subunits, strands passage, and DNA rejoining. Two families of antibiotics are typical inhibitors of DNA gyrase. Coumarins inhibit gyrase by competing with ATP for binding to the enzyme. Quinolones act by trapping a gyrase-DNA intermediate, which is revealed by addition of a protein denaturant such as SDS.

MccB17 produces the same effect as quinolones when assayed "*in vitro*" and "*in vivo*". "*In vitro*",

MccB17 induces double-strand cleavage of plasmid DNA. This reaction is dependent on gyrase as shown by the fact that no cleavage was detected when microcin-resistant gyrase was used in these assays. "*In vivo*" MccB17 induces massive fragmentation of DNA in susceptible cells, but has an insignificant effect on resistant cells. Altogether, the above results indicate that MccB17 blocks DNA gyrase by trapping an enzyme-DNA cleavable complex. A complete account of the work concerning the action of MccB17 on DNA gyrase has been presented in the paper from Vizán *et al.* (1991).

Acknowledgements

This research has been supported by the Fondo de Investigaciones Sanitarias and the Comisión Interministerial de Ciencia y Technologia of Spain.

References

Baquero F, Bouanchaud D, Martínez MC, Fernández C (1978) Microcin plasmids: a group of extrachromosomal elements coding for low molecular weight antibiotics in *Escherichia coli.* J Bacteriol 135:342-347

Cozzarelli NR (1980) DNA gyrase and the supercoiling of DNA. Science 207:953-960

Davagnino J, Herrero M, Furlong D, Moreno F, Kolter R (1986) The DNA replication inhibitor microcin B17 is a forty-three-amino acid protein containing sixty percent glycine. Proteins 1:230-238

Gellert M (1981) DNA topoisomerases. Annu Rev Biochem 50:879-910

Groisman EA, Castilho BA, Casadaban MJ (1984) "*In vivo*" DNA cloning and adjacent gene fusing with a mini-Mu-*lac* bacteriophage containing a plasmid replicon. Proc Natl Acad Sci USA 81:1480-1483

Herrero M, Kolter R, Moreno F (1986) Effects of microcin B17 on microcin B17-immune cells. J Gen Microbiol 132:403-410

Herrero M, Moreno F (1986) Microcin B17 blocks DNA replication and induces the SOS system in *Escherichia coli.* J Gen Microbiol 132:393-402

Laviña M, Pugsley AP, Moreno F (1986) Identification, mapping, cloning and characterization of a gene *(sbmA)* required for microcin B17 action on *Escherichia coli* K12. J Gen Microbiol 132:1685-1693

Mayo O, Hernández-Chico C, Moreno F (1988) Microcin B17, a novel tool for preparation of maxicells: identification of polypeptides encoded by the IncFII minireplicon pMccB17. J Bacteriol 170:2414-2417

Moreno F, San Millán JL, del Castillo I, Gómez JM, Rodríguez-SáinzMC, González-Pastor JE, Díaz-Guerra L (1991) *Escherichia coli* genes regulating the production of microcins B17 and C7. This volume.

Pugsley AP (1984) The ins and outs of colicins. Part I. Production and translocation across membranes. Part II. Lethal action, immunity and ecological implications. Microbiol Sciences 1:168-178

San Millán J., Kolter R, Moreno F (1987) Evidence that colicin X is microcin B17. J Bacteriol 169:2899-2901

Vizán JL, Hernández-Chico C, del Castillo I, Moreno F (1991) The peptide antibiotic microcin B17 induces double strand cleavage of DNA mediated by gyrase. EMBO J 10:467-476

THE STRUCTURE AND MATURATION PATHWAY OF MICROCIN B17

Peter Yorgey, Jonathan Lee and Roberto Kolter
Department of Microbiology and Molecular Genetics
Harvard Medical School
200 Longwood Avenue
Boston, MA 02115
U.S.A

The DNA replication inhibitor microcin B17 (MccB17) is a forty-three amino acid peptide antibiotic produced by diverse strains of *Escherichia coli* (Baquero & Moreno, 1984; Davagnino *et al.*, 1986; Herrero & Moreno, 1986; Vizán *et al.*, 1991). Seven genes, designated *mcbABCDEFG*, responsible for MccB17 production and immunity have been found in several large, single-copy plasmids, the most well studied of these being pMccB17 (San-Millán *et al.*, 1985; Garrido *et al.*, 1988; Genilloud *et al.*, 1989). Four of these genes, *mcbABCD*, are essential for the production of MccB17. Initially, we began studying MccB17 as a small protein amenable to structure/function analysis by using extensive mutagenesis followed by characterization of the mutant proteins. But as we began to characterize MccB17, we became more and more interested in its unique structure and in the pathway of its synthesis. In this paper we present what we know about the structure of MccB17. We also present a working model of the maturation pathway of MccB17 and our evidence which supports this model. Lastly, we briefly discuss some interesting observations we have made concerning the regulation of MccB17 maturation.

The structure of MccB17

Initial genetic analysis demonstrated that *mcbABCD* were essential for production of MccB17 (San-Millán *et al.*, 1985). Subsequent sequencing of these four genes revealed four open reading frames predicting proteins of 69, 296, 273, and 386 amino acids, corresponding well to the observed sizes of the gene products identified using minicells and maxicells (Genilloud *et al.*, 1989).

Amino acid analysis of purified MccB17 identified *mcbA* as the structural gene (Davagnino *et al.*, 1986). N-terminal sequencing of MccB17 showed that in the mature protein the N-terminal twenty-six amino acids are removed, leaving a forty-three amino acid peptide. Figure 1A shows the amino acid sequence of McbA and Figure 1B shows a schematic representation of the predicted amino acid sequence of McbA. Mature MccB17 begins at valine 27. An immediately evident and striking feature of MccB17 is its high glycine content, sixty percent. We refer to the N-terminal twenty-six amino acids as the leader peptide and the remainder of the peptide as the mature part of the peptide. The designation as leader is not meant to imply any particular function.

NATO ASI Series, Vol. H 65
Bacteriocins, Microcins and Lantibiotics
Edited by R. James, C. Lazdunski and F. Pattus

A

	Met	Glu	Leu	Lys	Ala	Ser	Glu	Phe
Gly	Val	Val	Leu	Ser	Val	Asp	Ala	Leu
Lys	Leu	Ser	Arg	Gln	Ser	Pro	Leu	Gly
Val	Gly	Ile	Gly	Gly	Gly	Gly	Gly	Gly
Gly	Gly	Gly	Gly	Ser	Cys	Gly	Gly	Gln
Gly	Gly	Gly	Cys	Gly	Gly	Cys	Ser	Asn
Gly	Cys	Ser	Gly	Gly	Asn	Gly	Gly	Ser
Gly	Gly	Ser	Gly	Ser	His	Ile		

B

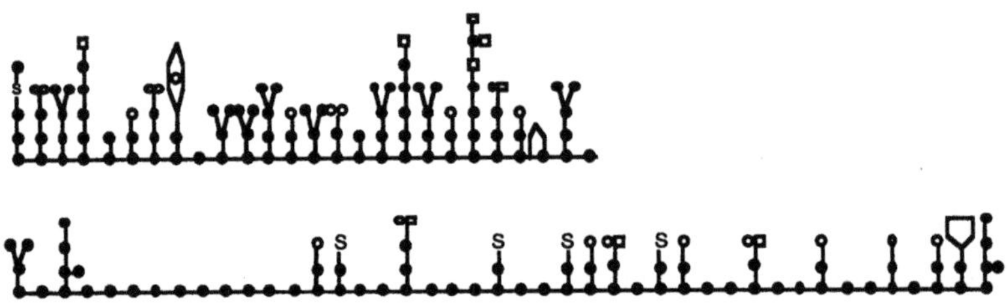

Figure 1. The predicted amino acid sequence of McbA (Davagnino *et al.*, 1986). A: presented as a three letter code. B: Schematic representation of the side chains. The top represents the leader residues, the bottom the mature peptide residues.

Amino acid analysis of mature MccB17 revealed that all four cysteine residues and four of the six serine residues are missing (Davagnino *et al.*, 1986). Performic acid oxidation and subsequent amino acid analysis yielded no cysteic acid, confirming that no free cysteine or disulfide bridges are present in MccB17. Since MccB17 can be labeled with ^{35}S-cysteine, it is clear that the cysteine sulfur is retained. Amino acid analysis also showed two unknown amino acid derivatives, one major and one minor.

When we first recognized the lack of four cysteines and four serines, we thought that these residues were modified to yield lanthionine (Ingram, 1969). To test for lanthionine we compared the acid hydrolysate of MccB17 to a lanthionine standard. Neither of the two unknown amino acid derivatives found in MccB17 comigrated with the lanthionine standard. This result and all subsequent results are inconsistent with the presence of lanthionine in MccB17.

The remainder of the structural information on MccB17 has been obtained using NMR, UV, and fluorescence spectroscopy. The one dimensional NMR spectrum of MccB17 in DMSO indicates that the β-methylene protons from four serines and four cysteines are missing in MccB17. In addition, eight conspicuous proton resonances are present in the aromatic region of the spectrum (see Figure 2, top spectrum). These "aromatic" resonances are uncharacteristic for any of the predicted amino acids in MccB17.

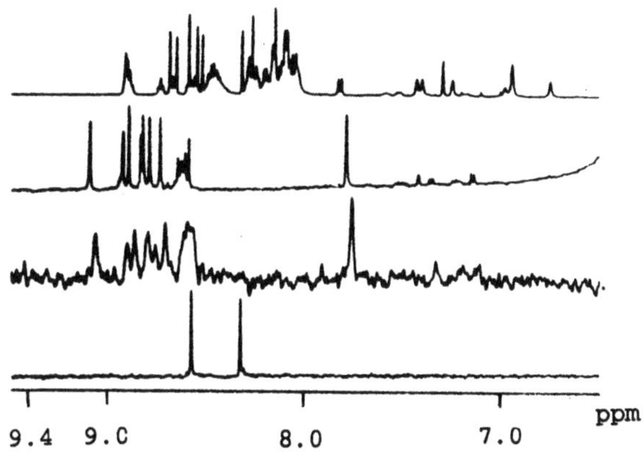

Figure 2. Expanded view of four one dimensional NMR spectra of MccB17. Each individual spectrum is explained in the text.

The second spectrum shown in Figure 2 is that of MccB17 after the addition of deuterated water. Since deuterium is not seen on NMR, only those protons which do not easily exchange remain visible (Wutrich, 1986). The eight "aromatic" protons do not exchange, suggesting that they are attached to carbon or they are buried in the protein and are inaccessible to solvent.

The third spectrum is the result of a "carbon-filtered" experiment (Wutrich, 1986), a technique which subtracts all protons except those which are attached to carbon. The eight "aromatic" resonances remain, confirming that those protons are bound to carbon.

These data combined with the location of the resonances and the singlet character of their spectral lines suggest that they may be part of a heterocyclic structure with aromatic character.

In order to isolate the putative ring structure for further characterization, we generated proteolytic fragments of MccB17 with Proteinase-K. These fragments were produced by extensive proteolysis, with excess Proteinase-K and incubation times of up to 10 days. Four different fragments were isolated; each contained one pair of the four pairs of "aromatic" resonances. The spectrum of one of these fragments is shown in the bottom of Figure 2. We interpret these results as indicating that each MccB17 molecule contains four ring structures, with each ring structure contributing two "aromatic" protons to the MccB17 spectrum.

The resonances of these "aromatic" protons and the putative heterocyclic ring structures are characteristic of ultraviolet absorbing chromophores. To investigate the possibility that MccB17 might indeed contain such a chromophore, the ultraviolet absorption spectrum of one of the proteolytic fragments was determined. (Figure 3). The fragment showed an ultraviolet absorption maximum at a wavelength of 274 nm, indicating the presence of a chromophore. The shape of the absorption peak suggests that the structure may have multiple rings.

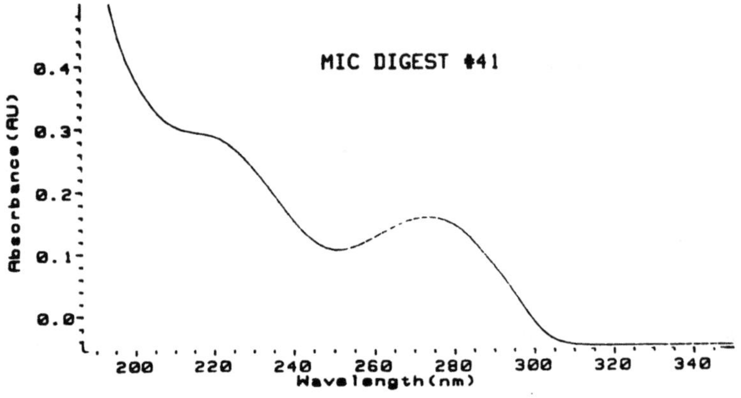

Figure 3. Ultraviolet absorption spectrum of a proteolytic fragment of MccB17.

The same proteolytic fragment was also analyzed by fluorescence spectroscopy. The fragment exhibited a fluorescence maximum at a wavelength of 338 nm. The maximum fluorescence was achieved at an excitation wavelength equal to the ultraviolet absorption maximum. These results strongly support the presence of a ring structure chromophore in MccB17.

The final structural information we present comes from two-dimensional NMR spectra of MccB17. The most unexpected result from these data is that forty-two distinct amide protons have been identified. This indicates that in DMSO MccB17 possesses a highly ordered structure, despite its high glycine content. Even the three N-terminal and three C-terminal residues of the protein are sufficiently ordered to have been identified and structurally defined in the spectra, results which are unexpected for any protein.

In summary, from the NMR, UV, and fluorescence spectroscopy studies we conclude that:

1) The β-methylene protons of four cysteines and four serines are missing in MccB17. This is supportive evidence that these residues are modified.

2) There are eight "aromatic" protons which are uncharacteristic of the predicted amino acids in MccB17. These protons are attached to carbon and are most likely present in four heterocyclic chromophores.

3) Despite its high glycine content, MccB17 has a highly ordered structure.

Maturation Pathway of MccB17

a. The model

The discovery that MccB17 contains forty-three residues, several unusual side-chain modifications, and a highly ordered structure led us to ask the following question. Starting from the sixty-nine residue primary translation product of *mcbA*, what is the pathway to the forty-three amino acid mature molecule with four chromophores?

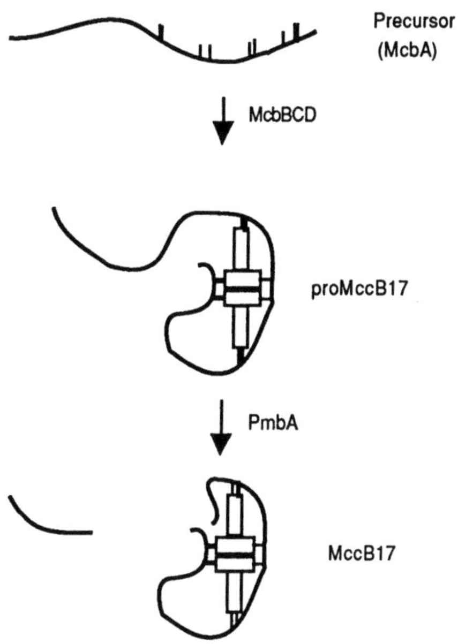

Figure 4. Model of the maturation pathway of MccB17. The model is explained in the text.

There are two steps in our model: modification and processing. In the first step the sixty-nine amino acid precursor, McbA, is modified by the McbBCD proteins to yield proMccB17. The second step or processing step involves removal of the leader peptide by the action of the chromosomal product PmbA thereby producing mature MccB17.

b. Evidence for the model: modifications.

The first step in understanding this pathway came from comparing the MccB17 peptide produced in the presence and absence of the *mcbBCD* genes. Figure 5 shows that when the *mcbA* gene is expressed in the presence of *mcbBCD* a peptide with an apparent molecular of 5.5 kDa is clearly visible in SDS-PAGE (lanes 3 and 4). However, in the absence of *mcbBCD*, a peptide with an apparent molecular weight of 8.5 kDa, which we designate as "precursor", is seen (lanes 5 and 6). Thus, the *mcbBCD* gene products are facilitating a large change in apparent molecular weight of the microcin peptide.

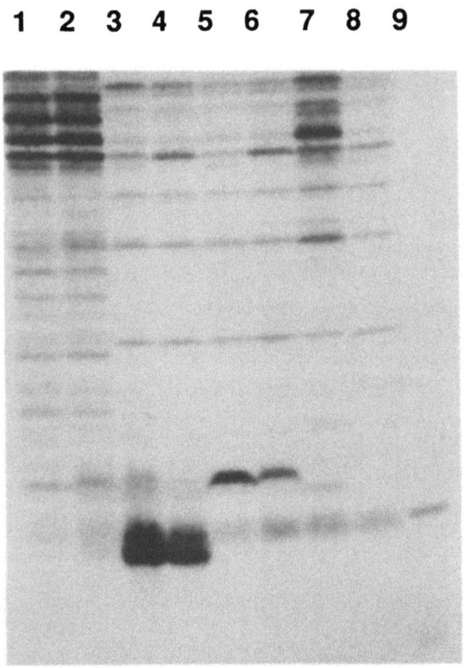

Figure 5. Identification of the microcin precursor, McbA. Unless otherwise noted the MccB17-producing strains used in these and subsequent labelings contained a pBR322-based plasmid harboring the seven MccB17 genes (Yorgey, 1991). The standard conditions for labeling involved incubation of cells with ^{3}H-glycine for two minutes. For total protein gels, cell pellets were then resuspended in Laemmli sample buffer and boiled before loading on a 20% Laemmli protein gel as modified for good resolution of small proteins (Laemmli, 1970; Thomas and Kornberg, 1978). The samples in lanes 1 through 4 were labeled at different times in the growth phase of the MccB17 producing strain (log, late log, and one and three hours into stationary phase, respectively). The results show that in the high copy plasmid used, like in the wild-type single copy plasmid, the MccB17 genes are expressed maximally in stationary phase (Connell *et al.*, 1987; Hernández-Chico *et al.*, 1986). Lanes 5 and 6 contain proteins from a strain lacking *mcbBCD*, labeled at one and three hours into stationary phase. Lanes 7 and 8 contain proteins from a strain harboring a *mcbA::IS1* mutation. In lane 9 are relevant molecular weight standards of 6 and 3.5 kDa.

Two possibilities seemed likely to us to account for the decrease in apparent molecular weight mediated by the *mcbBCD* gene products. Either these gene products facilitated the removal of the leader peptide or they facilitated modifications which somehow altered the gel mobility of the product. To address this question we tested whether the 5.5 kDa form of MccB17 contained the leader peptide. This was made simple by the fact that the leader peptide contains five leucines, while the mature molecule contains none. Using ^{3}H-leucine, we labeled and compared proteins from MccB1 7-producing and non-producing cells (Figure 6).

In the producing strain both the precursor and the 5.5 kd forms of MccB17 are labeled with ^{3}H-leucine (Figure 6, lanes 3, 4, and 5). This shows that the 5.5 kDa form of MccB17 contains the leader peptide. We concluded from this result that the 5.5 kDa form must therefore be modified in some way to account for the change in mobility. We designated this 5.5 kDa form, which is modified but still contains sixty-nine residues, proMccB17. We hypothesize that the *mcbBCD* gene products facilitate modifications which fold proMccB17, perhaps by crosslinking the peptide, and that this folding accounts for the decrease in apparent molecular weight.

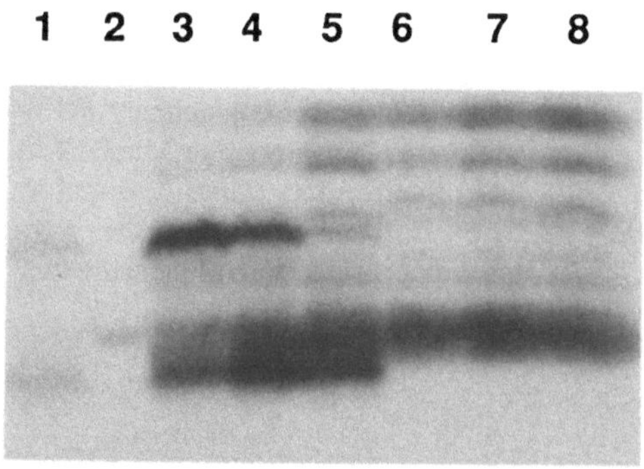

Figure 6. Leucine labeling of MccB17 precursor and proMccB17. Lane 1, ^{3}H-glycine labeled proteins as in Figure 5 for 5.5 and 8.5 kDa references. Lane 2, molecular weight standards as in Figure 5. Lanes 3, 4 and 5 contain proteins from the producing strain while lanes 6, 7 and 3 contain proteins from the strain with the insertion in the *mcbA* gene. Each set of three lanes from left to right represent increasing times of incubation with ^{3}H-leucine.

Additional information regarding the nature of these modifications came from experiments with antibodies raised to mature MccB17. A strain containing the *mcbA* gene alone and strains containing mutations in either *mcbB*, *mcbC*, or *mcbD* were labeled with ^{3}H-glycine and compared to a wild-type MccB17-producing strain. Half of each labeled sample was run as total protein (Figure 7, top panel) while the other half was immunoprecipitated with anti-MccB17 polyclonal antibodies (Figure 7, bottom panel). Polyclonal antibodies raised against mature MccB17, which immunoprecipitated proMccB17, do not recognize the precursor nor any of the comigrating bands from the *mcbB*, *C* and *D* mutants.

We find it amazing that polyclonal antibodies raised against the mature molecule do not recognize any epitope on the precursor. This suggests that either the modifications are the overwhelming antigenic determinants in mature MccB17 or they contribute to unique folding of the molecule (or both). The results shown in Figure 7 indicate that in the absence of any one of the three genes (*mcbBCD*), the precursor is not modified at all or, if it is modified, the partial modifications do not generate a recognizable epitope nor are they sufficient to facilitate folding of the molecule.

We have generated several dozen mutant forms of McbA which contain single amino acid replacements (Yorgey, 1991). To test whether any of these mutants were blocked in the first step of maturation from precursor to proMccB17, they were all screened by pulse-chase labeling with ^{3}H-glycine.

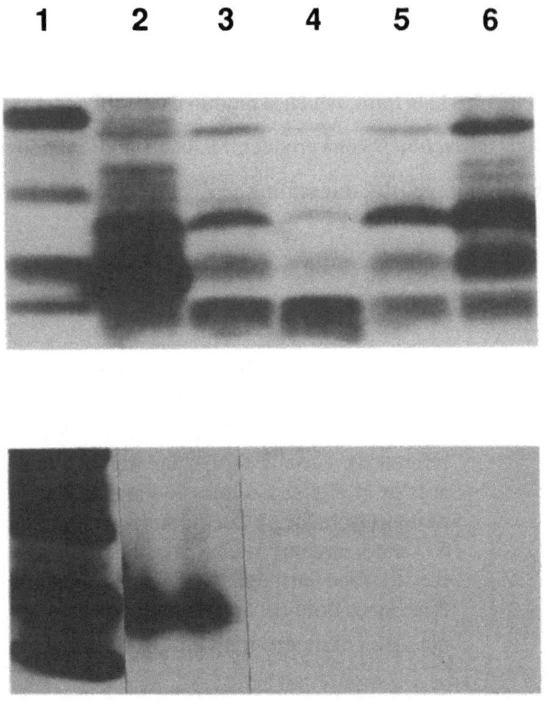

Figure 7. Immunoprecipitation of pulse labeled proMccB17 in wild-type and mutant backgrounds. In the top are total protein samples. Lane 1 has molecular weight standards. Lane 2 contains proteins from the MccB17-producing strain and shows both the precursor and proMccB17. Lane 6 has proteins from the strain containing only the *mcbA* gene, showing only the precursor and no proMccB17. Each of the other independent *mcbB*, *mcbC*, and *mcbD* mutants also exhibits a band which comigrates with the precursor and no proMccB17 or any other MccB17-related intermediate (lanes 3, 4 and 5). The bottom panel shows the immunoprecipitations of the same samples.

Two mutants, which show no antibiotic activity, accumulated precursor peptide which did not chase into proMccB17. Both mutations are glycine to aspartic acid changes in adjacent residues at positions 49 and 50. These two residues are flanked by cysteines 48 and 51. Surprisingly, both mutant peptides were recognized by anti-MccB17 antibodies, even though they comigrate with precursor (Figure 8)

The fact that we can immunoprecipitate the mutant precursor peptides suggests that modifications can occur prior to folding. We postulate that the modifications are added first and they facilitate the rapid cross-linking and/or folding of the polypeptide chain. The G49D and G50D mutations may trap a normally transient, modified but unfolded, intermediate by electrostatically or sterically interfering with cross-linking and/or folding.

c. Evidence for the model: N-terminal processing

The final step required in the maturation of MccB17 is the removal of the N-terminal twenty-six amino acids from proMccB17 (see Figure 4). N-terminal sequencing of an McbA-LacZ fusion protein expressed in the absence of the *mcbBCD* genes showed that about 50% of the molecules lacked the leader.

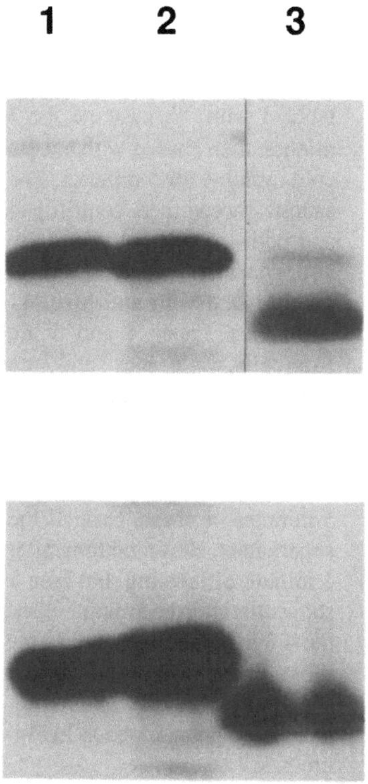

Figure 8. SDS-PAGE of mutant forms of McbA after pulse labeling with ^{3}H-glycine. In the top panel are total protein samples. Lanes 1 and 2 contain proteins from the G49D and G50D mutants and show a band comigrating with precursor. Lane 3 has proteins from the strain producing MccB17, showing mostly proMccB17. The bottom panel shows the immunoprecipitations of the same samples.

From this result several conclusions can be drawn. First, the processing is not mediated by the McbBCD proteins, nor is it dependent on the modifications that these proteins mediate. Second, the processing is not dependent on export since the MccB17 export genes *mcbEF* were not present; therefore processing occurs in the cytoplasm. Third, these results suggest that a chromosomally encoded, cytoplasmic peptidase carries out the processing. The best candidate for this peptidase is the product of the *pmbA* gene (Rodríguez-Sáinz *et al.*, 1990).

Pulse-chase experiments with the wild-type single copy plasmid pMccB17 have shown that label incorporated into proMccB17 in a one minute pulse can be chased into a lower molecular weight band of 4.8 kDa, which we believe to be mature MccB17 (Figure 9).

This result, however, has not been as reproducible when the labeled strain harbors a high copy plasmid with the *mcb* genes. While in a few experiments we have seen the chasing of proMccB17 into the 4.8 kDa product, most of the time this is not the case. Rather, most of the time the results we obtain suggest that the pulse-labeled proMccB17 is degraded during the chase (Figure 10).

Our present hypothesis to explain these observations is that when MccB17 is expressed from a high copy plasmid, the chromosomal processing protease, which we believe is PmbA, becomes the rate limiting step in the maturation pathway. We postulate that proMccB17 would accumulate, but it is rapidly degraded by cellular proteases. We hope to overexpress PmbA to test this hypothesis and to see if it will correct this problem.

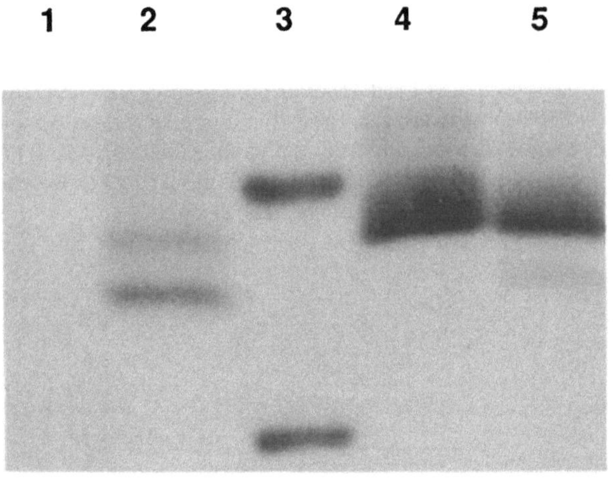

Figure 9. Pulse-chase labeling of a strain harboring the single-copy pMccB17 plasmid. Cells were labeled with ^{35}S-cysteine for 1 minute, then chased with excess cold cysteine for 5 minutes. The samples were then centrifuged and both the cell pellet and the supernatants. were immuno precip-itated with anti-MccB17 antibodies. Lanes 4 and 5 are from cells, while lanes 1 and 2 are from supernatant. Lane 4 shows proMccB17 after 1 minute of labeling and lane 5 shows a small amount of a 4.8 kDa product after 5 minutes of chase. Lane 1, the supernatant, shows nothing after 1 minute of labeling, but lane 2 shows that after the 5 minute chase the 4.8 kDa band appears as the most prominent band in the supernatant. Lane 3 contains molecular weight markers as in Figure 5.

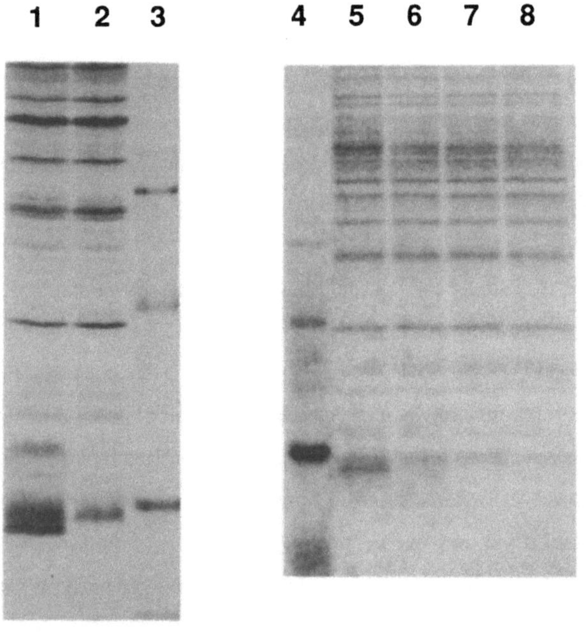

Figure 10. Pulse-chase labeling of strains harboring a multi-copy plasmid with the *mcb* genes. The panel on the left contains labeled proteins from the cell pellet. Lane 1 is a two minute pulse labeling showing a clear proMccB17 band. Lane 2 is a 30 minute chase of the same sample, the proMccB17 band is gone, no new band appears. The panel on the right contains labeled proteins from both cells and supernatant. Lane 5 shows proMccB17 after a 2 minute labeling, lanes 6, 7 and 8 represent chase periods of 5, 30 and 180 minutes. Even after 5 minutes of chase both the proMccB17 is gone and no 4.8 kDa peptide is visible. Lanes 3 and 4 contain molecular weight standards as in Figure 5.

Regulation of Maturation

We have made two interesting observations concerning the regulation of MccB17 maturation. Figure 11 shows a ^{3}H-glycine pulse-chase experiment from several different points in the growth curve of a MccB17 producing strain. In a two minute labeling the precursor is detected only at the onset of stationary phase. Thereafter all the label is incorporated into proMccB17 in two minutes. This suggests that the relative rate of modification to proMccB17 changes.

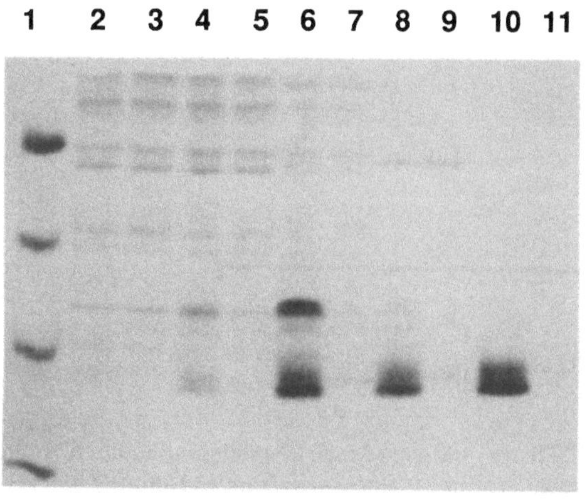

Figure 11. Pulse-chase labeling along the growth curve. Each pair of samples represents a two minute pulse and a thirty minute chase. Samples were labeled at the following time points: mid-log (lanes 2 and 3), late-log (lanes 4 and 5), onset of stationary phase (lanes 6 and 7), two hours into stationary phase (lanes 8 and 9), and ten hours into stationary phase (lanes 10 and 11). Lane 1, molecular weight standards.

Finally, the transcription of the *mcb* genes and the appearance of immunoprecipitable proMccB17 are uncoupled by about four hours when the *mcb* genes are present in the single copy, wild-type plasmid (Figure 12). The results from the Northern blot are consistent with previous results which indicated that *mcb* RNA expression is induced at the cessation of growth (Connell *et al.*, 1987; Hernández-Chico *et al.*, 1986). However, it is surprising that proMccB17 synthesis is delayed until four to six hours into stationary phase. This synthesis peaks quickly and shuts off within about three hours. We offer two explanations for these observations. It is possible that the *mcbA* transcript is made but not translated until several hours later. Alternatively, the precursor may have been made continually since the onset of stationary phase but the McbBCD-mediated modification may have not been added for several hours. We are currently carrying out experiments to distinguish between these two alternative explanations.

Summary and Perspectives

Isn't microcin wonderful?

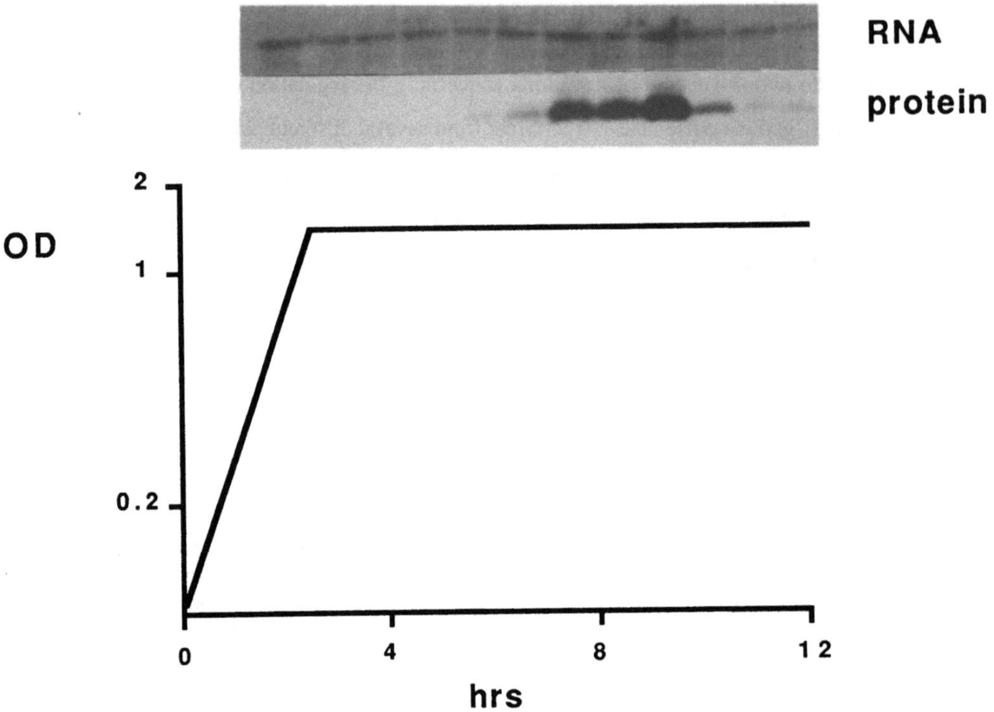

Figure 12. Production of *mcb* RNA and proMccB17 in the wild-type MccB17-producing strain. A strain harboring the wild-type plasmid pMccB17 was grown in minimal glucose medium and allowed to reach stationary phase. Samples were taken at regular intervals and: i) RNA was prepared and ii) cells were pulse-labeled with ^{35}S-cysteine and immunoprecipitated with anti-MccB17 antibodies. On top of the culture's growth curve is displayed a Northern blot where the probe was a DNA fragment harboring *mcbA*. Below the Northern blot are the results from the immunoprecipitation for each corresponding sample. The location of each sample indicates the time when it was taken.

Acknowledgements

We are grateful to Juan Davagnino for laying the foundation for much of the work presented here. We also wish to express our thanks to Felipe Moreno and members of his laboratory for their multi-faceted help in this project. Support for the work presented here came from a grant from the NIH (A125944) to R.K. R.K. is the recipient of an American Cancer Society Faculty Research Award. P.Y. is the recipient of a Ryan Foundation Predoctoral Fellowship.

References

Baquero F, Moreno F (1984) The microcins. FEMS Microbiol Lett 23:117-124

Connell N, Han Z, Moreno F, Kolter R (1987) An *E. coli* promoter induced by the cessation of growth. Mol Microbiol 1:195-201

Davagnino J, Herrero M, Furlong D, Moreno F, Kolter R (1986) The DNA replication inhibitor microcin B17 is a forty-three amino acid protein containing sixty percent glycine. Proteins 1:230-238

Garrido MC, Herrero M, Kolter R, Moreno F (1988) The export of the DNA replication inhibitor microcin B17 provides immunity for the host cell. EMBO J 7:1853-1862

Genilloud O, Moreno F, Kolter, R (1989) DNA sequence, products, and transcriptional pattern of the genes involved in production of the DNA replication inhibitor microcin B17. J Bacteriol 171:1126-1135

Hernández-Chico C, San Millán JL, Kolter R, Moreno F (1986) Growth phase and OmpR regulation of transcription of the Microcin B17 genes. J Bacteriol 167:1058-1065

Herrero M, Moreno F (1986) Microcin B17 blocks DNA replication and induces the SOS system in *Escherichia coli*. J Gen Microbiol 132:393-402

Ingram LC (1969) Synthesis of the antibiotic nisin: formation of lanthionine and β-methyl-lanthionine. Biochim Biophys Act 184:216-219

Laemmli UK (1970) Cleavage of structural proteins during the assembly of the head of bacteriophage T4. Nature 227:680-685

Rodríguez-Sáinz MC, Hernández-Chico C, Moreno F (1990) Molecular characterization of *pmb*A, an *Escherichia coli* chromosomal gene required for the production of the antibiotic peptide MccB17. Molec Microbiol 4:1921-1932

San-Millán JL, Kolter R, Moreno F (1985) Plasmid genes involved in microcin B17 production. J Bacteriol 163:1016-1020

Thomas JO, Kornberg RD (1978) The study of histone-histone associations by chemical cross-linking. In: Stein, S, Stein, J and Kleinsmith, LJ (ed) Methods in Cell Biology , Vol. 18, pp. 429-440. Academic Press, New York

Vizán JL, Hernández-Chico C, del Castillo I, Moreno F (1991) The peptide antibiotic microcin B17 induces double-strand cleavage of DNA mediated by *E. coli* DNA gyrase. EMBO J 10:466-467

Wutrich, K (1986) NMR of proteins and nucleic acids. John Wiley and Sons, New York.

Yorgey, P (1991) Unpublished constructs.

BACTERIOCINS OF GRAM-POSITIVE BACTERIA: AN OPINION REGARDING THEIR NATURE, NOMENCLATURE AND NUMBERS

J. R. Tagg
Department of Microbiology
University of Otago
Dunedin
New Zealand

A diverse variety of chemical substances, released extracellularly by bacteria, may function to inhibit the growth of other bacteria competing for a particular ecological niche (Tagg, Dajani & Wannamaker, 1976). These substances include:

(a) low molecular weight metabolic by-products such as ammonia, hydrogen peroxide and lactic acid
(b) so-called "classical" antibiotics, like bacitracin and polymyxin B, synthesized by multi-enzyme pathways
(c) small peptide antibiotics formed as ribosomally synthesized precursors and then post-translationally modified (e.g. lantibiotics and microcins)
(d) protein antibiotics in the molecular weight range ca. 50,000-100,000 daltons, that are defined as bacteriocins (e.g. colicins)
(e) bacteriolytic enzymes of the muramidase type (e.g. lysozyme-like enzymes)
(f) protein exotoxins (e.g. hemolysins, lecithinases)
(g) defective bacteriophage particles (some pyocins)

The term bacteriocin was originally defined quite specifically to refer to the colicin-type of protein antibiotics, the synthesis of which was "lethal" to the producing cell and the adsorption of which was dependent upon the presence of specific receptors on the sensitive bacterium (Jacob *et al.*, 1953). Other distinguishing features of the colicin-type of inhibitory substances include their relatively high molecular weights, their narrow spectra of inhibitory activity (generally restricted to other strains of the *Enterobacteriaceae*) and the plasmid association of their genetic determinants.

It has now become clear however, that most of the so-called bacteriocins produced by Gram-positive bacteria do not fit the classical colicin mold. Rather, they tend to be more broadly active against strains of Gram-positive species, with little evidence of their action being mediated by specific receptor molecules or their release from producer cells being enhanced by the action of lysins or bacteriocin release proteins. Obviously, the absence of an outer membrane in Gram-positive bacteria excludes any possibility of a

NATO ASI Series, Vol. H 65
Bacteriocins, Microcins and Lantibiotics
Edited by R. James, C. Lazdunski and F. Pattus

modulating effect of receptor molecules in the manner which applies to the interaction of colicins with sensitive bacteria. Rather, the potentially lethal interaction of "bacteriocins" of Gram-positive bacteria with sensitive cells appears to be dependent upon a more general compatibility between surface charges and hydrophobic domains of the interacting molecules. Another difference is that the level of immunity of the producing strain to its own inhibitory product is generally less strong for "bacteriocins" of Gram-positive bacteria than it is for the colicins.

It seems that if we are to continue to apply the term bacteriocin to the peptide and proteinaceous antibiotics produced by Gram-positive bacteria there are two options available. The term "bacteriocin" could be redefined to take into account the non-colicin-like characteristics of many of the more recently described inhibitors. Alternatively, the original definition could be retained and inhibitors which are broadly similar to the colicins could be referred to as bacteriocin-like inhibitory substances (BLIS). It is my recommendation that the latter course be adopted and that the acronym BLIS be used as a simple, catchy and expressive term for bacterially-produced antibiotic substances that have some similarities to the prototype bacteriocins, the colicins.

BLIS may be defined as bacterial peptide or protein molecules, released extracellularly, that in low concentrations are able to kill certain other closely related bacteria by a mechanism against which the producer cell exhibits a degree of specific immunity. This definition is not restrictive to substances that have plasmid-borne genetic determinants. Nor does it imply any requirements for lethal biosynthesis, specific receptors or a particularly narrow spectrum of inhibitory activity, all of which are key features of the originally-defined bacteriocin category.

What is the extent of BLIS production by Gram-positive bacteria? A survey conducted in 1976 (Tagg, Dajani & Wannamaker, 1976) indicated that representative strains of most Gram-positive bacteria could be found to produce BLIS. It was suggested that cross-testing of 100 or more strains of any one species should be rewarded by detection of BLIS producers. In our own subsequent studies we have used a single set of nine indicator bacteria to screen a wide variety of streptococcal species for BLIS production (Tagg, 1991). Only one small group of nutritionally variant streptococci have failed to yield BLIS producers using this test system. It has been our experience that by using additional indicator bacteria and by adoption of a variety of incubation conditions the apparent incidence of BLIS producers in any group of strains can be dramatically increased. Furthermore, it could be anticipated that some strains that appear to be BLIS-negative *in vitro* may actually produce BLIS when growing at the slower growth rates and under the conditions of nutritional depletion which characterize the "real" microbial world. It is my suggestion that the incidence of BLIS-positive bacteria in fresh isolates of any species from mixed natural populations may closely approach 100%.

REFERENCES

Jacob F, Lwoff A, Siminovitch A, Wollman E (1953) Définition de quelques termes relatifs à la lysogénie. Ann Inst Pasteur Paris 84: 222-224

Tagg JR (1991) BLIS production in the Genus *Streptococcus*. (This volume)

Tagg JR, Dajani AS, Wannamaker LW (1976) Bacteriocins of gram-positive bacteria. Bacteriol Rev 40: 722-756

MOLECULAR PROPERTIES OF *LACTOBACILLUS* BACTERIOCINS

T.R. Klaenhammer, C. Ahn, C. Fremaux and K. Milton
Department of Food Science
Southeast Dairy Foods Research Center
North Carolina State University
Raleigh, NC 27695-7624
U.S.A

INTRODUCTION

Among the members of lactic acid bacteria, the lactobacilli represent the most diverse genus. Individual species are ubiquitous in the environment while still occupying a multitude of specialized ecological niches. These include fermenting vegetables, meat, cereals, dairy products, and the intestinal tract of man and animals. Their metabolic capabilities and physiological characteristics are equally diverse and proper classification of many strains has occurred only recently following the development of more sophisticated biochemical and genomic classification systems. These bacteria derive their energy via either homo- or heterofermentative catabolism of carbohydrates in nutritionally complex environments. As the major end product of their metabolism, organic acids function directly as antagonists (Kashet, 1987) and lower the pH of the environment. Since lactobacilli are often more acid-tolerant than other competing bacteria, including other lactic acid bacteria, acidification of the environment promotes their ability to compete within and ultimately dominate fermenting ecosystems. In addition, the lactobacilli produce a variety of chemical and proteinaceous antimicrobials. These include hydrogen peroxide, various ill-characterized compounds (Vincent *et al.*, 1959; Hamdan & Mikolajcik, 1974; Silva *et al.*, 1987), antimicrobials (Talarico & Dobrogosz, 1989), and most notably, bacteriocins (Klaenhammer, 1988, 1990).

The lactic acid bacteria are well-recognized for their production of antimicrobial proteins (Klaenhammer, 1988; Lindgren & Dobrogosz, 1990). The lactobacilli are, however, most often cited to produce bacteriocins. Since the first descriptions of "bacteriocins" produced by homofermentative and heterofermentative lactobacilli (de Klerk & Coetzee, 1961; de Klerk, 1967), nineteen different bacteriocins have been reported (Table 1). These include the bacteriocins produced by a heterofermentative, non-aciduric *Lactobacillus, Carnobacterium piscicola*, which is naturally associated with meats (Ahn & Stiles, 1990).

The bacteriocins of *Lactobacillus* species are generally active against closely-related types that occupy similar ecological niches (Table 1). For example, lacticins A and B produced by *L. delbrueckii*

NATO ASI Series, Vol. H 65
Bacteriocins, Microcins and Lantibiotics
Edited by R. James, C. Lazdunski and F. Pattus

Table 1. Bacteriocins of Lactobacillus species.

Producer Strain	Bacteriocin	Spectrum of Activity	Characteristics	References*
L. acidophilus	lactacin B	Lactobacillus delbrueckii Lactobacillus helveticus Listeria monocytogenes (?)	high molecular weight complex, purified to 6.3 kDa; chromosomal determinants	a, b, c, d
	lactacin F	Lactobacillus fermentum Enterococcus faecalis Lactobacillus delbrueckii Lactobacillus helveticus Aeromonas hydrophila Staphylococcus aureus	high molecular weight complex, purified, cloned, expressed, and sequenced; 57 amino acids, 6.3 kDa peptide, 18 amino acid N-terminal extension; heat stable at 121 C for 15 min; episomal and conjugative genetic determinants	e, f, g, d
	acidophilucin A	Lactobacillus delbrueckii Lactobacillus helveticus	proteinaceous,inactivated by trypsin and actinase; heat labile at 60 C for 10 min	h
L. brevis	brevicin 37	Pediococcus damnosus Lactobacillus brevis Leuconostoc oenos	proteinaceous; heat stable, 121 C for 1 h; active over pH ranges 2-10; inactivated by chloroform	i
L. casei	caseicin 80	Lactobacillus casei	40 kDA protein; pI=4.5; heat stable, 60 C for 10 min; active over pH range of 3-9; inducible by mitomycin C	i
L. carnis‡	bacteriocin(s)	Lactobacillus Carnobacterium Pediococcus Enterococcus Listeria	proteinaceous; heat stable, 100 C for 30 min; pH stable, pH 2-11; membrane active; plasmid-linked	j, k
L. delbrueckii	lacticin A lacticin B	L. delbrueckii subsp. lactis L. delbrueckii subsp.bulgaricus L. delbrueckii subsp. delbreuckii	proteinaceous; heat labile at 60 C for 10 min	l
L. fermenti	bacteriocin	Lactobacillus fermenti	macromolecular complex; lipocarbohydrate moiety	m
L. gasseri	gassericin A	Lactobacillus acidophilus Lactobacillus delbrueckii Lactobacillus helveticus Lactobacillus casei Lactobacillus brevis	proteinaceous, trypsin sensitive; stable at 120 C for 20 min; characteristics similar to lactacin B and F	n
L. helveticus	lactocin 27	Lactobacillus acidophilus Lactobacillus helveticus	protein-lipopolysaccharide complex at >200 kDa, purified to 12.4 kDa	o
	helveticin J	Lactobacillus helveticus Lactobacillus delbrueckii subsp. lactis and bulgaricus	complex aggregate at >300 kDa, purified to 37 kDa, cloned, expressed, sequenced	p

L. plantarum	plantaricin A	Lactobacillus plantarum Lactobacillus spp. Leuconostoc spp. Pediococcus spp. Lactococcus lactis Enterococcus faecalis	proteinaceous; >8 kDa; stable at 100 C for 30 min; active over pH range of 4-6.5	q
	plantacin B	Lactobacillus plantarum Leuconostoc mesenteroides Pediococcus damnosus	proteinaceous	r
	bacteriocin	Leuconostoc Lactobacillus Pediococcus Lactococcus Streptococcus	sensitive to proteases, α-amylase, and lipase; heat stable at 100 C for 30 min; phenotype unstable, suggesting plasmid-borne determinants	s
strains 75 & 592	bacteriocin	Lactobacillus sake Lactobacillus curvatis Lactobacillus plantarum Lactobacillus divergens Listeria monocytogenes Clostridium botulinum spores Aeromonas hydrophila Staphylococcus aureus	proteinaceous	t, d, u, v
L. sake	sakacin A	Carnobacterium piscicola Enterococcus spp. Lactobacillus sake Lactobacillus curvatus Leuconostoc paramesenteroides Listeria monocytogenes Aeromonas hydrophila Staphylococcus aureus	plasmid-borne; 28kb; proteinaceous	t, d
	lactocin S	Lactobacillus Leuconostoc Pediococcus	33 amino acid protein; active over pH 4.5-7.5; 50% non-polar residues; lactocin S production and immunity linked to 50 kb plasmid pCIM1	w, x, y, z

‡ Current taxonomy is Carnobacterium piscicola

* a=Barefoot, Klaenhammer (1983); b=Barefoot, Klaenhammer (1984); c=Harris et al. (1989); d=Lewus et al. (1991); e=Muriana, Klaenhammer (1987); f=Muriana, Klaenhammer (1991a); g=Muriana, Klaenhammer (1991b); h=Toba et al. (1991b); i=Rammelsberg, Radler (1990); j=Ahn, Stiles (1990); k=Schillinger, Holzapfel (1990); l=Toba et al. (1991c); m=de Klerk, Smit (1967); n=Toba et al. (1991a); o=Upreti, Hinsdill (1975); p=Joerger, Klaenhammer (1986); q=Daeschel et al. (1990); r=West, Warner (1988); s=Jimenez-Diaz et al. (1990); t=Schillinger, Lucke (1989); u=Okereke, Montville (1991); v=Schillinger et al. (1991); w=Mortvedt, Nes (1990); x=Mortvedt et al. (1991a); y=Mortvedt et al. (1991b); and z=McCormick, Savage (1983)

target other delbrueckii-related subspecies associated with fermenting dairy products (Toba *et al.*, 1991c), while plantaricin A is bactericidal to numerous lactic acid bacteria normally associated with fermenting vegetables (Daeschel *et al.*, 1990). Some extension in the host range has been noted for bacteriocins which kill *Enterococcus* species, *Listeria monocytogenes*, *Leuconostoc* species, *Pediococcus* species, and more recently, *Clostridium botulinum*, *Staphylococcus aureus*, and *Aeromonas hydrophilia* (see Table 1 and Lewus *et al.*, 1991; Okereke & Montville, 1991). Antagonism of food-borne pathogens by selected *Lactobacillus* bacteriocins is highly significant in light of recent efforts to define natural antimicrobial systems that can contribute to the safety of minimally processed foods.

BIOCHEMICAL TYPES AND CLASSES

The majority of bacteriocins produced by lactobacilli have been characterized by the initial definition of a proteinaceous inhibitor, crude estimation of molecular weight (via retention in dialysis membranes or ultrafiltration), and determination of susceptible strains. Recent efforts to purify and biochemically characterize these compounds are beginning to provide information about their structure, processing, and mechanism of action. The lactic acid bacteria produce three general classes of antimicrobial proteins which include (I) lantibiotics, (II) small hydrophobic heat-stable peptides (< 13,000 daltons), and (III) large heat-labile proteins (> 30,000 daltons). Foremost among the characterized bacteriocins is the lantibiotic nisin produced by *Lactococcus lactis* subsp. *lactis*. This small 34 amino acid peptide contains two sulfur-containing amino acids, lanthionine and beta-methyllanthionine. Within the *Lactobacillaceae*, lantibiotics had not been described prior to this conference. However, Mortvedt *et al.* (1991a) reported at this symposium that the *L. sake* bacteriocin, lactocin S, contains lanthionine residues. In addition, Gudmundsdottir & Stoffels (1991) also implied the presence of lanthionine residues in a bacteriocin of 4632 daltons (33-35 amino acids) produced by a *Carnobacterium* species isolated from fish. These are very exciting preliminary reports in light of the broader host range typically observed for lanthionine-containing peptides (for example nisin, subtilin, and epiderminin). In general, the lactobacilli produce a variety of non-lanthionine bacteriocins. For those bacteriocins that have been purified and characterized, the antimicrobial proteins can be categorized on the basis of size, hydrophobic propensity, and heat stability (Table 2).

The first *Lactobacillus* bacteriocin purified and characterized was a lipocarbohydrate-protein macromolecular complex produced by *L. fermenti* (de Klerk & Smit, 1967). The bacteriocin was relatively heat-stable (96 C for 30 min) and contained a high percentage of glycine (11.1%) and alanine (13.4%) residues. The absence of net charge and a lack of mobility in an electrical field provided evidence for the involvement of a hydrophobic protein or suggested that the active component was associated with a lipid-

Table 2: **Lactobacillus Bacteriocins Biochemical Types / Classes**

- **Lantibiotics** ■ Lactocin S (?)
- **Small hydrophobic peptides**

  moderately heat stable (>30 min @ 100C - 15 min @ 121C)
  molecular weight < 13 kDa

  100C
  - Lactocin 27 ---------12.4 kDa, glycoprotein
  - Lactocin S ---------- 3.7 kDa (33aa)
  - Carnobacteriocins --- 4.9 kDa

  121C
  - Lactacin B ---------- 6.3 kDa
  - Lactacin F ---------- 6.3 kDa (57aa)
  - Brevicin 37

- **Large heat-labile proteins**

  inactivated within 10-15 min @ 60C - 100C

  Helveticin J 37 kDa (334aa)
  Acidophilucin A
  Lacticin A & B

like material. A number of additional heat-stable hydrophobic proteins have been purified subsequently. Lactocin 27, produced by *L. helveticus*, is a small heat-stable glycoprotein (Upreti & Hindsdill, 1973,1975). Similar to the *L. fermenti* bacteriocin, lactocin 27 was initially isolated as a large molecular weight complex (greater than 200,000 daltons) while the active peptide was defined only at 12,400 daltons and contained unusually high concentrations of glycine and alanine residues (15.1% and 18.1%, respectively). The action of lactocin 27 against the indicator strain *L. helveticus* LS18 was characterized as bacteriostatic even though a two log reduction in colony-forming units per ml occurred after treatment with the highest concentration of lactocin 27. Gross physical damage to the bacterial membrane was not detected, but the bacteriocin did cause a leakage of potassium ions, allowed an influx of sodium ions, and halted protein synthesis. Upreti and Hindsdill (1975) concluded that the target of lactocin 27 was, therefore, the cytoplasmic membrane.

Some of the bacteriocins from lactobacilli which have been characterized are reported to be heterogeneous in their chemical composition (de Klerk & Smit, 1967; Upreti & Hinsdill, 1975; Jimenez-Diaz *et al.*, 1990; Toba *et al.*, 1991b). If such compounds exist, they warrant a separate classification. However, at this time these bacteriocins are not represented as a separate class in Table 2 since it has not yet been established if the carbohydrate and lipid moieties are intrinsic to the primary structure and composition of the active bacteriocin. It is possible that the non-proteinaceous components may represent contaminating materials that associate with the bacteriocin during purification. Noting the hydrophobic propensity of many bacteriocins, interactions with lipids and other materials could be expected and have been shown to occur (Muriana & Klaenhammer, 1991a).

A number of heat-stable peptides have recently been identified and purified from *Lactobacillus* species. Table 3 presents the amino acid composition for those bacteriocins that have been characterized to date. These include lactocin S produced by *L. sake* (Mortvedt *et al.*, 1991b) and a number of carnobacteriocins produced by *Carnobacterium piscicola* (Ahn & Stiles, 1990). Both bacteriocins withstand 30 min of heating at 100 C, although lactocin S is more heat-labile and shows a 50% reduction in activity after 1 h. Lactacins B and F produced by *L. acidophilus* N2 and 11088, respectively, are highly heat-stable and withstand autoclaving for 15 min at 121 C. The antimicrobial peptides within this group are small, ranging in size between 3.7 kDa to 6.3 kDa. For example, molecular characterization of lactacin F (Muriana & Klaenhammer, 1991a, 1991b) determined that the peptide was composed primarily of hydrophobic and polar neutral residues (87.3%) which included glycine (21.6%), alanine (15.8%), and valine (8.8%). Analysis of the amino acid composition and N-terminal sequencing of lactocin S revealed that 50% of the approximated 33 amino acids were hydrophobic and nonpolar residues of alanine, valine, and glycine (Mortvedt *et al.*, 1991b). The lower glycine content and hydrophobic propensity of lactocin S compared to lactacin F correlates with their relative heat stabilities (lactacin F > lactocin S). Preliminary

Table 3. Amino acid composition of _Lactobacillus_ bacteriocins.

Amino acids	lactacin F	lactocin S	lactocin 27	helveticin J	_L. fermenti_ bacteriocin
Hydrophobic					
Alanine	15.8	24.0	15.1	5.7	13.4
Isoleucine	7.0	0	5.2	8.1	3.8
Phenylalanine	0	3.0	3.3	3.0	2.8
Leucine	3.5	12.0	5.1	6.3	6.4
Methionine	1.7	3.0	0	1.2	0.5
Proline	5.3	6.0	3.3	2.7	4.7
Valine	8.8	15.0	8.3	4.8	5.6
Tryptophan	3.5	0	0.8	1.5	nd
Charged					
Arginine	3.5	0	3.3	3.6	2.9
Aspartate	0	3.0	8.9 (a)	6.3	10.0 (a)
Glutamate	0	3.0	6.8 (b)	6.6	9.2 (b)
Histidine	1.7	6.0	2.3	3.0	1.9
Lysine	3.5	6.0	5.8	6.9	5.7
Polar neutral					
Asparagine	5.3	0	nd	8.7	nd
Cysteine	3.5	0	0	0.6	nd
Glycine	21.0	3.0	18.1	8.7	11.1
Glutamine	1.7	0	nd	4.2	nd
Serine	1.7	0	5.3	7.2	10.1
Threonine	10.6	3.0	5.8	5.1	8.7
Tyrosine	1.7	6.0	2.9	5.7	3.1
Estimated size in k-daltons	7.5	3.3	12.4	37.5	nd

nd: not determined; a: including asparagine; b: including glutamine.

studies have implicated the cytoplasmic membrane as the target for the action of lactocin S, lactacin F, and the carnobacteriocins (Ahn & Stiles, 1990; Mortvedt *et al.*, 1991b; Muriana & Klaenhammer, 1991a). These observations correlate well with the general size and hydrophobic character of this group of small heat-stable peptides. Two other recent reports on brevicin 37 (Rammelsberg & Radler, 1990) and gassericin A (Toba *et al.*, 1991a), which are also stable at 121 C for 20 min, suggest that these may be within the general class of small heat-stable peptides which appear to be common within lactobacilli and lactic acid bacteria.

A third general class of *Lactobacillus* bacteriocins include heat-labile proteins of large molecular weight. To date only helveticin J (37,000 daltons) has been purified and characterized at the genetic level. It is a heat-sensitive protein (inactivated at 100 C within 30 min) which does retain activity after treatment with various dissociating agents (Joerger & Klaenhammer, 1986). Recent reports of large heat-labile bacteriocins suggest there are numerous members of this class. Acidophilucin A, lacticin A and B, and caseicin 80 appear to be large proteins since they are inactivated within 10-15 min at 60 C (Rammelsberg & Radler, 1990; Toba *et al.*, 1991b, 1991c). The biochemical properties and mechanisms of action of many larger bacteriocins remain to be investigated. Noting their size and heat-lability, the bactericidal activities of these proteins are likely to be affected by changes in conformation and secondary structure.

COMPOSITIONAL ANALYSIS

Analysis of the amino acid composition for those *Lactobacillus* bacteriocins purified thus far indicates a group of highly heterogeneous proteins (Table 3). The smaller hydrophobic peptides retain a high proportion of non-polar residues (predominately alanine, glycine, valine). However, amino acid sequence comparisons between lactacin F and the available portions of lactacin B, lactocin S, and the carnobacteriocins have not yet revealed conserved or identical regions (Henkel *et al.*, 1991; Mortvedt *et al.*, 1991b; Muriana & Klaenhammer, 1991b; Nettles *et al.*, 1991). This is illustrated by a comparison of the N-terminus of the peptides lactacin B (Nettles *et al.*, 1991) and lactacin F (Muriana & Klaenhammer, 1991a):

lactacin F [N-Arg-Asn-Asn-Trp-Gln-Thr-Asn-Val-Gly-Gly-Ala..]

lactacin B [N-Arg-Gln-Pro-Gly-Phe-Ile-Leu-Phe-Pro-Thr-Val..]

Both are 6,300 dalton, heat-stable peptides that inhibit the same basic indicator group. Lactacin B is produced by a majority of *L. acidophilus* strains (Barefoot & Klaenhammer, 1983). This bacteriocin kills *L. delbrueckii* subsp. *lactis* and *bulgaricus* strains. Lactacin F is produced by the *L. acidophilus* strain VPI11088 (a member of the B2 homology group of Johnson *et al.*, 1980) and inhibits the same group of *L. delbrueckii* indicators plus *L..acidophilus*, *L. fermentum*, and *Enterococcus faecalis*. Neither DNA homology nor amino acid similarities have been detected between these two bacteriocins (Nettles *et al.*, 1991; Milton & Klaenhammer, unpublished). These two bacteriocins are categorized within the group of small, heat-stable hydrophobic peptides which are suggested to act at the cytoplasmic membrane (Ahn & Stiles, 1990; Mortvedt *et al.*, 1991b; Muriana & Klaenhammer, 1991a). Their general hydrophobic propensity and suspected membrane-active functions could be provided by a variety of hydrophobic and non-polar amino acids without conservation of specific DNA or amino acid sequences per se.

In our laboratory we have cloned and sequenced two *Lactobacillus* bacteriocins, one each from the two primary classes grouped on the basis of size and heat-stability. The following summarizes our molecular analysis of the helveticin J and lactacin F systems and compares their similarities and differences with bacteriocins produced by other gram-positive bacteria.

HELVETICIN J

Lactobacillus helveticus 481 produces a 37,000-dalton protein, helveticin J, that represents the least predominant group of bacteriocins produced by lactic acid bacteria. These bacteriocins are larger in size and more susceptible to heat treatment. No mechanisms of action have been determined. However, their size and heat-susceptibility suggest that these bacteriocins are more structurally complex. Helveticin J is bactericidal against closely-related species including *L . helveticus* and *L. delbrueckii.* This bacteriocin was selected for further molecular analysis in our laboratory since it was unique relative to other bacteriocins of lactic acid bacteria (Joerger & Klaenhammer, 1986, 1990).

The structural gene for helveticin J was cloned by creating gene fusions in the expression vector lambda gtll (Huynh *et al.*, 1984) and immunoscreening of the library with polyclonal antibodies specific for the bacteriocin (Joerger & Klaenhammer, 1990). Two recombinant phage produced helveticin J-lacZ fusion proteins and contained DNA inserts of 350 and 600 bp which were homologous to each other and specific for the strain producing helveticin J. These inserts were found subsequently to be internal to the helveticin J structural gene, but were too small to encode the entire helveticin J structural gene. Therefore, the 600-bp cloned sequence was used to screen a second *L. helveticus* 481 genomic library prepared in lambda EMBL3. The DNA hybridization analysis revealed that the helveticin J region was located on a 5.5-kb *Hin*dIII fragment. Attempts to clone this fragment into high copy number vectors in *E. coli* were unsuccessful. Plasmid deletions and rearrangements occurred frequently suggesting that the helveticin J region was either unstable or its gene products lethal in *E. coli.* A smaller 4-kb *Bgl*II fragment contained within the 5.5-kb *Hin*dIII fragment was ligated into the *Bcl*I site of pGK12, a low copy number vector in *E. coli* (Kok *et al.*, 1984). The recombinant plasmid (pTRK135) was electroporated into a number of

Lactobacillus host backgrounds but only one host, NCK64, expressed helveticin J. This is the first example in which a bacteriocin from lactic acid bacteria was expressed in a heterologous host.

Expression of helveticin J appeared to be dependent on the genetic complement of the expression host, which harbored some functional genetic determinants for bacteriocin production and immunity. *L. acidophilus* NCK64 is a derivative from the lactacin F producer (*L. acidophilus* 11088). NCK64 does not produce the bacteriocin but retains immunity (Muriana & Klaenhammer, 1987). A related expression host, NCK89, which does not produce lactacin F nor bear the immunity phenotype did not retain intact pTRK135 or express helveticin J. NCK89 has incurred a deletion in the lactacin F region of at least 2.2 kb (Muriana & Klaenhammer, 1991b). Alternatively, NCK64 retains this region suggesting that loss of lactacin F production resulted from a point mutation in the structural gene which did not disturb lactacin F-related processing or immunity functions. The failure to express helveticin J in NCK89 suggests that all the essential processing and immunity information for the production of helveticin J are either not present on the fragment cloned in pTRK135, or not expressed properly in *L. acidophilus*. In contrast, since the *L. acidophilus* NCK64 (pTRK135) expresses helveticin J, some essential function (processing, export, or immunity) must be provided *in trans* which results in the heterologous expression of helveticin J. There are numerous questions that remain to be answered concerning the processing systems of the different classes of bacteriocins produced by lactic acid bacteria. Noting the differences in the organization of genes within the helveticin J and lactacin F operons (see below) it will be very interesting to determine the similarities, differences, and complementarity in the processing and immunity systems for these bacteriocins. This type of information is vital to any future effort to direct the expression of bacteriocins in heterologous hosts for applications in food fermentation or preservation.

MOLECULAR ORGANIZATION OF THE HELVETICIN J OPERON

The fragment cloned in pTRK135 that encodes expression of helveticin J was sequenced initially over 2,655 bp (Joerger & Klaenhammer, 1990). Two complete open reading frames (ORF2 and ORF3) were identified originally and implicated in the production of helveticin J. Subsequently, the helveticin J region was resequenced and its flanking regions characterized (Fremaux & Klaenhammer, unpublished). The expansion of these sequence data will be presented elsewhere, but three corrections are noted in the published helveticin J sequence (Joerger & Klaenhammer, 1990; GenBank M30121):

@ nt-1141 GGATCAT instead of GGATCcAT
@ nt-1196 CGGcTT instead of CGGTT
@ nt-3029 GCGAT instead of CCGgAT

Figure 1 shows the overall organization of the region and defines the putative helveticin J operon. First,

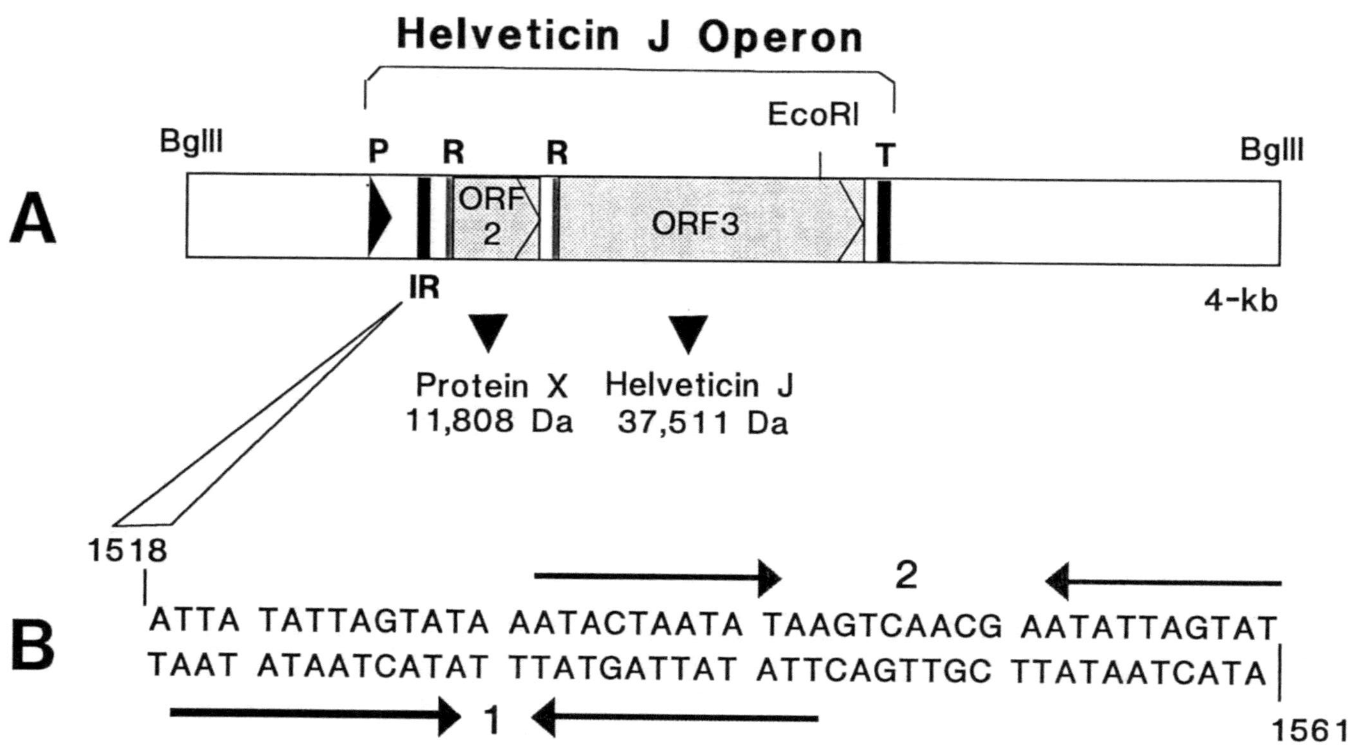

Figure 1A: Organization of the helveticin J operon within a 4 kb *Bgl*II fragment cloned from *L. helveticus* 481 into pGK12. P, promoter; T, a rho-independent terminator; R, ribosomal binding site; IR shows the position of two inverted repeats which could form two stem loop structures.
Figure 1B: Sequence of the region defining two inverted repeats within the helveticin J operon. The free energy of folding for hairpin #1 is -11.4 kcal and hairpin 2 is -8.4 kcal.

ORF3 (999 bp) has an excellent coding probability (92%) and could encode a protein of the expected size of purified helveticin J, 37,511 daltons (Joerger & Klaenhammer, 1986). Second, the reactive lambda gtll clones which contained inserts of 600 and 350 bp from the helveticin J producer, *L. helveticus* 481, were internal to ORF3. These data further implicated ORF3 as the structural gene for helveticin J, designated *hlvJ*. Thirty base pairs upstream from the *hlvJ* gene ends ORF2 (315 bp), which could encode an 11,808 dalton protein. The function of the putative ORF2 protein remains unknown at this time, but its organization relative to *hlvJ* suggests that it is an important component of an apparent bacteriocin operon. Surrounding these two open reading frames are the following putative expression signals:

(1) promoter-like sequences upstream from ORF2;
(2) ribosomal binding sites upstream from ORF2 and *hlvJ*;
(3) a rho-independent terminator 37 bp after the *hlvJ* stop codon, and;
(4) two inverted repeats located between the putative promoter and ribosomal binding site of ORF2.

The general organization of this region and the location of putative promoter, ribosomal binding sites, and terminator strongly suggest that ORF2 and *hlvJ* are organized in an operonlike structure. In addition to ORF2, four additional complete open reading frames have also been defined that surround the helveticin J operon (Fremaux & Klaenhammer, unpublished). It remains to be determined what roles, if any, these gene products play in the immunity, processing, or expression of helveticin J.

An additional feature of the promoter region are the two inverted repeats positioned between the putative promoter and ribosomal binding site upstream from ORF2 (Figure 1). Inverted repeats that serve as SOS-boxes have been defined in the regulatory regions of a number of inducible *E. coli* bacteriocins (van den Elzen *et al.*, 1982; Akutsu *et al.*, 1989). These regions serve as binding sites for the LexA protein which blocks transcription of the bacteriocin operons. Under "inducible" conditions that activate the SOS system, the LexA protein is cleaved and the bacteriocin operon is transcribed. Holo *et al.* (1991) have also defined an inverted repeat in the regulatory region of lactococcin A (LCN-A), a bacteriocin produced by *Lactococcus lactis* subsp. *cremoris*. In the case of helveticin J, it is not clear what role, if any, the inverted repeats play in the regulation or expression of helveticin J. Neither helveticin J nor lactococcin A have been reported to be inducible. In fact, little information is available on the inducibility of bacteriocins in lactic acid bacteria. Noting the presence of the inverted repeats in the suspected regulatory regions of these two bacteriocins, the inducibility of bacteriocins in lactic acid bacteria may be worth re-examination in future efforts to screen strains for new or more effective antimicrobials.

LACTACIN F

In a survey for bacteriocins produced by *L. acidophilus*, four strains were identified that produced a bacteriocin with a spectrum of activity that extended beyond the *L. delbrueckii* species to include two other residents of the gastrointestinal tract, *L. fermentum* and *Enterococcus faecalis* (Barefoot & Klaenhammer, 1983). *L. acidophilus* 11088 (NCK88) was selected for further study and its bacteriocin, lactacin F, has been characterized both biochemically and genetically. One of the more interesting features of lactacin F is its heat-stability. Since it retains full activity after autoclaving for 15 min at 121 C, the bacteriocin can be easily used as a selective component in bacteriological media. Lactacin F was employed to select transconjugants in matings where the genetic determinants for production (Laf^+) and immunity (Laf^r) were transferred by an episomal element (Muriana & Klaenhammer, 1987).

Lactacin F has been purified and biochemically characterized as a 6,300-dalton hydrophobic peptide (Muriana & Klaenhammer, 1991a). Compositional analysis originally estimated the peptide to be 54-57 amino acids in length and protein sequencing defined 25 amino acids from the N-terminus of the active bacteriocin. A 63-mer oligonucleotide probe was deduced and used to select a clone bearing a 2.2-kb *Eco*RI DNA fragment from a genomic library of the lactacin F producer, *L. acidophilus* 11088 (Muriana & Klaenhammer, 1991b). This fragment was subcloned onto an *E. coli* - *Lactobacillus* shuttle vector and the recombinant plasmid (pTRK162) introduced into two expression hosts that were deficient in lactacin F production, NCK64 (Laf^- Laf^r) and NCK89 (Laf^- Laf^s). With the introduction of pTRK162, transformants of both strains produced lactacin F. However, the lactacin F-producing colonies of NCK89 were small, variable in size, and produced significantly less bacteriocin than pTRK162 transformants of NCK64. Recent work by C. Ahn in our laboratory (unpublished) has shown that repeated propagation of the Laf^+ NCK89 (pTRK162) clones under bacteriocin-producing conditions (pH of 7.6) eventually results in loss of culture viability. In contrast, NCK89 (pTRK162) shows healthy colony formation and can be propagated continuously under conditions where lactacin F is not produced (pH 5.5 or less). The Laf^r NCK64 expression host supports excellent lactacin F production and can be propagated without adverse effects under optimal conditions for bacteriocin production. Collectively, these observations indicate that the 2.2-kb fragment which encodes genetic determinants for lactacin F production does not express, or fully express, immunity to the bacteriocin. It remains possible, however, that the immunity gene product is present and expressed in NCK89 but fails to function properly in this host.

The complete DNA sequence of the lactacin F structural gene determined previously (Muriana & Klaenhammer, 1991b) has been extended to confirm and define flanking upstream and downstream regions; these data will be published elsewhere (Ahn & Klaenhammer, unpublished). Collectively, DNA sequence analysis has currently identified four open reading frames within the 2.2-kb *Eco*RI fragment that are organized in an apparent operon-like structure (Figure 2). First, a putative promoter and ribosomal binding site are found upstream from the ORF that encodes the lactacin F prepeptide. The DNA sequence predicts that the gene product from the LAF/ORF is a 75 amino acid peptide with the following sequence:

1 Met Lys Gln Phe Asn Tyr Leu Ser His Lys Asp Leu Ala
Val Val Val Gly Gly* **Arg Asn Asn Trp Gln Thr Asn Val**
Gly Gly Ala Val Gly Ser Ala Met Ile Gly Ala Thr Val
Gly Gly Thr Ile Cys Gly Pro Ala Cys Ala Val Ala Gly
Ala His Tyr Leu Pro Ile Leu Trp Thr Gly Val Thr Ala
Ala Thr Gly Gly Phe Gly Lys Ile Arg Lys 75

The sequence of the first 25 amino acids of mature lactacin F, as determined from purified bacteriocin, was identified following the Arg (19) residue (underlined above). Lactacin F was the first non-lanthionine bacteriocin characterized from lactic acid bacteria where both DNA and protein sequence information were available.

The initial data supported the following conclusions:

(1) lactacin F is translated as a prepeptide with an 18 amino acid N-terminal extension;
(2) the bacteriocin is post-translationally processed by cleavage at a specific site [Val-Val-Gly-Gly* Arg (+1)], and;
(3) the mature hydrophobic peptide is 57 amino acids in length and predictions indicate formation of transmembrane helices.

The 18 amino acids comprising the N-terminal extension of the lactacin F prepeptide have several features that are characteristic of signal sequences (von Heijne, 1986): a positively-charged residue at the N-terminus (Arg), a central region of hydrophobic amino acids (Leu-Ala-Val-Val-Val), and a C-terminal cleavage site that shows small uncharged amino acids at positions -1 (Gly) and -3 (Val). It is not clear, however, if the N-terminal extension functions as a signal sequence. Secondary structure predictions via Garnier *et al.* (1978) identified an alpha-helical structure at the N-terminus and a beta turn at the processing site which could place the splicing region into juxtaposition with a peptidase. The "processed" lactacin F peptide retains two hydrophobic beta-sheets that are predicted to form a transmembrane helix. Hydropathy plots based on the methods of Kyte and Doolittle (1982) and Klein *et al.* (1985) are shown in Figure 3 and illustrate the hydrophobic character of the leader sequence and the lactacin F peptide itself. The processing site, defined by the arginine residue at position 19, is denoted by the negative hydropathy index in this region. The small size of the lactacin F peptide, the extent of beta structure, and its excellent potential to form a transmembrane helix implicate the cell membrane as the probable target of this antimicrobial peptide. Purified lactacin F was active against *Lactobacillus* and *Enterococcus* protoplasts in preliminary experiments (Muriana & Klaenhammer, 1991a) and we are currently investigating this further.

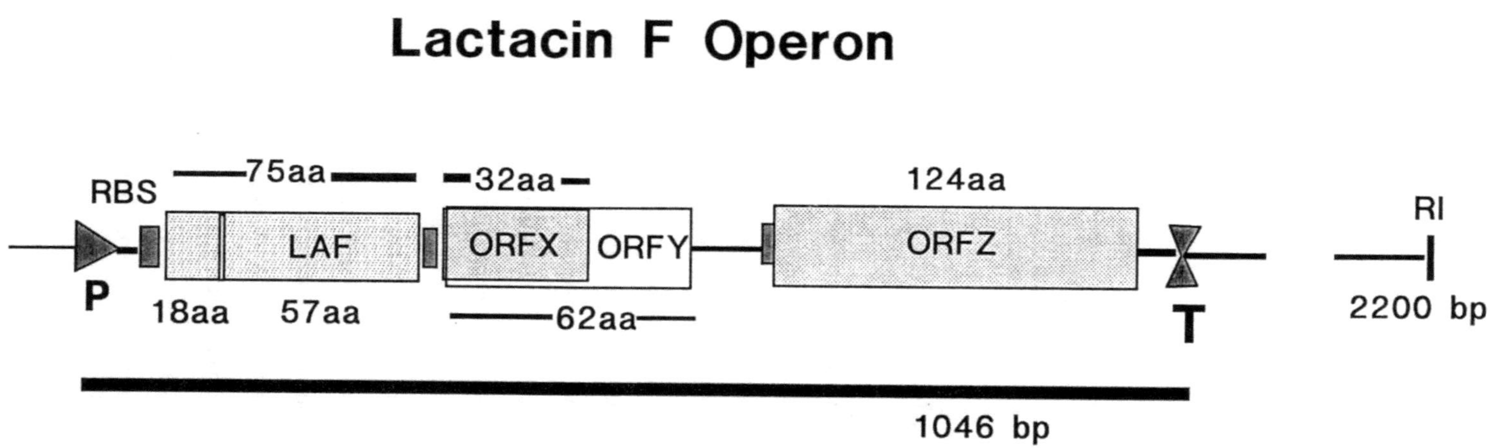

Figure 2: Organization of the lactacin F operon within a 2.2-kb fragment cloned from *L. acidophilus* 11088: P, promoter; T, a rho-independent terminator; RBS, ribosomal binding sites. LAF is the structural gene for pre-lactacin F showing the 18 amino acid leader sequence and 57 amino acid mature peptide.

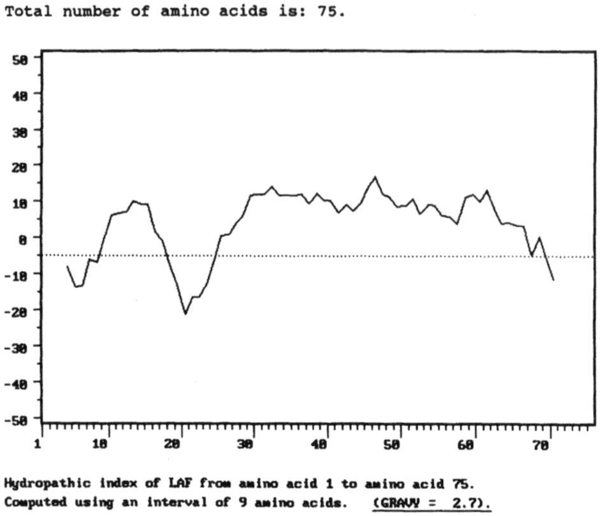

Figure 3: Hydropathic indices of the pre-lactacin F peptide computed using the PC GENE based programs of Kyte and Doolittle (1982) and Klein *et al.* (1985).

Following the gene encoding pre-lactacin F are two superimposed open reading frames, ORFX (32 amino acids) and Y (62 amino acids). The actual translation products produced from one or both ORFs have not yet been evaluated. Recent subcloning and expression studies have shown that disruption of ORFX/Y eliminates the Laf^+ phenotype (Ahn & Klaenhammer, unpublished data). Therefore, although their specific roles remain to be identified, one or both of these gene products are essential for expression of the Laf^+ phenotype. Expanded sequencing efforts downstream from ORFX/Y have recently identified a fourth open reading frame (ORFZ) that could encode a protein of 124 amino acids (Ahn & Klaenhammer, unpublished). The role of ORFZ remains unknown at this time. ORFZ is followed by a rho-independent terminator. The general organization of the genes, expression signals, and terminator over 1046 bp establishes an operon-like structure. This structure is most similar to the lactococcin determinants (now designated *lcnMa, lcnMb, lcnM*; van Belkum et al., 1992) organized in an operon-like structure of ORF-Al, ORF-A2, and ORF-A3 (van Belkum *et al.*, 1991). As with lactacin F, both ORF-Al and ORF-A2 were required for lactococcin production. In contrast to lactacin F, ORF-A3 appeared to be responsible for bacteriocin immunity. Additional experiments are being conducted to determine the roles of ORFX/Y and ORFZ in lactacin F production, immunity, and processing. The evidence to date strongly suggests that lactacin F, like helveticin J and the lactococcins, is organized in an operon structure and a number of gene products from the operon are likely involved in the expression, processing, export, and immunity functions of this *Lactobacillus* bacteriocin.

COMMON PROCESSING SITES IN PEPTIDE BACTERIOCINS

There are six peptide bacteriocins from lactic acid bacteria where the DNA or protein sequences are now described. These include the published sequences for lactacin F (Muriana & Klaenhammer, 1991b) and the lactococcins (van Belkum *et al.*, 1991, 1992), pediocin PA-1 (Gonzalez & Kunka, 1987; Marugg *et al.*, 1991; van Belkum, 1991), and leucocin A-UAL 187 produced by *Leuconostoc gelidum* UAL 187 (Hastings *et al.*, 1990; Hastings *et al.*, 1991; ME Stiles, personal communication). A comparison of the N-terminal extensions of lactacin F, leucocin A, pediocin PA-1, and lactococcins A, B, and Ma is shown in Figure 4. The arrow denotes the processing site for the bacteriocins where both the DNA sequence and amino acid sequence have been independently determined. The residues boxed in the lactacin F sequence are found in the identical position in one or more of the other five peptide bacteriocins. In all six cases there are N-terminal extensions of either 18 or 21 amino acids in length with a methionine and lysine at the amino terminus of the leader peptide. The DNA sequence of leucocin A (ME Stiles, personal communication) harbors two methionine residues that could start two possible ORFs generating either a 24 or 21 amino acid extension. Within all six prepeptides are two glycine residues in the -1 and -2 positions of the processing site. This strongly indicates that the Gly-Gly residues are a common feature of the processing site for prepeptide bacteriocins in lactic acid bacteria. van Belkum (1991) has also noted this feature and suggested that there may be a general processing mechanism for the maturation of small hydrophobic bacteriocins in lactic acid bacteria. This remains to be experimentally borne out, however, through expression of structural bacteriocin determinants in heterologous strains that harbor the general prepeptide processing mechanisms.

In the four cases where both the DNA sequence and N-terminal amino acid sequence of the processed peptide are known, the +l position is occupied by a postively-charged amino acid (see Figure 4). The N-terminal extension and processing sites of lactacin F are consistent with those of signal sequences (von Heijne, 1986). The leader peptide of lactacin F also exhibits good hydropathicity within its core region (Figure 3), supporting a possible membrane interaction. However, the 18 amino acid N- terminal extension is short relative to most signal peptides. The other prepeptides bearing the Gly-Gly processing site do not show features characteristic of signal peptides. For example, lactococcins A, B, Ma, and leucocin A exhibit low hydropathic indices over their leader regions (data not shown). Isoleucine and serine occupy the +l position in lactococcins Ma and B, respectively, suggesting that a charged amino acid is not essential for processing at the Gly-Gly site; however, the N- terminus of these bacteriocins has not yet been determined by amino acid sequencing of the mature peptide. The -3 position of lactococcins A and Ma is also occupied by a large and polar amino acid, asparagine, which is not found in sites processed by signal peptidases (von Heijne, 1986). Although the N-terminal extension of lactacin F shares some important features which are characteristic of signal peptides, the data accumulated thus far suggests that these small prepeptide bacteriocins are subject to a processing mechanism that is distinct of signal peptidases.

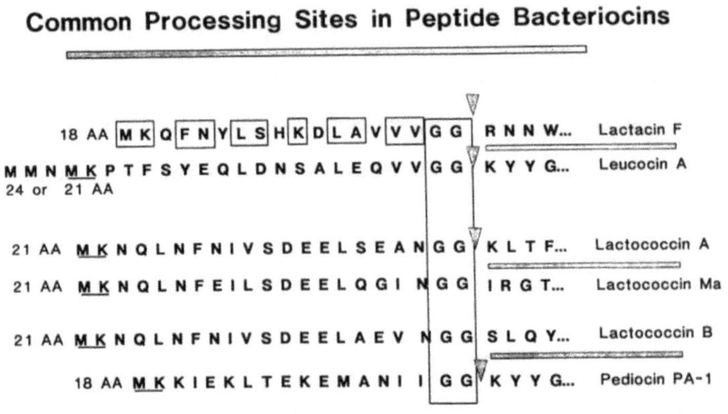

Figure 4: N-terminal amino acid sequences of six prepeptide bacteriocins produced by different lactic acid bacteria representing the genera *Lactobacillus* (lactacin F), *Leuconostoc* (leucocin A), *Lactococcus* (lactococcin A, Ma, B) and *Pediococcus* (pediocin). The processing sites, indicated by the inverted triangles, were determined by comparison of the DNA sequences with the known amino acid sequences of the mature peptides. Boxed residues illustrated in lactacin F are found in one or more of the other prepeptides at the identical position within the N- terminal extension.

The common features of prepeptide bacteriocins from lactic acid bacteria characterized to date are the following:

(1) an N-terminal extension;

(2) the -1 and -2 positions from the processing site are occupied by glycine-glycine;

(3) the two residues at the N-terminus of the leader peptide are methionine and lysine, and;

(4) the +1 position is occupied by a positively-charged amino acid (in those cases where the N-terminus of the mature peptide has been determined).

CONCLUSIONS

The lactobacilli produce an array of antimicrobial proteins which likely contribute to their competitive ability in a variety of nutritionally complex environments. The long-standing beneficial association of the lactic acid bacteria with man and his food supply provides added significance to the characterization of their bacteriocins, since research efforts may ultimately result in practical applications for these antimicrobials in food and fermentation systems. Although numerous genes and operons are now defined, the specific roles of the proteins produced from these operons remain to be investigated in depth. The most exciting developments will come from understanding the regulation and controls for bacteriocin production and immunity in lactic acid bacteria. This knowledge will provide new insight into the competitive roles of lactic acid bacteria in fermenting ecosystems and facilitate the design of bacteriocin systems that might be turned on or off by environmental signals at specified points in a fermentation process. Lastly, our understanding of the gene systems, regulation, and processing steps will elucidate the significant structure-function relationships for these bacteriocins. This in turn is expected to fuel genetic approaches to design new antimicrobial proteins for use in food systems using native bacteriocin systems which are available in lactic acid bacteria.

ACKNOWLEDGEMENTS

The research efforts on lactacin F and helveticin J systems have been funded in part by the National Dairy Promotion and Research Board, Arlington, VA, and Nestle, Ltd., Switzerland. We gratefully thank Dr. Susan Barefoot of Clemson University, Dr. Mike Stiles and Dr. John Vederas of the University of Alberta, Canada, for sharing their unpublished results on lactacin B, leucocin A-UAL 187, and the carnobacteriocins for comparison within this presentation. We also acknowledge and thank Dr. P. Muriana and Dr. M. Joerger for their original contributions which have led to further genetic characterization of the lactacin F and helveticin J systems.

Paper number FS91-50 of the Journal Series of the Department of Food Science, North Carolina State University, Raleigh, NC 27695-7624. The use of trade names in this publication does not imply endorsement by the North Carolina Agricultural Research Service of the products named, nor criticism of similar ones not mentioned.

REFERENCES

Ahn C, Stiles ME (1990) Plasmid-associated bacteriocin production by a strain of *Carnobacterium piscicola* from meat. Appl Environ Microbiol 56:2503-2510

Akutsu A, Masaki H, Ohta T (1989) Molecular structure and immunity specficity of colicin E6, an evolutionary intermediate between E-group colicins and cloacin DF13. J Bacteriol 171:6430-6436

Barefoot SF, Klaenhammer TR (1983) Detection and activity of lactacin B, a bacteriocin produced by *Lactobacillus acidophilus*. Appl Environ Microbiol 45:1808-1815

Barefoot SF, Klaenhammer TR (1984) Purification and characterization of the *Lactobacillus acidophilus* bacteriocin lactacin B. Antimicrobial Agents Chemother 26:328-334

Daeschel MA, McKenney MC, McDonald LC (1990) Bactericidal activity of *Lactobacillus plantarum* C11. Food Microbiol 7:91-98 de Klerk HC (1967) Bacteriocinogeny in *Lactobacillus fermenti*. Nature (London) 214:609

de Klerk HC, Coetzee JN (1961) Antibiosis among lactobacilli. Nature (London) 192:340-341

de Klerk HC, Smit JA (1967) Properties of a *Lactobacillus fermenti* bacteriocin. J Gen Microbiol 48:309-316

Garnier J, Osguthorpe DR, Robson B (1978) Analysis of the accuracy and implications of simple methods for predicting the secondary structure of globular proteins. J Mol Biol 120:97-120

Gonzalez CF, Kunka BS (1987) Plasmid-associated bacteriocin production and sucrose fermentation in *Pediococcus acidilactici*. Appl Environ Microbiol 53:2534-2538

Gudmundsdottir A, Stoffels G (1991) Characterization of a bacteriocin isolated from a psychrotrophic lactic acid bacterium. EMBO-FEMS-NATO Symposium Poster, September (1991), FRANCE

Hamdan IY, Mikolajcik EM (1974) Acidolin: an antibiotic produced by *Lactobacillus acidophilus*. J Antibiotics 27:631

Harris LJ, Daeschel MA, Stiles ME, Klaenhammer TR (1989) Antimicrobial activity of lactic acid bacteria against *Listeria monocytogenes*. J Food Prot 52:384-387

Hastings JW, Sailer M, Johnson K, Roy KL, Vederas JC, Stiles ME (1991) Characterization of leucocin A-UAL 187 and cloning of the bacteriocin gene from *Leuconostoc gelidum*. J Bacteriol 173:7491-7500

Hastings JW, Sailer M, Vederas JC, Stiles ME (1990) Antibiosis of a *Leuconostoc* sp. isolated from meat. FEMS Microbiol Rev 87:P86

Henkel T, Sailer M, Vederas JC, Worobo RW, Quandri L, Stiles ME (1991) Purification and characterization of bacteriocins produced by *Carnobacterium piscicola* LV17. EMBO-FEMS-NATO Symposium Poster, September (1991), FRANCE

Holo H, Nilssen O, Nes IF (1991) Lactococcin A, a new bacteriocin from *Lactococcus lactis* subsp. *cremoris*: isolation and characterization of the protein and its gene. J Bacteriol 173:3879-3887

Huynh TV, Young RA, Davis RW (1984) Constructing and screening cDNA libraries in lambda-gtl0 and lambda-gt11. In: Glover D (ed) DNA cloning techniques: a practical approach. IRL Press, Oxford, p:46-78

Jimenez-Diaz R, Piard JC, Ruiz-Barba JL, Desmazeaud MJ (1990) Isolation of a bacteriocin-producing *Lactobacillus plantarum* from a green olive fermentation. FEMS Microbiol Rev 87:P91

Joerger MC, Klaenhammer TR (1986) Characterization and purification of helveticin J and evidence for a chromosomally determined bacteriocin produced by *Lactobacillus helveticus* 481. J Bacteriol 167:439-446

Joerger MC, Klaenhammer TR (1990) Cloning, expression, and nucleotide sequence of the *Lactobacillus helveticus* 481 gene encoding the bacteriocin helveticin J. J Bacteriol 171:6339-6347

Johnson JL, Phelps CP, Cummins CS, London J, Gasser F (1980) Taxonomy of the *Lactobacillus acidophilus* group. Intern J Sys Bacteriol 30:53-68

Kashet ER (1987) Bioenergetics of lactic acid bacteria: cytoplasmic pH and osmotolerance. FEMS Microbiol Rev 46:233-244

Klaenhammer TR (1988) Bacteriocins of lactic acid bacteria. Biochimie 70:337-349

Klaenhammer TR (1990) Antimicrobial and bacteriocin interactions of the lactic acid bacteria. In: Heslot H, Davies J, Florent J, Bobichon L, Durand G, Penasse L (ed) Proceedings of the 6th International Symposium on Genetics of Industrial Microorganisms. Aug (1990). Societe Francaise de Microbiologie, France

Klein P, Kanehisa M, DeLisi C (1985) The detection and classification of membrane-spanning proteins. Biochimica et Biophysica Acta 815:468-476

Kok JJ, van der vossen JMBM, Venema G (1984) Construction of plasmid cloning vectors for lactic streptococci which also replicate in *Bacillus subtilis* and *Escherichia coli*. Appl Environ Microbiol 48:726-731

Kyte J, Doolittle RF (1982) A simple method for displaying the hydropathic character of a protein. J Mol Biol 157:105-132

Lewus CB, Kaiser A, Montville TJ (1991) Inhibition of food-borne bacterial pathogens by bacteriocins from lactic acid bacteria isolated from meat. Appl Environ Microbiol 57:1683-1688

Lindgren SE, Dobrogosz WJ (1990) Antagonistic activities of lactic acid bacteria in food and feed fermentations. FEMS Microbiol Rev 87:149-164

Marugg J, Chikindas M, Toonen M, Zoctmulder L, Ledeboer A, van Wassnenaar D, Henderson J, Vandenbergh P (1991) Molecular characterization and sequence analysis of genes involved in production of pediocin PA-1, a bacteriocin from *Pediococcus acidilactici*. EMBO-FEMS-NATO Symposium Poster, September (1991), FRANCE

McCormick EL, Savage DC (1983) Characterization of *Lactobacillus* sp. strain 100-37 from the murine gastrointestinal tract: ecology, plasmid content, and antagonistic activity toward *Clostridium ramosum* H1. Appl Environ Microbiol 46:1103-1112

Mortvedt CI, Nes I (1990) Plasmid-associated bacteriocin production by a *Lactobacillus sake* strain. J Gen Microbiol 136: 1601-1607

Mortvedt CI, Nissen-Meyer J, Nes IF (1991a) Lactocin S, a new lanthionine containing bacteriocin from *Lactobacillus sake*, purification and properties. EMBO-FEMS-NATO Symposium Poster, September (1991), FRANCE

Mortvedt CI, Nissen-Meyer J, Sletten K, Nes IF (1991b) Purification and amino acid sequence of lactocin S, a bacteriocin produced by *Lactobacillus sake* L45. Appl Environ Microbiol 57: 1829-1834

Muriana PM, Klaenhammer TR (1987) Conjugal transfer of plasmid encoded determinants for bacteriocin production and immunity in *Lactobacillus acidophilus* 88. Appl Environ Microbiol 53:553-560

Muriana PM, Klaenhammer TR (1991a) Purification and partial characterization of lactacin F, a bacteriocin produced by *Lactobacillus acidophilus* 11088. Appl Environ Microbiol 57:114-121

Muriana PM, Klaenhammer TR (1991b) Cloning, phenotypic expression, and DNA sequence of the gene for lactacin F, an antimicrobial peptide produced by *Lactobacillus* spp. J Bacteriol 173:1779-1788

Nettles CG, Barefoot SF, Bodine AB (1991) Purification and partial sequence of the *Lactobacillus acidophilus* bacteriocin lactacin B. Proceedings of the Annual Meeting of the Society for Industrial Microbiology, Philadelphia PA, (August 3-9). Abstract to be published in SIM News

Okereke A, Montville TJ (1991) Bacteriocin inhibition of *Clostridium botulinum* spores by lactic acid bacteria. J Food Protection 54:349-353

Rammelsberg M, Radler F (1990) Antibacterial polypeptides of *Lactobacillus* species. J Appl Bacteriol 69:177-184

Schillinger U, Holzapfel WH (1990) Antibacterial activity of carnobacteria. Food Microbiol 7:305-310

Schillinger U, Lucke FK (1989) Antibacterial activity of *Lactobacillus sake* isolated from meat. Appl Environ Microbiol 55: 1901-1906

Schillinger U, Kaya M, Lucke FK (1991) Behavior of *Listeria monocytogenes* in meat and its control by a bacteriocin-producing strain of *Lactobacillus sake*. J Appl Bacteriol 70:473-478

Silva M, Jacobus NV, Deneke C, Gorbach SL (1987) Antimicrobial substance from a human *Lactobacillus* strain. Antimicrobial Agents Chemother 31:1231-1233

Talarico TL, Dobrogosz WJ (1989) Chemical characterization of an antimicrobial substance produced by *Lactobacillus reuteri*. Antimicrobial Agents Chemother 33:674-679

Toba T, Yoshioka E, Itoh T (1991a) Potential of *Lactobacillus gasseri* isolated from infant faeces to produce bacteriocin. Letters in Appl Microbiol 12:228-231

Toba T, Yoshioka E, Itoh T (1991b) Acidophilucin A, a new heat labile bacteriocin produced by *Lactobacillus acidophilus* LAPT 1060. Letters in Appl Microbiol 12:106-108

Toba T, Yoshioka E, Itoh T (1991c) Lacticin, a bacteriocin produced by *Lactobacillus delbrueckii* subsp. *lactis*. Letters in Appl Microbiol 12:43-45

Upreti GC, Hinsdill RD (1973) Isolation and characterization of a bacteriocin from a homofermentative *Lactobacillus*. Antimicrobiol Agents Chemother 4:487-494

Upreti GC, Hinsdill RD (1975) Production and mode of action of lactocin 27: bacteriocin from a homofermentative *Lactobacillus*. Antimicrobial Agents Chemother 7:139-145

van den Elzen PJM, Maat J, Walters HHB, Velkamp E, Nijkamp HJJ (1982) The nucleotide sequence of the bacteriocin promoters of plasmids Col DF13 and Col EI: role of *lexA* repressor and cAMP in the regulation of promoter activity. Nucleic Acids Res 10:19131928

van Belkum M (1991) PhD Thesis, Lactococcal bacteriocins: genetics and mode of action. University of Groningen, Department of Genetics, The Netherlands

van Belkum M, Hayema BJ, Jeeninga RE, Kok J, Venema G (1991) Organization and nucleotide sequences of two lactococcal bacteriocin operons. Appl Environ Microbiol 57:492-498

van Belkum M, Kok J, Venema G (1992) Cloning, sequencing, and expression in *Escherichia coli* of *LCNB*, a third bacteriocin determinant from the lactococcal bacteriocin plasmid p9B4-6. Appl Environ Microbiol (to be published)

Vincent JG, Veomett RC, Riley RF (1959) Antibacterial activity associated with *Lactobacillus acidophilus*. J Bacteriol 78:477-484

von Heijne G (1986) A new method for predicting signal sequence cleavage sites. Nucleic Acids Res 14:4683-4690

West C, Warner PJ (1988) Plantacin B, a bacteriocin produced by *Lactobacillus plantarum* NCDO 1193. FEMS Microbiology Letters 49:163-165

LACTOCOCCAL BACTERIOCINS: GENETICS AND MODE OF ACTION

M.J. van Belkum[1], B.J. Hayema[1], J. Kok[1], G. Venema[1], H. Holo[2], I.F. Nes[2], W.N. Konings[3], and T. Abee[3]
Department of Genetics[1] and Department of Microbiology[3]
University of Groningen
Kerklaan 30, 9751 NN Haren
The Netherlands

INTRODUCTION

Lactic acid bacteria produce a variety of antimicrobial substances which are important in food fermentation and preservation. In several instances, the inhibitory activity results from metabolic end products such as hydrogen peroxide, diacetyl, and organic acids (Lindgren & Dobrogosz, 1990). In addition, the bactericidal activity of many strains appeared to result from bacteriocin production (Klaenhammer, 1988). Although bacteriocins of lactic acid bacteria have been the subject of many studies, only little is known about their chemical structure, their mode of action, and their genetic determinants. In recent years, the increasing interest in bacteriocins produced by these organisms has resulted in the cloning and genetic characterization of several bacteriocin determinants (Joerger & Klaenhammer, 1990; Marugg, 1991; Muriana & Klaenhammer, 1991). From lactococci, the structural gene for the lantibiotic nisin has been cloned and sequenced by several groups (Buchman *et al.*, 1988; Kaletta & Entian, 1989; Dodd *et al.*, 1990). The genetic determinant for bacteriocin production by *Lactococcus lactis* subsp. *lactis* WM_4 was shown to be associated with the 131-kb plasmid pNP2 (Scherwitz *et al.*, 1983). Cloning experiments in *L. lactis* identified an 18.4-kb DNA region containing the bacteriocin determinant (Scherwitz-Harmon & McKay, 1987).

In an extensive program, Geis *et al.* (1983) have screened 280 strains of lactococci for bacteriocin production. On the basis of secretion in liquid medium of substances that were sensitive to proteolysis, precipitable with ammonium sulfate, and which could inhibit the growth of closely related bacteria, 16 strains were identified which produced bacteriocin. Based on their heat stability at different pH values and their spectrum of inhibition, the bacteriocins could be classified into eight groups. Bacteriocin production and immunity of four of these bacteriocin-producing strains were transferred by conjugation to a plasmid-free *L. lactis* recipient (Neve *et al.*, 1984). One of these conjugative plasmids was p9B4-6 of 60 kb from *L. lactis* subsp. *cremoris* 9B4. This plasmid was used in our study to clone the genes specifying antagonistic activity.

Laboratory of Microbial Gene Technology[2], P.O.Box 51, N-1432 ÅS-NLH, Norway

NATO ASI Series, Vol. H 65
Bacteriocins, Microcins and Lantibiotics
Edited by R. James, C. Lazdunski and F. Pattus

CLONING OF THE GENETIC DETERMINANTS FOR LACTOCOCCIN A, B AND M PRODUCTION AND IMMUNITY

A restriction enzyme map of p9B4-6 was made (Figure 1) and about 70% of this plasmid was subcloned in *Escherichia coli*. The recombinant plasmids were subsequently transferred to *L. lactis* subsp. *lactis* IL1403 to screen for antagonistic activity (van Belkum *et al.*, 1989). Two distinct regions on the 60-kb plasmid were identified and cloned which specified bacteriocin production as well as immunity: a 7.9-kb fragment with low antagonistic activity (pMB200), and a 15-kb fragment specifying high antagonistic activity (pMB500).

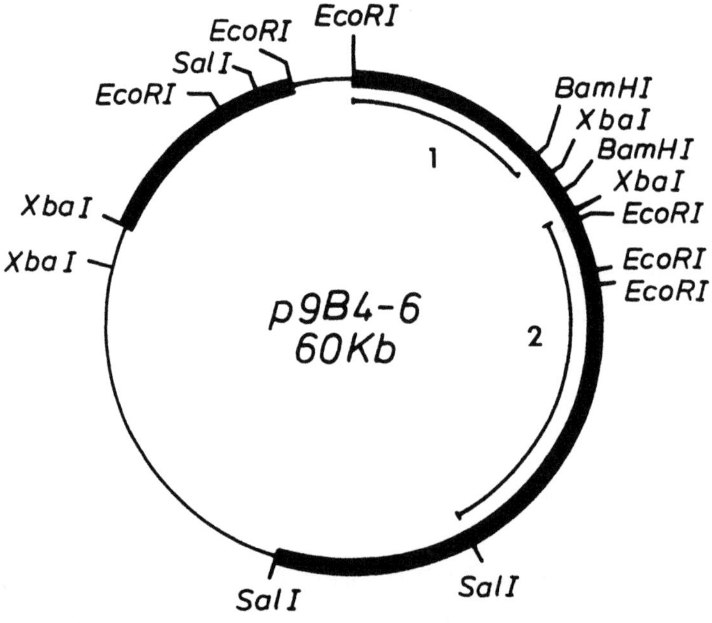

Figure 1. Restriction enzyme map of the bacteriocin plasmid p9B4-6. The thick bars represent the parts of p9B4-6 cloned in *E. coli*. Only those parts associated with bacteriocin activity are further specified as follows: 1, 7.9-kb *Eco*RI-*Bam*HI fragment in pMB200; 2, 15-kb *Sal*I-*Xba*I fragment in pMB500.

The inhibitory substances produced by the two clones were sensitive to proteolysis, confirming their proteinaceous nature. By deletion analyses, the determinants for high and low antagonistic activity could be further confined to a 1.3-kb *Sca*I-*Hin*dIII fragment and a 1.8-kb *Sca*I-*Cla*I fragment, respectively. Both fragments were sequenced completely and analyzed by deletion and mutation analyses (Figure 2) (van Belkum *et al.*, 1991). On the 1.8-kb fragment, three genes were identified that were transcribed as an operon. The first two genes, *lcnM* and *lcnN*, could encode polypeptides of 69 and 77 amino acids, respectively, and were involved in the production of the bacteriocin designated lactococcin M. The third gene, *lciM*, contained 154 codons and specified immunity towards lactococcin M. On the 1.3-kb fragment an operon containing two genes were present encoding polypeptides of 75 and 98 amino acids, respectively (Figure 2). The first gene, *lcnA*, specified bacteriocin activity, the second encoded the corresponding immunity *(lciA)*. Downstream of *lciA* two inverted repeats were identified which could act as rho-independent terminators. The bacteriocin specified by *lcnA* was termed lactococcin A. Using primer extension analyses, a promoter upstream of *lcnM* and *lcnA* was identified. As the nucleotide sequences upstream of *lcnM* and *lcnA* as well as the first 20 bp of both genes appeared to be identical, the difference in antagonistic activity we observed could not be caused by a difference in promoter strength. Lactococcin M and A appeared to have different specificities.

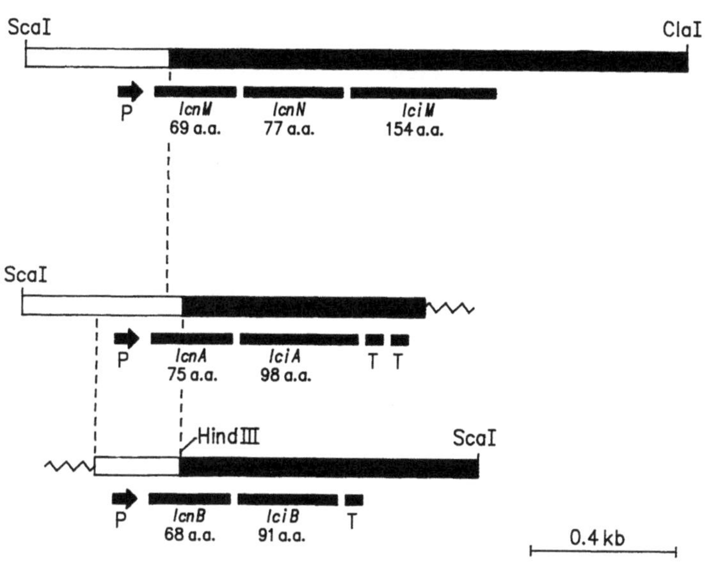

Figure 2. Schematic representation of the lactococcin M operon on the 1.8-kb *ScaI-ClaI* fragment and the lactococcin A and B operons on the 2.1-kb *ScaI* fragment. The region of homology between the nucleotide sequence upstream of *lcnM, lcnA* and *lcnB* and the 5' part of these genes is represented by the open bar. Abbreviations: P, promoter; T, putative rho-independent terminator; a.a., amino acids.

Downstream of the lactococcin A operon, sequence analysis revealed the presence of an additional bacteriocin determinant. The DNA region adjacent of the lactococcin A operon was cloned from the 15-kb fragment and analyzed (van Belkum *et al.*, 1992). Also in this case, two genes were present in an operon structure (Figure 2). The first gene, *lcnB*, contained 68 codons and specified lactococcin B. The second gene, *lciB*, contained 91 codons and was responsible for lactococcin B immunity. Downstream of *lciB* a putative rho-independent terminator was identified. The specificity of lactococcin B was different from that of lactococcin A and M. Also in this case, a nucleotide sequence upstream of, and encompassing the 5' part of the bacteriocin structural gene, showed similarity to the equivalent sequences of the other two bacteriocin operons. The similarity of these regions may suggest that recombinational events may have assembled these genes on p9B4-6.

L. lactis cells carrying a 2.1-kb *ScaI* fragment containing both the lactococcin A and B operon (Figure 2) showed a somewhat reduced level of antagonistic activity as compared to that produced by *L. lactis* cells containing the 15-kb fragment from which the 2.1-kb fragment was derived. No further bacteriocin determinants were present on the 15-kb fragment, indicating that some other additional information on this fragment is responsible for optimal expression of the bacteriocin phenotype.

LACTOCOCCIN A IS PRODUCED AS A PRECURSOR WITH AN N-TERMINAL EXTENSION

Recently, Holo *et al.* (1991) purified and analyzed a bacteriocin from *L. lactis* subsp. *cremoris* LMG 2130 which appeared to be encoded by a gene identical to *lcnA*. In their strain of *L. lactis* subsp. *cremoris*, the gene appeared to be located on a 55-kb plasmid. Lactococcin A was found to be effective against lactococci only. The mature lactococcin A polypeptide is composed of 54 amino acids and its amino acid sequence is located in the C-terminus of the 75 amino acid polypeptide encoded by *lcnA*. Apparently, *lcnA* encodes a precursor of lactococcin A from which an N-terminal extension of 21 amino acids is proteolytically cleaved off. Comparison of the primary translation products of *lcnA*, *lcnM* and *lcnB* revealed that the N-terminal 21 amino acids of these polypeptides are nearly identical (Figure 3), suggesting that the latter two are also produced as precursors and cleaved at equivalent sites.

```
lcnM   (69 a.a.)   M K N Q L N F E I L S D E E L Q G I N G G | I R G T G K
lcnA   (75 a.a.)   M K N Q L N F N I V S D E E L S E A N G G | K L T F I Q
lcnB   (68 a.a.)   M K N Q L N F N I V S D E E L A E V N G G | S L Q Y V M
laf    (75 a.a.)         M K Q F N Y L S H K D L A V V V G G | R N N W Q T
ORFl   (62 a.a.)         M K K I E K L T E K E M A N I I G G | K Y Y G N G
```

Figure 3. The N-terminal part of the polypeptides encoded by *lcnM*, *lcnA*, *lcnB*, *laf*, and ORFl (the open reading frame encoding the precursor of pediocin PA-1). The cleavage site at the C-terminal side of the two adjacent glycine residues is indicated by a dashed line.

Cleavage of the precursor of lactococcin A, and probably also those of lactococcins B and M, occurs at the C-terminal side of two adjacent glycine residues. Identical cleavage sites have also been identified in the precursors of two other bacteriocins, namely lactacin F *(laf)* from *Lactobacillus acidophilus* 11088 (Muriana & Klaenhammer, 1991)) and pediocin PA-1 (ORF 1) from *Pediococcus acidilactici* PAC1.0 (Henderson *et al.*, 1991; Marugg, 1991) (Figure 3). These observations suggest that a general processing mechanism underlies the maturation of small hydrophobic bacteriocins in lactic acid bacteria. The assumption that the *lcnA* and *lcnB* products are processed at identical positions seems to be confirmed by the fact that the mature lactococcin A and B polypeptides migrate to the same position as judged from their biological activity on an SDS-polyacrylamide gel. The antagonistic activity of lactococcin M was too low to detect a zone of inhibition on an SDS-polyacrylamide gel. The function of *lcnN* in lactococcin M activity remains unclear. A possible explanation might be that lactococcin M is a heterodimer of the *lcnM* and *lcnN* gene products, or that *lcnN* is involved in maturation and/or secretion of the bacteriocin. Using the vector pT580, in which the T7 RNA polymerase-specific promoter of pT712 directed the expression of the lactococcin B operon in *E. coli* BL21(DE3), we were able to demonstrate antagonistic activity in extracts of this host (Figure 4) (van Belkum *et al.*, 1992). However, the rate of migration in an SDS-polyacrylamide gel of the polypeptide produced by *E. coli* was less than that of the mature lactococcin B, suggesting that *E. coli* is not able to process the bacteriocin precursor.

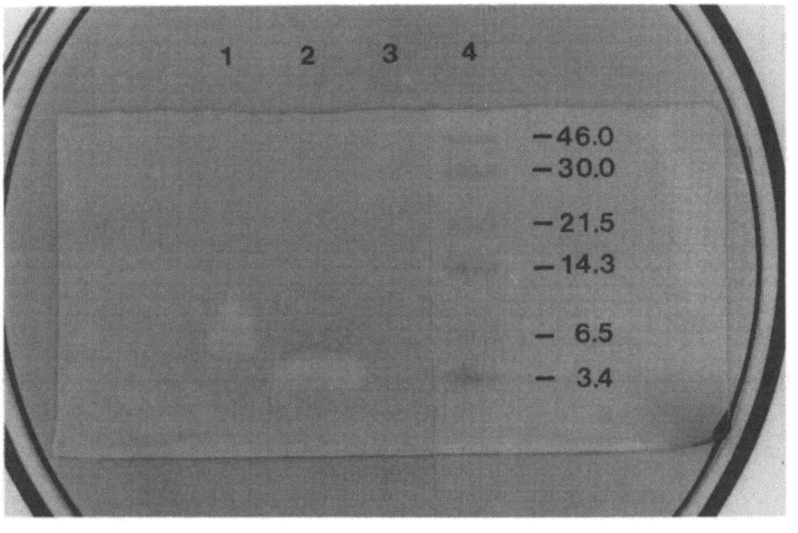

Figure 4. Detection of bacteriocin activity in a tricine-SDS-polyacrylamide gel by the overlay test using *L. lactis* IL1403 as indicator. Lanes: 1, and 3, lysates of *E. coli* BL21(DE3) containing pT580, and pT712, respectively; lane 2, supernatant of *L. lactis* cells producing lactococcin B; lane 4, Molecular size standard (Rainbow protein molecular weight marker; Amersham International, Amersham, England): molecular sizes (in kilodaltons) are shown on the right.

MODE OF ACTION OF LACTOCOCCIN A

Since lactococcins are small hydrophobic polypeptides, a possible target for their action might be the cytoplasmic membrane. The purified lactococcin A was used to study its mode of action on whole cells

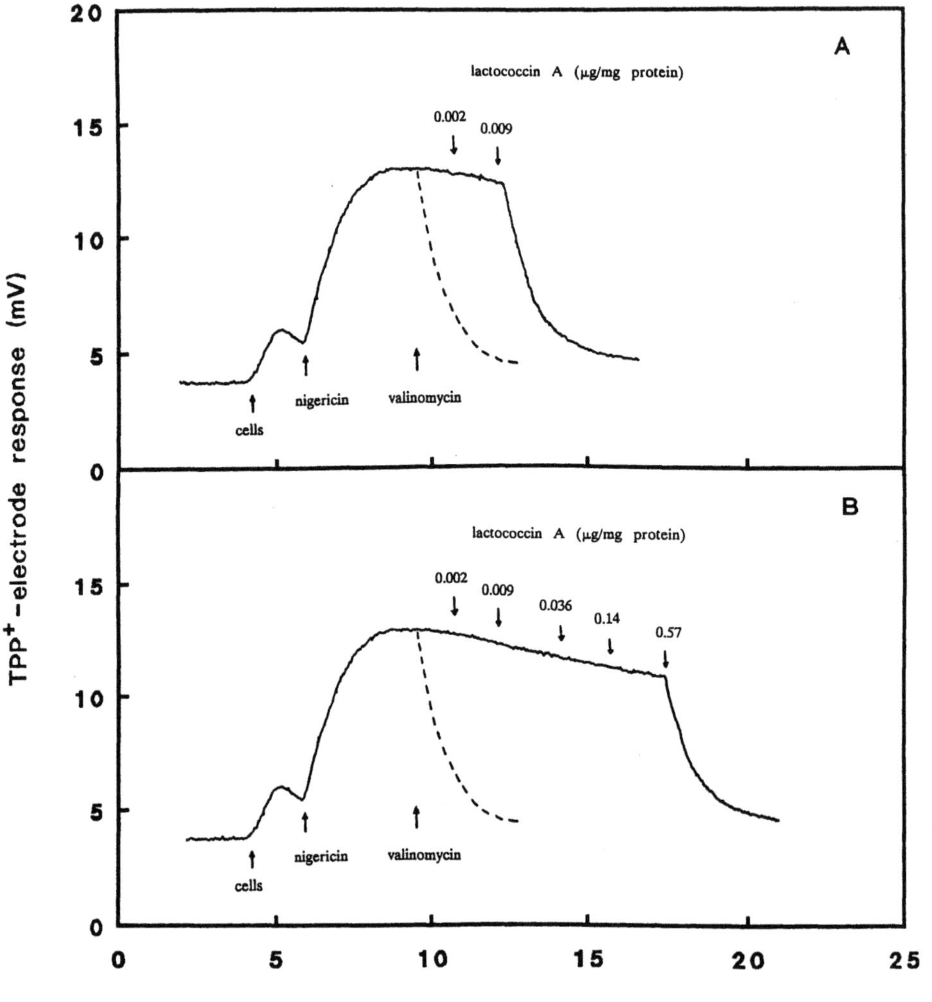

Figure 5. Effect of lactococcin A on the membrane potential of energized cells of *L. lactis* IL1403 (A) and *L. lactis* IL1403(pMB563) (B). The membrane potential was measured in the presence of nigericin with a TPP^+-ion selective electrode. The effect of valinomycin (1 µM) is indicated with the broken lines. Cells of IL1403 and IL1403(pMB563) were treated with increasing concentrations of lactococcin A. Taken from van Belkum *et al.* 1991)

and membrane vesicles of sensitive and immune lactococcal strains (van Belkum *et al.*, 1991). Lactococcin A dissipated the membrane potential of sensitive *L. lactis* IL1403 cells, as was monitored by the distribution of the lipophilic cation tetraphenylphosphonium (TPP^+) by using a TPP^+-selective electrode (Figure 5). A similar collapse of the membrane potential was observed after the addition of the ionophore valinomycin. A significantly higher concentration of lactococcin A was needed to dissipate the membrane potential of IL1403 containing plasmid pMB563 which carries the lactoccocin A-specific immunity gene (van Belkum *et al.*, 1992). Lactococcin A at a concentration which was enough to dissipate the membrane potential of IL1403 but not of strain IL1403(pMB563) was used to study its effect on the uptake of glutamate. Glutamate is taken up by an ATP-dependent unidirectional uptake process in *L. lactis* and is, therefore, not driven by proton motive force (Poolman *et al.*, 1987). The addition of the ionophores valinomycin and nigericin to cells of strain IL1403 did not result in efflux of glutamate. However the addition of lactococcin A to IL1403 resulted in inhibition of glutamate uptake (Figure 6B) and induced leakage of accumulated glutamate (Figure 6A), indicating that lactococcin A affected the permeability of the cytoplasmic membrane.

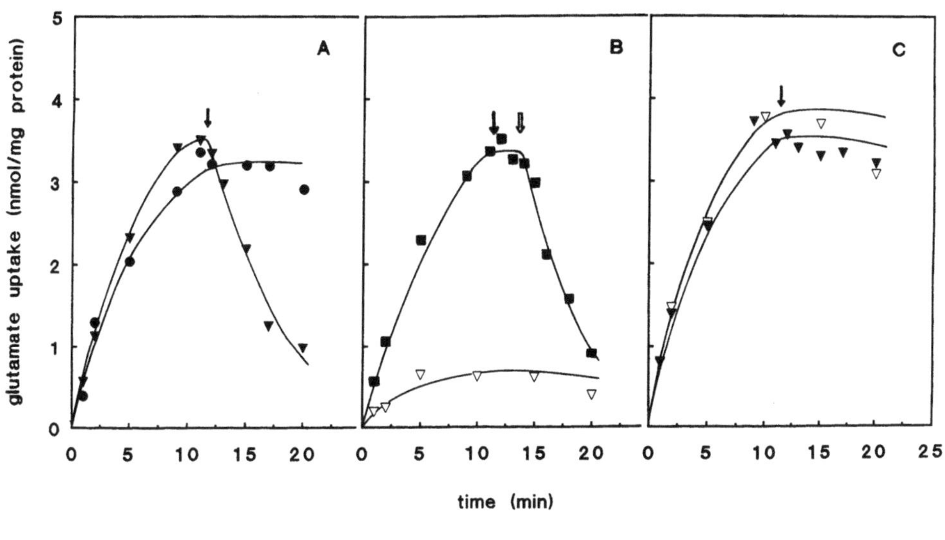

Figure 6. Effect of lactococcin A on L-glutamate uptake in energized cells of *L.lactis* IL403 (A and B) and *L.lactis* IL403(pMB563) (C). In seperate experiments valinomycin (1 μM) plus nigericin (0.5 μM) (●), or lactococcin A (0.029 μg/mg protein) (▼) were added to the cell suspensions at the times indicated by an arrow. Uptake was also monitored in an experiment (B) in which first valinomycin (1 μM) plus nigericin (0.5μM) and subsequently lactococcin A (0.029 μg/mg protein) were added to the cell suspension (indicated with the closed and open arrow respectively) (■). Uptake in cells of IL403 and IL403(pMB563) preincubated for 3 min with lactococcin A (0.029 μg/mg protein) is indicated with the symbol (▽). The assays were started by the addition of 1.75 μM of ^{14}C-labeled glutamate. Taken from van Belkum *et al.* 1991.

The activity of lactococcin A on whole cells was proton motive force-independent as lactococcin A-induced glutamate efflux still took place when cells of strain IL1403 were treated with valinomycin and nigericin to dissipate the proton motive force before lactococcin A was added (Figure 6B). This is in contrast to the other pore-forming bacteriocins which have a voltage-dependent activity (Schein *et al.*, 1978; Konisky, 1982; Ruhr & Sahl, 1985; Pressler *et al.*, 1986; Kordel *et al.*, 1988; Schüller *et al.*, 1989; Bourdineaud *et al.*, 1990; Wilmsen *et al.*, 1990; Gao *et al.*, 1991). The same concentration of lactococcin A did not affect the uptake of glutamate by cells of strain IL1403(pMB563) (Figure 6C).

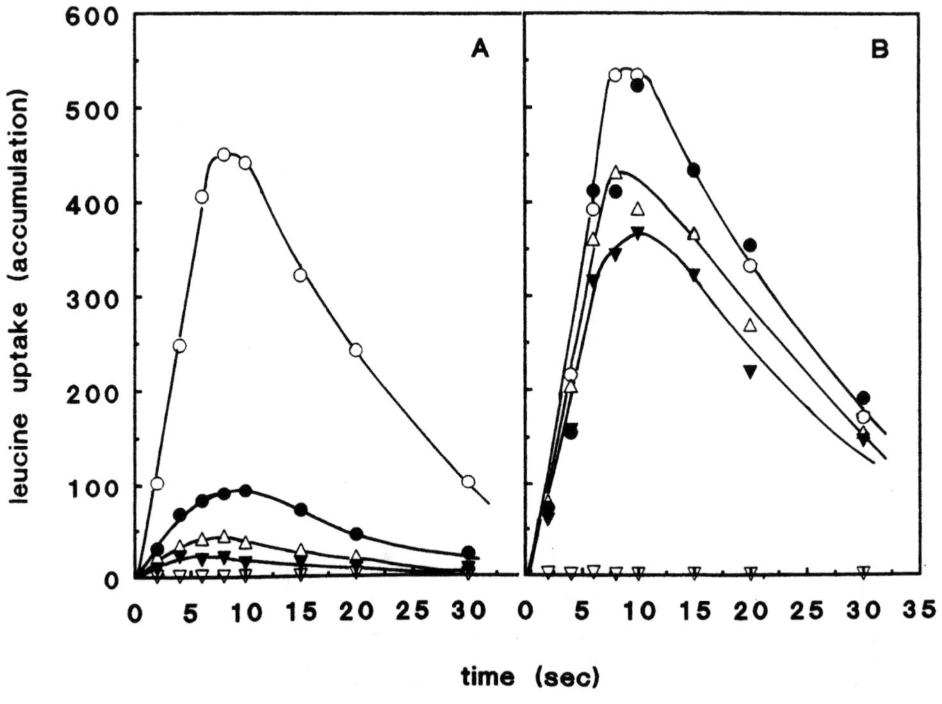

Figure 7. Effect of lactococcin A on the uptake of leucine driven by an artificially imposed proton motive force in cytoplasmic membrane vesicles derived from *L. lactis* IL1403 (A) and *L. lactis* IL1403(pMB563) (B). Proton motive force-driven uptake was started by diluting the K-Acetate-loaded, valinomycin-treated membrane vesicles 100-fold in 20 mM NaP_i (pH 6), 100 mM NaPipes, and 2 mM $MgSO_4$ containing 1.6 μM of [^{14}C]-labeled leucine. Symbols: ○ , membrane vesicles not preincubated with lactococcin A; uptake after 10 min preincubation of the membrane vesicles with lactococcin A at concentrations of 0.05 (●), 0.12 (△), and 0.25 (▼) μg/mg protein, respectively. Uptake of leucine in the absence of an imposed proton motive force was done by diluting the membrane vesicles in 20 mM KP_I (pH 6), 100 mM K-Acetate, 2 mM $MgSO_4$, and [^{14}C]-labeled leucine (▽). Taken from van Belkum *et al.* (1991).

The hypothesis that the cytoplasmic membrane is the primary target for lactococcin A was confirmed by the observation that proton motive force-driven uptake of leucine by membrane vesicles from strain IL1403 was strongly inhibited when the membrane vesicles were preincubated with lactococcin A (Figure 7A). In contrast, the uptake of leucine by membrane vesicles derived from IL1403(pMB563) was hardly affected (Figure 7B). This result also indicates that the immunity of IL1403(pMB563) is associated with the cytoplasmic membrane. Furthermore, lactococcin A inhibited leucine counterflow in membrane vesicles of the sensitive strain, indicating that lactococcin A also increased the permeability of membrane vesicles in the absence of a proton motive force (data not shown). The efflux of essential substrates by sensitive cells can explain the bactericidal activity of lactococcin A. Membrane vesicles of *Clostridium acetobutylicum, Bacillus subtilis* and *E. coli* were not affected by lactococcin A, nor were liposomes derived from phospholipids of *L. lactis* (data not shown) . These results reflect the narrow spectrum of activity of lactococcin A and indicate that a membrane-associated protein specific for lactococci is required for lactococcin A to be effective.

CONCLUSIONS

Molecular analysis of the lactococcal bacteriocin plasmid p9B4-6 has revealed the presence of three different bacteriocin determinants each of which have been cloned and sequenced. The fact that upstream of the three bacteriocin operons similar DNA regions are present which contain the 3' end of an ORF (preliminary results), suggests that the lactococcin operons are (or was in the case of the lactococcin B operon) part of a larger functional region whose combined presence on p9B4-6 is the result of recombinational events. The production of three different bacteriocins endowes the producer organism with a selective advantage over other bacteriocin producing lactococci. More study will be needed to unravel the mechanism of bacteriocin secretion, the bacteriocin receptor/target sites of the sensitive recipient and the exact mechanism by which the immunity system operates which is specific for each bacteriocin. The analysis of the amino acid sequence and the mode of action of lactococcin A has rapidly increased our knowledge of lactococcal bacteriocins. This may be of help to improve the antagonistic effectiveness of bacteriocins by molecular genetic techniques. Furthermore, bacteriocin determinants could be used in the development of food-grade gene cloning and expression vectors which are selectively retained in the bacterial population.

REFERENCES

Bourdineaud JP, Boulanger P, Lazdunski C, Letellier L (1990) *In vivo* properties of colicin A: Channel activity is voltage dependent but translocation may be voltage independent. Proc Natl Acad Sci USA 87:1037-1041

Buchman, GW, Banerjee S, Hansen JN (1988) Structure, expression, and evolution of a gene encoding the precursor of nisin, a small protein antibiotic. J Biol Chem 263:16260-16266

Dodd HM, Horn N, Gasson MJ (1990) Analysis of the genetic determinant for the production of the peptide antibiotic nisin. J Gen Microbiol 136:555-566

Gao FH, Abee T, Konings WN (1991) The mechanism of action of the peptide antibiotic nisin in liposomes and cytochrome *c* oxidase proteoliposomes. Appl Environ Microbiol 57:2164-2170

Geis A, Singh J, Teuber M (1983) Potential of lactic streptococci to produce bacteriocin. Appl Environ Microbiol 45:205-211

Henderson JT, Chopko AL, van Wassenaar P.D (1991) Purification and primary structure of the bacteriocin PA-1 produced by *Pediococcus acidilactici* PAC-1.0. Manuscript submitted

Holo H, Nilssen O, Nes IF (1991) Lactococcin A, a new bacteriocin from *Lactococcus lactis* subsp. *cremoris:* isolation and characterization of the protein and its gene. J Bacteriol 173:3879-3887

Joerger MC, Klaenhammer TR (1990) Cloning, expression, and nucleotide sequence of the *Lactobacillus helveticus* 481 gene encoding the bacteriocin helveticin J. J Bacteriol 172:6339-6347

Kaletta C, Entian K-D (1989) Nisin, a peptide antibiotic: cloning and sequencing of the *nisA* gene and posttranslational processing of its peptide product. J Bacteriol 171:1597-1601

Klaenhammer TR (1988) Bacteriocins of lactic acid bacteria. Biochimie 70:337-349

Konisky J (1982) Colicins and other bacteriocins with established modes of action. Ann Rev Microbiol 36:125-144

Kordel M, Benz R, Sahl H-G (1988) Mode of action of the staphylococcinlike peptide Pep5: voltage-dependent depolarization of bacterial and artificial membranes. J Bacteriol 170:84-88

Lindgren SE, Dobrogosz WJ (1990) Antagonistic activities of lactic acid bacteria in food and feed fermentations. FEMS Microbiol Reviews 87:149-164

Marugg J (1991) Personal communication

Muriana PM, Klaenhammer TR (1991) Cloning, phenotypic expression, and DNA sequence of the gene for lactacin F, an antimicrobial peptide produced by *Lactobacillus* spp. J Bacteriol 173:1779-1788

Neve H, Geis A, Teuber M (1984) Conjugal transfer and characterization of bacteriocin plasmids in group N (lactic acid) streptococci. J Bacteriol 157:833-838

Poolman B, Smid EJ, Konings WN (1987) Kinetic properties of a phosphate-bond-driven glutamate-glutamine transport system in *Streptococcus lactis* and *Streptococcus cremoris*. J Bacteriol 169:1460-1468

Pressler U, Braun V, Wittman-Liebold B, Benz R (1986) Structural and functional properties of colicin B. J Biol. Chem 261:2654-2659

Ruhr E, Sahl H-G (1985) Mode of action of the peptide antibiotic nisin and influence on the membrane potential of whole cells and on cytoplasmic and artificial membrane vesicles. Antimicrob Agents Chemother 27:841-845

Schein SD, Kagan B, Finkelstein A (1978) Colicin K acts by forming voltage dependent channels in phospholipid bilayer membranes. Nature (London) 276:159-163

Scherwitz KM, Baldwin KA, McKay LL (1983) Plasmid linkage of a bacteriocin-like substance in *Streptococcus lactis* subsp. *diacetylactis* strain WM4: transferability to *Streptococcus lactis*. Appl Environ Microbiol 45:1506-1512

Scherwitz-Harmon KM, McKay LL (1987) Restriction enzyme analysis of lactose and bacteriocin plasmids from *Streptococcus lactis* subsp. *diacetylactis* WM4 and cloning of *BclI* fragments coding for bacteriocin production. Appl Environ Microbiol 53:1171-1174

Schüller F, Benz R, Sahl H-G (1989) The peptide antibiotic subtilin acts by formation of voltage-dependent multi-state pores in bacterial and artificial membranes. Eur J Biochem 182:181-186.

van Belkum MJ, Hayema BJ, Geis A, Kok J, Venema G (1989) Cloning of two bacteriocin genes from a lactococcal bacteriocin plasmid. Appl Environ Microbiol 55:1187-1191

van Belkum MJ, Hayema BJ, Jeeninga RE, Kok J, Venema G (1991) Organization and nucleotide sequences of two lactococcal bacteriocin operons. Appl Environ Microbiol 57:492-498

van Belkum MJ, Kok J, Venema G, Holo H, Nes IF, Konings WN, Abee T (1991) The bacteriocin lactococcin A specifically increases permeability of lactococcal cytoplasmic membranes in a voltage-independent, protein-mediated manner. J Bacteriol 173:7934-7941

van Belkum MJ, Kok J, Venema G (1992) Cloning, sequencing, and expression in *Escherichia coli* of *lcnB*, a third bacteriocin determinant from the lactococcal bacteriocin plasmid p9B4-6. In press.

Wilmsen HU, Pugsley AP, Pattus F (1990) Colicin N forms voltage- and pH-dependent channels in planar lipid bilayer membranes. Eur Biophys J 18:149-158

INTRODUCTION TO THE LANTIBIOTICS SESSION

H.G. Sahl
Institute of Medical Microbiology and Immunology
University of Bonn
Sigmund-Freud-Str. 25
5300 Bonn-Venusberg
Germany

Lantibiotics are a group of antibacterial peptides which are distinguished from other bacteriocins of Gram-positive bacteria by unique structural features. The designation **lantibiotics** was proposed as abbreviation for **lan**thionine-containing peptide **antibiotics** to indicate their characteristic properties, i.e. the content of the thioether amino acid lanthionine and the antibacterial activity (Schnell *et al.*, 1988). Besides lanthionine and its analogue 3-methyllanthionine all lantibiotics contain the α,β unsaturated amino acids didehydroalanine and/or didehydrobutyrine. Individual lantibiotics may have further unusual residues such as S-aminovinyl-cysteine, *erythro*-3-hydroxyaspartic acid and lysinoalanine. A detailed description of these residues and the structures of lantibiotics is given by S. Freund and G. Jung in this session.

So far, lantibiotics are exclusively found to be produced by Gram-positive bacteria. Currently we distinguish two groups, type A lantibiotics including elongated peptides of molecular masses of 2100 Da to 3500 Da, and type B lantibiotics which are of globular shape and molecular masses of 1800 Da to 2100 Da. Type A lantibiotics are produced by lactococci, lactobacilli, staphylococci, streptococci and *Bacillus* species; the most prominent member of this group is the widely used food preservative nisin. Their antibacterial activity is primarily based on the formation of voltage-dependent, short-lived pores (see H.G. Sahl, this session). Solution structures of type A lantibiotics, as recently deduced from 1-D and 2-D NMR experiments and molecular dynamics simulation, provide the molecular basis for channel formation; the peptides proved to be screw-shaped, strongly amphiphilic and of sufficient length to span a membrane (see S. Freund and G. Jung, this session).

Type B lantibiotics are produced by streptomycetes only and have little antibacterial activity. However, they display other interesting biological activities; e.g. cinnamycin and the duramycin family are potent inhibitors of phospholipase A_2 and, therefore, promising substances for treatment of inflammation and allergy; ancovenin inhibits the angiotensin-converting enzyme and could be used for blood pressure regulation.

How are the unusual amino acid residues introduced into lantibiotics? Research in the past few years clearly demonstrated that lantibiotics derive from ribosomally synthesized prepeptides for which distinct

NATO ASI Series, Vol. H 65
Bacteriocins, Microcins and Lantibiotics
Edited by R. James, C. Lazdunski and F. Pattus

structural genes were detected. The prepeptides consist of hydrophilic, net negatively charged leader peptides and a positively charged propeptide region which is subject to several post-translational modifications to yield the active lantibiotic. These modifications include dehydration of serine and threonine residues to α,β-didehydroamino acids, addition of cysteine-sulphhydryl groups to form the thioether amino acids lanthionine and 3-methyllanthionin, further individual modifications and finally, proteolytic cleavage of the leader peptide. The sequence of the modification reactions was demonstrated by the isolation of prepeptides in different stages of modification (see H.G. Sahl, this session). The modifications are catalyzed by enzymes, the genes for which were found by complementation of mutants and DNA sequencing to be located adjacent to the lantibiotic structural genes. Generally, the biosynthesis genes are organized in operons, although at present the individual steps of biosynthesis cannot be definitely attributed to the identified genes. The current status of the molecular genetical studies on the biosynthesis of lantibiotics is summarized by K.D. Entian *et al.* (this session).

There has been some controversy on whether lantibiotics should be classified as peptide antibiotics or bacteriocins. The ribosomal synthesis (followed by modification) clearly separates lantibiotics from peptide antibiotics which are enzymatically synthezised. Moreover, particularly the type A lantibiotics have much in common with many bacteriocins from Gram-positive bacteria that do not contain modified amino acids. Such criteria include size (2000 to 4000 Da), a positive net charge, amphiphilicity, the mode of action (pore formation), a specific immunity system and an activity spectrum which is broader than that of the prototype bacteriocins, the colicins, but restricted to Gram-positive bacteria and therefore, more narrow than that of most antibiotics. It is tempting to speculate about a functional analogy of the channel-forming colicins and the membrane-depolarizing bacteriocins (including lantibiotics) of Gram-positive bacteria.

The narrow action spectrum of colicins is a consequence of the architecture of the Gram-negative cell wall which requires receptor-binding and complex translocation of the bacteriocin to the cytoplasmic membrane. The Gram-positive cell wall is a wide network of peptidoglycan strands and negatively charged teichoic or teichuronic acids through which peptides of the size of most Gram-positive bacteriocins can easily penetrate. There is no need for receptor-binding and translocation, and consequently these domains are not present in type A lantibiotics and similar bacteriocins. In this context, the recent finding that the voltage-gating of the colicin El channel can be attributed to a 36 amino acid, positively charged segment of the channel domain (Merrill & Cramer, 1990) is very interesting because the pore-forming bacteriocins of Gram-positive bacteria are of similar size and charge. It seems that the latter may be regarded as functional analogues to colicins being strictly reduced to the very minimum size required for pore formation with reflection to the different architecture of the Gram-positive cell wall. This analogy is further extended by the detection of an immunity peptide against the lantibiotic Pep5 (see H.G. Sahl, this session) which may have a functional organization similar to that found for immunity proteins of the channel-forming colicins.

For these reasons it was recommended at the first workshop on lantibiotics (Bad Honnef, 1991), where conventions on the nomenclature of lantibiotics were made (de Vos *et al.*, 1991), to designate lantibiotics as bacteriocin-like peptides. Moreover, at the Bendor conference it became clear that the lantibiotics have much in common with microcin B17 (see the session on Microcins) and colicin V (see the contribution of R. Kolter).

The future research on lantibiotics aims at understanding completely the molecular events that lead to production of an active lantibiotic. Currently, we do not know the precise role of the gene products encoded in the biosynthetic gene clusters; the catalytic mechanism of dehydration and thioether formation remains to be elucidated; the export mechanism is unclear, although MDR-transporters may be involved in excretion of either the active lantibiotic or modified but unprocessed prepeptides; the role of the unusual leader peptide in biosynthesis and/or export is obscure; the influence of the thioethers on the spatial structure of lantibiotics is not fully understood, and structure-function studies have to be performed to gain insight into the mechanism of channel formation. Once the fundamental questions of the biosynthesis of lantibiotics are answered there is hope to construct lanthionine-bridged peptides with unique properties regarding conformation and stability for use in agro-food industries and medical applications.

References

de Vos WM, Jung G, Sahl HG (1991) Definitions and nomenclature of lantibiotics. In: Jung G, Sahl HG (eds) Nisin and novel lantibiotics. ESCOM, Leiden, p 457

Merrill AR, Cramer WA (1990) Identification of a voltage-responsive segment of the potential-gated colicin El channel. Biochem 29:8529-8534

Schnell N, Entian KD, Schneider U, Gotz F, Zahner H, Kellner R, Jung G (1988) Prepeptide sequence of epidermin, a ribosomally synthesized antibiotic with four sulphide-rings. Nature 333:276-278

LANTIBIOTICS : AN OVERVIEW AND CONFORMATIONAL STUDIES ON GALLIDERMIN AND PEP5

Stefan Freund and Günther Jung
Institut für Organische Chemie,
Eberhard-Karls-Universität Tübingen,
Auf der Morgenstelle 18
7400 Tübingen, F. R. G.

Lantibiotics are polycyclic bacteriocins with the thioether amino acids lanthionine and β - methyllanthionine. In addition to these intrachain sulfide bridges, unsaturated amino acids like α,β-didehydroalanine and α, β - didehydrobutyric acid occur. Members of the lantibiotics family are for example nisin, an important food preservative, epidermin, a highly specific therapeutic drug against acne, a series of enzyme inhibitors and immunologically interesting peptides such as the duramycins. In recent years, an increasing number of novel lantibiotics has been discovered. Lantibiotics are produced via ribosomal synthesis from inactive precursor proteins (prelantibiotics), which are post-translationally converted to the biologically active polycyclic peptides. The enzyme involved herein catalyzes the dehydration of the serine and threonine residues to yield unsaturated residues. This step is followed by a stereospecific addition of the cysteine thiol groups to these double bonds leading to the formation of sulfide bridges. The final step is the release of the active peptide by proteolytic cleavage of the leader peptide. Conformational analysis of the lantibiotics, as well as their prepeptides, provide information about their mode of action and the steps of biosynthesis. Based on the knowledge of the lantibiotic biosynthesis, and by using the isolated novel enzymes, a gene technological construction of a variety of similarly modified polypeptides will be available in the near future.

Lantibiotics - an Overview

Heterodet polycyclic antibiotics possessing the thioether amino acids meso - lanthionine (Lan) and (2S,3S,6R) - 3 - methyllanthionine (MeLan) are called lantibiotics [Schnell *et al.*, 1988]. Most of them contain additional unusual residues such as the α, β - unsaturated residues 2,3 - didehydroalanine (Dha) or (Z) - 2,3 - didehydrobutyrine (Dhb) or they are further bridged via (2S, 9S) - lysinoalanine (LysN-Ala) or S - [(Z) - 2 - aminovinyl] - cysteine (Cys (Avi)). Other unusual components are *erythro* - 3 - hydroxyaspartic acid (HyAsp) and the N - terminally occuring 2 - oxobutyryl - group (Ob) (Figure 1, Table 1). A recent review article [Jung, 1991a] and a book [Jung & Sahl, 1991] summarize the results obtained on lantibiotics in different fields so far. This article follows, in parts, these reviews.

NATO ASI Series, Vol. H 65
Bacteriocins, Microcins and Lantibiotics
Edited by R. James, C. Lazdunski and F. Pattus

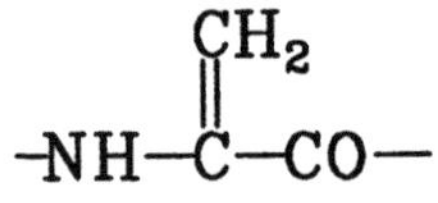

2.3-Didehydroalanine
Dha

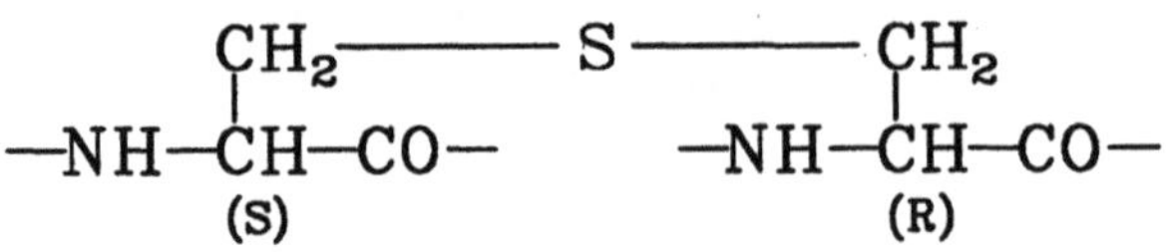

(2S,6R)-Lanthionine
meso-Lanthionine
Lan

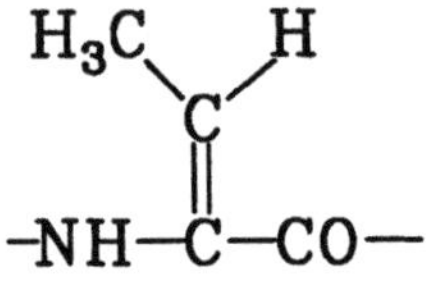

(Z)-2.3-Didehydrobutyrine
Dhb

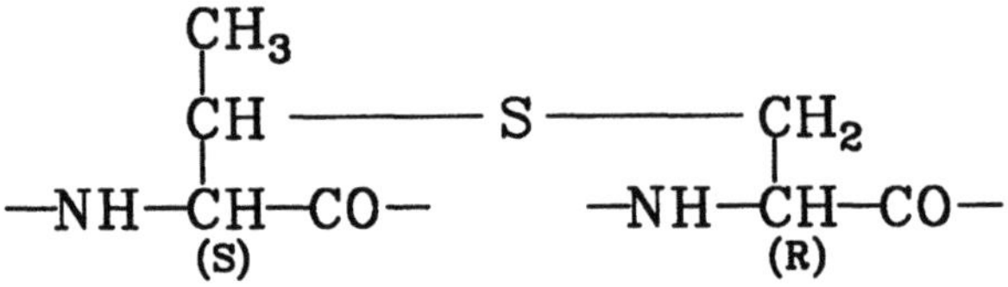

3-Methyllanthionine
MeLan

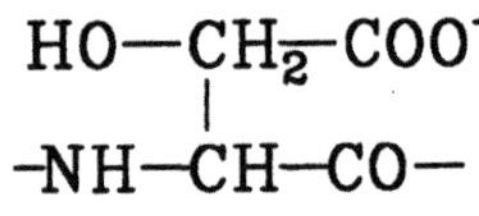

erythro-3-Hydroxy-L-aspartic acid
HyAsp

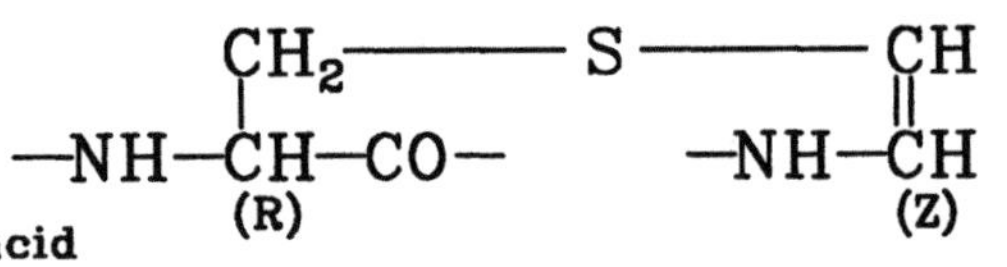

S-[(Z)-2-Aminovinyl]-D-cysteine
Cys(Avi)

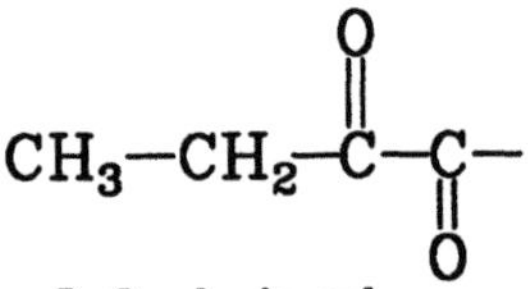

2-Oxobutyryl
Ob

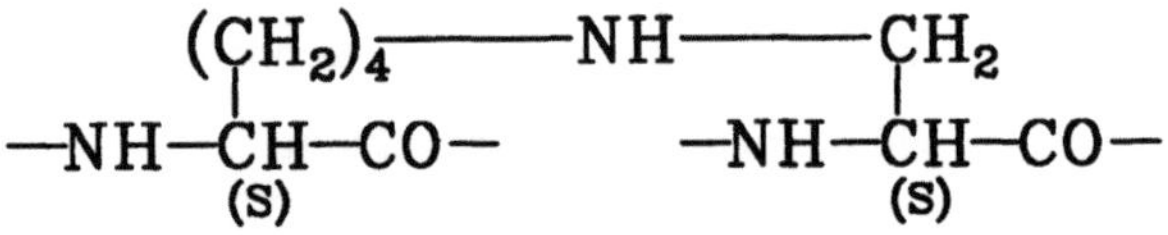

(2S,9S)-Lysinoalanine
LysN-Ala

Figure 1. Structural formula of unusual amino acid residues found in lantibiotics.

Table 1 : Lantibiotics of Type A and B

Lantibiotic	Mass	Naa[a]	Unusual amino acids[b]				
	[Da]		Lan	MeLan	Dha	Dhb	Others
Type A							
Pep5	3488	34	2	1	-	3	-
Nisin	3353	34	1	4	2	1	-
Subtilin	3317	32	1	4	2	1	-
Epidermin	2164	22	2	1	-	1	Cys (Avi)
Gallidermin	2164	22	2	1	-	1	Cys(Avi)
Mersacidin	1825	20	-	4	1	-	Cys(Avi)
Actagardine	1890	19	1	3	-	-	-
Type B							
Cinnamycin[c]	2041	19	1	2	-	-	HyAsp, LysN-Ala
Duramycin[d]	2012	19	1	2	-	-	HyAsp, LysN-Ala
Duramycin B,C	2012	19	1	2	-	-	HyAsp, LysN-Ala
Ancovenin	1959	19	1	2	1	-	-

[a] Total number of amino acids (Naa) in the precursor peptides, e. g. in Pep5, Thr in position 1 was counted instead of the N-terminal 2 - oxobutyryl residue.

[b] see Figure 1 for structural formula

[c] identical with Ro 09-0198, Lanthiopeptin

[d] identical with Leucopeptin

Lantibiotics differ in their overall structure, their molecular mass and their charges. They can be classified according to these criteria into two subtypes (Table 2). Pep5 [Sahl & Brandis, 1981; Sahl 1985; Kellner *et al.*, 1989] is the largest member of subtype A (Figure 2) with a molecular mass of 3488 Daltons, whereas duramycin type lantibiotics (subtypeB) Shotwell *et al.*, 1958; Navorro *et al.*, 1985; Kinne-Safran & Kinne, 1986; Chen & Thai , 1987; Choung *et al.*, 1988a,b; Dunkley *et al.*, 1988; Clejan *et al.*, 1989; Sokolve *et al.*, 1989; Fredenhagen *et al.*, 1990] have a molecular mass smaller than 2100 Dalton. The most important type A lantibiotic is nisin [Gross & Morell, 1971; Gross & Brown, 1976], a generally used food preservative. Subtype A lantibiotics are highly charged, with net positive charges between +2 and +7. Exceptions are the recently found mersacidin [Limbert *et al.*, 1991] and actagardine [Kettenring *et al.*,

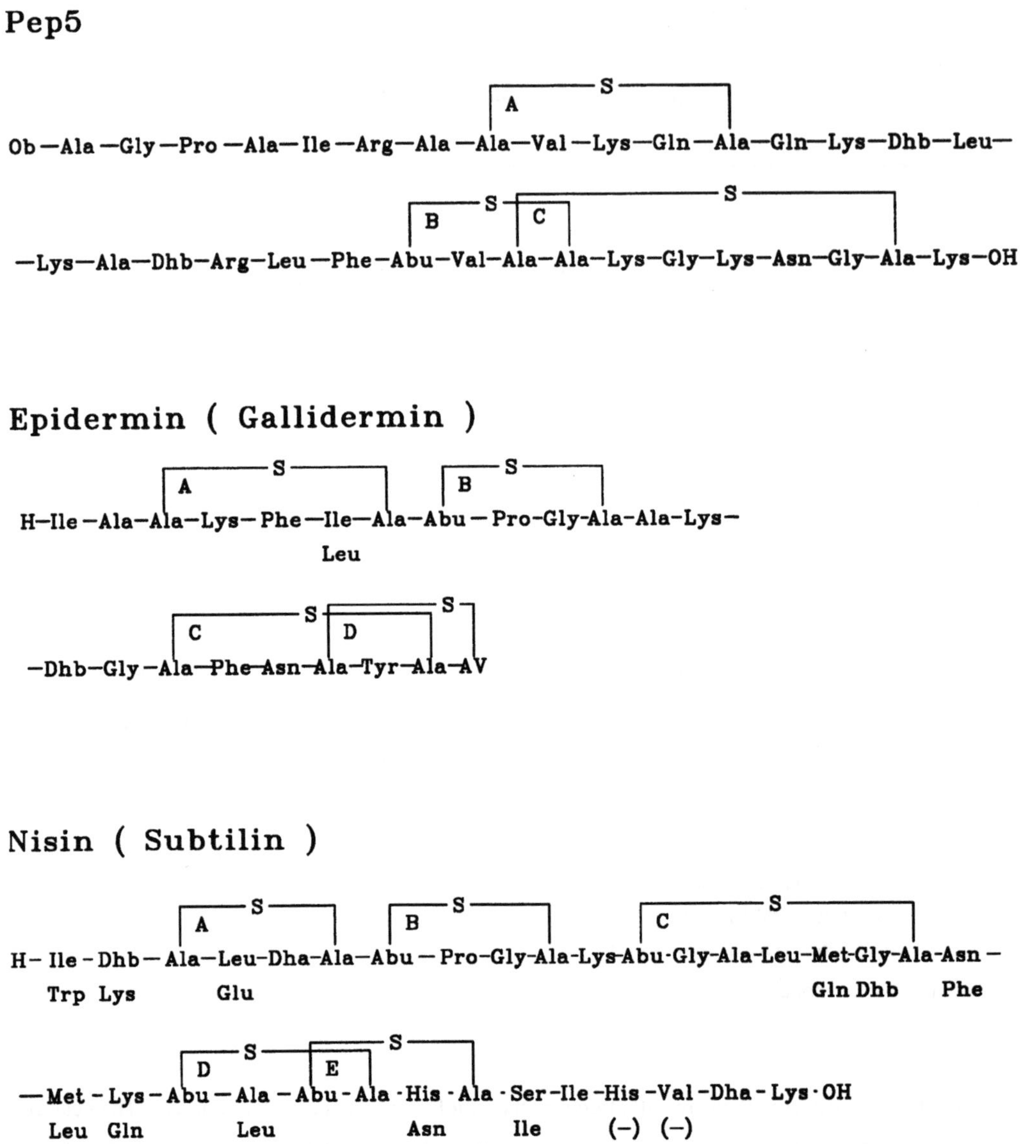

Figure 2. Primary Sequences of a few examples of Lantibiotics Type A.
The abbrevations of the unusual residues are defined in Figure 1. Gallidermin and subtilin show a close homology to epidermin and nisin respectively. The exchanged residues are marked below the primary sequence, deletions are shown as (-).

1990; Malabarba *et al.*, 1990] with net charges of 0 and -1 respectively. The properties of Lantibiotics Type A and B are different but similar within the subtypes. Only mersacidine and actagardine are difficult to classify.

Table 2: Classification of lantibiotics according to their charges, conformation and activity

Lantibiotic	**Charges**			**Properties**
	positive	negative	net	
Type A				
Pep5	8	1	7+	long stretched,
Nisin	4	1	3+	screw like / helical,
Subtilin	4	2	2+	cationic, amphiphilic
Epidermin	3	0	3+	voltage dependent
Gallidermin	3	0	3+	membrane pore-formers
Mersacidin	3	1	0	amphiphilic
Actagardine	1	2	1-	hydrophobic
Type B				
Cinnamycin[a]	3	2	1+	compact, almost neutral
Duramycin[b]	3	2	1+	amphiphilic
Duramycin B	3	2	1+	enzyme inhibitors
Duramycin C	2	2	0	immunologically
Ancovenin	2	2	0	active

[a] identical to Ro-09-0198, Lanthiopeptin
[b] identical to Leucopeptin

Judging from the primary sequence of lantibiotics of type A and from previous conformational studies, they are rod shaped with a screw - like or helical arrangement of the backbone. Although different in chain length, they show a clear structural relationship through homology and have a similar ring structure. The configuration of the individual thioether amino acids within subtype A is always D in the N - terminal, and L in the C - terminal amino acid, again with the exception of mersacidin and actagardine. Besides the unusual residues Lan and MeLan, only Dha and Dhb are found in this subtype A. Epidermin [Allgaier *et al.*, 1985; Allgaier *et al.*, 1986] and gallidermin [Kellner *et al.*, 1988] are further modified at the C- terminus (Cys(Avi)). Whereas subtype A producing strains are staphylococci, lactococci and bacilli, lantibiotics of the duramycin type (subtype B) are mainly produced by streptomycetes [Sahl, 1992]. They show a high homology among each other with molecular masses smaller than 2100 Daltons. In contrast to the basic lantibiotics of subtype A (nisin type), they have no or only +1 positive net charge. Apart from the two MeLan sulfide bridges and one Lan ring, which are located in the same positions, a lysinoalanine bridge between the sidechains of Lys^{19} and Dhb^{6} is a special feature of the duramycins (Figure 3). In the case of ancovenin, this bridge remains unformed, leaving an unsaturated dehydroalanine at position 6. An additional sulfide bridge between positions 1 and 18 further emphasizes the globular structure of the members of subtype B. All the duramycins show an *erythro* - 3 - hydroxyaspartic acid at position 15, again with the exception of ancovenin, which is not hydroxylated at this position. Further common features within the subtypes A and B can be found in connection with their biosynthesis and their conformations.

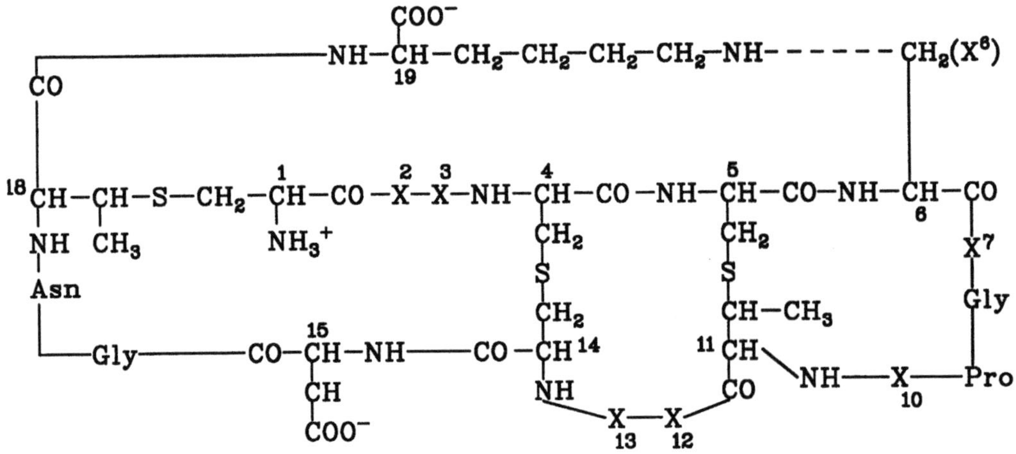

	Duramycin[a]	Cinnamycin[b]	Duramycin B[c]	Duramycin C[d]	Ancovenin
X^2	Lys	Arg	Arg	Ala	Val
X^3	Gln	Gln	Gln	Asn	Gln
X^6	Ala	Ala	Ala	Ala	Dha
X^7	Phe	Phe	Phe	Tyr	Phe
X^{10}	Phe	Phe	Leu	Leu	Leu
X^{13}	Phe	Phe	Phe	Trp	Trp
R	OH	OH	OH	OH	H

Figure 3. Bridging scheme of the lantibiotics Type B (duramycins).
a other name : leucopeptin
b other names : Ro 09-0198, lanthiopeptin
c proposed structures according to Fredenhagen 90
d 'Ala'bridged with Lys19 as lysinoalanine

Biosynthesis

The biosynthesis of lantibiotics will be discussed in detail by H.-G. Sahl and K-D. Entian in other articles in this volume. Only a brief summary of the steps involved in the formation of the mature lantibiotic will be presented here (Figure 4). The biosynthesis of lantibiotics differs from mechanisms involved in the formation of other peptide antibiotics, such as the thiotemplate mechanism [Kleinkauf & von Döhren, 1982]. As postulated for epidermin [Allgaier *et al.*, 1985, Allgaier *et al.*, 1986; Jung *et al.*, 1987] and also proven [Schnell *et al.*, 1988], the biosynthesis of all other lantibiotics consists of a ribosomal synthesis followed by several modification steps. The ribosomal synthesis yields linear precursor proteins consisting of a N-terminal leader sequence and a prolantibiotic part with Cys and Ser/Thr residues. The hydroxy-amino acid residues of the prolantibiotic are dehydrated enzymatically in a first modification step to the unsaturated residues dehydroalanine and dehydrobutyrine. The Thr/Ser residues in the hydrophilic and highly charged leader sequence remain unmodified, no Cys residues were found in these sequences.

The thioether bridged *meso* - lanthionine and (2S,3S,6R) - methyllanthionine are formed through an addition of the thiol groups of the Cys residues to the unsaturated residues, in the prolantibiotic part still attached to the leader sequence. The configuration is L - in the part stemming from the Cys residues, whereas the other half, which originates from the hydroxy - residues is inverted to D. In some cases, unsaturated residues such as the α,β - (Z) - didehydrobutyrine in epidermin remain unmodified, which do not encounter a Cys residue for addition. An additional decarboxylation of the C-terminal Cys residue yields S - (2-aminovinyl)-D-cysteine. A further N-terminal modificaton takes place in Pep5, where 2-oxobutyryl is formed via desamination of a Thr residue. The leader sequence remains unchanged in these modification steps. The matured lantibiotic is cleaved via a signal peptidase and translocated through the membrane.

Conformational studies on prelantibiotics

From several points of view it was interesting to study the conformation of leader sequences as well as the whole prelantibiotics. Conformational studies should allow to answer questions which arose from the biosynthesis pathways. First of all, why do the modification steps take place in the prolantibiotic part and not in the leader sequence ? An EPICON89 prediction plot [Tröger, 1990] as well as a helical wheel representation reveals, that all so far known very hydrophilic and strongly charged leader sequences of lantibiotics have a high tendency to form amphiphilic helices. In contrast, the prolantibiotic part is more lipophilic and lacks α - helical regions but has a higher tendency to form turn like structures. Is there an interaction between these two parts to prevent modifications and to enable the formation of the sometimes

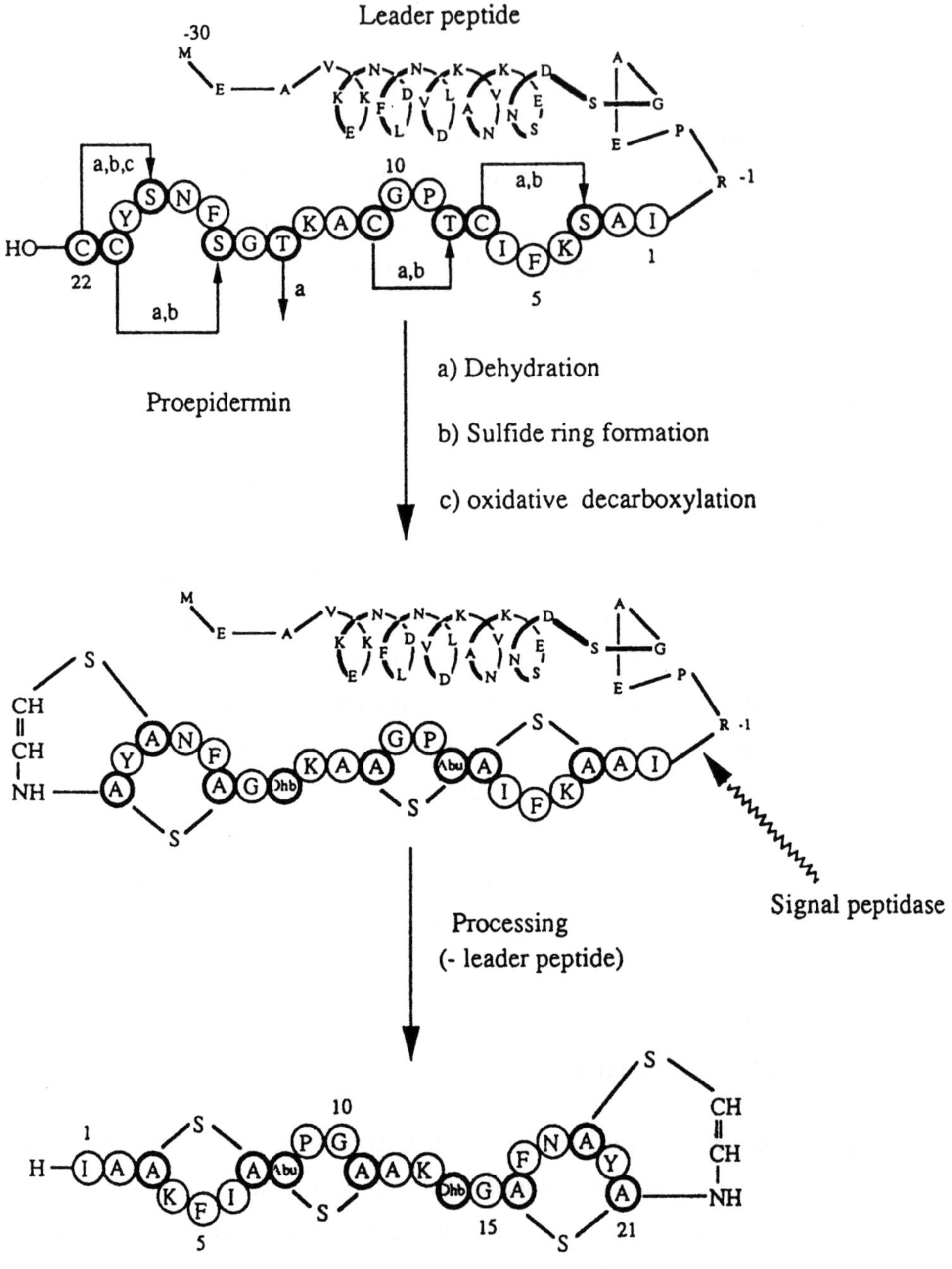

Figure 4. Postulated and proven events of the epidermin biosynthesis by modification of the precursor protein, pre-epidermin. The amino acids are marked in single letter code.

intertwined ring structures ? Studies of the conformation of nisin [van de Ven *et al.*, 1991a,b; Lian *et al.*, 1991] as well as Pep5 [Freund *et al.*, 1991a] suggest, that the formation of some of the cyclic structures in both peptides can be explained as a side chain linkage between residues one turn separated in a helical or turn like arrangement during the modification steps.

Merrifield solid phase methods utilizing Fmoc chemistry were used to synthesize prelantibiotics [Beck-Sickinger & Jung, 1991; Bycroft *et al.*, 1991] as well as leader sequences and parts of the prolantibiotics. Currently, the conformation of prenisin is studied by nmr. Preliminary results in water show a high flexibility of the whole molecule at least in a hydrophilic environment and did not reveal a tendency to form a preferred conformation. The circular dichroism of several leader peptides was measured in solvents of increasing lipophilicity. These studies proved the assumption of leader peptides being helical at least for environments of high lipophilicity. The similar results obtained for all leader peptides so far, can be interpreted in terms of a similar functional role of all leader peptides. Nevertheless, no interactions between leader and prolantibiotic have been found.

Conformational studies on lantibiotics

Conformational studies of lantibiotics in solution were carried out by several research groups based on two dimensional nmr spectroscopy. The strategy used is to determine a set of conformational parameters from proton nmr spectra after the corresponding proton signals have been assigned [Wüthrich, 1986]. The data set consists of dihedral angles, through space distances and solvent / temperature dependencies of mainly backbone amide protons. These data provide an experimental basis for molecular dynamics (MDS) or distance geometry calculations (DG).

The lantibiotic of major commercial interest nisin was studied by three independent groups [Slijper *et al.*, 1989; Chan *et al.*, 1989a,b; Palmer *et al.*, 1989, 1990; van de Ven *et al.*, 1991a,b, Lian *et al.*, 1991; Goodman *et al.*, 1991]. Other lantibiotics which are being investigated are gallidermin / epidermin [Freund *et al.*, 1991a,b] and Pep5 [Freund, 1991c], both of subtype A, and the duramycins Ro-09-0198 [Kessler *et al.*, 1987, 1988, 1991; Wakamatsu *et al.*, 1990] and duramycin B. These studies revealed, that although lantibiotics are polycyclic and therefore restricted in conformational freedom, their conformational analysis is not a simple task. The essential sequential assignment of the proton signals is hampered by the severe overlap of the C_β - proton signals of the numerous Lan and MeLan - bridges. In particular, attempts to determine the primary sequence solely by nmr were ambiguous [Kessler *et al.*, 1988]. Primary sequence assignments were reported for actagardine [Kettenring *et al.*, 1990], mersacidin [Kogler *et al.*, 1991] and Ro-09-0198 [Kessler *et al.*, 1987].

Studies of nisin and Pep5 in water revealed that lantibiotics of subtype A are far more flexible than it was originally assumed. Only the very tight ringsystems seem to be fixed in one conformation whereas the linear segments are often not definable. The lack of conformational parameters to determine the conformation of the latter indicates exchange processes or conformational equilibria. One reason for these findings could be the existence of sulfide bridges, which do not allow the formation of a regular secondary structure such as an α - helix. Nisin was also studied in dimethylformamide [Palmer *et al.*, 1989; Goodman *et al.*, 1991]. In this solvent, Nisin adopts a longstretched and screw like conformation which is bent in the middle region. These results are in agreement with results obtained for gallidermin [Freund *et al.*, 1991a,b].

In contrast to the lantibiotics type A, the duramycins (subtype B) are highly rigid as revealed by nmr - restrained molecular dynamics calculations of Ro-09-0198 in several solvents and preliminary molecular dynamics studies of the homologuous duramycin B. In dimethylsulfoxide as well as in water, Ro-09-0198 adopts almost globular, very packed structure. A certain amphiphilicity can be found as well, with all hydrophilic residues on one side, and the hydrophobic residues including the three phenylalanine residues on the other side. The interaction of Ro-09-0198 with lipids is highly specific. A nmr study [Wakamatsu *et al.*, 1990] of the interaction with lysophosphatidyethanolamine has shown that there is a specific interaction between the phospholipids and the hydrophobic as well as the hydrophilic parts of the lantibiotic. Judging from these highly specific interactions, Ro-09-0198 seems to have 'active' membrane disrupting functions. The specific interaction of duramycin B, duramycin C and actagardine with different phosphorlipids and conformational consequences were studied [Märki *et al.*, 1991]. They found a high selectivity towards phosphatidylethanolamine. Circular dichroism investigations revealed that these interactions induce conformational changes.

Conformations of epidermins

Conformational studies on gallidermin, the Leu^6 analogue of epidermin and of epidermin were carried out by 2D ^1H 500 MHz nmr in water, dimethylsulfoxide and mostly in trifluoroethanol / water (9:1) [Freund *et al.*, 1991a,b]. In general, different solvent mixtures are used to unravel the conformations or conformational changes of molecules in environments of different polarity. In particular, trifluoroethanol / water mixtures with a high percentage of the alcohol can serve as a model environment of high lipophilicity and can thus mimic the situation in biological membranes. Such an environment predominantly stabilizes intramolecular hydrogen bonds.

The nmr data set obtained for gallidermin in trifluoroethanol / water consisted of 190 through space distances, which were determined from NOESY experiments [Neuhaus & Williamson, 1989] (Figure 5) with different mixing times, of 14 dihedral angles for the backbone NH-CHa angles from coupling constants extracted from DQF COSY - spectra and of temperature dependencies of the amide protons in

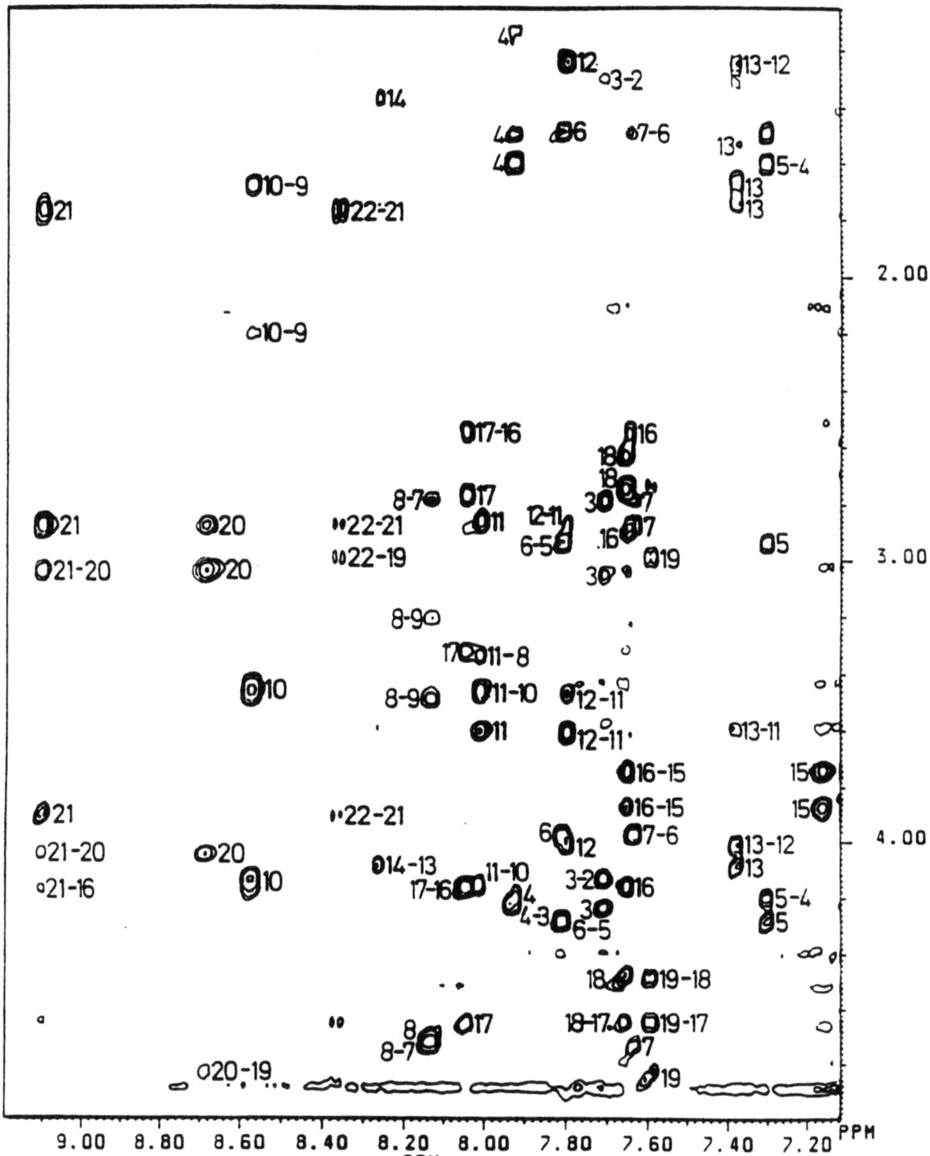

Figure 5. Expansion of a NOESY spectrum with a mixing time of 200 msec for gallidermin in trifluoroethanol / water (9 : 1). The found crosspeaks reveil intra - as well as interresidual contacts and can be correlated to the distance between the two protons, which give rise to these signals.

the range of 300 - 320 K. The latter can be used as a measure for the solvent accessibility or the involvement in hydrogen bonds of these amide protons. Small values for dd / dT (< 3 ppB/K) can be correlated to the hydrogen donor function or a solvent shielded orientation of these protons. The computer calculations based on this data set involved both restrained energy minimization (EM) and molecular dynamics runs (MD) using a modified version of the AMBER programme. The nmr restrained molecular dynamics studies / distance geometry runs yielded a set of 5 structures, which were in good accordance with the experimental data. A high definition was only obtained for the cyclic ring systems, whereas the linear segment Ala^{12} to Gly^{15} could not be defined precisely. It seems reasonable, that this 'hinge' region exhibits conformational flexibility.

Figure 6 shows the obtained averaged structure of gallidermin in trifluoroethanol / water. According to this model, gallidermin adopts a conformation with an overall screw - like arrangement of the peptide backbone. Its overall length is about 30 Å, with 8 - 10 Å in diameter. Only the conformation of ring B is a regular secondary structure, a ß - turn type II, which is in agreement with results for the homologuous ring B in nisin. All other cyclic substructures as well as the linear 'hinge' region could not be classified. The mds suggested a turn like motif for residues 11 - 14. The C - terminal part of gallidermin can be described as a cage formed by the small and rigid ringsystem 19 - 22, with an additional handle like bridge from D - Ala^{16} to Ala^{21}. The temperature dependencies of the amide signals correlate very well with the orientation of these labile protons. The mds suggests a hydrogen donor function for the amide signals Ala^{3}, Ala^{11}, Gly^{15} and AV^{22}, all of them exhibit low temperature dependencies. Especially for the C-terminal cage, we found an internal orientation for the residues Asn^{18}, Ala^{19}, Ala^{21} and AV^{22}, thus forming a hydrophilic core inside the rather hydrophobic cage, whereas the amide protons of the two aromatic residues Tyr^{20} and Phe^{17} exhibit unusual large temperature dependencies and are therefore externally oriented.

The biological mode of action of the epidermins is to form voltage - dependent pores in bacterial membranes [Kordel *et al.*, 1988; Schüller *et al.*,1989; Sahl, 1985]. Electrophysiological studies in black lipid membranes [Benz *et al.*, 1991] have shown, that especially the epidermins induce longlasting (up to 30 sec) voltage dependent ion channels even at very low membrane potentials (50 mV). These results indicate, that only little energy is required for the insertion of these lantibiotics into the membrane. Requirements for the formation of such ion channels were reported for other membrane peptides such as alamethicin [Schwarz *et al.*, 1983; Rizzo *et al.*, 1985]. The structural features for gallidermin in trifluoroethanol / water are consistent with the criteria found essential for membrane pore formation. The membrane spanning length and the screw - like amphiphilic structure, as well as the high dipolar moment of approximately 80 Debye of the overall structure can explain the electrophysiological and biochemical behaviour. Thus, stable ion channels with an estimated diameter of about 1 nm are formed as aggregates of several gallidermin / epidermin molecules. These results are similar to those for amphiphilic helices and prove, that the epidermins are able, although they possess rather conformationally demanding ring structures, to be easily inserted into lipid membranes. Another feature studied was the enzymatic cleavage of epidermin by trypsin. Molecular dynamics simulations based on the nmr results obtained for gallidermin in water have shown, that the linear segment 11 - 14 fits well into the active site of trypsin. These studies are the basis for gene technological modified mutants, which are currently under investigation, with the aim to yield an enzymatic stable epidermin analogue.

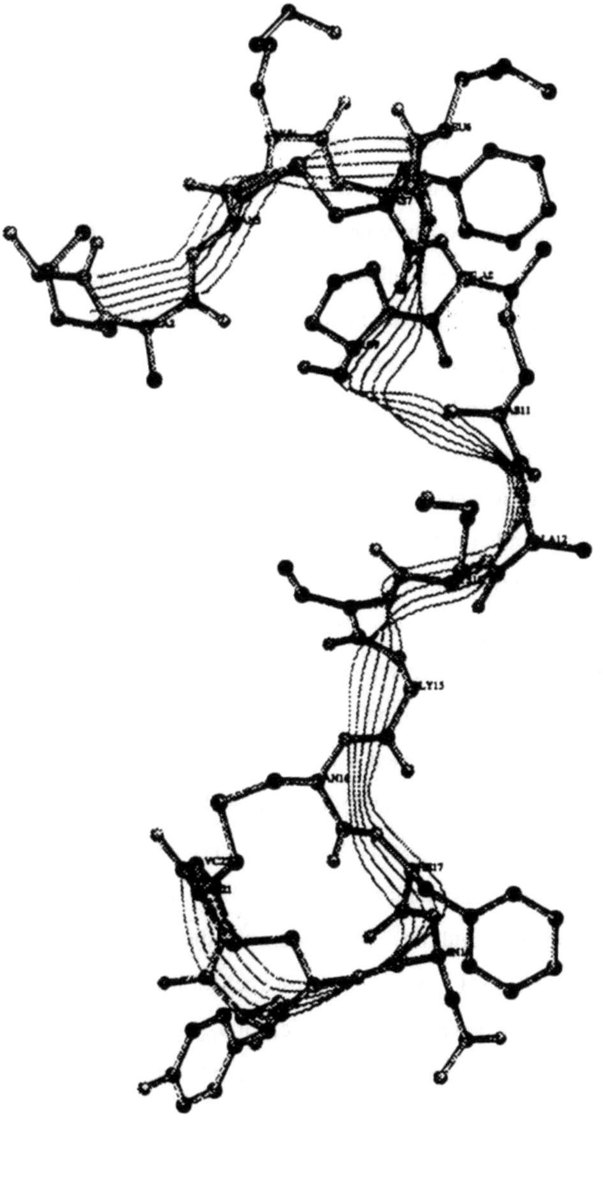

Figure 6. Ball and stick representation of the averaged structure of gallidermin in trifluoroethanol / water (9:1) with the N-terminus at the top. The positions of the sulfide atoms are indicated by dots. The peptide backbone is emphasized by a ribbon.

Conformation of Pep5

Pep5 is the largest member of subtype A with a molecular mass of 3488 Daltons (Table 1). Its isolation and the determination of its primary sequence was carried out by the groups of Sahl and Jung [Sahl & Brandis, 1981; Kellner *et al.*, 1989, 1991]. Like gallidermin it possesses three sulfide ringsystems, but larger stretches of linear sequences. Pep5 is highly hydrophilic due to a net positive charge of +8. The isolation and sequential analysis of pep5 was hampered by the strong tendency of Pep5 to form dimers.

Our interest is to study the conformation of Pep5 in aqueous and lipophilic environments and to compare these results with the ones obtained for gallidermin. The solvent dependency of Pep5 from water to trifluoroethanol was investigated by circular dichroism to reveil conformational changes in this solvent system [Freund *et al.*, 1991c]. The CD - spectrum obtained in trifluoroethanol / water (9:1) ressembled spectra of α - helical polypeptides. With increasing percentages of water, the helix content decreases in favour of a more random coil like shape of the spectra. However, a quantitative interpretation of these results is difficult due to additional contributions to the Cotton effects from unusual aminoacids and the number of D-aminoacids. Summarizing these results, we can state, that the conformation of Pep5 is strongly dependent on the polarity of its environment. An isodichroic point furthermore suggests the occurence of a two state equilibrium between the above mentioned helical and random coil arrangement. Nmr studies were carried out in water and in trifluoroethanol / water (9:1). The strategy was similar to the one used for gallidermin. For the definition of the overall fold, long range non sequential NOE's are of major importance. Similar to results found for nisin [van de Ven *et al.*, 1991a,b], only short range distances could be determined for Pep5 in water, which did not reveal a regular secondary structure. The lack of long range contacts has to be interpreted in terms of a rather high flexibility of the whole molecule with the exception of the small ringsystems. A fixed local conformation was found for the residues preceding and following the Dhb residues in the linear 'hinge' region 14 - 23. This suggests, that the role of such unsaturated residues is to stabilize local structural elements in agreement with results found for gallidermin. There Dhb stabilizes a turn like motif at the position of the cleavage site 11 - 14. Results obtained for model peptides with unsaturated residues revealed a high tendency to form β - turns. Nevertheless, the obtained data set was not sufficient, so far, to determine the overall structure of Pep5 in water. It has to be assumed, that Pep5 exists as an ensemble of several conformations in water, with local structural rigidity only for the tight ring systems. First conformational studies were also carried out on Pep5 in trifluoroethanol / water (9:1). In this solvent mixture, additional NOE's were obtained especially for the linear middle region 14 -22. Especially the found NH_i-NH_{i+1} contacts suggest a helical arrangement of this region involving the two unsaturated Dhb's. A distinction between α - helical or 3_{10} helical arrangement was not possible so far. Nevertheless, such an element of fixed secondary structure induces an at least partly rodlike shape of the whole molecule. A helical wheel representation furthermore reveals an amphiphilicity of this segment, with the three Lys residues at positions 11, 15 and 18 located on one side of the helix, whereas the unsaturated Dhb's are located on the other side. Pep5 therefore fulfills at least partly, similar criteria as the ones described for gallidermin and other channel forming peptides. Electrophysiological studies showed that Pep5 is only able to induce short - lived pores in the range of nsec when a high membrane potential is applied (> 80mV). This suggests, that the insertion of the highly charged lantibiotic into membranes is the rate limiting step of the formation of ion channels of Pep5. Further investigations are in progress to obtain sufficient data sets for molecular dynamics or distance geometry studies.

Summary

The conformations of lantibiotics as well as their precursor peptides have been investigated by nmr and circular dichroism in different solvents. In water, most of the studies revealed only local structural elements for lantibiotics of subtype A as for example nisin and Pep5. Also gallidermin seems to exhibit conformational freedom in water at least in the linear 'hinge' region 11-15. In contrast, lantibiotics of the duramycin family seem to be rather rigid and limited in their conformational flexibility due to the high number of sulfide - and lysinoalanine bridges. Interactions of the duramycins with phospholipids are highly specific and also site selective. Circular dichroism studies revealed that these interactions also alter locally their conformations. Most of them are inhibitors of phospholipase A_2. Further conformational studies could answer the question of whether there is a direct or a lipidmediated interaction between these lantibiotics and phospholipase. Further structural analysis of different duramycins could also correlate structural changes to their different biological activities.

The biological acitvity of lantibiotics of Type A is to form voltage dependent pores in bacterial membranes. The biochemical and electrophysiological results correlate to the found conformations in lipophilic environments. Further nmr experiments have to be carried out in micelles and simulated lipid bilayers. The conformations determined so far especially for gallidermin, will provide a structural basis for gene technological modified mutants. The function of unsaturated residues in Lantibiotics is not fully understood. Similar questions arise about the function of the N - terminal 2 - Oxobutyryl group in Pep5 and the hydroxoaspartic acid in the duramycins. To explain the biosynthetic mechanisms, the conformation of precursor and leader sequences is of major interest. The hypothesis of an interaction between leader and prolantibiotic part could still not be experimentally verified.

References

Allgaier H, Jung G, Werner RG, Schneider U, Zähner H (1985) Elucidation of the structure of epidermin a ribosomally synthesized, tetracyclic heterodetic polypeptide antibiotic. Angew Chem Int Ed Engl 24:1052 -1055

Allgaier H, Jung G, Werner RG, Schneider U, Zähner H (1986) Epidermin, sequencing of a heterodet tetracyclic 21-peptide amide antibiotic. Eur J Biochem 160:9-22

Beck-Sickinger AG, Jung G (1991) Synthesis and conformational analysis of lantibiotic leader-, pro- and pre-peptides. In : Nisin and Novel Lantibiotics, G. Jung & H.-G. Sahl (eds) pp 218-230, Escom, Leiden

Benz R, Jung G, Sahl H-G (1991) Mechanism of channel-formation by lantibiotics in black lipid membranes. In : G. Jung and H.-G. Sahl (eds), Nisin and Novel Lantibiotics, pp 359-372, Escom, Leiden

Boheim G, Hanke W, Jung G (1983) Alamethicin pore formation : Voltage dependent flip-flop of a-helical dipoles. Biophys Struct Mech 9:181-191

Bycroft BW, Chan WC, Roberts GCK (1991) Synthesis and characterization of pro- and prepeptides related to nisin and subtilin. In : G. Jung and H.-G. Sahl (eds), Nisin and Novel Lantibiotics, pp 204-217, Escom, Leiden

Chan WC, Bycroft BW, Lian L-Y, Roberts GCK (1989a) Isolation and characterization of two degradation products derived from the peptide antibiotic nisin. FEBS Lett 252:29-34

Chan WC, Lian L-Y, Bycroft BW, Roberts GCK (1989b) Confirmation of the structure of nisin by complete ^{1}H nmr resonance assignment in aqueous and dimethyl sulfoxide solution. J Chem Soc Perkin Trans 1:2359-2367

Chen L, Thai PC (1987) Effects of antibiotics and other inhibitors on ATP-dependent protein translocation into membrane vesicles. J Bacteriol 169:2373-2379

Choung S-Y, Kobayashi T, Inoue J-I, Takemoto K, Ishitsuka H, Inoue K (1988a) Hemolytic activity of a cyclic peptide Ro-09-0198 isolated from streptoverticullum. Biochim Biophys Acta 940:171-179

Choung S-Y, Kobayashi T, Takemoto K, Ishitsuka H, Inoue K (1988b) Interaction of a cyclic peptide, Ro-09-0198, with phosphatidylethanolamine in liposomal membranes. Biochim Biophys Acta 940:180-187

Clejan S, Guffanti AA, Cohen MA, Krulwich TA (1989) Mutation of *Bacillus firmus* OF4 to duramycin resistance results in substantial displacement of membrane lipid phosphatidylethanolamine by its plasmalogen form. J Bacteriol 171:1744-1746

Dunkley EA, Clejan S, Guffanti AA, Krulwich TA (1988) Large decreases in membrane phosphatidylethanolamine and diphosphatidylglycerol upon mutation to duramycin resistance do not change the protonophore resistance of Bacillus subtilis. Biochim Biophys Acta 943:13-18

Fredenhagen A, Fendrich G, Märki F, Märki W, Gruner J, Raschdorf F, Peter HH (1990) Duramycins B and C, two new lanthionine containing antibiotics as inhibitors of phospholipase A2. Structural revision of duramycin and cinnamycin. J Antibiotics 43:1403-1412

Freund S, Gutbrod O, Folkers G, Gibbons WA, Jung G (1991a) The solution structure of the lantibiotic gallidermin. Biopolymers 31:803-811

Freund S, Jung G, Gutbrod O, Folkers G, Gibbons W (1991b) The three-dimensional solution structure of gallidermin determined by nmr-based molecular graphics. In : G. Jung and H.-G. Sahl (eds), Nisin and Novel Lantibiotics, pp 91-102, Escom, Leiden

Freund S, Jung G, Gibbons WA, Sahl H.G (1991c) Nmr and circular dichroism studies on Pep5. In : G. Jung and H.-G. Sahl (eds), Nisin and Novel Lantibiotics, pp 103-112, Escom, Leiden

Goodman M, Palmer DE, Mierke D, Seonggu R, Nunami K, Wakamaya T, Fukase K, Horimoto S, Kitazawa M, Fujita H, Kubo A, Shiba T (1991) Conformations of nisin and its fragments using synthesis, nmr and computer simulations. In : G. Jung and H.-G. Sahl (eds), Nisin and Novel Lantibiotics, pp 59-76, Escom, Leiden

Gross E, Morell JL (1971) The structure of nisin. J Am Chem Soc 93:4634-4635q

Gross E, Brown JH (1976) Peptides with α,β-unsaturated and thioether amino acids, duramycin. In : Loffet, A. (ed) Peptides, pp 183-190, Editions de l'Universite de Bruxelles, Brüssel
Jung G, Allgaier H, Werner RG, Schneider U, Zähner H (1987) Sequence analysis of the antibiotic epidermin, a heterodet tetracyclic 21-peptide amide. In : Theodoropoulos, D. (ed) Peptides 1986, pp 243-246, de Gruyter, Berlin
Jung G, Sahl H-G (eds) (1991) Nisin and Novel Lantibiotics, Escom, Leiden
Jung G (1991a) Lantibiotics - ribosomally synthesized biologically active polypeptides containing sulfide bridges and α,β-didehydroamino acids. Angew Chem Int Ed Engl 30:1051-1068
Kellner R, Jung G, Hörner T, Zähner H, Schnell N, Entian K-D, Götz F (1988) Gallidermin : a new lanthionine containing polypeptide antibiotic. Eur J Biochem 177:53-59
Kellner R, Jung G, Josten M, Kaletta C, Entian K-D, Sahl H-G (1989) Pep5 : Structure elucidation of a large lantibiotic. Angew Chem Int Ed Engl 28:616-620
Kellner R, Jung G, Sahl H-G (1991) Structure elucidation of the tricyclic lantibiotic Pep5 containing eight positively charged amino acids. In : G. Jung & H.-G. Sahl (eds) Nisin and Novel Lantibiotics, pp 141-158, Escom, Leiden
Kessler H, Steuernagel S, Gillessen D, Kamiyama T (1987) Complete sequence determination and localization of one imino and three sulfide bridges of the nonadecapeptide Ro-09-0198 by homonuclear 2D-NMR spectroscopy. The DDF-RELAYED-NOESY experiment. Helv Chim Acta 70:726-741
Kessler H, Steuernagel S, Will M, Kellner R, Jung G, Gillessen D, Kamiyama T (1988) The structure of the polycyclic nonadecapeptide Ro-09-0198. Helv Chim Acta 71:1924-1929
Kessler H, Seip S, Wein T, Steuernagel S, Will M (1991) Structure of cinnamycin (Ro-09-0198) in solution. In : G. Jung and H.-G. Sahl (eds), Nisin and Novel Lantibiotics, pp 76-90, Escom, Leiden
Kettenring JK, Malabarba A, Vekey K, Cavalleri B (1990) Sequence determination of actagardine, a novel lantibiotic, by homonuclear 2D nmr spectroscopy. J Antibiotics 53:1082-1088
Kinne-Safran E, Kinne R (1986) Proton pump activity and MG-ATPase activity in rat kidney cortex brushborder membranes : effect of 'proton ATPase' inhibitors. Pfluegers Arch 407(Suppl. 2):180-185
Kleinkauf H, von Döhren H (eds) (1982) Peptide Antibiotics - Biosynthesis and Functions, de Gruyter, Berlin
Kogler H, Bauch M, Fehlhaber HW, Griesinger C, Schubert W & Teetz V (1991) Nmr-spectroscopic investigations on mersacidin. In : G. Jung and H.-G. Sahl (eds), Nisin and Novel Lantibiotics, pp 159-170, Escom, Leiden
Kordel M, Benz R, Sahl H-G (1988) Mode of action of the staphylococcinlike peptide Pep5 : Voltage dependent depolarization of bacterial and artificial membranes. J Bacteriol 170:84-88
Lian L-Y, Chan WC, Morley SD, Roberts GCK, Bycroft BW, Jackson D (1991) Nmr studies of the solution structure of nisin A and related peptides. In : G. Jung and H.-G. Sahl (eds), Nisin and Novel Lantibiotics, pp 43-58, Escom, Leiden
Limbert M, Isert D, Klesel N, Markus A, Seibert G, Chatterjee S, Chatterjee DK, Jani RH, Ganguli BN (1991) Chemotherapeutic properties of mersacidin in vitro and in vivo. In : G. Jung and H.-G. Sahl (eds), Nisin and Novel Lantibiotics, pp 448-456, Escom, Leiden
Malabarba A, Pallanza R, Berti M, Cavalleri B (1990) Synthesis and biological activity of some amide derivatives of the lantibiotic actagardine. J Antibiotics 53:1089-1097
Märki F, Hänni E, Fredenhagen A, van Oostrum J (1991) Mode of action of the lanthionine-containing peptide antibiotics duramycin, duramycin B and C, and cinnamycin as indirect inhibitors of phospholipase A2. Biochem Pharm 42:2027-2035
Navorro J, Chabot J, Sherill K, Aneja R, Zahler SA, Racker E (1985) Interaction of duramycin with artifical and natural membranes. Biochemistry 24:4645-4650
Neuhaus D, Williamson M (eds) (1989) The NOE - Effect in Structural and Conformational Analysis, VCH Publishers, New York
Palmer DE, Mierke DF, Pattaroni C, Goodman M, Wakamiya T, Fukase V, Kitazawa M, Fujita H, Shib, T (1989) Interactive nmr and computer simulation studies of lanthionine ring structures. Biopolymers 28:397-408

Rizzo V, Schwartz G, Voges K-P, Jung G (1985) Molecular shape and dipole moment of alamethicin-like synthetic peptides. Eur Biophys J 12:67-73

Sahl H-G, Brandis H (1981) Production, purification and chemical properties of an antistaphylococcal agent produced by *Staphylococcus epidermidis*. J Gen Microbiol 127:377

Sahl, H.-G. (1992) this volume

Sahl H-G (1985) Influence of the staphylococcin-like peptide Pep5 on membrane potential of bacterial cells and cytoplasmic membrane vesicles. J Bacteriol 162:833-836

Schnell N, Entian K-D, Schneider U, Götz F, Zähner H, Kellner R, Jung G (1988) Prepeptide sequence of epidermin, a ribosomally synthesized antibiotic with four sulfide-rings. Nature 333:276-278

Schüller F, Benz R, Sahl H-G (1989) The peptide antibiotic subtilin acts by formation of voltage dependent multi-state pores in bacterial and artificial membranes. Eur J Biochem.182:181-186

Schwarz G, Savko P, Jung G (1983) Solvent dependent structural features of the membrane active peptide trichotoxin A-40 as reflected in its dielectric dispersion. Biochim Biophys Acta 728:419-428

Shotwell OL, Stodola FH, Michael WR, Lindenfelser LA, Dworschak G, Pridham TG (1958) Antibiotics against plant disease. III. Duramycin, a new antibiotic from *Streptomyces cinnnamomeus* forma azacoluta. J Am Chem Soc 80:3912-3915

Slijper M, Hilbers CW, Konings RNH, van de Ven FJM (1989) NMR studies of lantibiotics. Assignment of the ^{1}H NMR spectrum of nisin and identification of interresidue contacts. FEBS Lett 252:22-28

Sokolve PM, Westphal PA, Kester MB, Wierwille R, Sikora-VanMeter K (1989) Duramycin effects on the structure and function of heart mitochondria. I. Structural alterations and changes in membrane permeability. Biochim Biophys Acta 983:15-22

Tröger W (1990) Ph.d thesis University of Tübingen, FRG

van de Ven FJM, van den Hooven HW, Konings RNH, Hilbers CW (1991a) The spatial structure of nisin in acqueous solution. In : G. Jung and H.-G. Sahl (eds), Nisin and Novel Lantibiotics, pp 35-42, Escom, Leiden

van de Ven FJM, van den Hooven HW, Konings RNH, Hilbers CW (1991b) NMR studies of lantibiotics. The structure of nisin in aqueous solution. Eur J Biochem 202:1181-1188

Wakamatsu TK, Choung S, Kobayashi T, Inoue K, Higashijma T, Miyazawa H (1990) Complex formation of peptide antibiotic Ro09-0198 with lysophosphatidyl-ethanolamine: 1H NMR analyses in dimethylsulfoxide solution. J Biochem 29:113-118

Wüthrich K (1986) in : NMR of Proteins and Nucleic Acids, J. Wiley, New York

BIOSYNTHESIS OF THE LANTIBIOTIC PEP5 AND MODE OF ACTION OF TYPE A LANTIBIOTICS

H.-G. Sahl
Institute of Medical Microbiology and Immunology
University of Bonn
Sigmund-Freud-Str. 25
5300 Bonn-Venusberg
Germany

Introduction

Lantibiotics are a group of bacteriocin-like peptides produced by Gram-positive bacteria which differ from all other bacteriocins by their content of didehydroamino acids (dehydroalanine and dehydrobutyrine) and thioether amino acids (lanthionine and 3-methyllanthionine). Currently, two types of lantibiotics are distinguished: Type A which comprises elongated, screw-shaped, amphipathic peptides with molecular masses of more than 2100 Da (Pep5, subtilin, epidermin, gallidermin and the widely used food preservative nisin) and type B lantibiotics which are of globular shape and have molecular masses of 1800 to 2100 Da (cinnamycin, several closely related duramycins and ancovenin). Type A lantibiotics are produced by bacilli, staphylococci, streptococci and lactococci and show strong antimicrobial activities based on formation of short-lived channels in energized membranes; type B lantibiotics are produced by streptomycetes and display less pronounced antibacterial activities but interesting other biological activities instead. For detailed reading on lantibiotics see Jung & Sahl (1991).

Pep5 is the largest (3488 Da) and, with a net charge of +7, the most basic lantibiotic detected so far (Figure 1, Kellner *et al.*, 1991); it is synthesized by *Staphylococcus epidermidis* 5 (Sahl & Brandis, 1981) and its production is associated with the 18.6 Kb plasmid pED 503 (Ersfeld-Dreßen *et al.*, 1984) which carries the structural gene *pepA*. Recently, we have isolated and characterized biosynthetic precursors of Pep5 in various stages of modification; as well as identified an immunity peptide. This work is summarized in the present contribution. Furthermore, an overview is provided on the mode of action of Pep5 and other type A lantibiotics.

Biosynthetic precursors of Pep 5

Lantibiotics are matured from ribosomally synthesized prepeptides for which defined structural genes can be identified. The prepeptides of type A lantibiotics share a characteristic organisation in that they consist of an unusual N-terminal leader peptide and a C-terminal propeptide (Figure 1). The leader

NATO ASI Series, Vol. H 65
Bacteriocins, Microcins and Lantibiotics
Edited by R. James, C. Lazdunski and F. Pattus

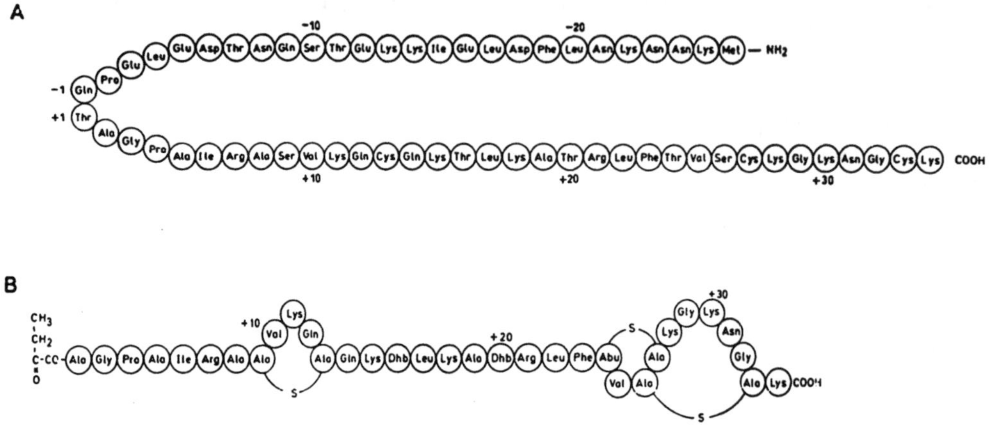

Figure 1. A: Amino acid sequence of the Pep5 prepeptide as predicted from the DNA sequence of the structural gene *pepA*. (Kaletta *et al.*, 1989). Amino acids -26 to -1 represent the leader peptide and amino acids 1-34 the propeptide. B: Structure of mature Pep5 derived from the propeptide domain of pre-Pep5 (Kellner *et al.*, 1991)

peptides are highly hydrophilic with a net negative charge and a predicted α-helix conformation; leader and propeptide region are linked by a turn region which contains a characteristic proteolytic cleavage site. The propeptide is modified by a series of reactions which include the dehydration of hydroxyamino acids, thioether formation and cleavage of the leader peptide; individual lantibiotics may have further modifications. Furthermore, it has become clear that the enzymes catalyzing these modification reactions are organized in an operon. Most progress has been obtained with epidermin, the production of which was heterologously expressed in *Staphylococcus carnosus* by transfer of an 8 kb DNA fragment carrying the structural gene *epiA* and seven further genes (Schnell *et al.* 1991). The essential role of the putative gene products of these genes was demonstrated by complementation of mutants (Augustin *et al.*, 1991). Currently, similar work is in progress with nisin and subtilin (Hansen *et al.*, 1991; Kaletta *et al.*, 1991), aiming at the elucidation of the lantibiotic biosynthesis on the molecular level. In addition to this molecular genetical approach we tried to obtain more information on how biosynthesis of lantibiotics proceeds by isolation and characterization of precursor peptides chosing Pep5 as a model system.

We grew the Pep5 producing strain, *S. epidermidis* 5, to the mid-log phase, disrupted the cells mechanically and separated the cell wall protein fraction, the membrane fraction and the cytoplasm by centrifugation (Weil *et al.*, 1990; Sahl *et a*l., 1991). All fractions were investigated as to the presence of pre-Pep5 by the Western-blot technique using a polyclonal rabbit antiserum raised against the synthetic Pep5 leader peptide. Positive blots were only obtained with the cytoplasmic fraction which was therefore chosen for purification of pre-Pep5. A rapid purification was achieved by two consecutive runs on reversed phase C18 HPLC. The cytoplasmic fraction was directly applied onto a semi-preparative column, and the blot positive protein fraction re-chromatographed on a second, analytical column (Weil *et al.*, 1990). This yielded a chromatographically pure 7 kDa peptide which strongly reacted with the anti-leader-peptide antibody but not with an antiserum raised against mature Pep5.

We then analysed this peptide with respect to amino acid composition, amino acid sequence and molecular mass. Amino acids were found in the relative amounts predicted from the structural gene, with the exception of serine and threonine which were 1.2 and 2.8 moles per mole of peptide instead of 3 moles and 6 moles, respectively; lanthionine and 3-methyllanthionine were not detected. Amino acid sequencing yielded exactly the expected sequence and proceeded through the entire leader peptide but stopped at the first residue of the propeptide region which is a threonine. This indicated that this residue was modified, most probably dehydrated, as dehydroamino acids were shown to block Edman degradation (e.g. Kellner *et al.*, 1991). Finally, we used mass spectrometry, applying the most accurate ion spray technique which yielded 6575 ± 1.7 Da. For the unmodified peptide a molecular mass of 6684 Da was calculated. The difference of approximately 109 Da indicated the absence of six water molecules from the prepeptide, and therefore dehydration of all serine and threonine residues in the propeptide region (Figure 2).

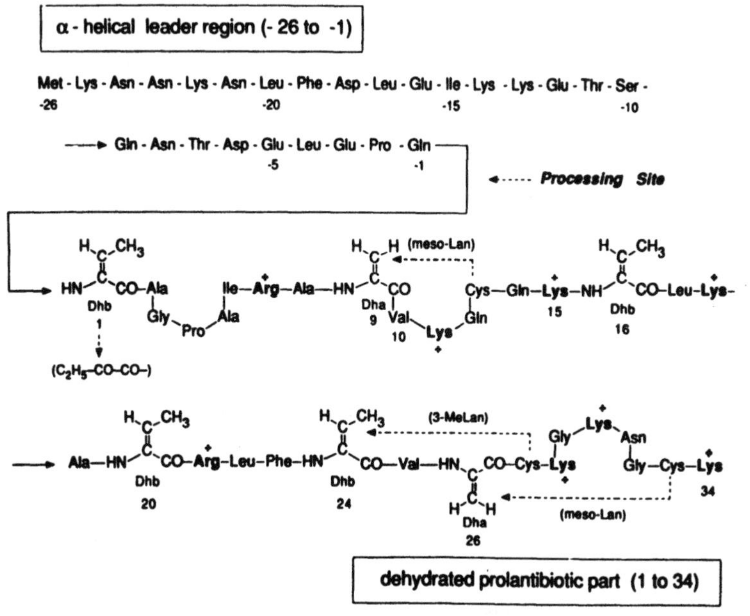

Figure 2. Structure of the isolated Pep 5 prepeptide as determined by amino acid analysis, amino acid sequencing and ion spray mass spectrometry (Weil *et al.*, 1990). Note that serine and threonine residues in the leader peptide are not dehydrated.

When we used a different protocol for purification of pre-Pep5, which included conventional gel-filtration and ion exchange chromatography, we also obtained dehydrated prepeptide. Moreover, careful analysis of the mass peaks obtained in ion spray mass spectrometry clearly demonstrated that the isolated prepeptide was a mixture of six fold dehydrated peptide as the main fraction, and, in decreasing amounts, five fold to one fold dehydrated peptides (Sahl *et al.*, 1991). However, in these preparations we also detected lanthionine and 3-methyllanthionine in somewhat varying but significant amounts. Thus, it seems possible that lanthionine formation occurs slowly after disruption of cells, when the cytoplasmic fraction is kept at physiological conditions during convential chromatography, while it is prevented when the cytoplasm is immediately subjected to reversed phase HPLC. Circular dichroism measurements demonstrated an extraordinarly high proportion of α-helix conformation in the isolated prepeptides (Beck-Sickinger & Jung, 1991), thus verifying prediction plots (Kaletta *et al.*, 1989).

We also isolated Pep5 prepeptides from a clone in which the wild-type plasmid pED503 had been replaced by vector pCUl into which the Pep 5 structural gene *pepA* (synthesized by means of the polymerase chain reaction) had been ligated (Reis & Sahl, 1991). This clone (*S. epidermidis* pMR 7; Figure 3) was expected to yield only unmodified primary translation product, as the biosynthetic enzymes necessary for Pep5 production were thought to be encoded on pED503. Indeed, we detected the primary translation product as shown by amino acid sequencing and mass spectroscopy (Sahl *et al.*, 1991). However, a similar amount of dehydrated peptide was also detected. This demonstrates that, although much slower than in the wild-type, dehydration takes place in this clone and that the dehydrating enzyme is likely to be found on the chromosome. In contrast to the wild-type strain, a significant portion of the prepeptide was associated with the cytoplasmic membrane in this clone. Thus, the export of pre-Pep5 seems to be blocked in the absence of intact pED503.

From the above results we can draw the following conclusions with respect to the sequence of events in the biosynthesis of Pep5, and possibly of lantibiotics in general:

biosynthesis of Pep5 is initiated by dehydration of hydroxyamino acids selectively in the propeptide domain;

the primary translation product has a short half-life and may be modified immediately after release from the ribosome;

thioether formation is not directly linked to dehydration and should be catalyzed by a separate enzyme or may occur spontaneously;

both modifications take place in the cytoplasm;

the modified prepeptide is exported via a specific transport protein, which seems to be encoded on pED503;

the final step of Pep5 biosynthesis is proteolytic processing, which is most likely to occur outside the cells.

We are currently testing the above outlined model by identification and localization of the enzymes involved in Pep5 biosynthesis.

Immunity towards Pep 5

It is a general feature of bacteriocin producing strains to protect themselves from the lethal action of their own products by specific immunity proteins (see several contributions elsewhere in this volume). Such self-protection was also observed with the Pep5-producing strain *S. epidermidis* 5 (Ersfeld-Dreβen *et al.*, 1984) which is at least 250-fold more sensitive after elimination of plasmid pED 503. In search of the Pep5 structural gene *pepA*, a 1.3 kb *Kpn*I fragment of pED503 was identified which hybridized with a *pepA* gene probe. This fragment was subcloned into a shuttle vector pCUl (kindly provided by F. Götz and coworkers) and transferred into a *S. epidermidis* 5 clone from which pED503 had been eliminated. The resulting clone, *S. epidermidis* 5 pMR 2, produced pre-Pep5 but not mature Pep5 as was observed with *S. epidermidis* 5 pMR 7 (see above). Moreover, this clone was fully immune, i.e. was insensitive towards externally added Pep5 to the same degree as the wild-type strain (Reis & Sahl, 1991). Therefore, the complete 1.3 kb *Kpn*I fragment was sequenced. Detection of single *Hae*III and *Hin*dIII restriction sites allowed further shortening of the fragment to 1.09 kb (clone pMR 11) without loss of immunity (Figure 3).

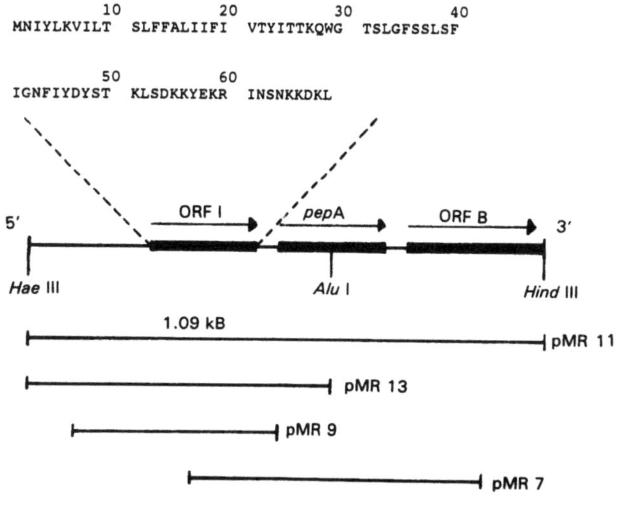

Figure 3. Genes involved in Pep5 immunity as analyzed by construction of mutants pMR 7, 9, 11 and 13. All DNA fragments were subcloned into pCU 1 and expressed in *Staphylococcus epidermidis* 5 Pep5⁻ devoid of the *pepA* carrying wild-type plasmid pED 503 (for further details see the text). The sequence of the ORF I protein is indicated; the pre-Pep5 sequence is given in Figure 1.

This clone contains two intact open reading frames (ORFs), ORF I and *pep*A; ORF B is incomplete; it starts downstream of *pepA* after a typical rho-independent terminator (Reis and Sahl, 1991) and extends to another 572 base pairs beyond the *Hin*dIII restriction site (Figure 3); the role of the putative 285 amino acid ORFB protein in the biosynthesis of Pep5, is currently being investigated (Iglesias-Wind & Sahl, unpublished).

In contrast to clone pMR 11, clone pMR 13, obtained by digestion with *AluI*, which cuts in the middle of *pepA* and removes the propeptide domain pre-Pep5, was fully sensitive. Also fully sensitive were clones pMR7 and pMR 9. Both contain the PCR-synthesized genes ORF I and *pepA*, respectively (Figure 3). Transcription of both genes was controlled by detection of an ORF I specific m-RNA in pMR 9 (Süling & Sahl, unpublished) and isolation of pre-Pep5 from pMR 7 (Sahl *et al.*, 1991; see also previous paragraph). Apparently, both gene products are necessary to express the immunity phenotype. The putative ORF I protein consists of 69 amino acids with a hydrophobic N-terminal segment and a highly charged C-terminus. Currently, we are studying how immunity may be provided by both gene products at the molecular level.

Mode of Action of Type A lantibiotics

Type A lantibiotics currently include nisin, subtilin, Pep5 and epidermin as well some closely related variants like nisin Z, and gallidermin. They are characterized by molecular masses of higher than 2200 Da and a net positive charge (Jung, 1991). Recent investigations of their solution structure by means of two dimensional ^{3}H-NMR and molecular dynamics simulation revealed that they have a strong tendency to adopt elongated, screw-like conformations with some flexibility in regions which are not included in thioether rings, and an overall amphiphilic character (van de Ven *et al.*, 1991; Goodman *et al.*, 1991; Lian *et al.*, 1991; Freund *et al.*, 1991 a,b). These structural properties of type A lantibiotics provide the basis for a mode of action model which had been proposed several years ago based on experiments with intact bacteria, membrane vesicles and artificial membranes: the formation of short-lived, potential-dependent pores by peptide aggregates in the cytoplasmic membrane of target bacteria (Sahl *et al.*, 1987; Kordel *et al.*, 1988). The experimental work leading to this model is summarized below.

Experiments with intact cells and cytoplasmic membrane vesicles

As early as 1960 nisin was reported to release UV-absorbing material from treated bacterial cells (Ramseier, 1960), indicating that nisin could affect membrane integrity. Subsequent studies did not make use of these findings; instead, other modes of action were postulated, such as interaction of dehydroamino acids with free sulfhydryl groups of enzymes (Groß & Morell, 1970) or inhibition of cell wall biosynthesis

(Reisinger *et al.*, 1980). We initially started mode of action studies with Pep5 and later included nisin, subtilin and epidermin, all of which gave basically similiar results.

Pep5 immediately inhibited incorporation of radiolabeled biosynthetic precursors such as thymidine, uracil, amino acids and glucose (Sahl & Brandis, 1982). These results pointed to the cytoplasmic membrane as the primary target of Pep5, which could either induce leakage of precursors or depolarize cells and cause depletion of energy for uptake and biosynthesis. We then treated cells of various Gram-positive strains with micromolar concentrations of type A lantibiotics after the cells had been loaded with radioactive markers (^{86}Rb, amino acids). In all cases we observed rapid efflux of these markers within one minute (Figure 4, Sahl & Brandis, 1983; Ruhr & Sahl, 1985; Schüller *et al.*, 1988).

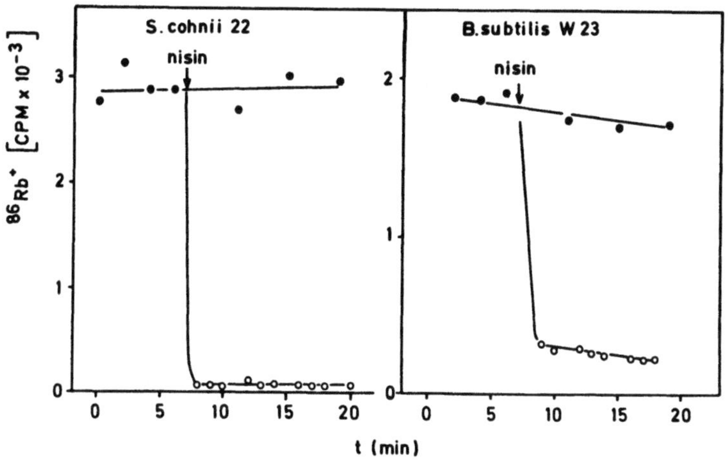

Figure 4. Efflux of $^{86}Rb^+$ from a *Staphylococcus* and a *Bacillus* strain after treatment with 3 μM nisin. Cells were grown on $^{86}Rb^+$ containing medium, resuspended in potassium phosphate buffer (10 mM, pH 7) and assayed for intracellular $^{86}Rb^+$ content by a filtration test.

ATP, but not larger molecules such as peptides, could also be rapidly detected outside the cells, indicating that pores of defined size are formed, rather than that a general disruption of the membrane takes place. Furthermore, we found that the efflux rates of markers were 5 to 10 times higher with energized cells than with starved cells (Sahl & Brandis, 1983) which suggested that the energization state of the membrane is of importance for lantibiotic mediated action. This interpretation was further supported by the observation that the phosphoenolpyruvate-dependent phosphotransferase system (PTS) remained functioning after treatment of cells with Pep 5 (Kordel & Sahl, 1986). Indeed, uptake of α-methylglucoside was stimulated by the lantibiotic to the same extent as with the protonophore carbonyl cyanide-m-chlorophenylhydrazone (CCCP). In contrast, the uptake of amino acids, which is driven by the proton motive force, was completely inhibited (Sahl & Brandis, 1983; Ruhr & Sahl, 1985). To test the hypothesis of potential-dependent pore formation we estimated the membrane potential ($\Delta\psi$) of *Staphylococcus simulans* 22 and simultaneously investigated the influence of the lantibiotics on ($\Delta\psi$). As expected, $\Delta\psi$ collapsed immediately after addition of the bacteriocins (Figure 5) and it became quite obvious that the rate of depolarizisation was directly correlated to the magnitude of the potential (Sahl, 1985; Ruhr & Sahl, 1985). The depolarization kinetics were similar to the efflux kinetics observed with K^+ and amino acids.

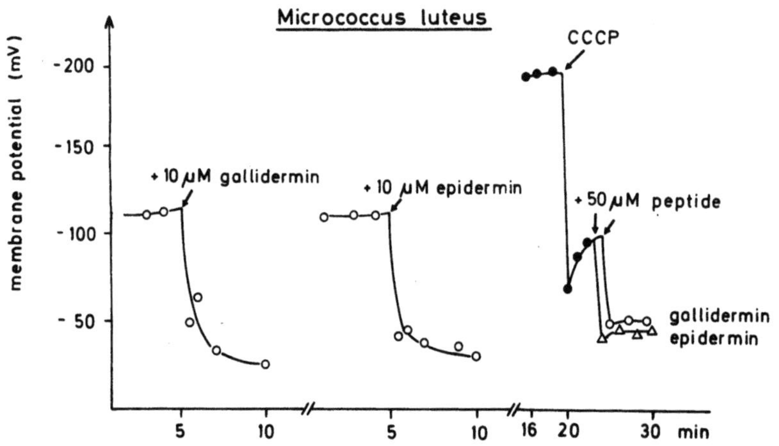

Figure 5. Influence of gallidermin and epidermin on the membrane potential of *Micrococcus luteus*. First two panels; starved cells suspended in potassium phosphate buffer (50 mM, pH 7). Third panel: growing cells after pre-depolarization with carbonylcyanide m-chlorophenylhydrazon (CCCP 50 μM).

We then tried by means of valinomycin-induced potassium diffusion potentials to identify the minimum level of energization (threshold potential), necessary to promote the action of lantibiotics (Sahl *et al.*, 1987; Kordel *et al.*, 1988). When micromolar concentrations of peptides were used threshold potentials were in the range of -80 to -100 mV for Pep5, subtilin and epidermin, while for nisin -50mV was sufficient (for a review see Sahl, 1991). However, threshold potentials varied with the bacterial indicator strain and pH. Nisin, in particular, was more active at pH 5.5, requiring considerably less than -50 mV. Gao *et al.* (1991) have recently shown, that ΔpH is sufficient as a driving force for nisin action, and thus can substitute for $\Delta\psi$ at low pH.

Experiments with cytoplasmic membrane vesicles provided further evidence for a mode of action model based upon formation of pores in energized membranes (e.g. Ruhr & Sahl, 1985). Such vesicles, prepared from different Gram-positive strains, actively accumulated amino acids when energized with ascorbate/phenazinemethosulfate. Addition of lantibiotics led to efflux of markers as seen with intact cells. Pre-incubation of vesicles with lipophilic uncouplers of phosphorylation (CCCP) completely prevented energization and no uptake was detected at any time in the course of an experiment. In contrast, preincubation with lantibiotics did not affect uptake for a short period after energization, then leading to efflux of the accumulated amino acid. Obviously, a certain level of energization has to be reached to induce pore formation.

Gram-negative bacteria are not susceptible to lantibiotics. This is obviously due to the outer membrane which provides an impermeable barrier for peptides of such size. Bypassing the outer membrane by osmotic shock or preparation of *E. coli* cytoplasmic membrane vesicles induced susceptibility of cells and membranes, respectively (Kordel & Sahl, 1986). In this study we found that eukaryotic cells such as lung fibroblasts or erythrocytes, as well as *Mycoplasma* cells, were also not susceptible at micromolar concentrations of lantibiotics. These cells are known to build up only small potentials across their cytoplasmic membranes ranging from -10 to -50 mV. It is in agreement with our results obtained with intact Gram-positive cells, that such a level of energization is not sufficient to induce pore formation. In millimolar concentrations nisin and Pep5 had some influence on eukaryotic cells presumably by physical destabilization of the membrane.

Experiments with liposomes and black lipid membranes

Pep5, nisin and subtilin did not induce any efflux of markers from non-energized liposomes composed of either dioleoylphostatidylcholine, dimyristoylphosphatidylcholine and dimyristoylphosphatidylserine, or defined mixtures of these phospholipids (Kordel *et al.*, 1988). However, it became obvious that the peptides do interact with these liposomes in the absence of a potential. Their effects on neutral membranes was rather negligable as studied by membrane fluidity, phase transition temperatures and carboxyfluorescein efflux, but acidic liposomes were affected more strongly, indicative of primarily electrostatic interaction with phospholipid head groups. Subtilin may slightly enter the hydrophobic core as suggested by tryptophan fluorescence quenching and liposome fusion experiments (Kordel *et al.*, 1988). Asolectin vesicles made from soybean lipid extracts were also not affected, even after application of valinomycin-induced diffusion potentials (Ruhr & Sahl, 1985; Sahl, 1985). However, Abee *et al.* (1991) showed that these potentials are rather short-lived and that vesicles were most likely depolarized before the lantibiotics were added.

Planar lipid bilayer experiments were most helpful to further substantiate the pore formation model (Sahl *et al.*, 1987; Kordel *et al.*, 1988; Schüller *et al.*, 1988; Benz *et al.*, 1991). With type A lantibiotics we observed three different types of behaviour in current-voltage recordings: (i) Pep5 and nisin did not induce membrane conductance when voltages smaller than 80 mV were applied; with neutral membranes even 100 mV were necessary. Above 80 and 100 mV, respectively, transmembrane currents increased exponentially (Figure 6A). Decreasing the voltage resulted in decreased membrane conductivity but we observed considerable hysteresis. Both peptides required a trans-negative orientation of the voltage, i. e. the negative pole had to be opposite to the site to which the peptides were added. (ii) In contrast, subtilin induced membrane conductivity at both polarities of the voltage; otherwise, the results were similar to those obtained with Pep5 and nisin.

(iii) Epidermin and its closely related variant gallidermin also induced membrane conductivity at both orientations of the voltage, however, membrane currents were already recorded at 50 mV (Benz *et al.*, 1991). These differences were also reflected in single channels experiments. With nisin, Pep5 and subtilin we did not observe well-defined, regular channels as have been reported for the channel-forming peptides melittin (Tosteson & Tosteson, 1984) and alamethicin (Boheim, 1974). Instead, we found rapid current fluctuations on the millisecond time scale including bursts and pulse-like spikes (Figure 6B). Increasing voltages increased the amplitude of the conductance fluctuations and their life times. Rough estimates of the pore diameters, based on the assumption of the formation of cylindric pores with a length corresponding to the membrane thickness, yielded pore sizes of approximately 1 nm for nisin (Sahl *et al.*, 1987) and Pep5 (Kordel *et al.*, 1988) and up to 2 nm for subtilin (Schüller *et al.*, 1989). Such pore diameters would allow passage of ions, amino acids and ATP as observed with intact cells (see above).

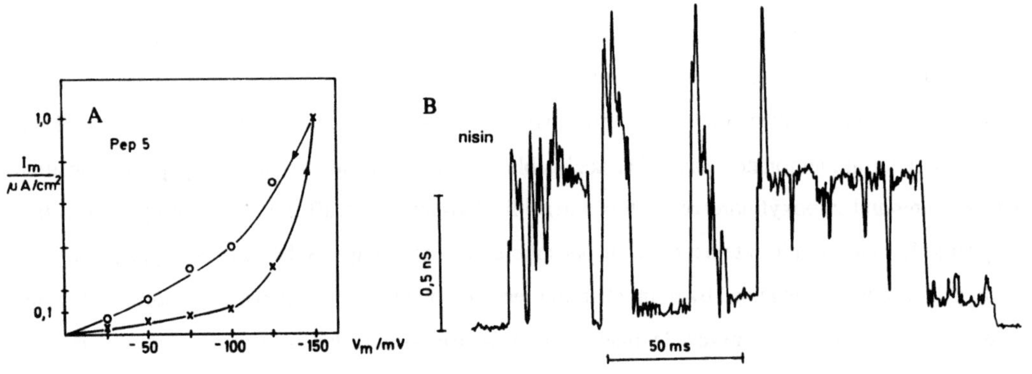

Figure 6. Black lipid bilayer experiments with type A lantibiotics.
A: Current-voltage plot with 0.3 μM Pep5 in 1 M KCl solution. The membrane was composed of dioleoyl phosphatidylcholine and phosphatidylserine (4:1). The voltage was increased stepwise by 25 V (crosses) and then decreased (open circles). Transmembrane current was only recorded with a trans-negative orientation of the voltage indicated by the sign of the voltage.
B: Single channel recordings with nisin (0.1 μM) at a constant voltage of -100 mV. Membrane composition was as in A. The highest conductance level corresponds to an estimated pore diameter of 0.9 nm. The time scale (horizontal bar) represents 50 milliseconds.

The conductance fluctuations obtained with epidermin and gallidermin were much more uniform and similar to those reported for alamethicin and melittin (Boheim, 1974; Hanke *et al.*, 1983). Statistical evaluation of single-channel conductances yielded an average value of 250 pS at 50 mV. However, the conductance was a function of the applied voltage, increasing to 500 pS at 80 mV; thus, pore sizes of epidermin and gallidermin are similar to those of nisin and Pep5. Moreover, epidermin- and gallidermin-induced pores were considerably more stable with average life times of several seconds (Benz *et al.*, 1991). The current-voltage recordings and single channel experiments indicate that considerably less energy is necessary to induce pore formation by epidermin and gallidermin than by the other lantibiotics.

Conclusions

The results obtained with intact cells, physiological and artificial membranes are in good agreement and strongly suggest that the primary mode of action of type A lantibiotics is the formation of pores in target membranes for which the following model is proposed. Pore formation requires energy which is provided by the membrane potential of energy transducing membranes. In the absence of a potential the peptides bind to membranes via ionic interactions of the positive charges on the peptides with the negative charges of phospholipid headgroups. This interaction induces or stabilizes the elongated screw-like solution structures found in NMR experiments under appropriate conditions (Jung & Sahl, 1991). In energized membranes the peptides are forced into a conducting orientation. The size of the lantibiotics excludes that channels of pore diameters of 1 nm are formed by a single peptide. Therefore, the channels must be composed of several molecules, the number of which may be varying as indicated by the single channels recordings with black lipid membranes. The amphipathic nature of lantibotics would suggest a channel architecture as outlined in Figure 7, where the hydrophobic sides of the peptides are facing the membrane core while the hydrophilic sides form the aqueous channel. Accumulation of positive charges may lead to repulsion of peptides and cause the instability of peptide aggregates. For channel formation in physiological membranes it may be important that lantibiotics could interact with integral membrane components as described by Reisinger *et al.* (1980). Such interaction could facilitate pore formation or could have an influence on the size, stability and average life time of pores, thus explaining the observed differences in susceptibility of Gram-positive bacteria. However, the above mentioned experiments should rule out that a specific interaction with a receptor-like molecule is essential for pore formation by type A lantibiotics.

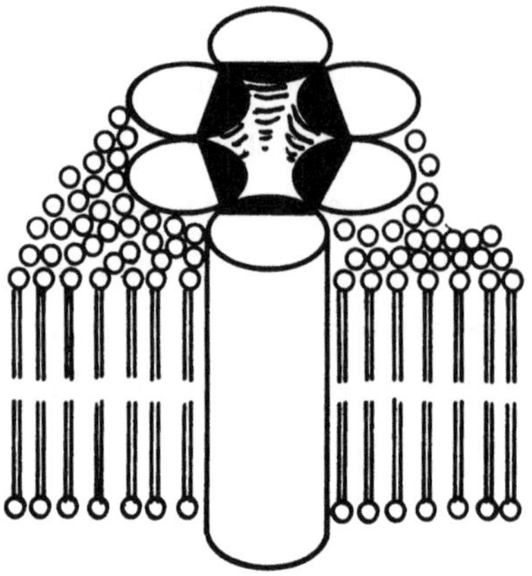

Figure 7. Simplified model for channels formed by amphipathic type A lantibiotics. The hydrophobic part of the molecules faces the membrane core while the hydrophilic sides form the aqueous channel. The number of peptides is arbitrary and may vary for individual channels. For further details see text and Sahl (1991).

Acknowledgement

The author gratefully acknowledges important contributions to the described work provided by the following coworkers: E. Bailly, G. Bierbaum, H. Ersfeld-Dreßen, M. Eschbach, M. Iglesias-Wind, M. Josten, M. Kordel-Bödigheimer, M. Reis, F. Schüller, J. Süling, C. Szekat and H. P. Weil. Much of the work would not have been possible without the excellent and exciting cooperation with R. Benz, Würzburg (black lipid membranes) and G. Jung, Tübingen and his co-workers A. G. Beck-Sickinger, S. Freund, R. Kellner, J. Metzger and S. Stevanovic, (peptide synthesis and sequencing, ion spray mass spectrometry and structure elucidation). The author obtained financial support from the Deutsche Forschungsgemeinschaft (Sa 292/2, Sa 292/4-2, Sa 292/5-1, 5-2, 5-3, Sa 292/6-1)

References

Abee T, Gao FH, Konings WN (1991) The mechanism of action of the lantibiotic nisin in artificial membranes. In: Jung G, Sahl H-G (eds) Nisin and Novel Lantibiotics. Escom, Leiden, p. 373

Augustin J, Rosenstein R, Kupke T, Schneider U, Schnell N, Engelke G, Entian K-D, Gotz F (1991) Identification of epidermin biosynthetic genes by complementation studies and heterologous expression. In: Jung G, Sahl H-G (eds) Nisin and Novel Lantibiotics. Escom, Leiden, p. 277

Beck-Sickinger A, Jung G (1991) Synthesis and conformational analysis of lantibiotic leaderpro- and prepeptides. In: Jung G, Sahl H-G (eds) Nisin and Novel Lantibiotics. Escom, Leiden, p. 218

Benz R, Jung G, Sahl H-G (1991) Mechanism of channel formation by lantibiotics in black lipid membranes. In: Jung G, Sahl H-G (eds) Nisin and Novel Lantibiotics. Escom, Leiden, p. 359

Boheim G (1974) Statistical analysis of alamethicin channels in black lipid membranes. J Membrane Biol 19: 277-303

Ersfeld-Dreβen H, Sahl H-G, Brandis H (1984) Plasmid involvement in production of and immunity to the staphylococcin-like peptide Pep 5. J Gen Microbiol 130: 3029-3035

Freund S, Jung G, Gutbrod O, Folkers G, Gibbons W (1991 a) The three-dimensional solution structure of gallidermin determined by NMR-based molecular graphics. In: Jung G, Sahl H-G (eds) Nisin and Novel Lantibiotics. Escom, Leiden, p. 91

Freund S, Jung G, Gibbons WA, Sahl H-G (1991 b) NMR and circular dichroism studies on Pep5. In: Jung G, Sahl H-G (eds) Nisin and Novel Lantibiotics. Escom, Leiden, p. 103

Gao FH, Abee T, Konings WN (1991) Mechanism of action of the peptide antibiotic nisin in liposomes and cytochrome c oxidase-containing proteoliposomes. Appl Environ Microbiol 57: 2164-2170

Goodman M, Palmer DE, Mierke D, Ro S, Nunami K, Wakamiya T, Fukase K, Horimoto S, Kitazawa M, Fujita H, Kubo A, Shiba T (1991) Conformation of nisin and its fragments using synthesis, NMR and computer simulations. In: Jung G, Sahl H-G (eds) Nisin and Novel Lantibiotics. Escom, Leiden, p. 59

Hanke W, Methfessel C, Wilmsen HU, Katz E, Jung G, Boheim G (1983) Melittin and a chemically modified trichotoxin form alamethicin-type multi-state pores. Biochim Biophys Acta 727: 108-114

Groβ E, Morell JL (1970) The structure of nisin. J Am Chem Soc 93: 4634-4635

Hansen JN, Chung, YJ, Steen MT (1991) Biosynthesis and mechanism of action of nisin and subtilin. In: Jung G, Sahl H-G (eds) Nisin and Novel Lantibiotics. Escom, Leiden, p. 287

Jung G (1991) Lantibiotics: a survey. In: Jung G, Sahl H-G (eds) Nisin and Novel Lantibiotics. Escom, Leiden, p. 1

Jung G, Sahl H-G (1991) (eds) Nisin and Novel Lantibiotics. Escom, Leiden.

Kaletta C, Entian K-D, Kellner R, Jung G, Reis M, Sahl H-G (1989) Pep5, a new lantibiotic: structural gene isolation and prepeptide sequence. Arch Microbiol 152: 16-19

Kaletta C, Klein C, Schnell N, Entian KD (1991) An operon-like structure of genes involved in subtilin biosynthesis. In: Jung G, Sahl H-G (eds) Nisin and Novel Lantibiotics. Escom, Leiden, p. 309

Kellner R, Jung G, Sahl H-G (1991) Structure elucidation of the tricyclic lantibiotic Pep5 containing eight positively charged amino acids. In: Jung G, Sahl H-G (eds) Nisin and Novel Lantibiotics. Escom, Leiden, p. 141

Kordel M, Sahl H-G (1986) Susceptibility of bacterial, eukaryotic and artificial membranes to the disruptive action of the cationic peptides Pep5 and nisin. FEMS Microbiol Lett 34: 139-144

Kordel M, Benz R, Sahl H-G (1988) Mode of action of the staphylococcinlike peptide Pep5: voltage-dependent depolarization of bacterial and artificial membranes. J Bacteriol 170: 84-88

Kordel M, Schuller F, Sahl H-G (1989) Interaction of the pore forming-peptide antibiotics PepS, nisin and subtilin with non-energized liposomes. FEBS LeK 244: 99-102

Lian L-Y, Chan WC, Morley D, Roberts GCK, Bycroft BW, Jackson D (1991) NMR studies of the solution structure of nisin A and related peptides. In: Jung G, Sahl HG (eds) Nisin and Novel Lantibiotics. Escom, Leiden, p. 43

Ramseier HR (1960) Die Wirkung von Nisin auf *Clostidium butyricum*. Arch Mikrobiol 37: 57-94

Reis M, Sahl H-G (1991) Genetic analysis of the producer self-protection mechanism ("immunity") against Pep5. In: Jung G, Sahl H-G (eds) Nisin and Novel Lantibiotics. Escom, Leiden, p. 320

Reisinger P, Seidel H, Tschesche H, Hammes WP (1980) The effect of nisin on murein synthesis. Arch Microbiol 127: 187-193

Ruhr E, Sahl H-G (1985) Mode of action of the peptide antibiotic nisin and influence on the membrane potential of whole cells and on cytoplasmic and artificial membrane vesicles. Antimicrob Agents Chemother 27: 841-845

Sahl H-G, Brandis H (1981) Production, purification and chemical properties of an antistaphylococcal agent produced by *Staphylococcus epidermidis*. J Gen Microbiol 127: 377-384

Sahl H-G, Brandis H (1982) Mode of action of the staphylococcin-like peptide Pep5 and culture conditions effecting its activity. Zbl Bakt Hyg, I Abt Orig A 252: 166-175

Sahl H-G, Brandis H (1983) Efflux of low-Mr substances from the cytoplasm of sensitive cells caused by the staphylococcin-like agent Pep 5. FEMS Microbiol LeK 16: 75-79

Sahl H-G, Kordel M, Benz R (1987) Voltage-dependent depolarization of bacterial membranes and artificial lipid bilayers by the peptide antibiotic nisin. Arch Microbiol 149: 120-124

Sahl H-G, Reis M, Eschbach M, Szekat C, Beck-Sickinger AG, Metzger J, Stevanovic S, Jung G (1991) Isolation of Pep5 prepeptides in different stages of modification. In: Jung G, Sahl H-G (eds) Nisin and Novel Lantibotics. Escom, Leiden, p. 332

Schnell N, Engelke G, Augustin J, Rosenstein R, Gotz F, Entian K-D (1991) The operon like organisation of lantibiotic epidermin biosynthesis genes. In: Jung G, Sahl H-G (eds) Nisin and Novel Lantibiotics. Escom, Leiden, p. 269

Schüller F, Benz R, Sahl H-G (1989) The peptide antibiotic subtilin acts by formation of voltage-dependent multi-state pores in bacterial and artificial membranes. Eur J Biochem 182: 181-186

Tosteson MT, Tosteson DC (1984) Activation and inactivation of melittin channels. Biophys J 45: 112-114

van de Ven FJM, van den Hooven HW, Konings RNH, Hilbers CW (1991) The spatial structure of nisin in aqueous solution. In: Jung G, Sahl H-G (eds) Nisin and Novel Lantibiotics. Escom, Leiden, p. 35

Weil H-P, Beck-Sickinger AG, Metzger J, Stevanovic S, Jung G, Josten M, Sahl H-G (1990) Biosynthesis of the lantibiotic Pep5. Isolation and characterization of a prepeptide containing dehydroamino acids. Eur J Biochem 194: 217-223

IDENTIFICATION OF GENES INVOLVED IN LANTIBIOTIC BIOSYNTHESIS

K. - D. Entian, C. Klein and C. Kaletta
Institute for Microbiology
Johann Wolfgang Goethe-Universitat
Theodor-Stern-Kai 7, Haus 75A
6000 Frankfurt/Main 70
Germany

SUMMARY

Using a hybridization probe specific for the structural gene of subtilin, *spaS*, the DNA regions adjacent to *spaS* could be isolated. Sequence analysis revealed several open reading frames with the same orientation as *spaS*. Upstream of *spaS* three reading frames, *spaB*, *spaC*, and *spaY* were identified (Klein *et al.*, 1992) which showed strong homology to genes identified near the structural gene of the lantibiotic epidermin (Schnell *et al.*, 1992). The SpaY protein derived from the *spaY* sequence was homologous to hemolysin B of *E. coli* which indicated its possible function in subtilin transport. Gene deletions within *spaB* and *spaC*, revealed subtilin negative mutants whereas *spaY*⁻ gene disruption mutants still produced subtilin (Klein *et al.*, 1992). Remarkably, the *spaY*⁻ colonies revealed a clumpy surface morphology on solid media and lost their viability in early stationary growth phase, possibly as a result of intracellular subtilin accumulation. Our results clearly proved that reading frames *spaB* and *spaC* are essential for subtilin biosynthesis whereas *spaY*⁻ mutants are probably deficient in subtilin transport.

INTRODUCTION

Lantibiotics are encoded by distinct genes and translated as prepeptides which are post-translationally matured (Schnell *et al.*, 1988). They can be divided into two subgroups (i) linear shaped and (ii) globular lantibiotics (G. Jung, pers. commun.). The group of linear lantibiotics includes nisin (Mattick & Hirsch 1944 and 1947, Rayman & Hurst, 1984), subtilin (Gross & Kiltz, 1973), epidermin (Allgaier *et al.*, 1985 and 1986) gallidermin (Kellner *et al.*, 1988), and Pep5 (Sahl & Brandis, 1981). Globular lantibiotics are cinnamycin (Benedict *et al.*, 1952; =Ro09-0198, Kessler *et al.*, 1987 and 1988, =lanthiopeptin, Naruse *et al.*, 1989), duramycin (Gross & Brown, 1976) and ancovenin (Wakamiya *et al.*, 1985) which are all synthesized by Streptomyces species. With respect to a possible industrial application as antibiotics, linear lantibiotics are of high interest whereas globular lantibiotics have only weak antimicrobial activities. Important linear lantibiotics with a very similar peptide structure and similar activity spectra are nisin from

NATO ASI Series, Vol. H 65
Bacteriocins, Microcins and Lantibiotics
Edited by R. James, C. Lazdunski and F. Pattus

Lactococcus lactis and subtilin from *Bacillus subtilis*. Nisin was described as early as 1928 in milk isolates (Rogers & Whittier, 1928) and occurs naturally in milk products. It may also replace nitrate as a food preservative thereby avoiding cancerous nitrite. As *B. subtilis* is much more suitable for genetic analysis than other lantibiotic producing species we have chosen *B. sublilis* as a model organism for the investigation of lantibiotic biosynthesis.

The general structure of lantibiotic genes is the same for all lantibiotic genes described so far. As first reported for epidermin (Schnell *et al.*, 1988) the primary transcript of lantibiotic genes is a prepeptide which consists of an N-terminal leader sequence followed by a C-terminal propeptide from which the lantibiotic is matured. This gene structure was also confirmed for subtilin (Buchmann *et al.*, 1988, Klein *et al.*, 1992), nisin (Kalettan & Entian, 1989, Banerjee and Hansen, 1988, Dodd *et al.*, 1990), gallidermin (Schnell *et al.*, 1989), and Pep5 (Kaletta *et al.*, 1989). The first gene structure of cinnamycin, a globular lantibiotic, (Kaletta *et al.*, 1991) revealed some differences between the prepeptides of linear and globular lantibiotics. The leader peptide was much longer than that of linear lantibiotics and the proteolytic processing site was different. Despite these differences precinnamycin had, as prepetides of linear lantibiotics, a high propensity for an α-helical leader sequence. Based on the results of Ingram (1969 and 1970) and the epidermin prepeptide sequence, Schnell *et al.* (1988) proposed an important role of the leader sequence during lantibiotic maturation with a dehydratase reaction at threonine and serine residues followed by sulfur addition from cysteine. This hypothesis was recently supported by the isolation of prepeptides containing dehydroalanine (Weil *et al.*, 1990).

Based on the common gene structure of linear lantibiotics it is very likely that the biosynthetic genes are also homologous in the different producing strains. We have identified three open reading frames upstream of the structural gene of subtilin and have genetically verified their importance for subtilin biosynthesis (Klein *et al.*, 1992). These open reading frames showed significant homologies to open reading frames identified near the structural gene of epidermin (Schnell *et al.*, 1992).

MATERIALS AND METHODS

Strains and media. *Bacillus subtilis* ATCC 6633 was used as a subtilin producing strain. Recombinant plasmids were amplified in *Escherichia coli* strain RR1 (F^- *hsd520 supE44 ara14 proA2 lacY1 galK2 rpsL20 xyl-5 mtl-l*). *E. coli* and *Micrococcus luteus* ATCC 9341 were grown on LB-media (Gibco, Neu-Isenburg, F.R.G.). *B. subtilis* was grown on TY-media [0.8% tryptone, 0.5% yeast extract (Difco, Detroit, USA), 0.1% glucose and 0.5% NaCl]. Where required for the selection of bacteria containing antibiotic resistance markers, 10 µg kanamycin/ml and 5 µg chloramphenicol/ml were added to the media.

For *B. subtilis* transformation HS medium contained 66.5 ml H_20, 10 ml 10x S-base (Spiziens salt), 2.5 ml 20% glucose, 5 ml filter sterilized tryptophan (1 mg/ml), 1 ml 2% casein hydrolysate (Gibco, Neu-Isenburg, F.R.G.), 5 ml yeast extract (Difco), and 10 ml of an 8% arginine, 0.4 % histidine solution. All components were autoclaved separately)

Molecular biology techniques. Established protocols were followed for molecular biology techniques (Maniatis *et al.*, 1990). DNA was cleaved according to the conditions recommended by the commercial suppliers of the restriction enzymes (Boehringer Mannheim, F.R.G., Pharmacia, Freiburg, F.R.G. and BRL, Neu-Isenburg, F.R.G.). Restriction digested DNA was eluted from 0.7% agarose gels by the freeze-squeeze method (Tautz & Renz, 1983).

RESULTS

Isolation of Genes Adjacent to *SpaS*

By using a 13-mer oligonucleotide specific to the subtilin structural gene, *spaS*, (Buchmann *et al.*, 1988) as a probe for Southern blot analysis (Southern, 1975) we determined the restriction sites near *spaS*. The *spaS* gene was localized on a 4.9 kb *Xba*I fragment (Figure 1). Assuming that genes for subtilin biosynthesis may be located near the subtilin structural gene, we cloned and sequenced the 4.9 kb *Xba*I fragment (Klein et *al.*, 1992).

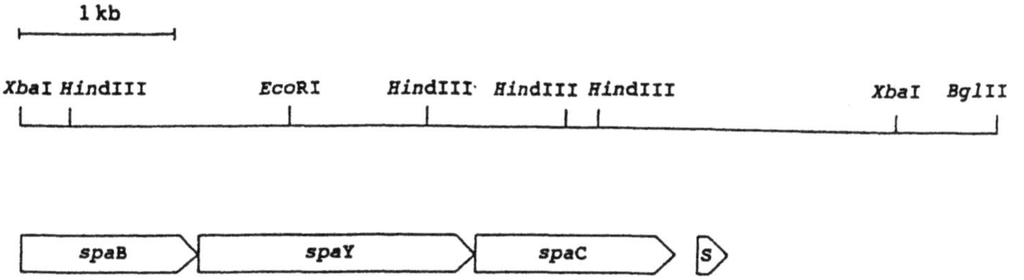

Figure 1. Genetic analysis of *B. subtilis* DNA adjacent to *spaS*.
A: Restriction map of *B. subtilis* DNA adjacent to *spaS*.
B: Open Reading frames derived from DNA sequencing of the 4.9 kb *Xba*I fragment.

Open reading frames upstream of *spaS*

An open reading frame, *spaC*, was found with the same orientation 26 base pairs 5' to *spaS*. This reading frame encodes a maximum of 441 amino acids, starts with an ATG codon for methionine and probably encodes a protein of 49.3 kDa. The *spaC* gene is preceded by a Shine Delgano sequence at an appropriate distance. No homologies have been found for the SpaC protein by searching protein sequence databases. SpaC, however, was homologous to EpiC, which is encoded by an open reading frame which was identified near *epiA*, the structural gene of lantibiotic epidermin (Klein *et al.*, 1992, Schnell *et al.*, 1992). The proteins derived from these genes have a 28.8% homology over the entire protein sequences (Figure 2).

```
          1                                                         60
SPAC      MERGTVSRIEVEIVKEMARQISNYDKVLEIVNQKDNFRSIGEVPLIPWKSTALSHGIPGI
               .: . :  .... . ..:  . :. . .:...  :.. .   :.. .:::: :::
EPIC      QKLKNISGMVVININNIKKILENKITFLSDI-EKATY-IIENQSESYWDPYTLSHGYPGI
         30                                                         87
SPAC      CMLYGELHAHFPEEGWDDIGHQYLSILVNEIKEKGLHTPSMFSGAAGIGLA-AICLSQRF
           .. .. .  :...  ... :::    ..   :.:... :.::: .:::.: .:  ....
EPIC      ILFLSASEKVFHKD-LEKVIHQYSIRKLGPYLESGIDGFSLFSGLSGIGFALDIASDKQY

SPAC      TYYNGLISDINEYLAETVPQLLTEFDQRQVCMSDYDVI-EGVSGIANYLL-LFQEDKAMG
          . :......:.. :.. : ..:.. :. .:  ..::.: .: ::...:::  .  . .
EPIC      SSYQSILEQIDNLLVQYVFDFLNN-DALEVTPTNYDIISQGFSGVGRYLLNRISYNYNAK

SPAC      DLLIDILKYLVRLTEDIIVDGEKVPGWHIPSQHQFTDIEKKAYPYGNFNMGLAHGIPG-P
          . : .::.:      ..:  ..:  .: ....::: ::.:...: ::.:.:::::: : :
EPIC      KALKHILNY-----FKTIHYSSK-DNWLVSNEHQFLDIDKQNFPSGNINLGLAHGILGSP

SPAC      ICVLSSALIQGIKVKGQEAAIEKMANFLLEFSEKEQDSLFWKGIISFEEYQYGSPPNAVN
          ... . . ..::...:.:. ......:::. . :. ..  :     :. :.  .   .
EPIC      LSLTALSKMNGIEIEGHEEFLQDFTSFLLKPEFKNSNNE-W-----FDRYDILENYIPNY

SPAC      FSRDAWCYGRPGVCLALVK-AGKALQNTELINIGVQNLRYTI-SDIRGIFSPTIC-HGYS
            :..:::: .:.  .:.. .::::.:..::... . :  .: ..  ...:::.: :: .
EPIC      SVRNGWCYGDTGIMNTLLSLSGKALNNEGLIKMSKNILINIIDKNNDDLISPTFCSHGLA

SPAC      GIGQILLAVNLLTGQEYFKEELQEIKQKIMSY-YDKDYIFGFHNYESMEGEEAVPLQYVG
          .   :. ..: . . .  .. ...: .::.:. :...  : :.. :   :..    . ::
EPIC      SHLTIIHQANKFFNLSQVSTYIDTIVRKIISHSYSEESSFMFQDIEYSYGQKIYKNK-VG
                                        441
SPAC      LLDGAVGV----GLGVLNMELGSKTDWTKALLI
          .:.:..::     :. .. .  :...:.. .::
EPIC      ILEGELGVLLSALLDYIDTQNQSRKNWKNMFLI
                                        465
```

Figure 2. Sequence homologies between *Spa*C and *Epi*C. Similar amino acid residues are indicated by dots and identical amino acid residues by colons.

```
                   130       140       150       160       170       180
HLYB        HIILITSRSSVTGKLAKFDFTWFIPAIIKYRRIFIETLVVSVFLQLFALITPLFFQVVMD
                                          :..:.. .  :..  ......  . : ...
SPAY        MEVKEQLKLKELLFIMKQMPKTFKLIFTLERSLFLKLIRFSIITGILPIVSLYISQELIN
                    10        20        30        40        50        60
                    190       200       210       220       230
HLYB        KVL-VHRGFSTLNVITVALSVVVVFEIILSGLRTYIFAHSTSRIDVELGAKLFRHLLALP
            ... .... :.. .: ..   :  :. ..: ....  ..   .:. .:. :....  .:.
SPAY        SLVTIRKEVSIVITIFLTYLGVSFFSELISQISEFYNGKFQLNIGYKLNYKVMKKSSNLA
                    70        80        90       100       110       120
          240       250       260           270        280       290
HLYB        ISYFESRRVGDTVARVRELDQIRNFLTGQA---LTSVLDLLFSLIFFAVMWYYSPKLTLV
            .. ::.. . :...:: .    . . . ::   .:. .  :.: : : . :  . .: :.
SPAY        LKDFENPEIYDKLERVTKEISYKPYQIIQAIITMTTSFVTLLSSIAFLMSWNPKVSLLLL
                   130       140       150       160       170       180
             300       310       320       330       340       350
HLYB        ILFSLPCYAAWSVFISPILRRRLDDKFSRNADNQSFLVESVTAINTIKAMAVSP-QMTNI
            ..  .. .   ..  . .. ..  .  .:..   :....   .....:  ...   ...
SPAY        VIPVISLFYFLKIGQEEFFIHWKRAGKERKSWYISYILTHDFSFKELKLYNLKDYLLNKY
                   190       200       210       220       230       240
              360       370           380       390       400       410
HLYB        WDKQLAGYVAAGFKV----TVLATIGQQGIQLIQKTVMIINLWLGAHLVISGDLSIGQLI
            :: .  ......  :.    :.:. : . ..::.  ....:.. ..:   ..:.. .:...
SPAY        WDIK-KSFIEQDTKILRKKTLLNLIYEIAVQLVGAVIIFIAI-MSA---FAGKIMVGNVM
                    250       260       270       280           290
                  420        430       440       450        460
HLYB        AFNMLAGQIVAPVIR-LAQIWQDFQQVGISVTRLGDVLNSPTESYHG-KLTLPEINGDIT
            .. . .. . ..  . .. :.  . . .. ...: . :.  .:. .: : .. :   ...
SPAY        SYIRSVSLVQNHSQSIMTSIYSIY-NSNLYMNQLYEFLELKEEKSQGHKKPIVEPIHSVV
              300       310        320       330       340       350
          470       480       490       500       510       520
HLYB        FRNIRFRYKPDSPVILDNINLSIKQGEVIGIVGRSGSGKSTLTKLIQRFYIPENGQVLID
            :.:..: :  ..  .:..::.:...:: ..:::..::::.:..::.  .:  ..:..::.
SPAY        FQNVSFIYPNQGEQTLKHINVSLHKGERVAIVGPNGSGKKTFIKLLTGLYEVHEGDILIN
               360       370       380       390       400       410
          530       540       550       560       570       580
HLYB        GHDLALADPNWLRRQVGVVLQDNVLLNRSIIDNISLANPGMSVEKVIYAAKLAGAH-DFI
            : ..   : .  ..:.....:: .  .... .::.... ..  ..    . :. .. ::.
SPAY        GINIKELDMDSYMNQIAALFQDFMKYEMTLKENIGFGQIDKLHQTNKMHEVLDIVRADFL
               420       430       440       450       460       470
           590        600          610       620       630       640
HLYB        SELRE-GYNTIVG---EQGAGLSGGQRQRIAIARALVNNPKILIFDEATSALDYESEHVI
            .. ..  ..: .:   ..:  X::::.:.::.:::  ..... :.::..::::  .:. .
SPAY        KSHSSYQFDTQLGLWFDEGRQLSGGQWQKIALARAYFREASLYILDEPSSALDPIAEKET
               480       490       500       510       520       530
               650       660       670       680       690       700
HLYB        MRNMHKICKGRTVIIIAHRLSTVKNADRIIVMEKGKIVEQGKHKELLSEPESLYSYLYQL
            . .. ...:... :.:.::: ..: :::::::.::.::. :.:.::X..  .::
SPAY        FDTFFSLSKDKIGIFISHRLVAAKLADRIIVMDKGEIVGIGTHEELLKTC-PLYKKMDES
               540       550       560       570       580        590

HLYB        QSD

SPAY        ENY
```

Figure 3. Sequence homologies between SpaY of *B. subtilis* and HlyB of *E. coli*. Similar amino acid residues are indicated by dots and identical amino acid residues by colons.

A further open reading frame, *spaY*, overlaps with the 5' end of *spaC* (Klein *et al.*, 1991). The SpaY protein starts with an ATG codon preceded by a 5' TGGTAG 3' sequence which may serve as a ribosomal binding site. The *spaY* gene probably encodes a protein of 614 amino acid residues with a molecular weight of about 71.2 kD. Furthermore, there is a strong homology between *spaY* and several transport proteins. The transport protein for hemolysin B of *E. coli* (Felmlee *et al.*, 1985) shares 22.2% homology with SpaY over the entire protein (Figure 3). Strong homologies also exist with the mouse "multi drug resistance" protein (Gros *et al.*, 1986) and the protein responsible for human cystic fibrosis desease (Riordan *et al.*, 1989). These two proteins arose from gene duplications, and therefore homology to SpaY occurs twice. Comparison of SpaY hydrophobicity plots with those of hemolysin B, mouse "multi drug resistance" protein, and human cystic fibrosis protein also revealed striking similarities (Klein *et al.*, 1992). For hemolysin B several membrane spanning helices were experimentally proved (Wang *et al.*, 1991). These membrane spanning helices are present within three hydrophobic peaks and are very similar in SpaY, hemolysin B, mouse "multi drug resistance" protein, and human cystic fibrosis protein, indicating a membrane location of all four proteins. Interestingly amino acid homologous sequences are strongly conserved at the C-terminal part of SpaY which corresponds to the ATP-binding site, whereas similarities in hydrophobicity profiles are very strong at the N-terminal part of SpaY which is important for membrane localization. These data strongly imply a similar function for SpaC, hemolysin B, mouse "multi drug resistance" protein, and human cystic fibrosis protein.

A reading frame, *spaB*, encoding the C-terminal 390 amino acid residues of another protein was located at the 5' of *spaY* were also identified (Klein *et al.*, 1992). The SpaB protein was not homologous to previously described proteins, but it is highly homologous to a protein encoded by the *epiB* gene (Schnell *et al.*, 1992) which is located 3' to the epidermin structural gene (Figure 4). The hydrophobicity plots also revealed strong similarities between both proteins, since a characterisitic change from a slightly hydrophobic to a strongly hydrophilic peak was observed. These similarities indicate an essential role of *spa*B and *epi*B in subtilin and epidermin biosyntheses.

Genetic Analysis of Open Reading Frames UFtream of *spaS*

To investigate the physiological importance of the identified reading frames in subtilin biosynthesis genes *spaS*, *spaC*, *spaY*, and *spaB*, essential parts of these genes were selectively destroyed by homologous recombination (Klein *et al.*, 1992). Therefore essential parts of these genes were replaced by either kanamycin or chloramphenicol resistance genes.

```
                                                                     60
SPAB      LEEIKSPFFEFEFHRTYELPQTFYIVNADNRLLIDIENDCTLDVFFWELKKTNHNQPLVA
                      .. ....:..  :. .X::::... ::  : ..  :::X .. . :
EPIB      SETENWLNRFATIREKWHIPKDVIIAFGDNRLLLNLLNDKHLIILKKELKKHG-------
                                                              677
SPAB      VEHDADALMDRNQNDYSGEIVVPLLRKQPEKPLYLPVLNAIEGSGSDRIKMPFEDWLFIK
               ..... ..:.   :::.:: .: . :       . :  .....    ..::. :.
EPIB      RIRILESFINESNNERMLEIVTPLYKKTSLKE-----QSFIIPKNRNKHFNNLKDWFSIH

SPAB      LYCKQTREEELIAFEIADFYNQIS-DQYPVRHFFMRYRDPKPHIRLRFNGKAEVLYSLFP
          :   .: ....:.  .  : .... ...   . :...... .  :.::.  ..:     ...
EPIB      LSIPKTYQDNFIQDYLLPFITELKVNNFINKFFYIKFKEDEDFIKLRLLREDEDYSQIYS

SPAB      QLLNWLK-SLREKGLVSESVITQYEREIERYGGLSLMEAAEQLFCEDSKVVEMMIRMHRM
           . :: . .: ...: . :.. .: .:. ::::  ..:. :..:      . . ...... .
EPIB      FIKNWKDYCLLNSELYDYSIV-DYVPEVYRYGGPHVIEDIENFF-----MYDSLLSINII

SPAB      K-DITISKEIAGMVSVIQFLEQFELTFEEQLTFLERNSLQNEYRT-EFKKDREMYIEICN
          . ...:.::.   .:.  .:. .:.. .:. ..: .:. .. ::. .... ...   .. :
EPIB      QSEFKIPKEFIVAISIDFLLDYLEINKSEKEEILINNA-EDLYRSNDIREYKNLLAKLTN

SPAB      SDRDWDNLKKTSDGGMLYETLKTRKMAAAHYAFLIKKAFDNKDEVYSRIGSIIHLHCNRL
          ...:.. :::.   .    .   :.. ... . . : :. ..... .   :::.::..:::.
EPIB      PKNDYEILKKEFPNLHEFLFNKISILENLKKT-LQKSLYTSRSRI---IGSFIHMRCNRI
                                          385
SPAB      FGTDRELENKILTLCRHSLYAQRYQKMNG
          ::...: :. .:.. ..    ...:
EPIB      FGINPEKEKFVLSIFNEITKTKKYWDGCD
                                    990
```

Figure 4. Sequence homology between SpaB and EpiB. Similar amino acid residues are indicated by dots, identical amino acid residues by colons and homologies of major importance by X.

The gene disruption mutants *spaS⁻*, *spaB⁻*, *spaY⁻*, and *spaC⁻* were tested for subtilin production by bioassay with *M. luteus* as a test organism (see Materials and Methods). A strong growth inhibition zone was observed with *spaS* wild type cells. As expected, no growth inhibition of *M. luteus* was observed in *spaS⁻* mutants. This clearly confirmed that growth inhibition of *M. luteus* was specifically the result of subtilin production by *B. subtilis 6633*. Surprisingly, *spaY⁻* mutants still inhibited the growth of *M.luteus* as wild type indicating that *SpaY* was not absolutely essential for subtilin production and that *spaS is* still expressed in *spaY⁻* mutants. Remarkably, *spaY⁻* mutant colonies clumped together compared to wild type colonies. Additionally, growth curves were different between *spaY* mutants and wild type. Wild type cells reached an optical density of 4.5 A_{578} units in stationary phase cultures, whereas in *spaY⁻* mutants growth ceased at 1.3 A_{578} units under the same growth conditions (Figure 5). Possibly, *spaY⁻* mutants were unable to secrete subtilin leading to an intracellular accumulation which may be responsible for the observed effects.

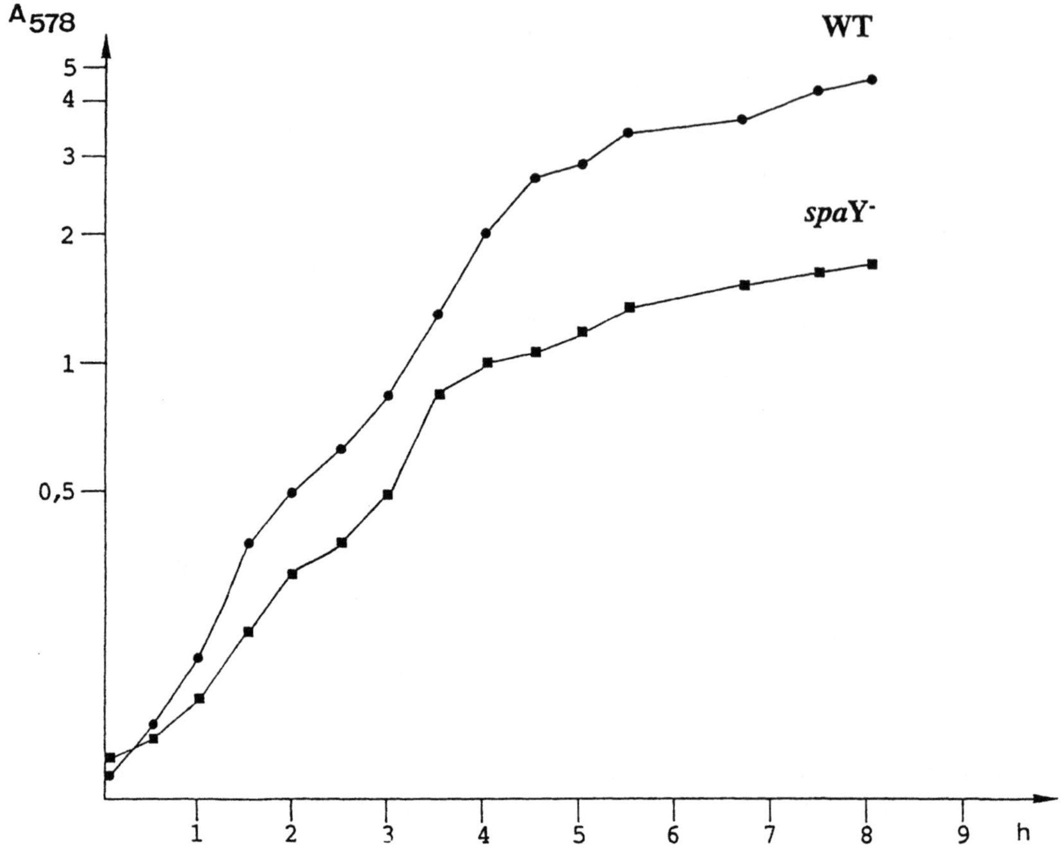

Figure 5. Growth curves of *B.subtilis* ATCC 6633 wild type and *spaY*⁻ mutant in TY media.

DISCUSSION

For the investigation of lantibiotic biosynthesis, the gene products involved must be identified. The conservative organisation of all linear lantibiotic structural genes indicates that the enzymes involved in lantibiotic maturation are probably homologous for the different producers. Assuming that genes involved in lantibiotic biosynthesis may be located near the respective structural genes we sequenced a DNA region adjacent to the subtilin structural gene *spaS*. Three open reading frames, *spaB*, *spaY*, and *spaC* were identified upstream to *spaS*. Indeed, we also found homologous open reading frames near *epiA*, the structural gene of epidermin which is produced by *Staphylococcus epidermidis* (Schnell *et al.*, 1992). By gene disruption we could obtain *spaS*⁻, *spaB*⁻, *spaY*⁻, and *spaC*⁻ mutants. The *spaS*⁻ mutants failed to inhibit growth of *M. luteus* and this result confirmed that growth inhibition was based on subtilin production. The *spaB⁻* and *spaC⁻* mutants also failed to inhibit growth of *M. luteus* and these data supported the conclusion that the respective gene products, SpaB and SpaC, are essentail for subtilin biosynthesis. The importance of these proteins for lantibiotic production was also confirmed by epidermin negative *S. epidermidis* mutants which could be complemented by using reading frames *epiB* and *epiC* (Augustin *et al.*, 1992).

The SpaY protein seems also to be of major importance for subtilin biosynthesis. Its strong homology to hemolysin B, the "multi drug" resistance protein, and the cystic fibrosis protein together with its hydrophobicity plot suggest a transport function for subtilin. We assume that a subtilin transport defect in *spaY⁻* mutants may cause a strong intracellular accumulation of subtilin which may be responsible for the phenotypes observed. Subtilin could be observed in the growth media of *spaY⁻* cultures about 1 h after agglutination and loss of viability and is possibly the result of cell leakiness.

A comparison of the organization of lantibiotic genes indicate that all essential genes are located near the respective structural gene and most biosynthetic genes are organized as operons. For epidermin biosynthesis the operon starts with *epiA* followed by *epiB*, *epiC*, and *epiD* (Schnell *et al.*, 1992). For subtilin biosynthesis the 3' end of the apparent operon was identified with genes *spaB*, *spaY*, and *spaC* (Klein *et al.*, 1992). A homologous reading frame for *spaY* is also found near the *epiA* structural gene. This reading frame, however, shows a reading frame shift resulting in two reading frames *epiY'* and *epiY''* (Schnell *et al.*, 1992) which are homologous to *spaY*. In contrast to *spaY*, reading frames *epiY'* and *epiY''* are not part of the *epiABCD* operon.

The gene disruptions in *B. subtilis* (Klein *et al.*, 1992), the analysis of epidermin-negative mutants (Augustin *et al.*, 1992), and the described homologies to genes necessary for epidermin production (Schnell *et al.*, 1992) clearly indicate that reading frames B and C are of major importance for lantibiotic biosynthesis. The present results also show that the genes for lantibiotic biosynthesis are similar, although lantibiotics are synthesized in strongly divergent organisms such as *S. epidermidis*, *B. subtilis* and *Lactococcus lactis*.

Acknowledgements

The authors would like to thank G. Engelke for stimulating discussions and H. Gruter for correcting the manuscript. This work was supported by Deutsche Forschungsgemeinschaft and Fonds der Chemischen Industrie. C. K was in receipt of a grant from the Fonds der Chemischen Industrie.

REFERENCES

Allgaier H, Jung G, Werner RG, Schneider U, Zähner H (1985) Angew Chem Int Ed Engl 24:1051-1053

Allgaier H, Jung G, Werne, RG, Schneider U, Zähner H (1986) Eur J Biochem 160:9-22
Anagnostopoulos C, Spizizen J (1961) J Bacteriol 81:741-746
Augustin J, Rosenstein R, Wieland B, Schneider U, Schnell N, Engelke G, Entian K-D, Götz F (1992) Eur J Biochem in press
Benedict RG, Dwonch W, Shotwell OL, Pridham TG, Lindenfelser LA (1952) Antibiot Chemother 2:591-594
Banerjee S, Hansen NJ (1988) J Biol Chem 263:9508-9514
Birnboim HC, Doly J (1979) Nucl Acid Res 7:1513-1518
Buchmann GW, Banerjee S, Hansen NJ (1988) J Biol Chem 263:16260-16266
Dodd HM, Horn N, Gasson MJ (1990) J Gen Microbiol 136:555-566
Dubnau D, Davidoff-Abelson R (1971) J Mol Biol 56:209-221
Felmlee T, Pellet S, Lee E., Welch RA (1985) J Bacteriol 163:94-105
Gros P, Croop J, Housman, D (1986) Cell 47:371-380
Gross E, Kiltz H (1973) Biochem Biophys Res Comm 50:559-565
Gross E, Brown JH (1976) In: (A. Loffet, ed.) Peptides 1976. Editions de l'Universite de Bruxelles, Brussels, p 183-190
Holmes DS, Quigley MA (1981) Anal Biochem 114:193-197
Ingram L (1969) Biochim Biophys Acta 184:216-219
Ingram L (1970) Biochim Biophys Acta 224:263-265
Kaletta C, Entian K-D (1989) J Bacteriol 171:1597-1601
Kaletta C, Entian K-D, Kellner R, Jung G, Reis M, Sahl H-G (1989) Arch Microbiol 152:16-19
Kaletta C, Entian K-D, Jung G (1991) Eur J Biochem 199:411-415
Kellner R, Jung G, Hörner T, Schnell N, Entian K-D, Götz F (1988) Eur J Biochem 177:53-59
Kessler H, Steuernagel S, Gillessen D, Kamiyama T (1987) Helv Chim Acta 70:726-741
Kessler H, Steuernagel S, Will M, Jung G, Kellner R (1988) Helv Chim Acta 71:1924-1929
Klein C, Kaletta C, Schnell N, Entian K-D (1992) Appl Environ Microbiol 58:132-142
Love PE, Lyle MJ, Yasbin RE (1985) Proc Natl Acad Sci USA 82:6201-6205
Maniatis T, Fritsch E, Sambrook J (1990) Molecular cloning. A laboratory manual. 2nd ed. Cold Spring Harbour Laboratory, Cold Spring Harbour New York
Naruse N, Tenmyo O, Tomita K, Konishi M, Myaki T, Kawaguchi H (1989) J.Antibiot 62:837-845
Mattick ATR, Hirsch A (1944) Nature 154: 551
Mattick ATR, Hirsch, A (1947) Lancet ii:5-12
Rayman K, Hurst A (1984) In: E.J. Vandamme (ed) Biotechnology of industrial antibiotics. Dekker New York, p607-628
Riordan JR, Rommens JM, Kerem BS, Alon N, Rozmahel R, Grzelczak Z, Zielenski Z, Lok S, Plavsic N, Chou JL, Drumm ML (1989) Science 245:1066-1073
Rogers, LA., and Whittier, E.O (1928) J Bacteriol 16: 321-325.
Sahl HG, Brandis H (1981) J Gen Microbiol 127:377-384
Sanger F, Nicklen S, Coulson AR (1977) Proc Natl Acad Sci USA 74:54635467
Schnell N, Entian K-D, Schneider U, Götz F, Zähner H, Kellner R, Jung G (1988) Nature 333:276-278
Schnell N, Entia, K-D, Götz F, Homer T, Kellner R, Jung, G (1989) FEMS Microbiol Lett 58:263-268
Schnell N, Engelke G, Augustin J, Rosenstein R, Ungermann V, Götz F, Entian K-D (1992) Eur J Biochem in press
Southern EM (1975) J Mol Biol 98:503-517
Tautz D, Renz N (1983) Anal Biochem 132:503-517
Wakamiya T, Ueki Y, Shiba T, Kido Y, Motoki Y (1985) Tetrahedron Lett 26:665-668
Wang R et al (1991) J Mol Biol 217:441-454
Weil H-P, Beck-Sockinger AG, Metzger J, Stevanovic S, Jung G, Josten M, Sahl HG (1990) Eur J Biochem 194:217-223

PORE FORMING BACTERIOCINS

D. Baty, F. Pattus* and A. Finkelstein+
CBBM, CNRS
31 chemin Joseph Aiguier
BP71, 13402 Marseille Cedex 9
France

INTRODUCTION

X-ray crystallography reveals that, in solution, the pore-forming domain of colicin A possesses a hydrophobic helical hairpin buried away from the solvent within the core of a compact globular protein. Presumably, the other pore-forming colicins have a similar structure. Despite our knowledge of this structure, however, the mechanism of membrane insertion, of channel induction by membrane potential, and the structure of the channel itself are still far from being well understood. This session dealt with two general experimental approaches to these problems.

Revival of *in vivo* studies to study the pore-formation and mechanism of membrane translocation

Measurement of the kinetics of K^+ release from *E. coli* cells provide a necessary link between conductance measurements in artificial membranes and the mode of action of the bacteriocins *in vivo* (Letellier *et al.*). They show that, as *in vitro*, the channel is voltage-dependent and can be closed and re-opened depending on the magnitude of the membrane potential. There is also evidence that one colicin kills one bacteria. This puts enormous constraints on attempts to model the colicin channel structure, given that no more than 180 amino-acid residues (and probably only 113 amino-acid residues) are sufficient to form the channel and that, apart from the hydrophobic hairpin, the sequence is largely hydrophilic.

The combination of site-directed mutagenesis and biophysical studies

Cysteine residues were introduced by site-directed mutagenesis at specific points on the colicin A pore-forming domain to introduce extrinsic fluorescent probes and do fluorescent energy transfer measurements with the soluble and the membrane bound form of the colicin (Lakey *et al.*). The results of these studies are consistent with the hydrophilic helices of the colicin lying parallel to the membrane

* EMBL, Postfach 10.2209, Meyerhofstrasse 1, 6900 Germany. + Albert Einstein College of Medecine of Yeshiva University, 1300 Morris Park Avenue, Bronx NY 10461, USA.

NATO ASI Series, Vol. H 65
Bacteriocins, Microcins and Lantibiotics
Edited by R. James, C. Lazdunski and F. Pattus

surface when the channel is in the closed state. Photolabeling experiments in phospholipid vesicles suggest the voltage-dependent insertion of a specific sub-domain of colicin El into the bilayer (Cramer *et al.*). Introducing or removing charged groups on the polypeptide chain were found to affect either selectivity or the gating kinetics of the colicin El channel in planar lipid bilayers (Slatin *et al.*). With the easy accessibility of both sides of the membrane, the planar bilayer system allows accurate determination of the sidedness of the effects of mutations introduced by genetic engineering.

SUMMARY

A consensus picture of the membrane-bound (channel closed) form of pore-forming colicins is emerging from the studies by different groups. There is an agreement that the hydrophobic hairpin spans the membrane bilayer with the other amphipathic helices parallel to the membrane surface. Application of a membrane potential sufficient to open the channel seems to promote the insertion of at least one other helical hairpin inside the bilayer.

There is still a long way to go before the overall molecular mechanism of colicin insertion, channel formation and gating by membrane potential will be elucidated. This meeting has shown that substantial progress is to be expected within the next few years.

IN VIVO PROPERTIES OF COLICIN A: CHANNEL ACTIVITY AND TRANSLOCATION

L. Letellier, C. Lazdunski*, H. Benedetti*, J.P. Bourdineaud*, P. Boulanger
Laboratoire des Biomembranes
URA. CNRS 1116
Université Paris Sud, Bat 433
91405 Orsay Cedex,
France.

Colicin A belongs to the group of colicins which form voltage dependent ionic channels in planar lipid bilayers. (Schein *et al.*, 1978; Pattus *et al.*, 1983). *In vivo*, the primary effects of these colicins are a leakage of cytoplasmic K^+ (Wendt, 1970) and small ions (Lusk and Nelson, 1972), a decrease of internal ATP, a collapse of the electrochemical gradient of protons ($\Delta \mu H^+$), and consequently an inhibition of the $\Delta\mu H^+$-driven active transport systems (for reviews see Luria, 1975; Konisky, 1978, 1982). On the basis of these *in vitro* and *in vivo* properties, it is generally admitted that the killing activity of these colicins results from the formation of channels in the cytoplasmic membrane .

The first part of this review summarizes recent experiments devoted to the characterization of a putative channel activity of colicin A in *E. coli* cells (Bourdineaud *et al.*, 1990). This channel activity has been inferred from the kinetic analysis of the K^+ efflux induced by colicin A. For this purpose, we used a valinomycin-K^+ selective electrode which allows a direct and quantitative measurement of the K^+ fluxes (Boulanger & Letellier, 1988) under conditions close to those of killing.

To kill *E.coli* cells, colicin A first binds to the BtuB/OmpF outer membrane receptors, then it is translocated through the envelope. This translocation involves the participation of bacterial proteins encoded by the *tolQ,R,A,B* gene cluster (Lazdunski *et al.*, 1988). Recent data obtained on the mechanism of translocation of colicin A are described in the second part of the review (Benedetti *et al.*, 1991).

COLICIN A-INDUCED K+ EFFLUX MAY BE THE CONSEQUENCE OF THE FORMATION OF VOLTAGE-DEPENDENT CHANNELS

Energized *E.coli* cells retain 450 to 500 mM cytoplasmic K^+. Addition of colicin A resulted in a net efflux of K^+ whose amplitude increased linearly with increasing number of added colicin molecules, up to 50 molecules/cell, and then saturated. A single colicin molecule per bacterial cell induced an efflux of

* Centre de Biochimie et de Biologie Moleculaire du CNRS, 31 Chcmin Joseph Aiguier. BP 71. 13402 Marseille Cedex 9, France

NATO ASI Series, Vol. H 65
Bacteriocins, Microcins and Lantibiotics
Edited by R. James, C. Lazdunski and F. Pattus

3×10^5 K^+ ions/sec^{-1}, a rate larger than the rate of efflux catalyzed by the different K^+ transport systems and also high when compared to the turnover rate for mobile carriers (10^4 ions/sec^{-1}) (Boulanger & Letellier, 1988). This suggests that colicin A causes a new route for K^+ efflux to open which is likely to be channel-mediated. The linear relationship between the initial rate of efflux and the amount of colicins suggests a lack of cooperation between colicin molecules. Presumably, each colicin forms a single channel.

Respiring *E.coli* cells generate a transmembrane electrical potential ($\Delta\psi$) of about 165 mV at pH 6.8. Addition of colicin A led to a partial depolarization of the inner membrane which occurred with the same kinetics as the K^+ efflux. The same steady state of $\Delta\psi$ (70 mV, negative inside) was attained whatever the number of added colicin molecules suggesting that the channel closes when this threshold is attained. This hypothesis is supported by the fact that a decrease of $\Delta\psi$ below 80 mV after the onset of the K^+ efflux by addition of the protonophore TCS[1] resulted in an immediate arrest of the efflux. Furthermore, when $\Delta\psi$ was re-increased above 80 mV by addition of BSA[2] which desorbs amphiphilic molecules from membranes, the efflux started immediately suggesting that the channel could be reopened at will in whole cells.

It has been long known that *E.coli* cells are protected against colicin K and E1 by strict anaerobiosis (Fields & Luria, 1969). Similarly, pretreatment of the bacteria with inhibitors of the respiratory chain such as cyanide or with protonophores prior to the addition of these colicins also prevents the occurrence of irreversible damage by these colicins (Okamoto, 1975, Jetten & Jetten, 1975). Since all these treatments are known to decrease or abolish $\Delta\psi$, it is likely that they prevent the insertion and/or opening of the channel in the inner membrane .

RELATIONSHIP BETWEEN CHANNEL ACTIVITY, RESPIRATION AND CELL KILLING/ RESCUE

The fact that the channel closes below 80 mV raises the question of how colicin A kills bacteria. The decrease of $\Delta\psi$ cannot alone explain the death since bacteria can retain viability with a reduced $\Delta\mu H^+$ (Kinoshita *et al.*, 1984) . We rather think that the primary event causing the death is the inhibition of the respiration which takes place as a consequence of the K^+ depletion. We have previously shown that the rate of respiration of *E.coli* decreases linearly with decreasing internal concentration of K^+ . This observation not only applies in the case where cells are depleted of K^+ by an hypoosmotic shock (Letellier & Boulanger, 1986) but when they are infected by phage (Boulanger & Letellier, 1988), or colicins. Indeed, colicin A-treated cells incubated in a low K^+ medium (less than lmM) retained low levels of internal K^+ and of respiratory activity, whereas when they were incubated in high K^+ medium (> 100 mM), the K^+ loss

[1] TCS: 3,3' ,4',tetrachlorosalicylanilide. [2] BSA: bovine serum alhumine

was reduced and their respiration maintained for a longer time (Letellier *et al.*, unpublished results). Since the generation of a $\Delta\mu H^+$ requires the functioning of the respiratory chain, it is likely that in low K^+ medium, $\Delta\mu H^+$ cannot be regenerated after colicin A addition and thus, that all the $\Delta\mu H^+$ -dependent transports are inhibited. Since the level of internal ATP is also lowered (Fields & Luria, 1969) there is no chance for the bacteria to remain viable, even if the colicin A channel closes when the threshold of 80 mV is attained. On the other hand, in high K^+, since the respiratory activity is not impaired, it is likely that the colicin-induced depolarization is compensated by a reincrease of $\Delta\mu H^+$ and thus that the channel switches from a closed to an open configuration. However, even under these conditions, the respiring bacteria should retain the main metabolic functions; this would explain why cells can be rescued from colicin action by addition of high K^+ to the incubation medium (Kopecky *et al.*, 1975).

UNFOLDING OF COLICIN A ACCELERATES ITS TRANSLOCATION

Colicin A was denatured in 8M urea. The kinetics of its renaturation was determined from the fluorescence of the tryptophans; colicin A elicits an initial rapid refolding step with a $t_{1/2} < 1$ min followed by a slow renaturation step lasting about 60 min. The lag time before K^+ efflux was reduced by a factor of two when colicin A was first caused to unfold in 8M urea and then directly diluted 200 fold in the cell suspension. A renaturation time of about 60 min was required for the lag time of the native colicin A to be recovered. Unfolding, on the other hand, had no effect on the channel characteristics (rate of efflux, membrane potential threshold for closing). These results suggest that one of the limiting steps in the translocation of colicin A may be the partial unfolding of the polypeptide chain.

A MEMBRANE POTENTIAL IS NOT REQUIRED FOR THE TRANSLOCATION AND /OR INSERTION OF COLICIN A IN THE CYTOPLASMIC MEMBRANE.

The eventual role of the membrane potential in the translocation of colicin A was studied by comparing the lag time preceding efflux in cells either energized or not. When colicin A was added to energized cells, the K^+ efflux occurred after a lag time of 3 min (at 25°C). This lag time was shortened by about 20-30 sec when "denaturated" colicin A was used (Letellier *et al.*, unpublished data). This indicates, that, at low temperature, the rate limiting step in the formation of the pore is the translocation rather than the unfolding of colicin A. In a parallel experiment, the cells were first depolarized with TCS below 80 mV. Colicin A was added 5 minutes later and the cells were incubated for 10 more minutes in order to allow the binding (which occurs normally in deenergized cells) (Jetten & Jetten, 1975) and eventually the translocation and insertion steps. No K^+ efflux occurred under these conditions. Then, BSA was added to trap TCS. Five seconds later, a K^+ efflux of the same amplitude as that of the control was observed. The absence of a lag time before K^+ efflux in the TCS-BSA treated cells, as compared to the delayed efflux in the control experiment, suggests that one of the steps following binding (i.e., translocation, insertion in the cytoplasmic membrane) occurs in deenergized cells. However, these experiments do not allow us to conclude if colicin A is translocated and inserted into the deenergized membrane, the threshold of $\Delta\psi$ only necessary for channel opening, or if $\Delta\psi$ serves both for insertion into the inner membrane and channel opening.

COLICIN A SPANS THE WHOLE ENVELOPE WHEN ITS PORE IS OPENED IN THE CYTOPLASMIC MEMBRANE.

Addition of trypsin (50 µg/ml), after the onset of the colicin A induced K^+ efflux induced an immediate and irreversible arrest of this efflux. A similar result was obtained with colicin B, which shares neither the same receptor nor the same translocation machinery as colicin A (Roos *et al.*, 1989). Efflux occurred normally when cells were first treated with trypsin, then with a trypsin inhibitor and few minutes later with colicin A. This indicates that trypsin had not damaged one essential bacterial component involved in its binding or translocation.

Trypsin did not gain access either to the periplasmic space or to the cytoplasmic membrane in the presence of colicin A, since we could not observe any degradation of OmpA, which is normally proteolysed by trypsin from the periplasmic side (Tomassen & Lugtenberg, 1984).

The following arguments further suggest that the arrest of the efflux corresponds to the closing of the channels inserted in the inner membrane. i) the experiments were performed at multiplicities (M<5O) at which the rate of K^+ efflux is proportional to the number of added colicin molecules: this implies that each colicin added forms a pore in the inner membrane; ii) trypsin was added 1 min after the onset of the K^+ efflux; since the rate of K^+ efflux is constant under these conditions, it is likely that all the colicins have reached their target in the inner membrane.

Taken as a whole these results suggest that the colicins remain accessible from the external medium when forming the pore in the inner membrane. Although colicins are elongated molecules, they are not long enough to traverse the whole envelope, it is thus likely that they are active in an unfolded state. These results yield new insights into the so called "trypsin rescue state". Early experiments led to the conclusion that the action of trypsin on the channel-forming colicins, K and Ib, is divided in stage 1 and stage 2 depending on whether or not the cell viability can be restored by adding trypsin (Nomura & Nakamura, 1962;

Levisohn *et al.*, 1968). These results were confirmed by Dankert *et al.* (1980) in the case of colicin El. They concluded that colicin El may acquire a trans-envelope conformation corresponding to an intermediate state in the penetration of colicin through the cell envelope. Our results suggest that colicins do span the envelope even when the pore has been formed. It is thus likely that stage 1 is that at which the pore can be closed by trypsin and the main transport activities required for the vital activity of the cells can be restored; stage 2 may take place when this closing occurs too long after the addition of trypsin, i.e. when the levels of internal K^+ and ATP (Fields & Luria, 1969) are too low for the cell to recover. It has been reported that bacteria treated with colicin E2 can also be rescued by trypsin after the colicin has begun to exert its lethal activity in the cytoplasm (Beppu *et al.*, 1972). In view of the above results it is thus tempting to conclude that cytoplasmic-acting colicins also span the envelope while exerting their lethal activity.

What triggers unfolding of colicins? It is likely that the interactions of colicins with their receptor trigger the first step of unfolding in the external membrane (Lazdunski *et al.*, 1988, Baty *et al.*, 1988, , Fourel *et al.*, 1990; Benedetti *et al.*, 1991a). Further unfolding into the periplasmic space may be favoured by interactions with their translocation machinery. It has been recently shown (Benedetti *et al.*, 1991b) that the N-terminal domain of colicin A interacts *in vitro* with the C-terminal domain of TolA, a protein of the translocation machinery associated with the inner membrane and extending in the periplasmic space (Levengood & Webster, 1989). This interaction may be relevant to the *in vivo* translocation mechanism.

ARE THERE COMMON PATHWAYS FOR THE TRANSLOCATION OF COLICINS AND BACTERIOPHAGES?

Colicins and bacteriophages share not only common receptors but also common translocation mechanisms; phage Tl and Φ80 and type "B" colicins used the TonB pathway whereas type "A" colicins and filamentous phages used the TolQRAB pathway (Smilowitz, 1974; Postle, 1990).

We have proposed, on the basis of the analysis of the K^+ efflux induced by several phages that the transfer of their DNA through the envelope would take place through specific channels (Boulanger & Letellier, 1988, 1989 and 1992 in press). The participation of phage T5 tail proteins in the formation of these channels has been recently suggested. These proteins, which, like colicins, are soluble proteins (Feucht *et al.*, 1990), were found essentially in the contact sites between the inner and outer membranes after fractionation of the cell envelope according to the protocol described by Yshidate *et al.*, (1986) (Guihard *et al.* 1992). Recent fractionation experiments suggest that colicin A is also preferentially associated with contact sites (Benedetti *et al.*, unpublished results).

The origin of these contact sites is still obscure. Their existence, which was first described by Bayer

(1968) has recently been challenged by Kellenberger (1990). Both, the observations by electron microscopy and the fractionation experiments, suggest that the number of contact sites increases after infection by phages or upon expression of a phage encoded proteins (Lopez & Webster, 1985 ; Walderich *et al.*, 1989, Guihard *et al.*, 1992). It is thus likely that such events create new contact sites or stabilize pre-existing ones. In view of the recent data on the location of the Tol proteins (Bourdineaud *et al.*, 1989, Webster, 1991), and the finding that colicin A spans the whole envelope, it is tempting to speculate that the contact sites are formed by the interaction of colicin A with its receptors and the Tol A (and other Tol?) proteins. It is likely that the interaction of other colicins with their receptors and Tol or TonB proteins would also lead to the formation of such contact sites (Letellier, in press).

REFERENCES

Baty D, Frenette M, Lloubès R, Géli V, Howard SP, Pattus F, Lazdunski C (1988) Functional domains of colicin A. Mol Microbiol 2:807-811

Bayer ME (1968) Areas of adhesion between wall and membrane of *E.coli*. J Gen Microbiol 53: 395-404

Bénédetti H, Frenette M, Baty D, Knibiehler M, Pattus F, Lazdunski C (1991) Individual domains of colicins confer specificity in colicin uptake, in pore properties and in immunity requirement. J Mol Biol 217:429-439

Bénédetti H, Lazdunski C, Lloubès R (1991) Colicins A and E 1 interact with TolA protein, a component of their translocation system. EMBO J 8:1989-1995

Bénédetti H, Lloubès R, Lazdunski C, Letellier L (1992) Colicin A unfolds during its translocation and spans the whole cell envelope when its pore is formed. EMBO J in press

Beppu T, Kawabata N, Arima K (1972) Specific inhibition of cell division by colicin E2 without degradation of DNA in a new colicin sensitive mutant of *E.coli*. J Bacteriol 110:485-493

Bourdineaud JP, Howard P, Lazdunski C (1989) Localization and assembly into the *E.coli* envelope of a protein required for entry of colicin A. J Bacteriol 171:2458-2461

Bourdineaud JP, Boulanger P, Lazdunski C, Letellier L (1990) *I n* vivo properties of colicin A: channel activity is voltage dependent but translocation may be voltage independent. Proc Natl Acad Sci USA 87:1037-1041

Boulanger P, Letellier L (1988) Characterization of ion channels involved in the penetration of phage T4 DNA. J Biol Chem 263:9767-9775

Boulanger P, Letellier L (1992) Ion channels are likely to be involved in the two steps of phage T5 penetration into *E.coli* cells. J Biol Chem in press

Dankert J, Hammond SM, Cramer WA (1980) Reversal by trypsin of the inhibition of active transport by colicin El. J Bacteriol 1 43:594-602

Feucht A, Schmid A, Benz R, Schwarz H, Heller KJ (1990) Pore formation associated with the tail-tip protein pb2 of bacteriophage T5. J Biol Chem 265:18561-18567

Fields KL, Luria SE (1969) Effects of colicin El and K on cellular metabolism. J Bacteriol 97:64-77

Fourel D, Hikita C, Bolla JM, Mizushima S, Pages JM (1990) Characterization of OmpF domains involved in *E.coli* K-12 sensitivity to colicins A and N. J Bacteriol 172:3675-3680

Guihard G, Boulanger P, Letellier L (1992) Involvement of phage T5 tail proteins and contact sites between the inner and outer membrane of *E.coli* in phage T5 DNA injection. J Biol Chem in press

Ishidate K, Creege, ES, Zrike J, Deb S, Glauner B, Mc Allister TJ, Rothfield LI (1986) Isolation of differenciated membrane domains from *E.coli* and *S. Typhimurium,* including a fraction containing attachment sites between the inner and outer membranes and the murein skeleton of the cell envelope. J Biol Chem 261:428-443

Jetten AM, Jetten ME (1975) Energy requirement for the initiation of colicin action in *E.coli*. Biochim Biophys Acta 387:12-22

Kellenberger E (1990) The "Bayer bridges" confronted with results from improved electron microscopy. Mol Microbiol 4:697-705

Kinoshita N, Unemoto T, Kobayashi H (1984) Protonmotive force is not obligatory for growth of *E.coli*. J Bacteriol 160:1074-1077

Konisky J (1978) The bacteriocins, in "The Bacteria" Vol VI, Academic Press

Konisky J (1982) Colicins and other bacteriocins with established modes of action. Ann Rev Microbiol 36:125-144

Kopecki A, Copelan, DP, Lusk JE (1975) Viability of *E.coli* treated with colicin K. Proc Natl Acad Sci 72:4631-4634

Lazdunski C, Baty D, Géli V, Cavard D, Morlon J, Lloubès R, Howard P, Knibiehler M, Chartier M, Varenne S, Frenette M, Dasseux JL, Pattus F (1988) The membrane-channel forming colicin A: synthesis, secretion, structure, action and immunity. Biochem BiophysActa 947:445-464

Letellier L (1991) Bacteriocin and bacteriophage channels in procaryotes in "Alkali cation transport systems in procaryotes". Ed. E.P. Bakker, CRC Press, in press

Letellier L, Boulanger P (1989) Involvement of ion channels in the transport of phage DNA through the cytoplasmic membrane of *E.coli*. Biochimie 71:167-174

Letellier L, Boulanger P (1986) Relationship between respiratory activity and intracellular concentration of K^+ in *E.coli* cells. Fourth European Bioenergetics Conference, Prague, p362

Levisohn R, Konisky J, Nomura M (1968) Interaction of colicin with bacterial cells. J Bacteriol 96:811-821

Lopez J, Webster RE (1985) Assembly site of bacteriophage fl corresponds to adhesion zones between the inner and outer membrane of the host cell. J Bacteriol 163:1270-1274

Luria SE (1973) Colicins. in "Bacterial membranes and walls" ed. L.Leive, Dekker, New York, Vol 1, pp. 293-320

Lusk JE, Nelson DL (1972) Effects of colicin El and K on permeability to magnesium and cobaltous ions. J Bacteriol 112:148-160

Levengood SK, Webster RE (1989) Nucleotide sequences of the *tolA* and *tolB* genes and localization of their products, components of a multistep translocation system in *E.coli*. J Bacteriol 171:6600-6609

Nomura M, Nakamura M (1962), Reversibility of inhibition of nucleic acid and protein synthesis by colicin K. Biochem Biophys Res Commun 7:306-314

Okamoto K (1975) Requirement of heat and metabolic energy for the expression of inhibitory action of colicin K. Biochem Biophys Acta 389:370-379

Pattus F, Martinez MC, Dargent B, Cavard D, Verger R, Lazdunski C (1983) Interaction of colicin A with phospholipid monolayers and liposomes. Biochemistry 22:5698-5707.

Postle K (1990) TonB and the Gram negative dilemma. Mol Microbiol 4:2019-2025

Roos U, Harkness RE, Braun V (1989) Assembly of colicin genes from a few DNA fragments. Nucleotide sequence of colicin D. Mol Microbiol 3:891-902

Schein SJ, Kagan BL, Finkelstein A (1978) Colicin K acts by forming voltage-dependent channels in phospholipid bilayer membranes. Nature 276:159-163

Smilowitz H (1974) Bacteriophage fl infection and colicin tolerance. J Virol 13: 100-106

Tommassen J, Lugtenberg B (1984) Amino terminus of outer membrane PhoE protein: localization by use of a *bla-phoE* hybrid gene. J Bacteriol. 157:327-329

Walderich B, Holtje JV (1989) Specific localization of the lysis protein of bacteriophage MS2 in membrane adhesion sites of *E.coli*.. J Bacteriol 171:3331-3336

Webster RE (1991) the *tol* gene products and the import of macromolecules into *E.coli*. Mol Microbiol 5:1005-1011

Wendt L (1970) Mechanism of colicin action: early events. J Bacteriol 104:1236-1241

SITE-DIRECTED FLUORESCENCE SPECTROSCOPY AS A TOOL TO STUDY THE MEMBRANE INSERTION OF COLICIN A

J.H. Lakey, D. Baty*, JM González-Mañas, D. Duché* and F. Pattus
European Molecular Biology Laboratory
Meyerhofstrasse 1
D-6900 Heidelberg
Germany

INTRODUCTION

Colicin A is a water soluble 60kD protein which can be produced in large amounts by *Citrobacter freundii* cells carrying the ColA plasmid. Upon induction of colicin synthesis via the *caa* gene, various processes occur which allow the protein to pass into the ambient medium.The plasmid-carrying cells are resistant to the effects of their own toxin due to the presence on the same plasmid of the *cai* gene which codes for the "immunity protein". These features will be fully discussed by other contributors to this meeting (see also Lazdunski *et al.*, 1989)

The colicin binds to a specific receptor in the outer membrane of the target cell (colicin A requires both the Vitamin B12 receptor and the porin OmpF) via the middle of its three domains. Translocation to the inner membrane appears to involve the N-terminal region and is dealt with in the paper by Letellier *et al.* in this book. Having arrived in the periplasmic space, the colicin A (like colicins N,B,E1,Ia, Ib) causes a large increase in the permeability of the cytoplasmic membrane leading to a loss of potassium ions, depolarisation of the inner membrane and cell death. As colicins of this group all form voltage-gated channels in planar bilayer membranes the increase in permeability in living cells has been thought to arise from the formation of such channels in the bacterial inner membrane. (Bourdineaud *et al.*, 1990).

That the C-terminal domain of 205 amino acids contains the entire channel forming activity of the protein has been demonstrated by using the fragment generated by thermolytic digestion of Colicin A. This thermolytic fragment is also water soluble and inserts spontaneously into negatively charged membranes. Upon application of a transmembrane potential difference (positive on the side of protein addition) the fragments form defined ion channels which are slightly anion selective and have a predicted diameter of 8Å. The water soluble form of this fragment has been crystalised and the structure solved to 2.7Å resolution (Figure 1). It consists of ten alpha helices folded in three layers so as to bury helices 8 and 9 completely within the structure. This complete burial of two very hydrophobic helices is possibly the key to this

* Centre de Biochimie et de Biologie Moleculaire du CNRS, 31 Chemin Joseph Aiguier, B.P.71, 13042 Marseille Cedex 9, France.

NATO ASI Series, Vol. H 65
Bacteriocins, Microcins and Lantibiotics
Edited by R. James, C. Lazdunski and F. Pattus

protein's ability to be both water soluble and membrane penetrating. The transition from the water soluble to membrane-bound to open channel configuration is currently the central puzzle in the study of channel forming colicins. A model for the membrane-bound form of colicin A has been proposed by Parker *et al.* (1990) and its predictions have been used to design a series of experiments aimed at building up a better picture of this so far uncrystalisable structure. In the model the molecule orients itself with respect to the membrane so that the hairpin structure of helices 8 and 9 is pointing towards the membrane. This orientation may be favoured by the presence of a ring of positive charges on this side of the molecule. Subsequently the insertion of the helices occurs within the confines of the surrounding amphipathic alpha helices so that no unfavourable contact with the polar phase occurs. The final result of this insertion is to fold up the amphipathic helices such that they lie on the surface of the membrane. This proposal has acquired the nickname "the umbrella model" as helices 8 and 9 appear to be supporting a canopy formed by the rest of the molecule (Figure 1).

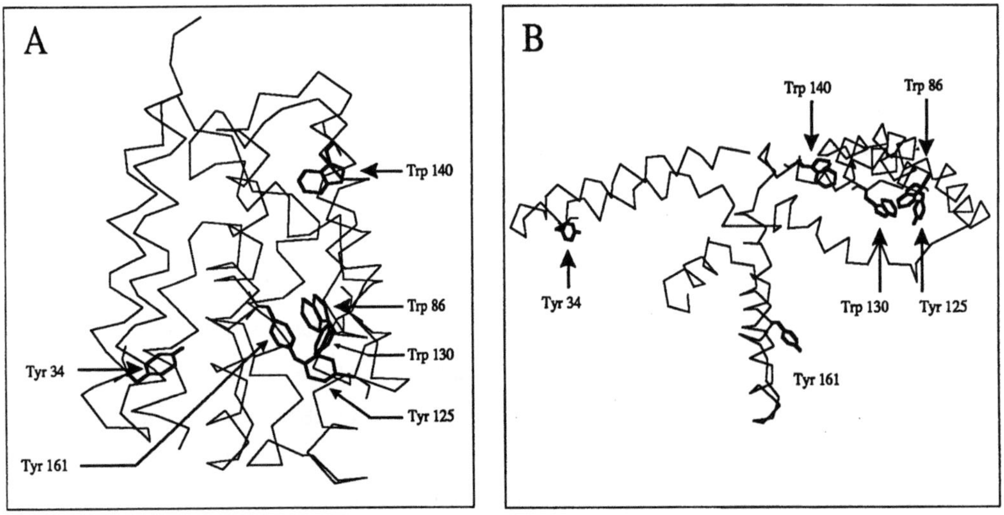

Figure 1. A peptide backbone representation of the colicin thermolytic fragment with aromatic residues attached. **A** Soluble protein structure determined by X-ray diffraction and **B** predicted model of the membrane bound form with two helices (8 & 9) projecting down into the bilayer.

The intrinsic chromophores of the thermolytic fragment gave evidence that was used in the design of the umbrella model (Lakey *et al.*, 1991b). The three tryptophans and three tyrosines provide an intrinsic fluorescence spectrum that blue shifts by only 1.4 nm on membrane insertion without a significant change in intensity. Hence, the apolar environment of the tryptophans is maintained in the membrane bound state. Accessibility studies with the polar quencher acrylamide show that the tryptophans are only slightly less accessible in the membrane than they are in the soluble protein, whilst quenching with trichloroethanol showed that the tryptophans accessibility to this hydrophobic probe goes up on membrane insertion thus indicating an enlarged hydrophobic pocket. The circular dichroism spectrum of the aromatic residues is quite complicated and indicates a number of fixed asymmetric orientations between the Tyr and Trp in the soluble protein. In the soluble form this characteristic signal flattens significantly due to a much more isotropic and therefore mobile relationship between the various chromophores. Nevertheless, the far UV spectrum, which provides an indication of the secondary structure of the protein, shows little change on insertion into the membrane, indicating that the proportion of alpha helix remains essentially the same. Thus, the intrinsic fluorescence of the colicin A thermolytic fragment gave initial clues as to the mode of membrane insertion but provided few possibilities to define the membrane bound or the open channel configurations. To this end we have exploited site-directed mutagenesis to increase the information available from steady-state fluorescence measurements (Lakey *et al.*, 1991a). The interested reader should also refer to the papers by Hubbel's group on site-directed spin label studies of colicin E1 (Todd *et al.*,. 1989) and other membrane proteins.

ATTACHMENT OF EXTRINSIC FLUOROPHORES TO CYSTEINE MUTANTS

Site-directed mutagenesis

Colicin A contains no natural cysteines so the replacement of any residue by a cysteine creates a uniquely reactive site on the the protein. Oligonucleotide-directed mutagenesis was carried out as previously described (Baty *et al.*, 1989) using a *Sma*I-*Hin*dIII fragment of pColA9 inserted into phage M13 vector mp9. Mutations were checked by using dideoxy sequencing and the *Bss*HII-*Bgl* II fragment, which contained the mutation, was then exchanged with the same fragment of pColA9. The first four mutants used were those designed to provide sites for heavy atom derivatives necessary to solve the X-ray structure (Tucker *et al.*, 1989). These sites were chosen to be on the exterior of the protein on the basis of structural predictions [Pattus *et al.*, 1985] and this turned out to be quite successful. Initially three mutants were designed but one (S16C) was found to be followed by a direct repeat of ten amino acids in some recombinants, perhaps due to DNA replication or repair errors. Eventually this mutant gave the best heavy atom derivative. Now that the structure is known, we are better placed to choose mutants which will provide useful information without impeding the activity of the protein. As cysteines are not capable of forming hydrogen bonds they should not replace residues which have these liaisons in the native structure. If possible the size change should be small and hence in theory alanines or valines are suitable targets for replacement. In our experience, see below, surface residues can be replaced without difficulty whilst the main problem with buried residues appears to be our ability to label them.

Of the wide range of cysteine specific fluorescent labels available we have tried three which represent different classes of fluorophore. 5-iodoacetamidofluorescein (5-IAF) is highly fluorescent with a very large extinction coefficient but has several disadvantages. There is significant non-specific binding of the probe to colicin, whether it binds to the hydrophobic center or to surface residues is not known but such binding manifests itself in two ways. Firstly, the probe and the protein do not comigrate on SDS gels run in the absence of β-mercaptoethanol and secondly, controls using the wild-type colicin show significant bound fluorescence. Another disadvantage is the juxtaposition of the excitation and emission bands (small Stokes shift) so that experiments in turbid conditions (i.e. vesicles) suffer significantly from light scattering artefacts.

1,5-IAEDANS (N'-(iodoacetyl)-N'-(5-sulfo-1-napthyl)ethylenediamine is a much better probe and is the one that we use routinely in our fluorescent labelling. It has a large Stokes shift (150 nm) so is ideal for use in vesicle studies. Also the emission wavelength is very environmentally sensitive thus giving information on the aqueous exposure of the probe. We have seen only specific binding of IAEDANS and yields are generally between 75 and 100% for surface exposed cysteines. A very important feature of IAEDANS is its ability to act as an acceptor for tryptophan fluorescence emission which enables us to measure intramolecular distances by resonance energy transfer.

Monobromobimane (MBB) is, potentially, a very useful probe in these systems. It carries no charge at any pH and is smaller than the above probes. It is very highly specific for cysteines and has a useful but not enormous Stokes shift. It does show environmental sensitivity but unfortunately it is of little use in energy transfer experiments.

Labelling procedure

To achieve good yields of specific binding to cysteine mutants the procedure outlined in figure 2 was developed. Variations of this method will still yield good labelling results if the following points are appreciated. Surface exposed cysteines such as those introduced by mutagenesis appear very sensitive to oxidation and they should be reduced just before the reaction. This reduction should be followed by rapid total removal of the reducing agent (DTT) by G-25 or G-50 gel filtration for example. From this point there is competition between the probe and oxygen; add the probe quickly and flush with N_2. Use the same G-25 to remove the excess probe. With buried or slowly reacting cysteines the IAEDANS or MBB can be left to react for 24 hours or more but a control with wild-type should be made in case of side reactions. It is better not to boil your samples before running the gel and not to use reducing agent in the sample buffer.

With buried cysteines such as I26C, we observed no labelling under these conditions and thus it was

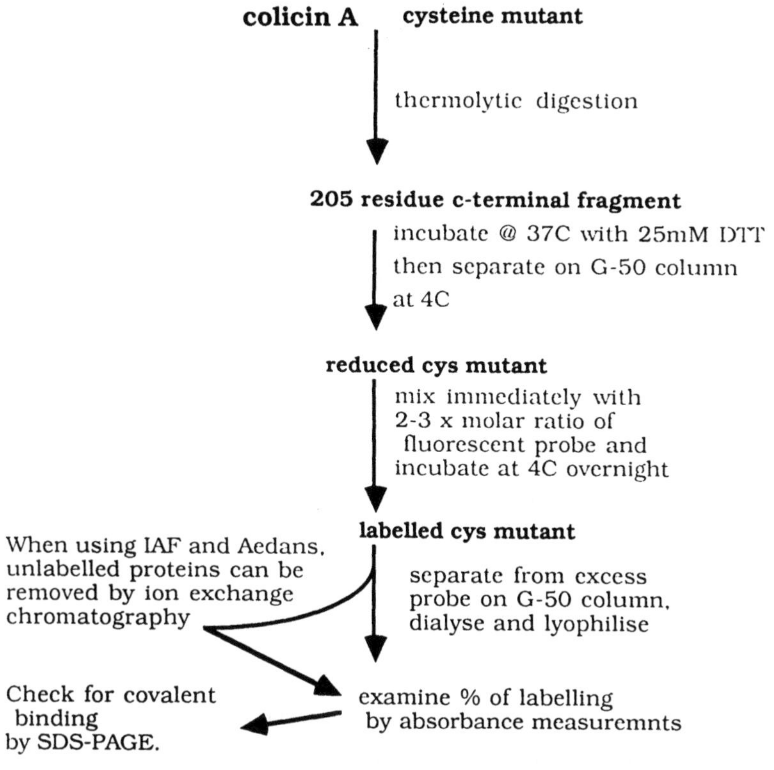

Figure 2. Outline method for the fluorescent labelling of colicin cysteine mutants.

necessary to unfold the protein to expose this residue. Similar work has been done before on the single natural cysteine of colicin E1 (Merrill *et al.*, 1990) which is probably on the equivalent of helix 9. This was denatured in 6M GdnHCl, labelled and renatured with complete recovery of voltage dependent activity. We have seen that at pH 5 only 1M GdnHCl is capable of flattening the near-UV circular dichroism spectrum of the colicin A thermolytic fragment and under these conditions we can label the I26C mutant very efficiently whereas at 0.5M there is no labelling. Spin label probes which have been attached to cysteines on colicin E1 (Todd *et al.*, 1989) penetrate proteins far more easily than IAEDANS and unfolding of the protein does not seem necessary. The stoichiometry of labelling is most easily measured by absorption measurements using the published values for the extinction coefficient of the protein and the probe.

In testing for activity we have no easy enzymatic assay to quantitate residual activity after labelling. It is difficult to use channel analysis in planar bilayer systems as the relationship between bulk concentration and bilayer conductivity is variable because it is difficult to achieve steady state conditions with colicin in such systems. In this case 10% unlabelled protein may give the impression of full activity. Merrill *et al.* (1990) used a vesicle assay which may avoid this problem but we prefer to use the entire colicin molecule to treat living, susceptible *E.coli* cells. The subsequent initial rate of potassium efflux is proportional to the activity and concentration of the colicin added and so slight differences in activity are apparent from these measurements. It is however, an overdemanding assay as it requires that the receptor binding, translocation and channel activity are unaffected by the mutation and labelling when we are only really interested in the channel forming ability of the C-terminal 205 amino acids. So far all the mutants labelled with IAEDANS have given >80% of wild type activity.

Fluorescence Spectroscopy

Details of the fluorescence methods used can be found in Lakey *et al.*, (1991a,b).In these experiments fluorescence spectra were recorded by an SLM 8000 spectrofluorometer operating in ratio mode with spectral bandwidths of 4 nm for both excitation and emission .

The first experiment is to record an emission spectrum of the IAEDANS label. This gives information on the yield and emission wavelength of the probe at this position. IAEDANS in water has an emission maximum near 512 nm whilst in hydrophobic environments this can be as short as 460 nm, hence its relative aqueous exposure can be estimated. The yield (brightness) of the probe is a less quantitative measurement and depends upon many parameters (hydrophobicity, quenching, energy transfer etc.) and may only be useful if it exhibits changes that are proportional to some other measurable parameter such as lipid phase change, etc.

Polarisation anisotropy measurements have been extensively discussed by Lakowicz (1983). This technique detects the degree of rotation of the fluorophore during the excited lifetime and thus is a measure of the mobility of the probe and the molecule to which it is attached. With Colicin A cysteine mutants we have shown that this measure of mobility is governed by the rotation of the thermolytic peptide in solution and is thus very low (0.035) for all mutants. When attached to lipid vesicles the probes show anisotropies which are site specific, we have thus ascribed this to the local mobility of the probe which becomes significant when the protein rotation is slowed down on the lipid vesicle. For example, the Lysine to Cysteine mutant (K39C) has a very high mobility when attached to vesicles but the (S16C) mutant is much lower. In the X-ray structure the serine is in the contact zone between helices 1 and 2 and thus some similar kind of steric effects must exist in the membrane bound form.

In studies of membrane proteins it is especially important to discover which parts of the polypeptide are exposed to solvent and which to membrane. Clearly the emission wavelength gives us a clue but another approach is more rigorous. I have mentioned earlier the quenching of the intrinsic tryptophan fluorescence by acrylamide and TCE. This can be applied to the localisation of IAEDANS but recently we have developed the use of brominated dioleoylphosphatidylglycerol for membrane specific quenching of our site-directed fluorescent probes. These brominated lipids (East & Lee,1982) carry a bromine atom at positions 9 and 10 on each fatty acyl tail. These powerful quenchers are thus situated 1/4 and 3/4 of the distance across the bilayer if the lipids pack in a normal bilayer conformation. The Br-DOPG has already been used to study the exposure of the tryptophan residues and has resulted in 80% quenching of these indicating their extensive contact with the acyl chains of the bilayer. On the other hand quenching of the IAEDANS probes on the surface is non-existent except for T127C-IAEDANS which shows 20% quenching. This residue is in a loop region between two helices and thus it must have contact with the membrane phase.

The clearest evidence to come out of the use of site-directed IAEDANS placement is through the measurement of intramolecular distances by resonance energy transfer (Lakey *et al.*, 1991a). This technique relies on the rate at which fluorescence is transferred from a donor fluorophore to an acceptor chromophore due to the overlap of the donor emission and the acceptor absorbance spectra (see Stryer, 1978). Förster showed that the rate of energy transfer depended on the distance (R) between donor and acceptor;

$$R = [E^{-1}-1]^{\frac{1}{6}} R_0 \qquad (1)$$

where R_0 is the Förster critical distance at which E is 50% (Förster 1951). This was calculated in the usual manner;

$$R_0 = (J \kappa^2 Q_o n^{-4})^{\frac{1}{6}} \times 9.7 \times 10^3 \text{ Å} \qquad (2)$$

where *n*, the refractive index of the solution, was taken as 1.4, Q, the donor quantum yield determined in the absence of acceptor [Q = 0.11 at 35°C and pH 5.0 for tryptophan in thermolytic fragment] and κ^2 is the orientation factor. J, the overlap integral (in $cm^3 M^{-1}$) is given by;

$$J = \frac{\int F(\lambda)\,\varepsilon(\lambda)\,\lambda^4 d\lambda}{\int F(\lambda) d\lambda} \qquad (3)$$

where F(λ) is the corrected fluorescence of the thermolytic fragment excited at 290 nm and ε is the extinction coefficient of attached IAEDANS expressed in $M^{-1}cm^{-1}$. J was numerically integrated at 1 nm intervals.

This leads to a sigmoid relationship between energy transfer efficiency and distance (see figure 3). Distance measurements are best around R_0, which is characteristic of each donor acceptor pair. Distances away from R_0 give poor correlations between energy transfer efficiency and measured distances due to the small change in energy transfer per unit distance. The poor overlap (J) of MBB with Trp showing up in a distance of 12Å for colicin A.

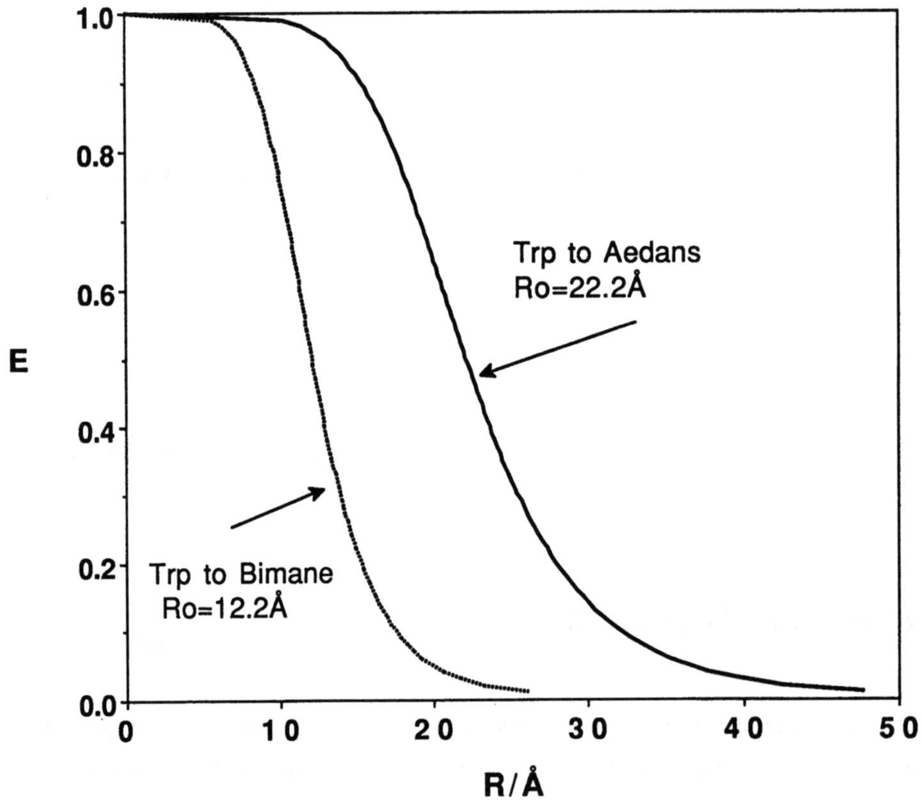

Figure 3. The relationship between the efficiency of energy transfer and Trp to acceptor distance for two fluorescent probes attached to colicin A thermolytic fragment.

The greatest source of inaccuracy in all measurements of this kind is the size of the orientation factor

κ^2, i.e. the measure of relative dipole orientation which cannot be determined uniquely. It can adopt values between 0 and 4 but if both donor and acceptor can rotate isotropically within times much shorter than the transfer time then $\kappa^2 = 2/3$. Although this condition is not always met, an estimate of the upper and lower bounds of κ^2 can be made from steady state polarisation measurements (Dale *et al.*, 1979). One advantage of the site- directed inclusion of fluorescent probes is that we can make measurements on several sites with presumably different orientations in the same part of the molecule. If they all indicate a similar distance then it is unlikely that the orientation factor differs significantly from 2/3.

As shown in figure 4 the energy transfer between the three tryptophans and K39C-IAEDANS falls significantly when lipids are added.

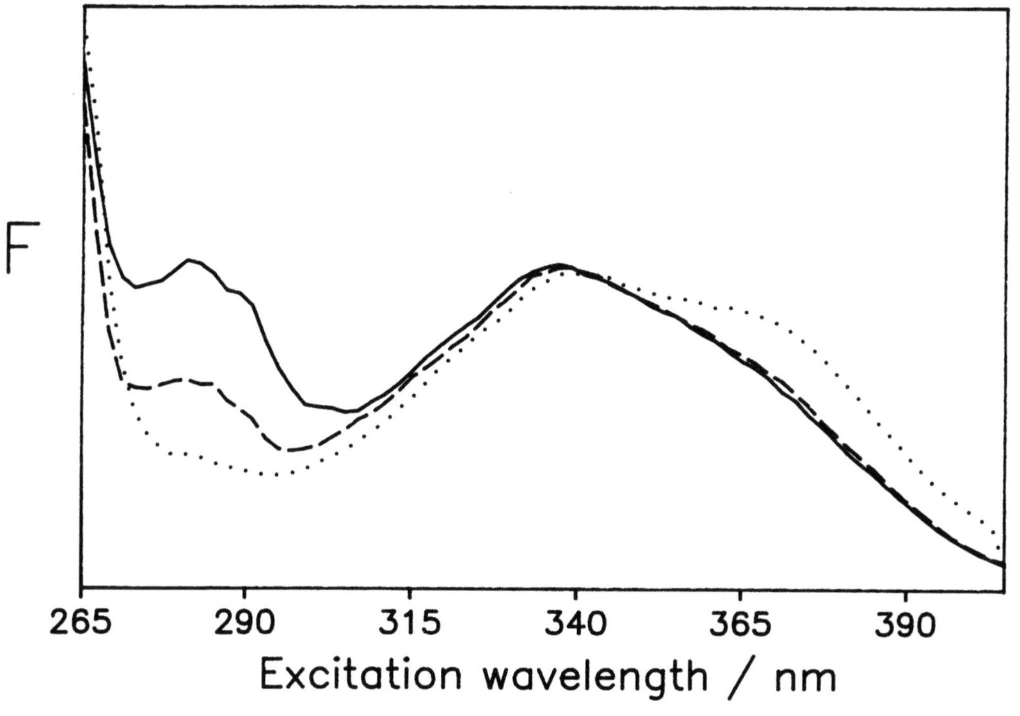

Figure 4. Fluorescence excitation spectra of the IAEDANS label on the K39C mutant. Emission wavelength = 490 nm. Solid line = protein in aqueous solution. Dashed line = protein in DMPG vesicles, indicating the loss of tryptophan contribution (at 290 nm) to the IAEDANS fluorescence. This is due to a reduction in distance dependent energy transfer. Dotted line = protein in 6M GdnHCl. All data normalised to the same value at 340 nm.

The same is true for S16C-IAEDANS whilst T127C-IAEDANS shows only a small decrease. These results indicate that, in accordance with the model of Parker, helices 1 and 2 move significantly farther away from helices 5,6 and 7 when binding to the membrane occurs (Lakey et al., 1991a). It should be noted here that intramolecular energy transfer measurements are sensitive to intermolecular transfer from neighbouring proteins especially in the membrane phase.This can be avoided by the addition of an excess of non tryptophan-containing (N-bromosuccinimide treated, Spande & Witkop, 1967) colicin or by sufficient dilution of the proteins by adding excess lipid.We are currently involved in characterising a series of mutants which like I26C are internal to the protein structure in solution and may therefore be exposed to the membrane phase when bound to the membrane. In addition we have labelled three cysteine mutants A169C, G166C, G176C which define the point of the hydrophobic hairpin. These will be used in distance and membrane exposure measurements to better define the position of this region in membrane-bound form.

MODIFICATION OF INTRINSIC FLUOROPHORES

The use of site-directed mutagenesis to place fluorescent amino acid residues at selected points on the structure of colicin A holds many advantages for the study of membrane insertion mechanisms. Most advantageous is the lack of labelling reactions required, hence making the purification of the mutant protein the only step before use. Secondly, it may be thought that the insertion of tryptophan or tyrosine groups may be more likely to maintain the original activity and conformation than the addition of artificial groups like IAEDANS. Finally, the ability to move the donors around the molecule makes distance measurement by energy transfer far more flexible.

The first stage in this process is to create a starting molecule which is non-fluorescent by removing the intrinsic tryptophan and tyrosine fluorescence. Tryptophan removal, which is the most important step as these are the best natural fluorophores to reinsert, has been successfully achieved for several proteins (Axelsen *et al.*, 1991; Loewenthal *et al.*, 1991) , most notably in the case of colicin E1 (Merrill *et al.*, 1991). The removal of the three tryptophans from Colicin A thermolytic fragment is still at an early stage but, in view of our knowledge of the three dimensional structure, is quite informative regarding the structural constraints of this process.We chose to replace the tryptophan residues by phenylalanine as this is the next most bulky aromatic group which does not introduce an additional hydrogen bonding constraint such as that present with the -OH of tyrosine. In fact no other residue can replace structural features of tryptophan due to its extreme bulkiness and the hydrogen bonding afforded by the ring nitrogen. This inequality may be the reason why the replacement of tryptophans is rather problematic.

W130 occurs in a hydrophobic pocket with weak contacts with Y125 (whose ring is perpendicular to that of W130) and Leu 178. W130F still has a toxic activity although its secretion from the cells is poorer. W86 is partially exposed to solvent and has a series of close contacts with other residues R118, L158, L179, a possible hydrogen bond with S121 and a parallel stacking reminiscent of base stacking in DNA with Y161. In spite of this, the W86F still has almost the same activity as the wild type. W140 is at the end of helix seven and is the most water exposed of all the tryptophans. It has a hydrogen bond with E136 and has contact with K113 and F154. W140F has negligible secretion from the producer cell and little or no activity. It is not resistant to thermolysin digestion so it cannot yield a stable C-terminal fragment. By inspecting the 3-D structure we believed that part of the problem may arise from the removal of the H-bond from E136 so that it seeks another link with a lysine further away. In the homologous region from Col E1 there is no tryptophan or glutamate so it was decided to remodel this region using replacements that mimic the Col E1 structure. Hence the constructions E136T/V137L/W140K and W140K/K113F which refilled the space made by the W140 substitution all show close to wild type activity. This is a very good example of the usefulness of aligned sequences in guiding the choices in site-directed mutagenesis studies. Such complicated replacement is only necessary to produce the initial W^- form upon which simple X to W mutations can be made. These will have to satisfy the activity criteria outlined above but hopefully a significant proportion of these will be successful if the 3-D information can be well interpreted.

CONCLUSIONS

The use of site-directed cysteine mutants has already shown promise in the elucidation of the membrane bound structure of the channel-forming region of colicin A. With improved quenching methods and tryptophan mutants we hope to fully characterise this structure and eventually also model the open channel configuration that exists under the influence of a membrane potential.

ACKNOWLEDGEMENTS

We thank N. Didat and S. Scianimanico for technical assistance, J.M.G was a recipient of a fellowship from the Spanish Ministry of Education and Science .

REFERENCES

Axelsen PH, Bajzer Z, Prendergast FG, Cottam PF, Ho C (1991) Resolution of fluorescence intensity decays of the two tryptophan residues in glutamine-binding protein from *Escherichia coli* using single tryptophan mutants. Biophys J 60:650-659

Bourdineaud J-P, Boulanger P, Lazdunski C, Letellier L (1991) *In vivo* properties of colicin A: channel activity is voltage dependent but translocation may be voltage independent. Proc Natl Acad Sci USA 87:1037-1041

Dale RE, Eisinger J, Blumberg WE (1979) The orientational freedom of molecular probes. The orientation factor in intramolecular energy transfer. Biophys J 26:161-194

East JM, Lee AG (1982) Biochemistry 21:4144-4151

Förster T (1951) Fluoreszenz organischer Verbindung, Vandenhoeck and Rupprecht, Göttingen.

Lakey JH, Baty D, Pattus F (1991a) Fluoresence energy transfer measurements using site-directed single cysteine mutants. The membrane insertion of colicin A. J Mol Biol 219:639-653

Lakey JH, Massotte D, Heitz F, Faucon J-F, Dasseux J-L, Parker MW, Pattus F (1991b) Conformation of the pore-forming domain of colicin A in its membrane-bound form. Eur J Biochem 196:599-607

Lakowicz JR (1983) Principles of Fluorescence Spectroscopy, Plenum Press, New York London

Lazdunski C, Baty D, Géli V, Cavard D, Morlon J, Lloubès R, Howard P, Knibiehler M, Chartier M, Varenne S, Frenette M, Dasseux J-L, Pattus F (1988) The membrane channel-forming colicin A: synthesis secretion, structure, action and immunity. Biochim Biophys Acta 947:445-464

Loewenthal R, Sancho J, Fersht AR (1991) Flourescence spectrum of Barnase: Contributions of the three tryptophan residues and a histidine related pH dependence. Biochemistry 30:6775-6779

Martinez MC, Lazdunski C, Pattus F (1983) Isolation, molecular and functional properties of the C-terminal domain of colicin A. EMBO J 2:1501-1507

Merrill AR, Cohen FS, Cramer WA (1990) On the nature of the structural change of the colicin E1 channel peptide necessary for its translocation-competent state. Biochemistry 29:5829-5836

Merrill AR, Szabo AG, Cramer WA (1990) Correlation of fluorescence parameters of single tryptophan mutants of a colicin E1 channel peptide with its translocation-competent state. Biophys J 59:458a

Parker MW, Pattus F, Tucker AD, Tsernoglou D (1989) Structure of the membrane-pore-forming fragment of colicin A. Nature 337:93-96

Parker MW, Tucker AD, Tsernoglou D, Pattus F (1990) Insights into membrane insertion based on studies of colicins. TIBS 15:126-129

Spande TF, Witkop B (1967) Determination of the tryptophan content of proteins with N-bromosuccinimide.In: Hirs CHW (ed) Methods in Enzymology, vol 11. Academic Press, New York London, p 506

Stryer L (1978) Fluorescence energy transfer as a spectroscopic ruler. Ann Rev Biochem 47:819-846

Todd AP, Cong J, Levinthal F, Levinthal C, Hubbel WL (1989) Site-directed mutagenesis of colE1 provides site specific attachment sites for spin labels whose spectra are sensitive to local conformation. Proteins 6:294-305

Tucker A D, Baty D, Parker MW, Pattus F, Lazdunski C, Tsernoglou D (1989) Crystallographic phases through genetic engineering: experiences with colicin A. Protein Eng 2:399-405

STRUCTURE-FUNCTION OF THE COLICIN E1 ION CHANNEL: VOLTAGE-DRIVEN TRANSLOCATION AND GATING OF A TETRA- (OR HEXA-) HELIX CHANNEL

W. A. Cramer, F. S. Cohen*, C. V. Stauffacher, Y.-L. Zhang, A. R. Merrill**, H. Y. Song and P. Elkins
Department of Biological Sciences
Purdue University
West Lafayette, IN 47907
U.S.A.

The problem of the formation and gating of the colicin El (or Ia, Ib, A, B, N) ion channel differs from that of the more traditionally studied eukaryotic ion channels because the toxin-like colicin molecules have two lives, one in solution and an after-life following passage into the membrane bilayer. Thus, the problem of formation of the colicin channel structure is related to problems of protein import into membranes. The basic properties of the membrane import and insertion processes for the channel-forming domain of the 522 residue colicin El molecule are: (a) the 178-187 residue COOH-terminal channel peptide has a predominantly (50-60%) α-helical conformation in solution (Brunden *et al.*, 1984); (b) the helical content, if not the same helical domains, is preserved in channel peptide-proteoliposomes (Rath *et al.*, 1991); (c) channel activity *in vitro* requires an acidic pH (Davidson *et al.*, 1985), (d) membranes with an acidic lipid composition or a small amount of added non-ionic detergent (Bullock & Cohen, 1986), and (e) a transnegative membrane potential (Schein *et al.*, 1978; Bullock *et al.*, 1983; Peterson & Cramer, 1987). The membrane potential requirement makes the channel "voltage-gated." This requirement was also inferred from studies on the *in vivo* cytotoxic action of the colicin molecule (Jetten & Jetten, 1975), and is reflected in such experiments by the decreased sensitivity of *E. coli* cells in the stationary phase of growth.

The Water Soluble Channel Peptide: Crystallization; the Nature of the Translocation-Competent State

Crystallization experiments have been done on the channel peptide of colicin El at both alkaline pH (soluble form, pH 7-8) and acidic pH (membrane binding form, pH 4-5). Channel peptide from proteolytically digested colicin, as well as a cloned COOH-terminal fragment of colicin El of the same length (Song, 1990), have both been crystallized from solutions of 22-26% PEG and sodium salts at alkaline pH values. These crystals grow in several morphologies, but with related crystal packing, in the

*Department of Physiology, Rush Medical College, Chicago, IL 60612, U.S.A. **Current address: Department of Chemistry and Biochemistry, University of Guelph, Guelph, Ontario, Canada NlG 2Wl.

NATO ASI Series, Vol. H 65
Bacteriocins, Microcins and Lantibiotics
Edited by R. James, C. Lazdunski and F. Pattus

space group I4 (Figure 1). Proteolytic fragments grow as long rods and the engineered fragment as rhombohedral prisms (Elkins *et al.*, 1990, and in preparation). Both crystal forms diffract to atomic resolution with the rod-shaped I4 crystals diffracting to 2.2Å and the rhombohedral ones to 2.6Å. Studies at low pH also indicate that crystals can be grown from these peptides, with small crystals produced from both, but in this case the engineered fragment appears to give more promising results.

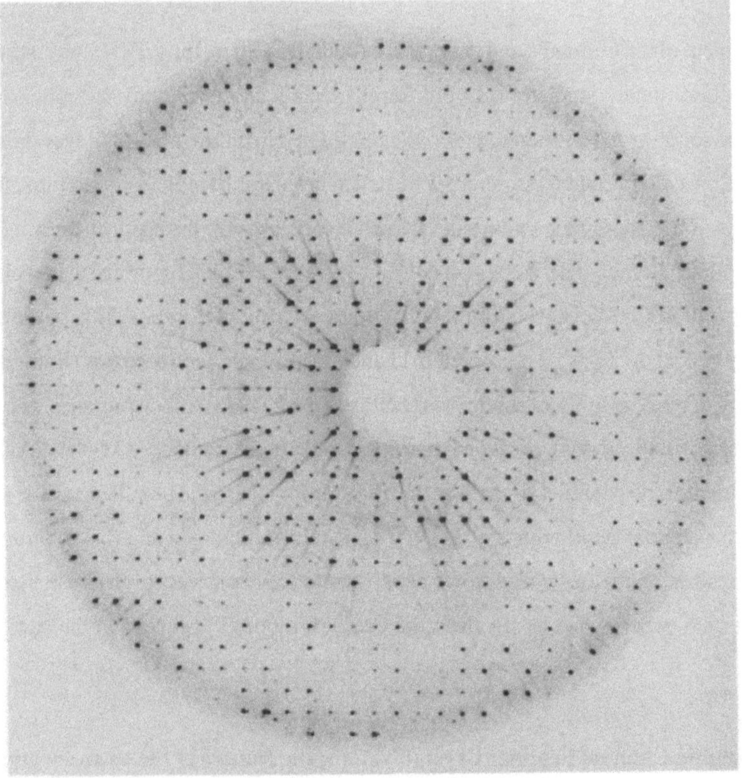

Figure 1. Diffraction pattern of the tryptic colicin E1 channel peptide along the four-fold axis of the I4 unit cell.

The translocation competent (pH ≤ 4) state of the channel peptide was initially inferred to be significantly unfolded by the criteria of increased protease accessibility (Randall & Hardy, 1986) and partition into non-ionic detergents (Olsnes *et al.*, 1988), that have commonly been used to infer unfolding and conformation change of translocatable proteins. However, the average hydrodynamic radius was found not to increase, but to decrease slightly, from 23Å to 21Å, as the pH was decreased from 6.0 to 3.5 (Merrill *et al.*, 1990). The peptide denatured in guanidine hydrochloride had a radius of 85Å. This suggests that translocation competence is not associated with significant unfolding, but perhaps with a "molten globule" state.

Sequence Information

Comparison of the amino acid sequences of the channel domains (last 202 residues of colicin El) of six channel forming colicins shows 28 conserved residues (Cramer *et al.*, 1990). The calculated pI of the colicin El tryptic COOH-terminal 187 residue channel domain is 9.4. This domain also contains (a) a pronounced hydrophobic domain extending approximately over residues 471-508, and (b) four 16-18 residue segments, Q357-D376, I383-N401, K420-K436, and S442-K461, of pronounced amphipathicity. The latter peptide segments are candidates for additional membrane spanning α-helices of the channel structure.

The Hydrophobic Membrane Anchor

The pronounced 35-40 residue hydrophobic region, A471-I508 in colicin El, is approximately 10 residues longer in the colicin A channel peptide. The existence of a hydrophobic sequence in colicin El, and a pre-formed hydrophobic α-helical hairpin in the crystallographic structure of the soluble colicin A peptide, led to the suggestions that this hydrophobic hairpin might spontaneously insert into the membrane to form an anchor for the channel (Cleveland *et al.*, 1983; Parker *et al.*, 1989). The COOH-terminus of the colicin E1 channel peptide inserted into artificial membranes was found to be exposed on the *cis*-side of the membrane, on the basis of the ability of carboxypeptidase Y added from this side to inhibit activity or to cleave the membrane-bound peptide (Cleveland *et al.*, 1983; Xu *et al.*, 1988). Further details on the topography of the colicin El channel hydrophobic hairpin in the membrane were obtained by saturation mutagenesis and insertion of positive charges into the 35-40 residue hydrophobic region (Song *et al.*, 1991). The pattern of activities of the resultant mutants implied the existence of a membrane spanning helical hairpin including a turn region covering residues 489-493 that includes 2 Gly and 2 Thr residues.

Search for Peptide Segments Involved in Voltage Gating

With no other information on the identity of membrane-spanning α-helices, and no other regions in the channel peptide that would be predicted to insert spontaneously, we decided to search for voltage-dependent insertion of segments of the channel domain into the membrane (Merrill & Cramer, 1989, 1990a,b).

The protocol, which may be of general utility for studies of potential-dependent protein import, is as follows: the 178 residue thermolytic channel peptide was added to large (0.4 μ average diameter) K^+-loaded membrane vesicles made of soybean lipid ("asolectin"). Labelling of potentially voltage responsive, importable peptide segments was assayed with two probes: (a) the lipid was doped with the lipophilic photoaffinity probe [^{125}I]TID [3-(trifluoromethyl)-3-(*m*-[^{125}I]iodophenyldiazirin]; (b) the water-soluble tyrosine-specific halide, [^{125}I] plus the catalyst Iodogen, was added to the aqueous phase external to the liposomes as a probe of the accessibility of the 9 Tyr residues in the thermolytic channel peptide. The two labelling systems provided complementary information on (i) increased peptide presence in the membrane bilayer as seen by the TID, or (ii) a decrease in peptide accessibility to the Iodogen from the outer aqueous phase. The iodine-radiolabelling of peptide extracted from the vesicle suspension, illuminated with UV light in the case of the TID labelling, was measured after incubation of the vesicle suspension in the presence and absence of a potassium diffusion potential ($\Delta\psi$ = -100 mV or 0 mV). This potential was produced by dilution of the vesicle suspension in the presence of valinomycin into a potassium-free solution. Labelling of the peptide by the [^{125}I]-TID was increased by 30-40% in the presence of the $\Delta\psi$, and the external Iodogen showed the same decrease. Statistically significant labeling changes were localized to the segment, Trp424-Trp460, by use of a tryptophan-specific cleavage reagent (BNPS-skatole) applied to channel peptide extracted from the liposomes. The peptide used in these cleavage experiments contained only a single tryptophan at positions 424, 460, or 495 after substitution of a phenylalanine at the other two positions. The data are summarized in Table I.

For the interpretation of this experiment, it was important to know that the channel peptide was bound tightly to the membrane vesicles in the absence of the membrane potential. The 178-190 residue COOH-terminal channel peptides are bound so tightly downstream of Lys382 that there is no cleavage by trypsin in the terminal 140 residues from Lys382 to Ile522-COOH in spite of the presence of 16 Lys in this segment (Xu *et al.*, 1988). It was found using other proteases, i.e., chymotrypsin, pepsin, and pronase E, that there is also no significant cleavage detectable downstream of Lys382 (Zhang & Cramer, in preparation). This implies that the COOH-terminal segment is bound very tightly, that the $\Delta\psi$ does not cause increased binding of the channel peptide, and that a segment including Trp424-Trp460, the "gating peptide," is inserted into the bilayer upon imposition of a membrane potential (Figure 2). Moreover, this insertion was found to be **reversible**. Upon dissipation of the $\Delta\psi$, the gating peptide was extruded from the membrane (Merrill & Cramer, 1990b). It was concluded that reversible insertion of the peptide segment into the membrane may account for channel gating.

Amphipathic Nature of Gating Hairpin

The segment of colicin E1, extending from Lys420 to Lys461, that is inserted into the membrane

Table I. Dependence on Membrane Potential ($\Delta\psi$) of the Labelling of the Channel Peptide from Two Colicin E1 Mutants Containing a Single Tryptophan [(1) W424-F460-F495 and (2) F424-W460-F495] by (A) [^{125}I]TID in the Membrane Bilayer and (B) [^{125}I]Iodogen in the External Aqueous Solution (Merrill and Cramer, 1990b).

	$\Delta\psi$ (mV)	(A) [^{125}I]TID Relative Labelling			*n*
(1) F424-W460-F495		Uncleaved[a]	I345-W460[b]	K461-I522[b]	7
	0	1.00 ± 0.07	1.00 ± 0.08	1.00 ± 0.11	
	-100	1.30 ± 0.08	1.57 ± 0.10	0.92 ± 0.12	
(2) W424-F460-F495		uncleaved[a]	I345-W424		3
	0	1.00 ± 0.09	1.00 ± 0.06		
	-100	1.34 ± 0.08	1.06 ± 0.13		

	$\Delta\psi$ (mV)	(B) [^{125}I]Iodogen Relative Labelling			*n*
(1) F424-W460-F495		Uncleaved[b]	I345-W460[b]	K461-I522[c]	4
membrane	0	1.00 ± 0.07	1.00 ± 0.06	1.00 ± 0.09	
	-100	0.76 ± 0.04	0.62 ± 0.05	1.14 ± 0.12	
soln (no liposomes)		1.46 ± 0.10	1.88 ± 0.09	1.07 ± 0.08	
(2) W424-F460-F495		Uncleaved[d]	I345-W424[e]		3
membrane	0	1.00 ± 0.07	1.00 ± 0.12		
	-100	0.65 ± 0.07	0.95 ± 0.15		
soln (no liposomes)		1.80 ± 0.12	2.16 ± 0.44		

[a]Uncleaved refers to intact peptide that was not fragmented during the tryptophan cleavage reaction. This was typically 20-25% of the total peptide subjected to cleavage. [b]The data in these columns were found to be significantly different with a confidence level of $p = 0.99$ for n trials. [c]The data are not significantly different: confidence level $p = 0.70$-0.80. [d] $p = 0.98$ indicates that these differences are significant. [e] $p = 0.95$ indicates that these differences are significant.

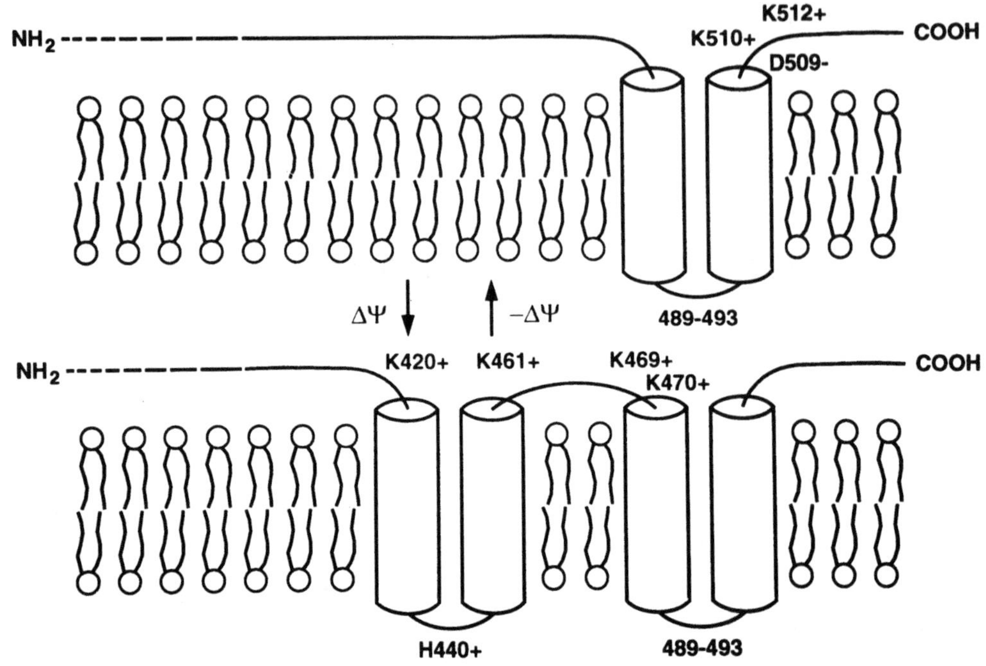

Figure 2. Four helix model of reversible potential-dependent insertion of colicin E1 gating peptide, K420-K461, into the membrane bilayer, after spontaneous insertion of the hydrophobic anchor helical hairpin, K470-D509. $\Delta\psi$, membrane potential.

bilayer in the "open" state can be depicted as two amphipathic peptides (helical wheel diagram, Figure 3A,B) extending from Lys420-Lys436 and from Ser442-Lys461. It is proposed that the gating peptide inserts into the membrane as an **amphipathic helical hairpin**. The amphipathic nature of these helices has been tested by mutagenesis of the non-polar amino acids of the hairpin to charged residues (Song, 1990). Together with the hydrophobic (anchor) hairpin described above, the open channel would then contain at least four transmembrane helices (Figure 4).

Role of Upstream (above Lys382) Region; Number of Helices in Channel

As noted above, the segments Gln357-Asp376 and Ile383-Asn401 are very amphipathic (hydrophobic moment, $\mu = 0.45$ and 0.49 for periodicities of 96° and 110°, respectively), and might be predicted on this basis to span the bilayer, generating a six helix channel. Such a six helix model would require that Lys382 be positioned on the trans-side of the membrane, where it would be inaccessible to proteases. This prediction contradicts the experimental data on trypsin to the channel peptide incorporated into membrane vesicles in the absence of a membrane potential ($-\Delta\psi$). These data showed a cleavage site after K381 and K382 (Xu *et al.*, 1988). Furthermore, the cleavage sites in the peptide inserted into liposomes ($-\Delta\psi$), caused by chymotrypsin and pepsin, were after residues Tyr367 or Leu374, and Phe355 or Glu373, respectively (Zhang & Cramer, in preparation).

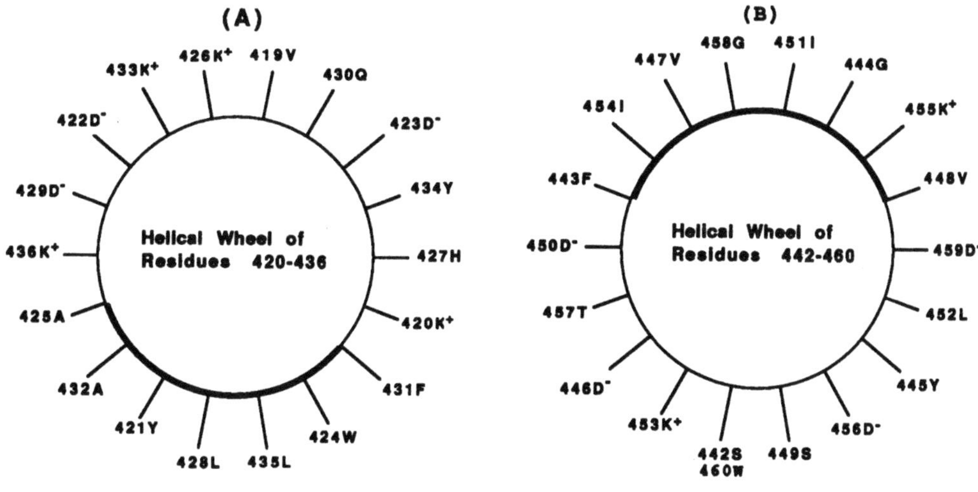

Figure 3. Helical wheel diagrams of amphipathic helices (A) Lys420-Lys436 and (B) Ser442-Lys461 that constitute the amphipathic "gating peptide. The non-polar surface is marked by a bold perimeter. Note that Lys455 violates the simple amphipathic character in (B).

We had suggested that another "tight" amphipathic helical hairpin from about Lys382 to Lys420 might exist in the membrane in the absence of a $\Delta\psi$, if four helices are not sufficient to form a channel with lumen $\leq$ 8-10Å diameter (Raymond *et al.*, 1985). Upon further reflection, it seems unlikely that a K382-K420 hairpin could insert spontaneously into the membrane in the absence of a $\Delta\psi$ because it contains 5 Lys, 1 Arg, 3 Asp, and 2 Glu residues. Perhaps this hairpin also inserts into the membrane by a potential-dependent mechanism and is also part of the "gating" mechanism, but requires a larger or longer lasting membrane potential than could be sustained in our liposome experiments by a potassium diffusion potential. On the other hand, there is precedent in studies on voltage- and ligand-gated ion channels for four helix models (Grove *et al.*, 1991). The proposed folding of a four helix model for colicin El is shown (Figure 4).

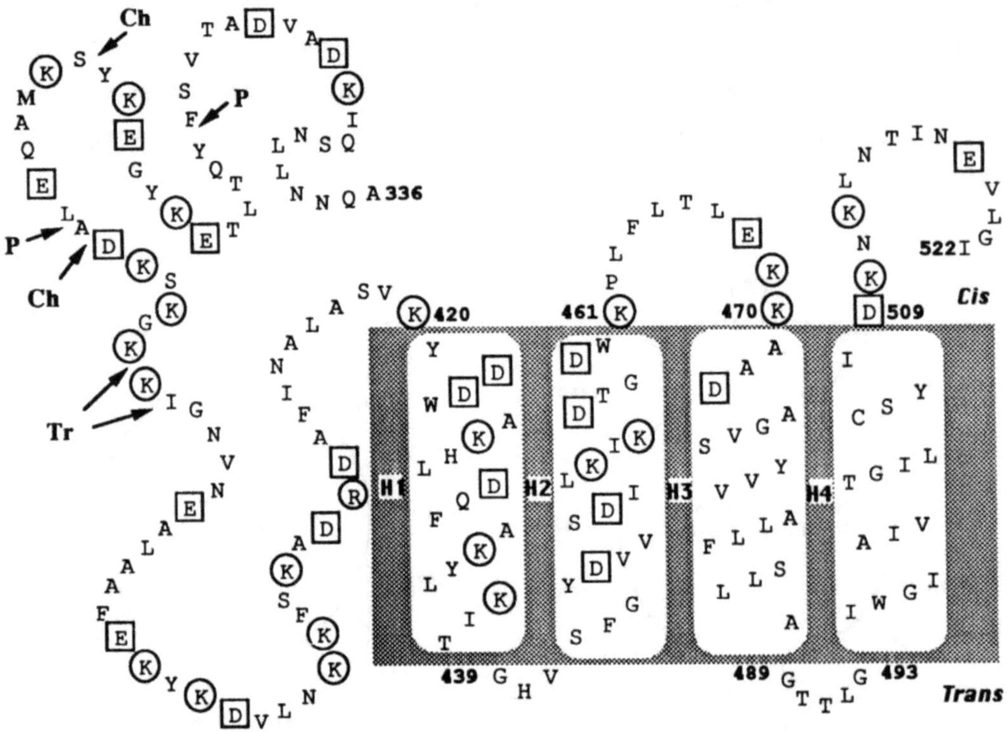

Figure 4. Folding of a four helix (H1-H4) model of the colicin E1 channel in the membrane bilayer. Sites of action of trypsin (Tr), chymotrypsin (Ch), and pepsin (P) are marked (↑). The model has the unusual features of somewhat short α-helices and short turns on the *trans* side.

Channel Models

A graphics model of a four helix channel was constructed according to the following criteria: (i) helices I-IV consisted of (I) Lys420-Lys436, (II) Ser442-Lys461, (III) Lys470-Ala488, and (IV) Ile494-Lys510, the former two helices constituting the amphipathic gating peptide helical hairpin, and the latter two the hydrophobic anchor hairpin. The charged residues included on the ends of the helices represent the "punctuating" residues where side chains are in the polar phase. (ii) Van der Waals and H bond distances between residues were individually checked. An overall energy minimization was not performed. (iii) The amphipathic helices I and II were oriented so that their polar sides faced the internal aqueous lumen; hydrophobic helix IV was oriented so that Thr501 and Gly502, whose mutation to Glu affected the ion selectivity of the channel (Shirabe *et al.*, 1989), would both also face the aqueous channel lumen.

Some conclusions that result from visualization (Figure 5A,B) of the four helix model are: (i) there are no obvious violations of thermodynamic and structural principles among the eighteen charged residues, 9 Lys and 9 Asp, except for the side chain of Lys455 in helix II that is directed into the lipid phase;

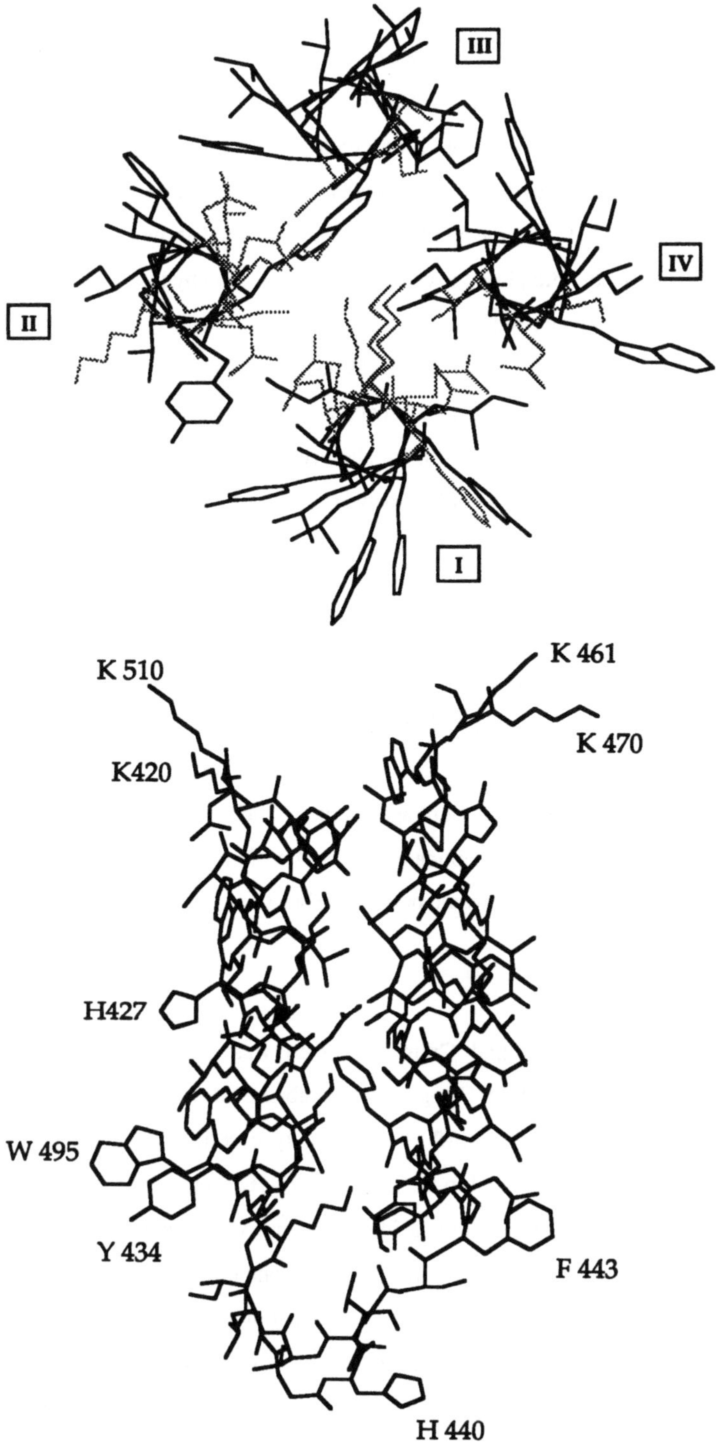

A

B

Figure 5. Graphics model of four helix colicin E1 channel viewed (A) along the axis perpendicular to the membrane plane and (B) from side, with transmembrane helices I, IV on left and II, III on right.

(ii) in fact, the charged residues used for the channel in this model are the more compact and more flexible Lys relative to Arg, and the smaller Asp relative to Glu; (iii) of the eleven aromatic residues, Y421, W424, F431, Y434, F443, Y445, W460, Y478, F484, W495, and Y507, all face outwardly into the lipid except for (a) W460 near the *cis*-side aqueous interface, which would partially occlude the central ion channel on the *cis* side if an α-helical structure is maintained, and (b) Phe484 which occupies a position in a hydrophobic pocket between helices III and IV; (iv) the amphipathic helices I and II show an interesting alteration of basic and acidic side chains extending into the channel. As can be seen by inspection of the amino acid sequence, the arrangement of charged residues along the channel side of helix I is (*cis* → *trans*) K420-D422-D423-K426-D429-K433-K436, and in (II) K461-D459-D456-K453-D450-D446; (v) it is difficult to determine the minimum diameter of the channel lumen, as it is dependent upon side chain conformation. However, it could readily be 8-10Å with minimal allowed changes of side chain torsion angles. In summary, this graphics exercise indicates that it is possible that the colicin El channel is a four helix monomer, although the position of Lys455 in the second amphipathic helix may pose a problem. On the other hand, if the channel is made of six helices, including the segment Lys382-Lys420 that would define the first two helices, the structure would have three very short turns on the *trans* side instead of two, and an additional very short turn on the *cis* side. These first two helices of a six helix model would differ from the other four by containing Glu and Arg (residue 409) residues. These helices would also require a membrane potential for insertion of the first two helices that was not detected in the $\Delta\psi$-dependence experiment of Merrill and Cramer (1990,a,b). One argument against the six helix, and for a four helix, model is the ability of a 136 residue COOH-terminal peptide of colicin A to form ion channels in lipid bilayers (Baty *et al.*, 1990). The latter channel peptide would not include helix one of the six helix model.

Electrophysiological Experiments

The "gating peptide" model of Merrill and Cramer (1990b) was further tested using electrophysiological methods by Abrams *et al.* (1991). Using planar membrane bilayers, they compared the voltage dependence of the rate of channel "turn-off" of the wild type colicin molecule and a His440→Cys mutant. His440 is believed to be located in the interhelix Hl-H2 loop (Figure 4) and, in contrast to the uncharged Cys440, would be charged at the acidic pH values where *in vitro* activity is assayed. The decreased voltage-dependence of "turn-off" of the Cys440 mutant implied that His440 was translocated across the membrane energy barrier in the presence of the gating potential. Although the deactivation appears to have more complicated kinetics than indicated by Abrams *et al.* (1991), the qualitative data that we have obtained by the application of this technique to several channel mutants implies: (i) Asp446 is not a key residue in pH-dependent deactivation kinetics; (ii) Lys453 and 455 have no role in determining the deactivation rate and are therefore on the *cis* side of the activation barrier, and a Lys433 → Met mutant shows a $\Delta\psi$- dependence slightly reduced relative to wild type, consistent with the channel folding model of Figure 4.

Acknowledgement

The studies discussed here were supported by grants from the NIH: GM18457 (WAC), GM27367 (FSC), and GM44001 (CVS). We thank Dr. T. J. Smith for advice on the construction of Figure 5, Dr. R Brasseur for communicating results prior to publication, Janet Hollister for her careful and aesthetic work on this manuscript. We note that it has not been possible because of space limitations to discuss or refer in the above to many of the important studies on other channel-forming colicins, particularly the large body of elegant work on colicin A, from the laboratories of C. Lazdunski and F. Pattus.

References

Abrams CK, Jakes KS, Finkelstein A, Slatin SL (1991) Identification of a translocated gating charge in a voltage-dependent channel. Colicin E1 channels in planar phospholipid bilayer membranes. J Gen Physiol 98:77-93

Baty D, Lakey J, Pattus F, Lazdunski C (1990) A 136 amino acid residue COOH-terminal fragment of colicin A is endowed with ionophoric activity. Eur J Biochem 189:409-413

Brunden KR, Uratani Y, Cramer WA (1984) Dependence of the conformation of a colicin El channel-forming peptide on the acidic pH and solvent polarity. J Biol Chem 259:7682-7687

Bullock JO, Cohen FS (1986) Octyl glucoside promotes incorporation of channels into planar phospholipid bilayers. Biochim Biophys Acta 856:101-108

Bullock JO, Cohen FS, Dankert JR, Cramer WA (1983) Comparison of the macroscopic and single channel conductance properties of colicin El and its C-terminal tryptic peptide. J Biol Chem 258:9908-9912

Cleveland BM, Slatin S, Finkelstein A, Levinthal C (1983) Structure-function for a voltage-dependent ion channel: properties of a COOH-terminal fragment of colicin El. Proc Natl Acad Sci USA 80:3706-3710

Cramer WA, Cohen FS, Merrill AR, Song HY (1990) Structure and dynamics of the colicin El channel. Molec Microbiol 4:519-526.

Davidson VL, Brunden KR, Cramer WA (1985) An acidic pH requirement for insertion of colicin El into artificial membrane vesicles: relevance to the mechanism of action of colicins and certain toxins. Proc Natl Acad Sci USA 82:1386-1390

Elkins P, Fisher A, Merrill AR, Cramer WA, Stauffacher CV (1990) Crystallization of colicin E1 pore forming fragment. Biophys J 57:419a.

Grove A, Tomich JM, Montal M (1991) A molecular blueprint for the pore-forming structure of voltage-gated calcium channels. Proc Nat Acad Sci USA 88:6418-6422

Jetten AM, Jetten MER (1975) Energy requirement for the initiation of colicin action in *E. coli*. Biochim Biophys Acta 387:12-22

Merrill AR, Cramer WA (1989) The mechanism of colicin El channel formation: pH-dependent unfolding and membrane potential dependent insertion into the lipid bilayer. Biophys J 55:247a

Merrill AR, Cramer WA (1990a) Protein translocation in the voltage gating mechanism of the colicin El ion channel. Biophys J 57:316a

Merrill AR, Cramer WA (1990b) Identification of a voltage-responsive segment of the potential-gated colicin El ion channel. Biochemistry 29:8529-8534

Merrill AR, Cohen FS, Cramer, WA (1990) On the nature of the structural change of the colicin E1 channel peptide necessary for its translocation-competent state. Biochemistry 29:5829-5836.

Olsnes S, Moskaug JO, Stenmark H, Sandvig K (1988) Diphtheria toxin entry: protein translocation in the reverse direction. Trends Biochem Sci 13:348-351

Parker MW, Pattus F, Tucker AD, Tsernoglou D (1989) Structure of the membrane-pore-forming fragment of colicin A. Nature 337:93-96

Peterson AA, Cramer WA (1987) Voltage-dependent monomeric channel activity of colicin El in artificial membrane vesicles. J Mem Biol 99:197-204

Randall LL, Hardy SJS (1986) Correlation of competence for export with lack of tertiary structure of the mature species: A study *in vivo* of maltose-binding protein in *E. coli.* Cell 46:421-428

Rath P, Bousché O, Merrill AR, Cramer WA, Rothschild KJ (1991) FTIR evidence for a predominantly α-helical structure of the membrane-bound channel forming C-terminal peptide of colicin El. Biophys J 59:516-522

Raymond L, Slatin SL, Finkelstein A (1985) Channels formed by colicin El in planar lipid bilayers are large and exhibit pH-dependent ion selectivity. J Mem Biol 84:173-181

Schein SJ, Kagan BL, Finkelstein A (1978) Colicin K acts by forming voltage-dependent channels in phospholipid bilayer membranes. Nature 276:159-163

Shirabe K, Peterson AA, Shiver JW, Cohen FS, Nakazawa A, Cramer WA (1989) Decrease of anion selectivity caused by mutation of Thr501 and Gly502 to Glu in the hydrophobic domain of the colicin El channel. J Biol Chem 264:1951-1957

Song HY (1990) Ph.D. Thesis, Purdue University. Membrane topography of *E. coli* ColEl gene products: the channel forming domain and the immunity protein of colicin El. 206 pp.

Song HY, Cramer WA (1991) Membrane topography of ColE1 gene products: (I) The hydrophobic anchor of the colicin El channel is a helical hairpin. J Bacteriol 173:2927-2934

Xu S, Peterson AA, Montecucco C, Cramer WA (1988) Dynamic properties of membrane proteins. Reversible insertion into membrane vesicles of a colicin El channel-forming peptide. Proc Natl Acad Sci, USA 85:7531-7535

VOLTAGE-DEPENDENT GATING OF COLICIN E1 CHANNELS IN PLANAR BILAYERS

S.L. Slatin, K.S. Jakes, C.K. Abrams & A. Finkelstein
Department of Physiology and Biophysics
Albert Einstein College of Medicine
Bronx, N.Y. 10461
USA

Introduction

The colicins are a group of plasmid-encoded bacteriocins expressed by various strains of *E. coli*. A subgroup of these toxins, which includes colicins E1, Ia, Ib, B, N, and A, form ion-permeable channels both in the inner membrane of the target bacterium and in lipid bilayer membranes (for reviews, see Slatin, 1988; Pattus *et al.*, 1990; Cramer *et al.*, 1990). Each of these channel-forming toxins consists of a single polypeptide chain of between 500 and 600 amino acids, with a single hydrophobic segment near the C-terminus (residues 474-508 in E1). For colicins E1 and A, it has been shown explicitly that only the carboxyl terminal third of the protein is required for channel formation. The addition of colicin E1 in nanomolar concentrations to the bathing solution on one side of a planar lipid bilayer (the *cis* side) induces a large, voltage-dependent conductance attributable to the gating of colicin channels. *Cis* positive voltages (all voltages refer to the *cis* compartment with respect to the *trans*, which is always at zero potential) tend to open channels, whereas *cis* negative potentials close them. The rate at which channels open and close is highly voltage dependent. Gating is faster at large absolute voltages than at voltages near zero. The rates are also pH dependent -- increasing the pH of the *cis* solution from 3.5 dramatically decreases the turn-on rate, whereas raising the *trans* pH dramatically decreases the turn-off rate.

A mechanism by which the colicin peptide interacts with the transmembrane voltage was suggested by the effect of the protease pepsin on colicin channels in planar bilayers (Slatin *et al.*, 1986). When added to the *cis* side of a colicin-doped membrane, pepsin destroys channels only when they are closed; open channels are unaffected. Pepsin added to the *trans* side does not destroy channels, whether they are open or closed. This result suggests that channels that are closed are more exposed to the *cis* solution than channels that are open, which are evidently protected by the membrane from pepsin digestion. If this is so, then positive voltage might cause the channel to open by driving a segment of the protein into the membrane. This voltage-sensitive segment, probably along with other parts of the protein, then forms the actual channel. To examine this model of colicin gating, we have carried out a series of experiments designed to locate specific regions of the protein with respect to the membrane when the channel is open

NATO ASI Series, Vol. H 65
Bacteriocins, Microcins and Lantibiotics
Edited by R. James, C. Lazdunski and F. Pattus

or closed. An alteration is made in the protein, usually by site-directed mutagenesis, and channels formed by the altered protein are studied in the planar bilayer system to try to place the altered site on one side of the membrane or the other. Often, we were able to exploit the highly pH-dependent nature of the colicin channel. Both the ion selectivity and the gating kinetics are sensitive to the pH on both sides of the membrane, and this dependence can be used to locate a titratable chemical group on the protein by measuring the channel's response to the bath pH on either side of the membrane (see below). Because the pH range in question is fairly low (3.5 - 6.5), the groups involved are probably carboxyl groups. However, the "gating charges" that are directly responsible for the postulated insertion of protein are probably positive, judging from the sign of the voltage dependence.

...-Asp-Val-Leu-Asn-Lys-Lys-Phe-Ser-Lys-Ala-Asp-Arg-Asp-Ala-Ile-Phe-Asn
398

-Ala-Leu-Ala-Ser-Val-Lys-Tyr-Asp-Asp-Trp-Ala-Lys-His-Leu-Asp-Gln-Phe-Ala
415

-Lys-Tyr-Leu-Lys-Ile-Thr-Gly-**HIS**-Val-Ser-Phe-Gly-Tyr-**ASP**-Val-Val-Ser-Asp
433 440 446

-Ile-Leu-Lys-Ile-Lys-Asp-Thr-Gly-Asp-Try-Lys-Pro-Leu-Phe-Leu-Thr-Leu-Glu
451

-Lys-Lys-Ala-Ala-**ASP**-Ala-Gly-Val-Ser-Tyr-Val-Val-Ala-Leu-Leu-Phe-Ser-Leu
469 473

-Leu-Ala-Gly-Thr-Thr-Leu-Gly-Ile-Trp-Gly-Ile-Ala-Ile-Val-Thr-Gly-Ile-Leu
487

-**CYS**-Ser-Tyr-Ile-Asp-Lys-Asn-Lys-Leu-Asn-Thr-Ile-Asn-Glu-Val-Leu-Gly-Ile
505 522

Figure 1. Sequence of the C-terminal region of colicin E1. The hydrophobic segment is underlined. Residues mentioned in the text are in bold face.

Asp 473

The ion selectivity of colicin E1 for small ions is such that the channel favors cations over anions at neutral pH and anions over cations at low pH. For example, in a 10:1 activity gradient of KCl, the reversal potential is 41 mV cation selective at pH 6.0 and 43 mV anion selective at pH 3.5 (Raymond *et al.*, 1985).

Experiments with unequal pH on the two sides of the membrane showed that the overall ion selectivity of the channel is controlled by titratable groups on the *cis* side, the *trans* side, and in the lumen. We used site-directed mutagenesis to remove particular aspartic acid residues from the protein in a search for the carboxyl groups that are titrated in the selectivity experiment. We usually replaced the wild type residue with cysteine, which besides neutralizing the negative charge, could subsequently serve as a substrate for chemical modification. [Colicin E1, and the other channel-forming colicins, are nicely set up for such chemistry, because (for E1) there is only one naturally occurring cysteine (at position 505) in the protein. We have changed it to an alanine, and found that the channel formed by this mutant (A505) is identical to the channel formed by the wild type. In general, then, most mutagenesis was carried out on the A505 mutant, so that introduced cysteines would be unique sites for chemical modification.]

We found, for example, that the aspartate at position 473 was involved in the control of selectivity (Jakes *et al.*, 1990). This residue is at the N-terminal end of the 35 amino-acid hydrophobic segment (Figure 1). At low pH (3.5) the reversal potential in a KCl gradient of C473/A505 is identical to that of A505, presumably because the carboxyl at position 473 is protonated at this pH. [In fact, none of the mutants which we made that removed a carboxyl group changed the reversal potential at pH 3.5.] However, as the pH is raised, the wild-type channel becomes more cation selective than the C473/A505 mutant. For example, at pH 5.5 in a 3:1 KCl gradient, the reversal potential of A505 is 10 mV cation-selective, whereas C473/A505 is 2 mV cation-selective. This reduced cation selectivity could be due directly to the elimination of one negative charge on the protein, or it could be the result of a different three-dimensional structure assumed by the mutant channel, which happens to be less cation selective than wild-type. If the loss of the charge were directly responsible for the effect, then restoring it should restore the wild-type phenotype. To do this, we reacted C473/A505 with iodoacetic acid, which adds a carboxyl group to sulfhydryls. This indeed restored the wild type phenotype, whereas adding an amide group by reacting with iodoacetamide had no effect (Jakes *et al.*, 1990). This result is consistent with the negative charge on asp 473 influencing the selectivity directly, and makes it possible to interpret the asymmetric pH experiments described below.

Having found one of the titratable groups influencing selectivity, we next wanted to determine its location in the open channel. With the *cis* pH 3.5 and the *trans* pH 5.5 (high enough to reveal a selectivity difference at symmetric pH), A505 and C473/A505 have identical selectivities. However, when the *cis* pH is 5.5 and the *trans* pH is 3.5, C473/A505 is again more anion selective than A505. These results suggest that asp 473, the residue responsible for the heightened cation selectivity of the wild type channel at high pH, does not "see" the pH of the *trans* solution; that is, it is either on the *cis* side or in the lumen.

If a titratable group is actually in the lumen of the channel, it will "see" a pH somewhere between the pH's of the *cis* and *trans* baths. While the exact pH profile within the lumen is unknown, it should be possible to manipulate it by controlling the buffering capacities of the two bathing solutions. In particular, if the buffering capacity of one solution is high while the other is low, and the lumen is accessible to the buffers, the pH of the high buffering-capacity side would be expected to dominate, effectively "clamping" the luminal pH at its value. In a series of experiments based on this effect, channels were incorporated at symmetric pH 4 and very low buffer concentration (1 mM) in a 3:1 KCl gradient. The reversal potentials for C473/A505 (14 mV anion selective) and A505 (16 mV anion selective) were essentially the same under these conditions. The *cis* pH was then increased to 5.5 with a minimal amount of buffer (so that the buffering capacity was still less than 1 mM buffer). Both the C473/A505 and the A505 channel became less anion-selective (going to 9 mV and 4 mV, respectively), but the buffering is so poor that the pH actually experienced by residues of the protein in the channel may be quite different than in better buffered solutions. Next, glycerate buffer at pH 4 was added to the *trans* compartment (which was still at pH 4) to a concentration of 20 mM. This shifted the reversal potential of both channels back toward higher anion selectivity. C473/A505 essentially returned back to its potential at symmetric pH 4 (14.6 mV anion selective), and A505 was only about 4 mV less anion selective than at symmetric pH 4 (ending up at 10.5 mV anion selelctive). The implication is that the greater cation selectivity of A505, compared to that of C473/A505, seen when the *cis* pH is raised, results from a titratable group present near the *cis* surface of A505 (but not of C473/A505) that is accessible to the low pH buffer in the lumen. Nevertheless, if residue 473 were "seeing" primarily the pH in the lumen, the reversal potential for A505 and C473/A505 would be the same whenever the pH at residue 473 was low enough to insure protonation of the aspartic acid. High luminal buffer concentrations at low pH, established with large permeant buffer capacity in the *trans* bath, brought the reversal potential of A505 close to that of C473/A505, but there remained a small difference. If the reversal potentials were identical under these conditions, we could unambiguously assign asp 473 to the lumen; the small difference that remains implies that asp 473 is located near the interface of the lumen with the *cis* bath. This assignment of location is consistent with the idea that the end of the hydrophobic segment ought to be at the membrane/solution interface.

His 440

The assignment of asp 473 to the *cis* side of the open channel [along with earlier work that suggested that the C-terminus of the protein is also on the *cis* side (Cleveland *et al.*, 1983; Xu *et al.*, 1988)] is consistent with the idea that the hydrophobic segment forms a pair of alpha helices [as it does in the crystal structure of the analogous protein, colicin A] that are buried in the membrane. Since the hydrophobic segment is uncharged, it is unlikely to be the segment postulated to insert in response to positive voltage. More likely, it inserts spontaneously and remains in the membrane when the channel is closed. Since there is spectroscopic evidence that the channel is largely alpha helical (Rath *et al.*, 1991) it seems natural to examine the region just upstream from asp 473 for sequences that could potentially form membrane-spanning alpha helices. As seen in Figure 2, the region from residue 420 to 460 can be modeled as a pair of amphipathic alpha helices, which (since the segment from 460 to 473 is too short to form a membrane-spanning helix) could only be inserted in the orientation that places 420 and 460 on the *cis* side. This orientation also places his 440 on the *trans* side, and, since histidine should carry a full positive charge at pH's where the channel gates well, implies that his 440 carries one of the postulated gating charges.

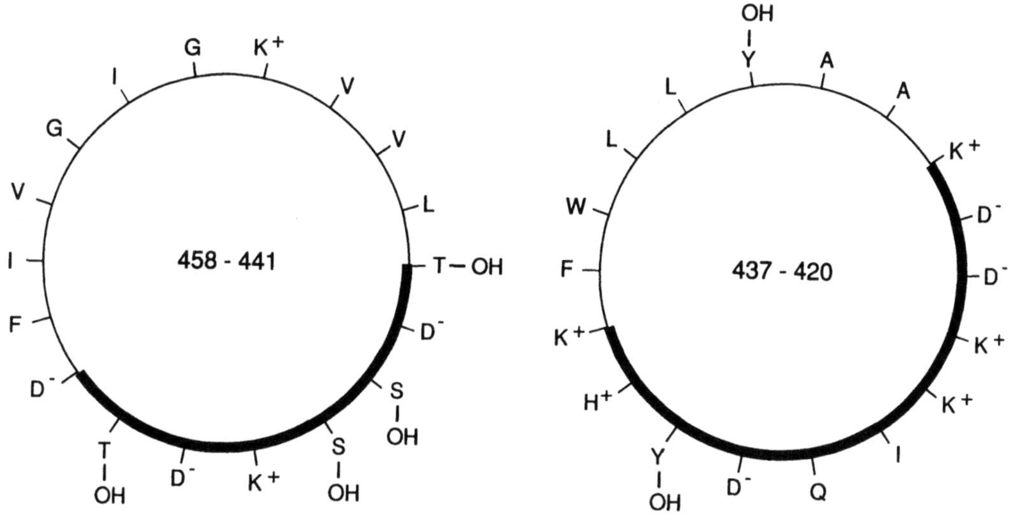

Figure 2. Helical wheel representations of residues 441-458 and 420 to 437. The hydrophilic portion of these amphipathic helices is highlighted.

The straightforward way to test this would be to construct a mutant with a neutral amino acid substituted for his 440 and measure its voltage dependence, which should be less than that of wild type if the model is right. Therefore, we made a mutant with his 440 replaced by cysteine (Abrams *et al.*, 1991). However, it is not possible to measure a steady-state voltage dependence (i.e., to directly measure the probability of the channel being open as a function of voltage) for colicin E1 channels, because the conductance does not reach a steady state fast enough. [In practical terms, the channels are always open at large positive voltages and always closed at large negative voltages, and at voltages between these extremes the gating is slower than the rate at which new channels come into the membrane.] It is, nevertheless, possible to measure the voltage-dependence of the channel by measuring rates of opening and closing, but the analysis is then model-dependent. Specifically, if one assumes a two state process for gating (which, for colicin E1, is not totally accurate)

$$C \underset{k_{-1}}{\overset{k_1}{\leftrightarrow}} O$$

one can write for the rate constants

$$k_1 = A \exp(-\Delta G_1/kT)$$
$$k_{-1} = A \exp(-\Delta G_{-1}/kT)$$

where the ΔG's are the activation free energy differences for the closed to open and open to closed transitions, respectively, A is a constant, k is the Boltzmann constant, and T is temperature. If gating involves the movement of n charges of unit charge q in a transmembrane electric field, one can further expand the ΔG's into chemical and electrical parts:

$$\Delta G = (\Delta G)_{chem} + nq\delta V$$

so that,

$$k_1 = A \exp\{-[nq\delta_{on}V + (\Delta G_1)_{chem}]/kT\}$$
$$k_{-1} = A \exp\{-[nq\delta_{off}V + (\Delta G_{-1})_{chem}]/kT\}$$

where $nq\delta_{on}$ and $nq\delta_{off}$ are the effective charges moved in opening and closing the channels, respectively, and the $(\Delta G)_{chem}$ is the chemical (i.e., non-voltage-dependent) contribution to the activation energy. (If a unit charge q moves completely across the membrane, then $n = 1$ and $q\delta_{on} + q\delta_{off} = q$; that is, δ_{on} and δ_{off} are then the electrical distances from the *cis* and *trans* sides, respectively, to the energy barrier for q, and $\delta_{on} + \delta_{off} = 1$.) We can then express the rate constants as

$$k_1 = A' \exp(-nq\delta_{on}V/kT)$$
$$k_{-1} = A'' \exp(-nq\delta_{off}V/kT)$$

where A' and A" include the chemical parts of the rate constants.

Figure 3 shows semilogarithmic plots of the closing of C440/A505 channels as a function of time and voltage. Closing rates are voltage dependent, and, at large negative voltages, the fraction of channels remaining open decreases expontentially with time with a single time constant τ_{off}. (At less negative voltages a second, slower process appears.) Semilogarithmic plots of τ_{off} vs. voltage yield straight lines (Figure 4). For wild-type channels, the voltage dependence of the off-rate, k_{-1}, is an e-fold change for 20 mV, making (since kT/q = 25.6 mV) $nq\delta_{off}$ equal to 1.28 effective charges; for the C440/A505 mutant channels, the voltage dependence of k_{-1} is an e-fold change for 41 mV or 0.62 effective charges. Thus, roughly 0.6 more effective charges move in closing the wild type channel than in closing the mutant channel. Assuming that the his 440 residue is translocated across the membrane in channel gating (i.e., it is on the *trans* side in the open state and the *cis* side in the closed state), this suggests that the barrier to its translocation is located an electrical distance of 0.6 from the *trans* side.

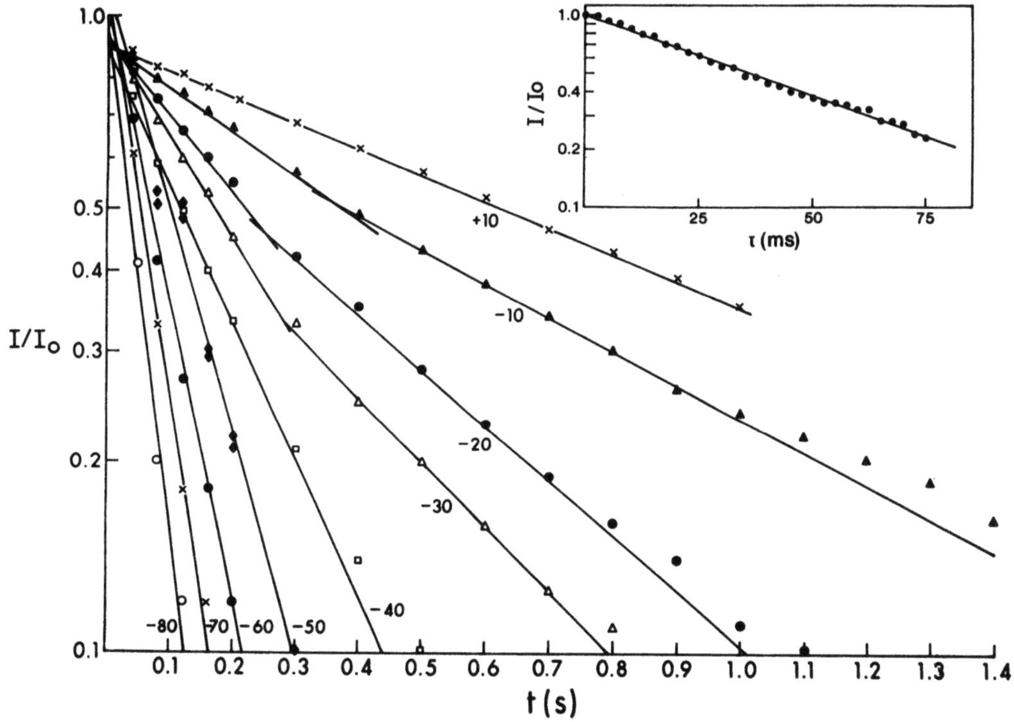

Figure 3. Turn off of C440/A505 channels at various voltages. Currents were digitized, background current was subtracted, and the normalized current I/I_o (where I_o is the current at t = 0) was plotted vs. time. Note that at voltages ≤ -50 mV the log (I/I_o) vs. time is linear, while at less negative voltages a slower component appears. At voltages for which there is one component, the slope of the line reflects the turn-off time constant, τ_{off}. The inset shows the first 75 ms of the -80 mV pulse plotted on an expanded scale.

If his 440 moves toward the *cis* side from the *trans* side during turn off, then it must move from *cis* to *trans* side during turn on, and removing the charge at 440, as is done in constructing C440/A505, ought to make the voltage dependence of turn-on less steep than that in the wild type. In fact, within experimental error, we find no difference between the voltage dependence of on rates of the wild type or mutant channels, but this is not inconsistent with the model. The voltage dependence of the on rate constant, k_1, for wild-type channels is e-fold for 10 mV. From the above equations this corresponds to 2.56 effective charges. If the electrical barrier to the histidine is located an electrical distance of 0.4 from the *cis* side, $nq\delta_{on}$ for C440/A505 channels would become 2.16 effective charges, and the voltage dependence of the on rate would then be an e-fold change for 12 mV, a difference of only 2 mV. This is too small a difference for our experiment to detect.

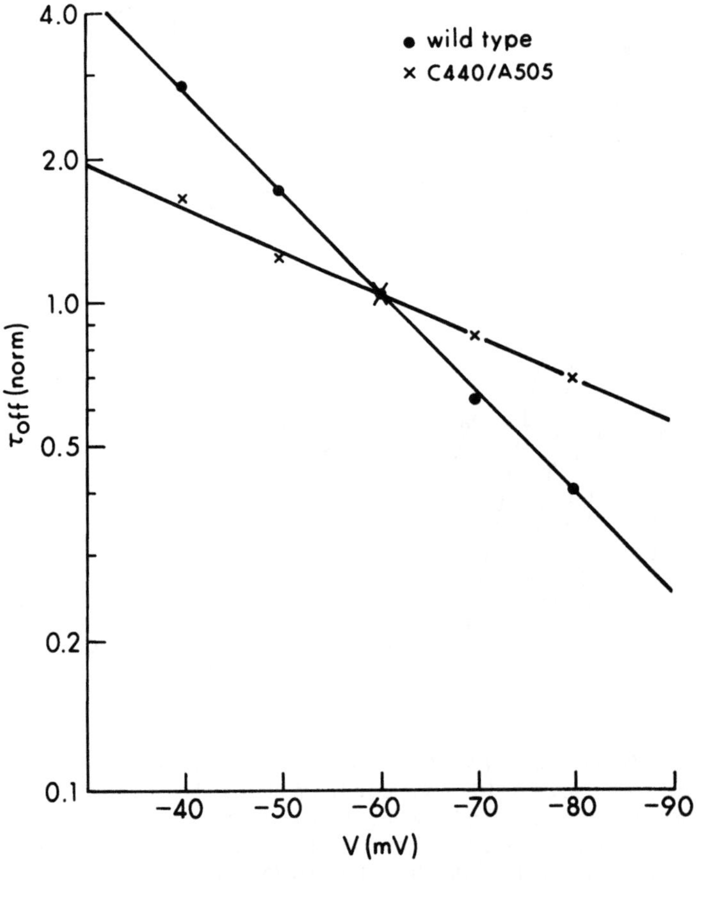

Figure 4. Voltage dependence of τ_{off} for wild type and C440/A505 channels. For each separate experiment (n = 4 for wild type, n = 6 for the mutant) the value of τ_{off} was normalized to its value at -60 mV.

These results support the view that his 440 is on the *trans* side in the open channel. It might be possible to use ion selectivity to test this inference if the charge on residue 440 affected selectivity. We found that while C440/A505 channels are only slightly less anion selective than wild type, when a carboxyl group was chemically added to cys 440, a significant selectivity difference could be seen. In a 10 -fold KCl gradient at symmetric pH 4.1, the carboxylated mutant channel was 23 mV anion selective, compared to 32 mV for wild type. When the *trans* pH was lowered to 3.5, this difference completely disappeared, probably because the introduced carboxyl is fully protonated at pH 3.5. If, however, the *trans* pH is raised to 5.0, the carboxylated mutant becomes 11 mV anion selective compared to 28 mV for wild type. This 17 mV difference is essentially maintained even if the channel-permeant buffer concentration in the *cis* compartment is raised to 100 times higher than that of the *trans* compartment. This result is strong evidence that residue 440 is on the *trans* side, as expected from the voltage dependence experiments. A model incorporating these ideas is shown in Figure 5.

The amphipathic nature of the two voltage-dependent helices suggests that they form (at least part of) an aqueous pore through the bilayer, with their hydrophobic faces contacting the lipid. In the closed state, they are pictured as lying parallel to the membrane surface, still with their hydrophobic sides in contact with lipid. [A very similar model was proposed by Merrill and Cramer (1990) based on quite different experiments.] Notice that no attempt is made here to arrange the inserted helices into a pore; we lack sufficient knowledge of the open state to attempt this. [Pattus *et al.* (1990) have proposed a model of the tertiary structure of the closed state of the colicin A channel.] Notice also that the N-terminal border of the channel has not been clearly defined. This is meant to indicate our ignorance of the upstream boundary of the channel-forming region of the protein (see Slatin, 1988, for a more detailed discussion of this point). In fact, it is not clear whether there are any transmembrane segments upstream of the the four helices shown in Figure 5. There is evidence that a channel can be formed by the helices shown in the model (Liu *et al.*, 1986; Merrill & Cramer, 1990); however, the experiment described below argues for the involvement of at least one more transmembrane segment.

The turn-off rate of colicin E1 is controlled by the *trans* pH. At a *trans* pH of 6.8 (in 1 M KCl), for example, the turn off time is immeasurably long even at -80 mV (conditions where τ_{off} is about 70 msec at pH 3.5). One can ask the question "how *trans* is this effect?" - that is, can the effect of high *trans* pH be reversed by clamping the pH of the lumen at a low value? We find that in a *cis* pH 3.5 / *trans* pH 6.8 gradient with a 50:1 ratio of channel-permeant buffering capacity, the turn off is still blocked. This says that the group(s) being titrated see(s) essentially only the *trans* pH, not the pH of the lumen. Comparing the four helix model and the sequence (figures 1 and 5), there are very few titratable groups predicted to be close enough to the *trans* solution to account for this effect. One obvious candidate is his 440, which we have shown to be on the *trans* side. However, we find that the C440/A505 mutant behaves like wild-

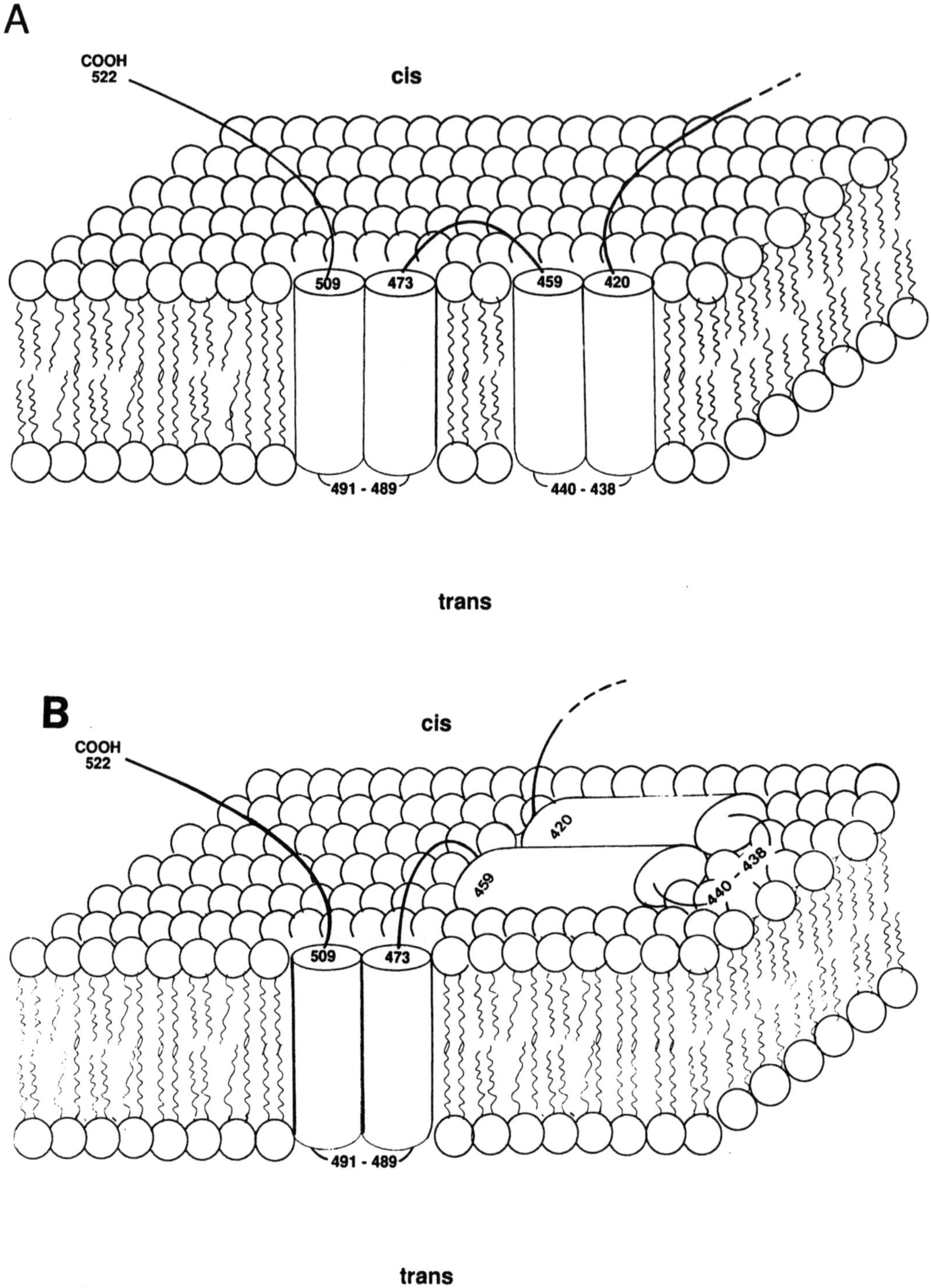

Figure 5. Representation of colicin E1 when the channel is open (A) and closed (B). Numbers refer to residues; cylinders are alpha helices.

type in this experiment, demonstrating that his 440 is not involved in the effect. Since there are no charged groups in the hydrophobic hairpin, we are left with the remaining titratable groups in the voltage-dependent hairpin. After his 440, asp 446 is predicted to be the closest to the *trans* side, so we constructed a mutant with asp 446 replaced by a cysteine, C446/A505. Channels formed by this mutant also fail to turn off in a pH 3.5 / pH 6.8 gradient (1 M KCl), even with a 100:1 gradient in buffering capacity (Figure 6).

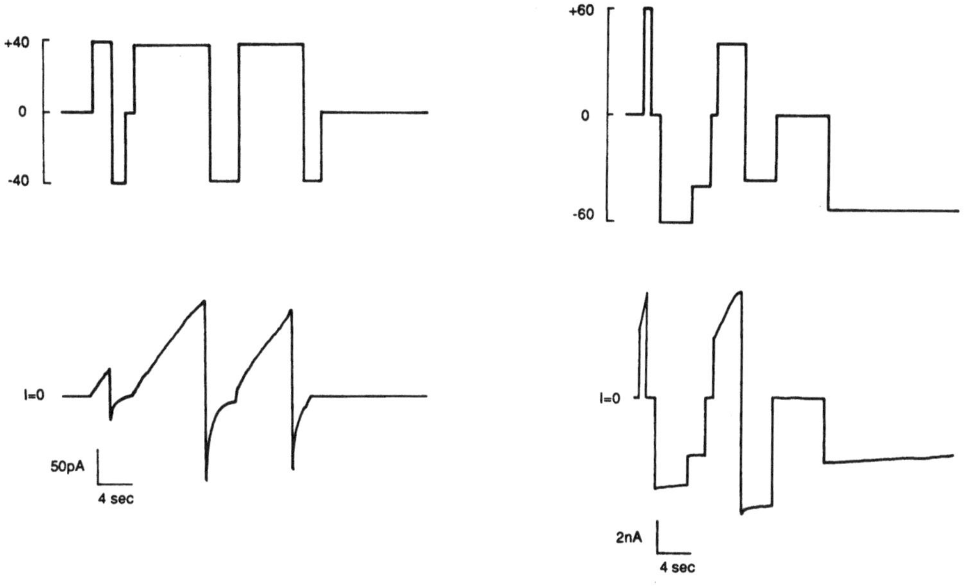

Figure 6. Gating of the C446/A505 mutant. The record on the left is from a membrane in symmetric pH 4.1, 1 M KCl. Notice that the current (lower trace) induced by the mutant colicin turns on and off in response to voltage pulses (upper trace) of positive and negative 40 mV, respectively. The conditions in the record on the right are similar except that the *cis* side contains 100mM glycerate buffer, pH 3.5, while the *trans* side contains 1mM bis tris propane buffer, pH 6.8.

Clearly, asp 446 is not responsible for the effect[1]. We are currently in the process of constructing mutants to remove the remaining titratable groups in the voltage-dependent hairpin, but, if the model is correct, none of them is very close to the *trans* solution, and it is unlikely that any of them will prove to be

[1] In 0.1 M KCl, colicin channels turn off at high *trans* pH faster than in 1 M KCl. Some of this difference could be due to the lower pH expected at the surface of the negatively charged asolectin membranes (used in these experiments) in 0.1 M KCl compared to 1 M KCl. However, the difference appears to be greater than can be accounted for by this effect, and so is probably partly due to a direct effect of lower salt concentration on the protein. Furthermore, in 0.1 M KCl, C446/A505 channels turn off significantly faster than wild-type channels at pH 6.2, although not as fast as they do at low pH. These data show that asp 446 does have some influence on the turn-off kinetics, although it does not account for the entire *trans* pH effect.

responsible for the *trans* pH effect on channel gating. This suggests that the effect must be due to a region of the protein upstream from residue 420, that is, that the four helix model is incomplete. Additional support for this possibility was given by Raymond *et al.* (1986), who argued that regions outside of the channel-forming domain were inserted in the membrane and were partly responsible for the *trans* pH effect. Experiments are currently in progress to search in the region upstream of residue 420 for the sensors of the *trans* pH.

REFERENCES

Abrams CK, Jakes KS, Finkelstein A, Slatin SL (1991) Identification of a translocated gating charge in a voltage-dependent channel. J Gen Phys 98:77-93

Cleveland MvB, Slatin S, Finkelstein A, Levinthal C (1983) Structure-function relationships for a voltage-dependent ion channel: properties of COOH-terminal fragments of colicin E1. Proc Natl Acad Sci USA 80:3706-3710

Cramer WA, Cohen FS, Merrill AR, Song HY (1990) Structure and dynamics of the colicin E1 channel. Mol Microbiol 4:519-526

Jakes KS, Abrams CK, Finkelstein A, Slatin SL (1990) Alteration of the pH-dependent ion selectivity of the colicin E1 channel by site-directed mutagenesis. J Biol Chem 265:6984-6991

Liu QR, Crozel V, Levinthal F, Slatin S, Finkelstein A, Levinthal C (1986) A very short peptide makes a voltage-dependent ion channel: the critical length of the channel domain of colicin E1. Proteins 1:218-229

Martinez MC, Lazdunski C, Pattus F (1983) Isolation, molecular and functional properties of the C-terminal domain of colicin A. EMBO J 2:1501-1507

Merrill AR, Cramer WA (1990) Identification of a voltage-responsive segment of the potential-gated colicin E1 ion channel. Biochemistry 29:8529-8534

Parker MW, Pattus F, Tucker AD, Tsernoglou D (1989) Structure of the membrane-pore-forming fragment of colicin A. Nature 337:93-96

Pattus F, Massotte D, Wilmsen HU, Lakey J, Tsernoglou A, Tucker A, Parker MW (1990) Colicins: prokaryotic killer pores. Experientia 46:180-192

Rath P, Bousche O, Merrill AR, Cramer WA, Rothschild KJ (1991) Fourier transform infrared evidence for a predominantly alpha-helical structure for the membrane-bound channel-forming COOH-terminal peptide of colicin E1. Biophys J 59:516-522

Raymond L, Slatin SL, Finkelstein A (1985) Channels formed by colicin E1 in planar lipid bilayers are large and exhibit pH-dependent ion selectivity. J Membrane Biol 84:173-181

Raymond L, Slatin SL, Finkelstein A, Liu QR, Levinthal C (1986) Gating of a voltage-dependent channel (colicin E1) in planar lipid bilayers: translocation of regions outside the channel-forming region. J Membrane Biol 92:255-268

Slatin SL, Raymond L, Finkelstein A (1986) Gating of a voltage-dependent channel (colicin E1) in planar lipid bilayers; the role of protein translocation. J Membrane Biol 92:247-254

Slatin SL (1988) Colicin E1 in planar lipid bilayers. Int J Biochem 20:737-744

Xu S, Cramer WA, Peterson AA, Hermodson A, Montecucco C (1988) Dynamic properties of membrane proteins: reversible insertion into membrane vesicles of a colicin E1 channel-forming peptide. Proc Natl Acad Sci USA 85:7531-7535

IMMUNITY TO COLICINS

Karen S. Jakes* and Claude Lazdunski+
*Department of Physiology and Biophysics
Albert Einstein College of Medicine
1300 Morris Park Avenue
Bronx, New York 10461 USA

In this introductory section, we attempt to provide some historical background to the subject of immunity to colicins, as well as to raise some of the questions that remain to be addressed with respect to immunity. This Introduction is not intended as a complete literature review of the subject, but rather as a brief overview. We therefore apologize for our failure to cite the work of all those who have contributed to our knowledge in this area.

The phenomenon of immunity was first described by Fredericq (1958). It is the property of colicinogenic cells that they are not killed by the particular colicin that they produce. It should not be confused with either of the more general properties of resistance to colicins conferred by loss of the outer membrane receptor, or tolerance to colicins, conferred by any of a number of target cell mutations which interfere with the translocation of colicins through the membranes of those cells. Immunity is specific for the colicin produced by a given strain, and the genes for immunity are borne on the same plasmids that specify the colicin (with certain exceptions, which will be discussed later).

Those colicins which act enzymatically on cytoplasmic targets, such as the ribonucleases, colicin E3 and cloacin DF13, and the DNases, colicin E2 and E8, present different problems for both the producing cells and the target cells, than do the channel-forming colicins exemplified by colicins A and El. Immunity to these two categories of colicins will therefore be discussed separately.

Immunity to colicin E3 and other enzymatic colicins

Boon (1971) and Bowman *et al.* (1971) showed that cytoplasmic extracts prepared from cells colicinogenic for E3 prevented the *in vitro* inactivation of ribosomes by colicin E3. Based upon those observations, the colicin E3 immunity protein was purified from cells carrying the ColE3 plasmid (Jakes *et al.*,1974; Sidikaro & Nomura, 1974). Subsequent purification and characterization of the immunity proteins for other E3-like and E2-like colicins, as well as the E3-like cloacin DF13, showed them to share essentially identical properties (De Graaf & Klaasen-Boor, 1974; Schaller & Nomura, 1976). They will therefore be discussed interchangeably.

+Centre de Biochimie et de Biologie Moleculaire, Centre National de la Recherche Scientifique
31 Chemin Joseph-Aiguier, B.P. 71, 13402 Marseille Cedex 9, France

NATO ASI Series, Vol. H 65
Bacteriocins, Microcins and Lantibiotics
Edited by R. James, C. Lazdunski and F. Pattus

The colicin E3 immunity protein is a highly acidic protein (pI 4.3) of molecular weight about 10,000 daltons. The addition of purified immunity protein prevents the inactivation of ribosomes by colicin E3 *in vitro* but not *in* vivo. Simply purifying the immunity protein almost immediately suggested the mechanism by which it prevents the action of the colicin. Sodium dodecyl sulfate polyacrylamide gels of purified colicin E3 run side-by-side with purified immunity protein showed a prominent band in the colicin which comigrates with free immunity protein. In fact, colicin E3 is released from producing cells as a heterodimer with its immunity protein. The complex can only be dissociated under strongly denaturing conditions (Jakes & Zinder, 1974; Hirose *et al.*, 1976). As expected, colicin E3 freed of bound immunity protein had greatly enhanced activity, as measured by inactivation of ribosomes *in vitro*. [Presumably, the *in vitro* ribosome inactivation observed with the colicin-immunity protein heterodimer is due to some low level of dissociation of the complex. Dissociation in the absence of denaturing agents was observed during diffusion of the complex on Ouchterlony plates (Jakes & Zinder, 1974). The complexes of both cloacin DF13 and colicin E2 with their respective immunity proteins have no activity *in vitro;* those toxins must be dissociated from the immunity protein in order to see activity (De Graaf & Klaasen-Boor, 1977; Schaller & Nomura, 1976)]. More surprisingly, the uncomplexed colicin had substantially reduced activity *in vivo,* when measured by its ability to kill sensitive cells. Results of Oudega *et al.* (1977) with cloacin DF13 suggested that the immunity protein in the complex somehow facilitates the translocation of the cloacin from the outer membrane receptor. Subsequent work by this group suggested that the immunity protein is removed from the cloacin after receptor binding occurs and translocation is initiated (Krone *et al.*, 1986). The question of precisely when and how immunity protein is dissociated from the enzymatic bacteriocins remains to be answered.

The binding site for the immunity protein on the colicin is at the carboxyl terminal end of the colicin. Trypsin treatment of either cloacin DF13 or colicin E3 yields a carboxyl-terminal fragment of the bacteriocin which is still complexed with the immunity protein and which still has *in vitro* activity on ribosomes (Mooi & De Graaf, 1976; Ohno *et al.*, 1977). The C-terminal portions of all of the enzymatic colicins are highly positively charged and thus have the potential for binding the highly acidic immunity proteins. The bound immunity protein protects the colicin from proteolysis, since trypsin treatment of the isolated cloacin results in its complete digestion (Mooi & De Graaf, 1976).

Since the enzymatic bacteriocins act on cytoplasmic targets (ribosomes or DNA) found in the cells where they are synthesized, as well as in susceptible cells, the synthesizing cells themselves must be protected from the lethal effects of these toxins. Immunity protein was found to be synthesized constitutively, at a relatively low level, by uninduced cultures of cells colicinogenic for E3 (Jakes *et al.*, 1974). Upon induction, the synthesis of immunity protein increased along with that of the colicin, with a molar excess of immunity protein synthesized at all times (Jakes *et al.*, 1974). Presumably, the role of this immunity protein is to protect producing cultures, both from colicin produced by neighboring cells and from colicin being produced by the cell itself. Nevertheless, Masaki and Ohta (1985) succeeded in constructing ColE3 plasmids which conferred a Col^{+} Imm^{-} phenotype in uninduced cultures. Those plasmids could only be maintained in host cells that were grown below 30°C in the presence of vitamin B12 or trypsin, to block receptors or degrade external colicin, or grown below 30°C in *recA* or *btuB* host cells. On the other hand, Jakes *et al.* (1988) were unable to isolate Col^{+} Imm^{-} plasmids from cells lacking the receptor for a fusion protein containing the enzymatic moiety of colicin E3. Their results argue that the immunity protein is, in fact, needed both to protect producing cells, as well as to protect those cells from exogenous colicin. Finally, the ability of James *et al.*, as discussed in their paper in this section, to create a hybrid colicin from E8 and E9, against which neither parent immunity works, also argues that the immunity protein is not absolutely required in the producing cell. Perhaps the colicin does not fold into a fully active conformation until it has both left the producing cell bound to immunity protein and then entered the target cell, freed from immunity protein by a mechanism which remains to be elucidated.

There is extensive amino acid homology between all of the enzymatic E colicins and their immunity proteins. In spite of this homology, immunity is highly specific; strains producing one colicin generally are not immune to closely related colicins. However, chimeric plasmids with the C-terminal and immunity portions of cloacin DF13 confer immunity to colicin E6 but not E3, despite the considerable sequence homology between the C-terminal bacteriocin and immunity domains of all three. By examining where the sequences differed, Akutsu *et al.* (1989) were able to identify specific residues as candidates for those amino acids which confer specificity on the interactions between the proteins. A related approach is taken by Richard James and his group in their careful dissection of the precise requirements for the interaction between the immunity protein and its cognate colicin domain, as discussed in their contribution to this section.

Further insights into the interactions of immunity protein and colicin will be gained when the X-ray crystal structure of the immunity protein and the complex of colicin and immunity protein are solved, as discussed in the contribution by Shoham in this section.

Immunity to pore-forming colicins

Since they act stoichiometrically, the immunity proteins for the enzymatic colicins become very abundant in induced cultures; the case is very different for the channel-forming colicins A, B, K, N, El, Ia, and Ib. The channel-forming colicins are not released from producing cells as dimers with their immunity proteins. In fact, their immunity proteins are not readily seen in induced colicinogenic cultures.

In order to identify the products of the immunity genes for these colicins, it has been necessary to clone the genes in appropriate overexpression vectors (Mankovich *et al.*, 1986) or under control of the colicin promoter itself (Geli *et al.*, 1988). Techniques such as these have made possible the demonstration that these immunity proteins are localized in the inner membranes of producing cells. Furthermore, Weaver *et al.* (1981) showed that membrane vesicles prepared from colicin Ia-immune cells could be depolarized by colicins El and Ib, but not by colicin Ia. Thus, rather than acting by directly binding to the active domain of the colicin, the immunity proteins work at the site of action of the colicins, the inner membrane.

The difference between immunity to the enzymatic colicins and to the channel-forming colicins is reflected in the regulation of their syntheses and in the arrangement of the various colicin operons. All colicin operons have the same basic arrangement of the genes—5'-colicin-immunity-lysis-3'. [It should be noted that on the ColE3-CA38 plasmid, there is an additional gene for immunity to colicin E8 which lies between the E3 immunity and lysis genes (Chak & James, 1984).] However, in the case of the enzymatic colicins, the immunity gene is transcribed in the same direction as the colicin and lysis genes (Masaki & Ohta, 1985; Cole *et al.*, 1985); for the channel-forming colicins, immunity is encoded on the opposite DNA strand and is thus transcribed in the opposite direction from the genes for colicin and lysis protein (Chan *et al.*, 1985; Mankovich *et al.*, 1986; Morlon *et al.*, 1988). The arrangements of the two types of operons is shown in Figure 1. While both the E3 and E8 immunity genes on ColE3-CA38 have their own promoters and terminators, induction of colicin synthesis from the SOS promoter is presumed to be powerful enough to overcome the effects of the independent immunity terminators (Chak & James, 1985; Jakes & Zinder, 1984). Induction of colicin synthesis thus results in an increase in the amount of immunity protein synthesized, as well as in relatively early synthesis of the lysis protein. On the other hand, induction of synthesis of any of the pore-forming colicins from its SOS promoter should not result in a concomitant increase in immunity protein; in fact, transcription from the immunity promoter on the opposite DNA strand should interfere with transcription of the downstream lysis gene. This gene arrangement probably explains the delay in lysis of El-colicinogenic cultures compared with E3-colicinogenic cultures (Jakes & Model, 1979; Lloubes *et al.*, 1986; Zhang *et al.*, 1988).

It is quite clear that, unlike the case for the enzymatic colicins, the immunity protein is not required to protect colicinogenic cells from endogenously produced colicin. Its role is simply to protect a given cell from colicin produced by its neighbors. This is not surprising, since the transmembrane potential inside the cell is of the opposite polarity to that required for colicin function. Thus, Pugsley (1984) and Geli *et al.* (1986) were able to isolate plasmids which conferred a $Col^{+}Imm$ phenotype for both ColN and ColA, although cells bearing these plasmids rapidly acquire mutations in the corresponding colicin receptor.

Examination of the available DNA sequences of the immunity genes for channel-forming colicins reveals that they fall into two potential structural classes. The sequences of the immunity proteins for

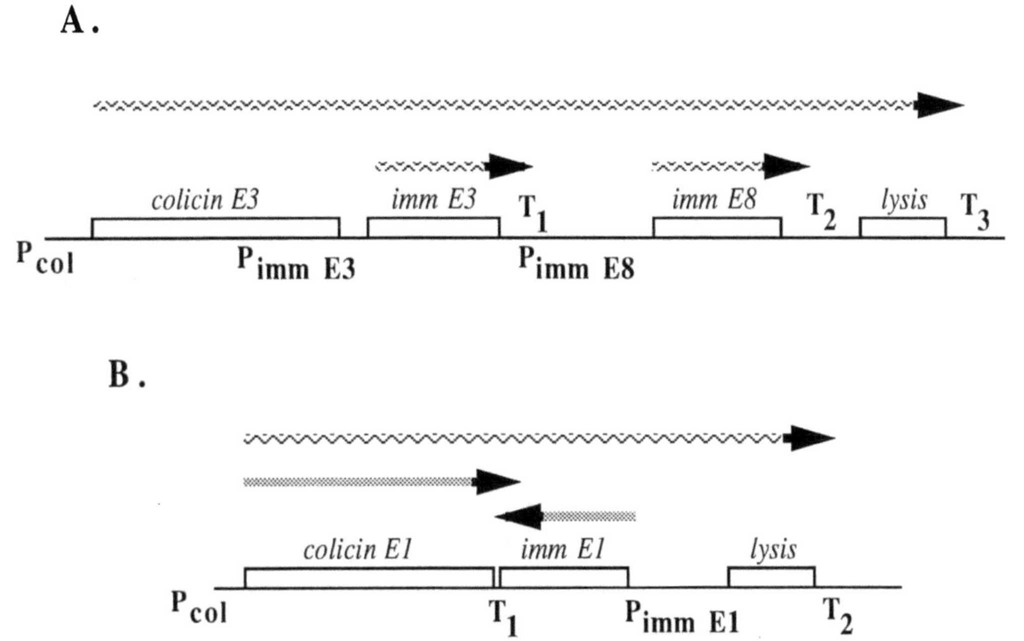

Figure 1. Arrangement of colicin operons. Boxes represent the genes for colicin, immunity protein(s) and lysis protein. **P**: represents promoters; **T**: represents terminators; arrows represent messenger RNA transcripts. **A**. Operon arrangement for ColE3-CA38 (Chak & James, 1985; Masaki & Ohta, 1985). Other enzymatic colicin operons are arranged the same way, although they do not contain the extra E8 immunity gene carried by ColE3-CA38 (van den Elzen *et al.*, 1980). **B**. Operon arrangement for the channel-forming colicins, shown here for ColE1. The same gene arrangement is shared by ColA, ColN, ColB, and ColIa and Ib (Lloubès *et al.*, 1986; Mankovich *et al.*, 1986; Pugsley, 1988; Schramm *et al.*, 1988; Zhang *et al.*, 1988).

colicins A, B, and N have four potential membrane-spanning regions (Lloubes *et al.*, 1984; Schramm *et al.*, 1988; Pugsley, 1988), while the genes for colicins El, Ia, and Ib appear to have only three potential membrane-spanning domains (Oka *et al.*, 1979; Mankovich *et al.*, 1986). Construction of fusion proteins between the immunity proteins of colicin A and colicin El and alkaline phosphatase have begun to confirm the conformations predicted by sequence analysis (Geli *et al.*, 1989; Song & Cramer, 1991).

As is the case for the enzymatic colicins, immunity to the channel-forming colicins is highly specific. Although colicins Ia and Ib are 60% identical in the C-terminal channel-forming domain, there is no cross-immunity between them (Mankovich *et al.*, 1986). Similarly, there is a high degree of sequence identity between colicins A and B and their immunity proteins, but there is no cross-immunity between them, either (Geli *et al.*, 1986; Schramm *et al.*, 1988). The determinants of this specificity have begun to be addressed using the approaches discussed by Geli in his contribution to this section.

To date, no specific interaction has been detected between any channel-forming colicin and its immunity protein. However, the recent results of Ölschläger *et al.* (1991) may bear on the question of whether such an interaction takes place. Colicin M has a unique mode of action; it causes lysis of susceptible cells by inhibiting peptidoglycan and lipopolysaccharide biosynthesis (Schaller *et al.*, 1982). Its immunity protein is found in the cytoplasmic membrane, as are the immunity proteins for the channel-forming colicins. Ölschläger *et al.* (1991) were able to demonstrate that trypsin added to spheroplasts of immune cells degraded the M immunity protein, but the prior addition of colicin M, but not colicin B, partially blocked the action of the trypsin on the surface-exposed immunity protein. Increasing transport of colicin M into immune cells resulted in immunity breakdown at lower colicin concentrations, supporting the notion of a direct interaction between the colicin M and its immunity protein. This is the first demonstration of any kind of interaction between a membrane-bound immunity protein and its corresponding colicin. It should be remembered, however, that colicin M has a completely different mode of action than the channel-forming colicins, and its immunity protein has only one transmembrane domain, rather than three or four. Whether such interactions occur between channel-forming colicins and their immunity proteins remains to be seen.

A number of other questions about the mode of action of the immunity proteins for pore-forming colicins remain to be answered. Among those questions are:

Does the immunity protein act at a specific site in the membrane where the colicin makes a channel?

If the immunity protein is acting by somehow blocking the site of channel-formation, how is specificity maintained?

What determines specificity of immunity between closely homologous colicins?

Are proteins from the susceptible cells involved in immunity?

Several of these issues are addressed by Geli in his contribution to this section.

REFERENCES

Akutsu A, Masaki H, Ohta T (1989) Molecular structure and immunity specificity of colicin E3, an evolutionary intermediate between E group colicins and cloacin DF13. J Bacteriol 171:6430-6436

Boon T (1971) Inactivation of ribosomes *in vitro* by colicin E3. Proc Natl Acad Sci USA 68:2421-2425

Bowman CM, Sidikaro J and Nomura M (1971) Specific inactivation of ribosomes by colicin E3 *in vitro* and mechanism of immunity in colicinogenic cells. Nature New Biol 234:133-137

Chak K-F, James R (1984) Localization and characterization of a gene on the ColE3-CA38 plasmid that confers immunity to colicin E8. J Gen Microbiol 130:701-710

Chak K-F, James R (1985) Analysis of the promoters for the two immunity genes present in the ColE3-CA38 plasmid using two new promoter probe vectors. Nucl Acids Res 13:2519-2530

Chan PT, Ohmori H, Tomizawa J-I, Lebowitz J (1985) Nucleotide sequence and gene organization of ColEl DNA. J Biol Chem 260:8925-8935

Cole ST, Saint-Joanis B, Pugsley AP (1985) Molecular characterisation of the colicin E2 operon and identification of its products. Mol Gen Genet 198:465-472

De Graaf FK, Klaasen-Boor P (1974) Purification and characterization of the cloacin DF13 immunity protein. FEBS Letters 40:293-296

De Graaf FK, Klaasen-Boor P (1977) Purification and characterization of a complex between cloacin and its immunity protein isolated from *Enterobacter cloacae* (Clo DF13). Eur J Biochem 73:107-114

Fredericq P (1958) Colicins and colicinogenic factors. Sympo Soc Exp Biol XII:104-122

Geli V, Baty D, Crozel V, Morlon J, Lloubes R, Pattus F, Lazdunski C (1986) A molecular genetic approach to the functioning of the immunity protein to colicin A. Mol Gen Genet 202:455-460

Geli V, Baty D, Lazdunski C (1988) Use of a foreign epitope as a "tag" for the localization of minor proteins within a cell: The case of the immunity protein to colicin A. Proc Natl Acad Sci USA 85:689-693

Geli V, Baty D, Pattus F, Lazdunski C (1989) Topology and function of the integral membrane protein conferring immunity to colicin A. Mol Microbiol 3:679-687

Hirose A, Kumagai J, Imahori K (1976) Dissociation and reconstitution of colicin E3 and immunity substance complex. J Biochem 79:305-311

Jakes K, Zinder ND, Boon T (1974) Purification and properties of colicin E3 immunity protein. J Biol Chem 249:438-444

Jakes KS, Model P (1979) Mechanism of export of colicin El and colicin E3. J Bacteriol 138:770-778

Jakes KS, Zinder ND (1974) Highly purified colicin E3 contains immunity protein. Proc Natl Acad Sci USA 71:3380-3384

Jakes KS, Davis NG, Zinder ND (1988) A hybrid toxin from bacteriophage fl attachment protein and colicin E3 has altered cell receptor specificity. J Bacteriol 170:4231-4238

Jakes KS, Zinder ND (1984) Plasmid ColE3 specifies a lysis protein. J Bacteriol 157:582-590

Krone WJA, De Vries P, Koningstein G, De Jonge AJR, De Graaf FK, Oudega B (1986) Uptake of cloacin DF13 by susceptible cells: removal of immunity protein and fragmentation of cloacin molecules. J Bacteriol 166:260-268

Lloubès R, Baty D, Lazdunski C (1986) The promoters of the genes for colicin production, release and immunity in the ColA plasmid: effects of convergent transcription and Lex A protein. Nucl Acids Res 14:2621-2636

Mankovich JA, Hsu C-H, Konisky J (1986) DNA and amino acid sequence analysis of structural and immunity genes of colicins Ia and Ib. J Bacteriol 168:228-236

Masaki H, Ohta T (1985) Colicin E3 and its immunity genes. J Mol Biol 1 82:217-227

Mooi FR, De Graaf FK (1976) Effect of limited proteolysis on bacteriocin activity *in vivo* and *in vitro*. FEBS Letters 62:304-308

Morlon J, Chartier M, Bidaud M, Lazdunski C (1988) The complete nucleotide sequence of the colicinogenic plasmid ColA. High extent of homology with ColEl. Mol Gen Genet 211:231-243

Ohno S, Ohno-Iwashita Y, Suzuki K, Imahori K (1977) Purification and characterization of active

component and active fragment of colicin E3. J Biochem 82:1045-1053
Oka A, Nomura N, Morita M, Sugisaki H, Sugimoto K, Takanami M (1979) Nucleotide sequence of small ColEl derivatives: Structure of the regions essential for autonomous replication and colicin El immunity. Mol Gen Genet 172:151-159
Ölschläger T, Turba A, Braun V (1991) Binding of the immunity protein inactivates colicin M. Mol Microbiol 5:1105-1111
Oudega B, Klaasen-Boor P, Sneeuwloper G, De Graaf FK (1977) Interaction of the complex between cloacin and its immunity protein and of cloacin with the outer and cytoplasmic membrane of sensitive cells. Eur J Biochem 78:445-453
Pugsley AP (1984) Genetic analysis of ColN plasmid determinants for colicin production, release, and immunity. J Bacteriol 158:523-529
Pugsley AP (1988) The immunity and lysis genes of ColN plasmid pCHAP4. Mol Gen Genet 211:335-341
Schaller K, Höltje J-V, Braun V (1982) Colicin M is an inhibitor of murein biosynthesis. J Bacteriol 152:994-1000
Schaller K, Nomura M (1976) Colicin E2 is a DNA endonuclease. Proc Natl Acad Sci USA 73:3989-3993
Schramm E, Ölschläger T, Tröger W, Braun V (1988) Sequence, expression and localization of the immunity protein for colicin B. Mol Gen Genet 211:176-182
Sidikaro J, Nomura M (1974) E3 immunity substance. J Biol Chem 249:445-453
Song HY, Cramer WA (1991) Membrane topography of ColEl gene products: the immunity protein. J Bacteriol 173:2935-2943
van den Elzen PJM, Konings RNH, Veltkamp E, Nijkamp HJJ (1980) Transcription of bacteriocinogenic plasmid CloDF13 *in* vivo and *in vitro:* Structure of the cloacin immunity operon. J Bacteriol 144:579591
Weaver C, Redborg AH, Konisky J (1981) Plasmid-determined immunity of *Escherichia coli* K-12 to colicin Ia is mediated by a plasmidencoded membrane protein. J Bacteriol 148:817-828
Zhang S, Yan L, Zubay G (1988) Regulation of gene expression in plasmid ColEl: Delayed expression of the *kil* gene. J Bacteriol 170:5460-5467

IMMUNITY PROTEIN TO PORE FORMING COLICINS

Vincent Géli & Claude Lazdunski
Centre de Biochimie et Biologie Moleculaire. C.N.R.S
31 chemin Joseph Aiguier BP71, 13402, cedex 9
Marseille, France.

Many levels of insensitivity to colicins are reported for bacteria (Davies & Reeves 1975). Transport of colicins are multistep processes and therefore mutations in genes encoding proteins involved in their entry into bacteria render the cells insensitive to their action. These mechanisms of insensitivity are not highly specific to the action of a particular colicin since many colicins share common components in their uptake system (see the Chapter: Uptake of Bacteriocins). In contrast, immunity is a highly specific insensitivity of bacteria to the action of a particular colicin (Konisky 1978). Determinants for colicin production, release, and immunity are carried by the same plasmid. As mentioned, immunity is restricted to the related plamid encoded colicin.

Identification of pore-forming colicins immunity gene products

Immunity genes of pore-forming colicins A, B, El, Ia, Ib, N have been identified (Morlon *et al.*, 1983; Lloubès *et al.*, 1984; Oka *et al.*, 1979; Goldman *et al.*, 1985; Weaver *et al.*, 1981a; Mankovich *et al.*, 1986; Ölschläger *et al.*, 1984; Schramm *et al.*, 1988; Pugsley 1984; Pugsley 1988). Immunity genes from colicins A, El, N are located in the intercistronic region between the genes coding for the colicin and their corresponding lysis protein (Figure 1) and are transcribed in the opposite direction (Cavard *et al.*, 1985; Lloubès *et al.*, 1986; Chan *et al.*, 1984; Pugsley 1987). In contrast to the high level of induced expression of colicins, immunity proteins are expressed constitutively at very low levels. Plasmids pColB, pColla, and pCollb do not have a lysis gene (Mankovich *et al.*, 1984; Schramm *et al.*, 1987). However, the organisation of the genes coding for the colicins and their immunity proteins is identical. Expression of the cloned immunity genes in maxicells or under T7 promoter control has allowed the identification of the ^{35}S-labeled gene products and their localization in the inner membrane (Crozel *et al.*, 1984; Schramm *et al.*, 1988; Goldman *et al.*, 1985; Mankovich *et al.*, 1986; Pugsley 1988).

Immunity proteins are localized in the inner membrane, the N-terminal part of the colicin A immunity protein is facing the cytopolasm and is neither required for insertion nor for function

The colicin A immunity protein (Cai) has been tagged in its N-terminal region with an epitope for

NATO ASI Series, Vol. H 65
Bacteriocins, Microcins and Lantibiotics
Edited by R. James, C. Lazdunski and F. Pattus

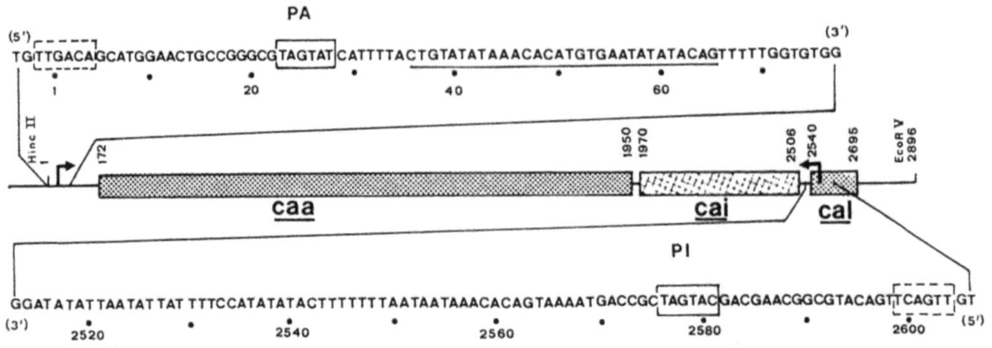

Figure 1. Organisation of the genes coding for colicin A (*caa*), its lysis protein (*cal*), and its immunity protein (*cai*) (from Lloubès *et al.*, 1986); *caa* and *cal* form an operon, *cai* is transcribed in the opposite direction. The recognition and binding sequences for RNA polymerase are boxed with dotted lines (-35 sequence) and solid lines (Pribnow box). Arrows above the start sites indicate the direction of transcription. The sequence is numbered as referenced (Lloubès *et al.*, 1986).

which a monoclonal antibody was available (Géli *et al.*, 1988). The gene coding for the modified immunity protein has been placed under the control of the inducible colicin A promoter. Induction kinetics showed a correlation between the appearance of the immunity phenotype for the induced cultures and the immunodetection of the modified immunity protein (Cai*). The hydrophylic tag was protected from proteolytic action in trypsin-treated inner membrane vesicles (right side out) of induced cultures. Thus, these results indicated that the N-terminal region of Cai* was directed toward the cytoplasm and not required for the immunity function (Géli *et al.*, 1988).

Immunity proteins are polytopic membrane proteins

The topology of the colicin A and El immunity proteins (Cai and Cei) was determined using fusions to alkaline phosphatase. Cai contains four transmembrane sequences with its N- and C- terminal regions directed toward the cytoplasm (Figure 2) (Géli *et al.*, 1989). The colicin El immunity protein has only 3 membrane spanning segments with its N-terminus directed toward the cytoplasm and the C-terminus located in the periplasm (Song & Cramer 1991).

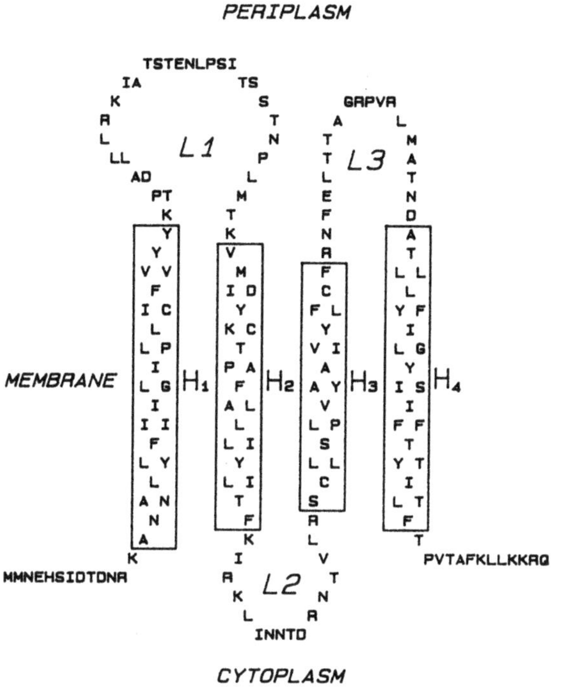

Figure 2. Model of Cai topology. Membrane spanning segments are indicated by boxes, the periplasmic and cytoplasmic loops are labeled Ll, L3 and L2 respectively.

Endogeneous colicins do not kill producing cells devoid of immunity

When the immunity and lysis genes are deleted from pColA in a colicin A producing strain, the cells grow normally, although they produce large amounts of endogeneous colicin, suggesting that colicin is not able to kill from inside. However, even in the absence of lysis protein, low amounts of colicin molecules can be released into the media resulting in a rapid selection of resistant mutants. Appearance of resistant mutants can be prevented by addition of trypsin into the media (Mankovich *et al.*, 1984; Géli *et al.*, 1986). A similar result was reported for colicin M which kills sensitive cells by inhibiting the synthesis of the peptidoglycan (Harkness & Braun, 1990).

Immunity proteins interact with the pore-forming domains of their cognated colicin

Strains harboring pColIa can bind as much colicin Ia as strains devoided of plasmid (Konisky & Cowell 1972). Immunity proteins thus do not modify the affinity of the colicin for its receptor. Colicins

Ia and Ib have the same pathway of uptake as suggested by the analysis of their tolerant mutants (Konisky 1982). Nevertheless, each of these colicins has its own immunity protein (Weaver *et al.*, 1981a). In consequence, it is unlikely that immunity proteins act on the translocation system. It is more likely that immunity proteins inactivate their corresponding colicin by directly interacting with it (Konisky 1982). Construction of hybrid proteins between colicins Ia and Ib and colicin A and El has demonstrated that the C-terminal pore-forming domain was responsible for the recognition of their immunity proteins by pore-forming colicins (Mankovich *et al.*, 1984; Bénédetti *et al.*, 1991). Similar conclusions were obtained with the purified pore-forming domain of colicin El. This peptide, devoid of its translocation and receptor binding domain, was able to kill osmotically shocked sensitive cells (with low efficiency) but unable to kill immune cells treated in the same manner (Bishop *et al.*, 1985). In another set of experiments, inner membrane vesicles prepared from colicin Ia immune cells, subjected to a freeze-thaw cycle in the presence of colicin Ib and El, leads to the inhibition in the ability of such vesicles to generate a potential. Such inhibition was not observed when cells were treated with colicin Ia (Tokuda & Konisky 1978; Weaver *et al.*, 1981a). Taken all together, these results indicate that immunity proteins of pore forming colicins interact with the pore-forming domains of their corresponding colicin at the level of the inner membrane. Their action should prevent the formation of the pore.

Immunity can be overcome when colicinogenic wild-type strains are treated with large amounts of colicin (Levisohn *et al.*, 1968; Pugsley 1988). Sensitivity of immune cells to their corresponding colicin decreases when the copy number of plasmids carrying the immunity gene increases. Immunity is overcome if the number of translocated colicins is greater than the number of immunity proteins necessary to neutralyze them - this phenomenon is called immunity breakdown (Konisky 1982). Nothing is known about the stochiometry of the colicin/immunity interaction. The Cai* and Cei proteins have been purified to homogeneity but immunity function has not so far been reconstituted with purified immunity proteins using *in vitro* model systems (Géli *et al.*, 1989; Cramer *et al.*, 1990; Shirabe *et al.*, 1991).

Functional regions of immunity proteins

Colicins can be divided into two groups based on the extent of homology of their pore-forming domain. The first group comprises colicins A, B, and N and the second group colicins El, Ia, and Ib (Pugsley 1987; Schramm *et al.*, 1987; Pattus *et al.*, 1990; Cramer *et al.*, 1990). These two groups correspond also to two groups of immunity proteins (Schramm *et al.*, 1988; Géli *et al.*, 1989; Song & Cramer 1991). Homologies in the cytoplasmic regions probably correspond to determinants required for the proper topology of the immunity proteins (Nilsson & von Heijne 1990). The role of the cytoplasmic positive charges of Cai as determinants of topology is currently being studied in our laboratory. It is reasonable to think that conserved amino-acids in the periplasmic regions may contribute to the protective function without nevertheless ruling out a possible role for the transmembrane regions (Figure 3).

```
1) E1 TYPE

ImmE1:MSLRYYIKNILFGLYCTLIYI...YLITKNSEGYYFLVSDKMLY...AIVISTILCPYSK
ImmIA:MNRKYYFNNMWWGWVTGGYML...YM....SWDYEF..KYRLLF.WCISLCGMVLYPVAK
ImmIB:MKLDISVKYLLKSLIPILIILTVFYLGWKDNQE.....NARMFYAFIGCIISAITFPFSM
      *.     . .  .       .   *.    . .    . ....     . . . * ..

ImmE1:YAIEYIAFNFIKKDFFERRKNLNNAPVAKLNLFMLYN....LLCLVLAIPFGLLGLFISI
ImmIA:WYIEDTALKFTRPDF..WNSGFFADTPGKMGLLAVYTGTVFILSLPLSMIYILSVIIKRL
ImmIB:RIIQKMVIRFTGKEF..WQKDFFTNPVGG.SLTAIFELFCFVISVPVVAIYLIFILCKAL
        *.  ...*.  .*  ..... ... . .*  ...    .... .   . .  .   .

ImmE1:KNN
ImmIA:SVR
ImmIB:SGK
      . .

2) A TYPE

Imm A:MMNEHSIDTDNRKANNALYLFIIIGLIPLLCIFVVYYKTPDALLLRKIATSTENLPSITS
Imm B:MTSNK...DKNKKANEILYAFSIIGIIPLMAILILRINDPYSQVLYYLYNKVAFLPSITS
Imm N:MHNT.......................LLEKIIAYLS...............LPGFHS
      * ..                        *. ..   .               **.. *

Imm A:SYNPLMTKVMDIYCKTAPFLALILYILTFKIRKLINNTDRNTVLRSCLLSP..LVYAAIV
Imm B:LHDPVMTTLMSNYNKTAPVMGILVFLCTYKTREIIKPVTRKLVVQSCFWGP..VFYAILI
Imm N:LNNPPLSEAFNLYVHTAPLAATSLFIFTHKELELKPKSSPLRALK..ILTPFTILYISMI
        .* ... .. * .***  .  ... * * ..    ..  ...  ..*  . * ..

Imm A:YLFCFRNFELTTAGRPVRLMATNDATLLLFYIGLYSIIFFTTYITLFTPVTAFKLLKKRQ
Imm B:YITLFYNLELTTAGGFFKLLSHNVITLFILYCSIYFTVLTMTYAILLMPLLVIKYFKGRQ
Imm N:YCFLLTDTELTLSSKTFVLIVKKR.SVFVFF..LYNTIYWDIYIHIFVLLVPYRNI....
      *   . . *** ..   *.  .  ......  .* ..   .*  ..  .   . .
```

Figure 3. Sequence alignment of the El and A-type immunity proteins (from Song & Cramer 1991). Sequences were obtained from Oka *et al.*, 1979; Mankovich *et al.*, 1986; Lloubès *et al*, 1984; Schramm *et al.*, 1988; and Pugsley 1988. * Match across all sequences, •conservative substitution.

The role in the protective function of the regions of Cai and Cei (as they were defined in the Cai topology model; see figure 2) has been investigated by site-directed mutagenesis. Introduction of single mutations (Arg to Asp), either in the second periplasmic loop (L3) or in the transmembrane regions (Hl, H4), of Cai abolish its function. In contrast, mutations introduced in the first periplasmic loop (Ll) did not have any significant effect (Géli *et al.*, 1989). The Cei protein tolerates a high degree of substitution. Only double lysine mutations, introduced either in Hl, H3 or Ll, caused a loss of function (Song & Cramer 1991). At this stage, the results from site-directed mutagenesis cannot be interpreted as indicating precisely how the immunity protein interacts with the pore-formmg domain of colicin A. What is the role of the periplasmic loops and of the transmembrane segments in the protective function? Is the membrane potential required for the colicin/immunity interaction ? What are the specific determinants of the colicin molecule that recognize the immunity protein? Does the interaction occur in the inner membrane or at the periplasmic side of the inner membrane? Does the immunity protein interact with other components, such as the Tol proteins (Webster 1991) or the TonB protein (Fischer *et al.*, 1989) or does the immunity protein diffuse freely within the membrane to interact with its cognate colicin? These are the questions we are trying to elucidate.

Very recently, we obtained insights into the specific determinants of the pore-forming domain responsible for the immunity recognition (Géli and Lazdunski 1991, manuscript submitted; see below).

The hydrophobic α-helices of the pore-forming domain are responsible for recognition of the immunity protein.

As mentioned above, colicins A, B, and N belong to the same group of colicins according to the homology in their pore-forming domains. The pore-forming domains of colicins A and B have 57% of identical residues plus 17% of structurally related amino-acids (Schramm *et al.*, 1987). Their two immunity proteins (Cai and Cbi) share 38% of identity and 38% of conservative substitutions (Schramm *et al.*, 1988). In order to identify the specific determinants for immunity recognition, we have generated a series of ColA/ColB and ColB/ColA chimeric pore-forming domains and tested the killing activity of these chimeric proteins on *immA* and *immB* indicator strains.

The hybrid proteins were generated by homologous recombination of their respective genes sharing 65 % of identity. Fusion sites in the hybrid pore-forming domains occured throughout the sequence, in or between the 10 α-helices defined by the resolution of the tridimensional structure of the colicin A pore-forming domain (Parker *et al.*, 1989; Parker *et al.*, 1990). All ColA/ColB and ColB/ColA hybrid colicins were dependent of the Tol system and the TonB system respectively. Extracts of each hybrid colicin were prepared and their activity was titrated against sensitive, *immA* and *immB* indicator strains. From the analysis of the killing activity of the hybrid proteins, the main specific determinant for immunity recognition was found to be located between residues L530 and D577 of the ColA sequence. This region corresponds mainly to the hydrophobic hairpin of the colicin A pore-forming domain which inserts spontaneously into the lipid bilayer in a potential independent-manner (Figure 4) (Lakey *et al.*, 1991).

Using the same approach, a set of ImmB/ImmA hybrid proteins has been constructed. The immunity phenotypes against ColA and ColB, of colonies harboring plasmids encoding the chimeric protein, were tested. Only the first transmembrane segment (Hl see figure 2) could be functionally exchanged between ImmA and ImmB (V. Géli, unpublished results).

In conclusion, colicin A recognizes its immunity protein through its hydrophobic helical hairpin and probably the interaction requires at least the hydrophobic H2, H3, and H4 helices of the immunity

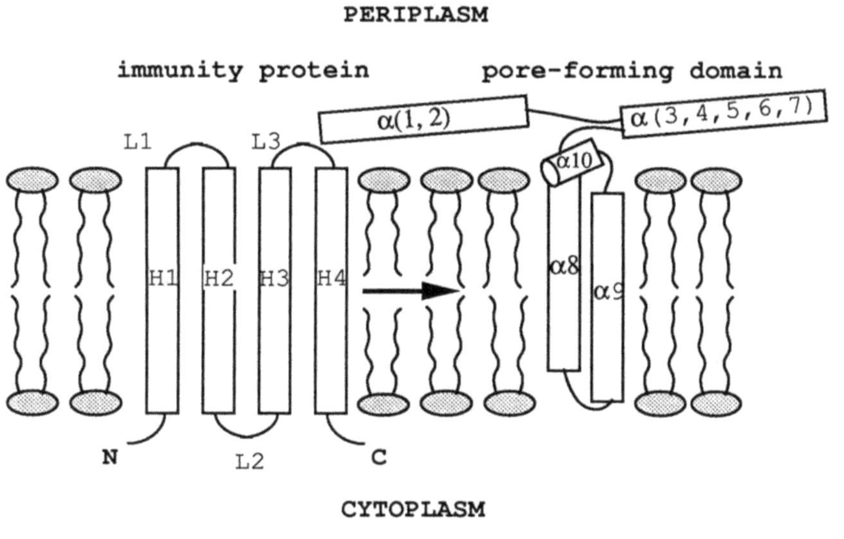

Figure 4. The hydrophobic α-helices of the colicin A pore-forming domain (helices 8 and 9) responsible for recognition of the immunity protein. The structure of the colicin A pore-forming domain and the predicted orientation of the α-helices with respect to the membrane phase are from Parker *et al.*, 1989 and Lakey *et al.*, 1991. It is not known whether the immunity protein freely diffuses within the membrane or interacts with a component common to the ColA and ColB uptake pathways.

protein in addition to the periplasmic loop as mentioned above. The second main conclusion was that immunity specificity was dependent only on the pore-forming domain of the hybrid colicin and not on its uptake pathway. Thus, immunity proteins either are not interacting with the colicin uptake system or immunity proteins are able to recognize both Tol and Ton systems through a component common to the two uptake pathways (Eick-Helmerich and Braun 1989; Braun 1989). In the light of these new data, we are now trying to set up an *in vitro* system to biochemically characterize the interaction between pore-forming colicins and their respective immunity proteins.

References

Bénédetti H, Frenette M, Baty D, Knibiehler M, Pattus F, Lazdunski C (1991) Individual domains of colicins confer specificity in colicin uptake, in pore-properties and in immunity requirements. J Mol Biol 217:429-439

Bishop LJ, Bjes ES, Davidson VL, Cramer WA (1985) Localization of the immunity protein-reactive domain in unmodified and chemically modified COOH-terminal peptides of colicin El. J Bacteriol 164:237-244

Braun V (1989) The structurally related *exc*B and *tol*Q genes are interchangeable in conferring *tonB*-dependent colicin, bacteriophage, and albomycin sensitivity. J Bacteriol 171:6387-6390

Chan PT, Ohmori H, Tomizawa J, Leibovitz J (1986) Nucleotide sequence and gene organisation of ColEl DNA. J Biol Chem 260:8925-8935

Cramer WA, Cohen S, Merrill AR, Song HY (1990) Structure and dynamics of the colicin El channel. Mol Microbiol 4:519-526

Crozel V, Lazdunski C, Cavard D (1983) Localization of genes responsible for replication and immunity to colicin A on plasmid ColACA31. Mol Gen Genet 192:500-505

Davies JK, Reeves P (1975) Genetic of resistance to colicins in *Escherichia coli* K-12. Cross-resistance among colicins of group B. J Bacteriol 123:96-101

Eick-Helmerich K, Braun V (1989) Import of biopolymers into *Escherichia coli*: nucleotide sequences of the *excB* and *exbD* genes are homologous to those of the *tolQ* and *tolR* genes, respectively. J Bacteriol 171:5117-5123

Fischer E, Günter P K, Braun V (1989) Involvement of ExbB and TonB in transport across the outer membrane of *E. coli*: phenotypic complementation of *exbB* mutants by overexpressed *tonB* and physical stabilization of TonB by ExbB. J Bacteriol 171:5127-5134

Géli V, Baty D, Crozel V, Morlon J, Lloubès R, Pattus F, Lazdunski C (1986) A molecular genetic approach to the functioning of the immunity protein to colicin A. Mol Gen Genet 202:455-460

Géli V, Baty D, Lazdunski C (1988) Use of a foreign epitope as a "tag" for the localization of minor proteins within a cell: the case of the immunity protein to colicin A. Proc Natl Acad Sci.USA 85:689-693

Géli V, Baty D, Pattus F, Lazdunski C (1989) Topology and function of the integral membrane protein conferring immunity to colicin A. Mol Microbiol 3:679-687

Géli V, Knibiehler M, Bemadac A, Lazdunski C (1989) Purification and reconstitution into liposomes of an integral membrane protein conferring immunity to colicin A. FEMS Microbiol Lett 60: 239-244.

Goldman K, Suit J, Kayalar C (1985) Identification of the plasmid encoded immunity protein for colicin El in the inner membrane of *Escherichia coli*. FEBS Lett 190:319-323.

Harkness R E, Braun V (1990) Colicin M is only bactericidal when provided from outside the cell. Mol Gen Genet 222:37-40

Konisky J, Cowell BS (1972) Interaction of colicin Ia with bacterial cells. Direct measurement of Ia-receptor interaction. J Biol Chem 247:6524-6529

Konisky J (1978) The bacteriocins, pp 71-136. In: L.N. Omston & J.R. Sokatch (eds). The Bacteria, vol. 6, Academic Press Inc., London.

Konisky J (1982) Colicins and other bacteriocins with established modes of action. Ann Rev Microbiol 36:25-144

Lakey JH, Massotte D, Heitz F, Faucon JF, Dasseux JL, Parker, MW, Pattus FP (1991) Conformation of the pore forming domain of colicin A in its membrane-bound form. Eur J Biochem 196:599-607

Levisohn R, Konisky J, Nomura M (1968) Interaction of colicins with bacterial cells. IV. Immunity breakdown studied with colicins Ia and Ib. J Bacteriol 96:811-821

Lloubès RP, Chartier JJ, Joumet AM, Varenne SG, Lazdunski C (1984) Nucleotide sequence of the gene for the immunity protein to colicin A. FEBS Lett 144:73-78

Lloubès R, Baty D, Lazdunski C (1986) The promoter of the genes for colicin production, release and immunity in the ColA plasmid: effects of convergent transcription and LexA. Nucl Acids Res 14: 2621-2636

Mankovich J.A, Lai PH, Gokul N, Konisky J (1984) Organization of the colicin Ib gene. J Biol Chem 259:8764-8768

Mankovich, JA, Hsu C, Konisky J (1986) DNA and amino acid sequence analysis of structural and immunity genes of colicins Ia and Ib. J Bacteriol 168:228-236

Nilsson I, Von Heijne G (1990) Fine-tuning the topology of a polytopic membrane protein: role of positively and negatively charged amino-acids. Cell 62:1135-1141

Morlon J, Lloubès R, Chartier M, Lazdunski C (1983) Complete nucleotide sequence of the structural gene for colicin A, a gene translated at non uniform rate. J Mol Biol 170:271-285

Oka A, Nomura N, Morita M, Sugisaki H, Sugimoto K, Takanami M (1979) Nucleotide sequence of small ColEl derivatives: structure of the regions essential for autonomous replication and colicin El immunity. Mol Gen Genet 172:151-159

Ölschläger T, Schramm E, Braun V (1984) Cloning and expression of the activity and immunity genes of colicins B and M on ColBM plasmids. Mol Gen Genet 196:482-487

Parker M, Pattus F, Tucker AD, Tsernoglou D (1989) Structure of the membrane-pore forming fragment of colicin A. Nature 337:93-96

Parker M, Tucker A, Tsernoglou D, Pattus F (1990) Insights into membrane insertion based on studies of colicin. TIBS 15:126-129

Pattus F, Massote D, Wilmsen HU, Lakey J, Tsernoglou D, Tucker A, Parker M. W (1990) Colicins: Prokaryotic killer pores. Experientia 46:180-192

Pugsley AP (1984) Genetic analysis of ColN plasmid determinants for colicin production, release and immunity. J Bacteriol 158:523-529

Pugsley AP (1987) Nucleotide sequencing of the structural gene for colicin N reveals homology between the catalytic C-terminal domains of colicin A and N. Mol Microbiol 1:317-325

Pugsley AP (1988) The immunity and lysis genes of ColN plasmid pCHAP4. Mol Gen Genet 211:335-341

Shirabe K, Yamada M, Nakazawa A (1991) Purification of colicin El immunity protein (in preparation)

Schramm E, Mende J, Braun V, Kamp R.(1987) Nucleotide sequence of the colicin B activity gene *cba*: consensus pentapeptide among TonB dependent colicins and receptors. J Bacteriol 169:3350-3357

Schramm E, Ölschläger T, Tröger W, Braun V (1988) Sequence, expression and localization of the immunity protein for colicin B. Mol Gen Genet 211:176-182

Tokuda H, Konisky J (1978) In vitro depolarization of *Escherichia coli* membrane vesicles by colicin Ia. J Biol Chem 253:7731-7737

Waleh NS, Johnson PH (1985) Structural and functional organisation of the colicin El operon. Proc Natl Acad Sci USA 82:8389-8393

Weaver CA, Kagan BL, Finkelstein A, Konisky, J (1981b) Mode of action of colicin Ib. Formation of ion-permeable membrane channels. Biochim Biophys Acta 645:137-142

Weaver CA, Redborg AH, Konisky J (1981a) Plasmid-determined imunity of *Escherichia coli* K-12 to colicin Ia is mediated by a plasmid encoded membrane protein. J Bacteriol 148:817-828

Webster R (1991) The *tol* gene products and the import of macromolecules into *Escherichia coli*. Mol Microbiol 5:1005-1011

SPECIFICITY DETERMINANTS OF THE INTERACTION BETWEEN COLICIN E9 AND ITS IMMUNITY PROTEIN

R. James[¶], M.D. Curtis[¶], R. Wallis[¶], M. Osborne[#], C. Kleanthous[¶] and G.R. Moore[#]
School of Biological Sciences[¶]
School of Chemical Sciences[#]
University of East Anglia
Norwich NR4 7TJ
Norfolk
U.K.

INTRODUCTION

E colicins are plasmid-encoded, protein antibiotics produced by *E.coli* which kill sensitive *E.coli* cells after binding to the *btuB*-encoded cell surface receptor (Di Masi et al., 1973). Each E colicin plasmid also codes for the production of a specific immunity protein which, on synthesis, binds to the C-terminal domain of its cognate colicin (with the exception of the colicin E1 immunity protein). This is assumed to protect the producing *E. coli* cells from being killed by intracellular colicin (Sidikaro & Nomura, 1974; Jakes *et al.*, 1974). The heterodimer of the colicin and immunity protein is secreted from the producing cell. The synthesis of excess immunity protein also protects against external colicin of the same immunity type. The E-group colicins have been sub-divided into colicins E1 to E9 on the basis of immunity tests (Watson *et al.*, 1981; Males & Stocker, 1982; Mock & Pugsley, 1982; Cooper & James, 1984). The cytotoxic activity of several of the E colicins (which is located in the C-terminal trypsin fragment) has been identified, and falls into one of three classes; colicin E1 is a **pore-forming** colicin (Tokuda & Konisky, 1979), colicins E2 (Schaller & Nomura, 1976), E7 (Chak *et al.*, 1991), E8 (Uchimara & Lau, 1987) and E9 (Chak *et al.*, 1991) are **DNAases,** whilst colicins E3 (Bowman *et al.*, 1971) and E6 (Lau & Condie, 1989) are **RNAases.** *E.coli* K12 strains carrying an E-colicin plasmid are sensitive to mitomycin C (MC) compared with plasmid-free isogenic strains (Herschman & Helinski, 1967). This is due to the SOS induction of a plasmid-encoded gene (*lys*) which is involved in release of the colicin/immunity protein complex from the producing cell.

Organization of the colicin E9 operon

The ColE9-J plasmid encodes the colicin structural and immunity genes in an operon (Chak & James, 1986). The E9*imm* gene has its own promoter, which is located within the colicin E9 structural gene. Transcription from this promoter results in constitutive expression of the immunity protein and provides

NATO ASI Series, Vol. H 65
Bacteriocins, Microcins and Lantibiotics
Edited by R. James, C. Lazdunski and F. Pattus

immunity to external colicin E9. Transcription from the SOS-inducible promoter, located upstream of the colicin structural gene, results in expression of all three genes of the operon and leads to the synthesis and secretion of the colicin/immunity protein complex (Figure 1).

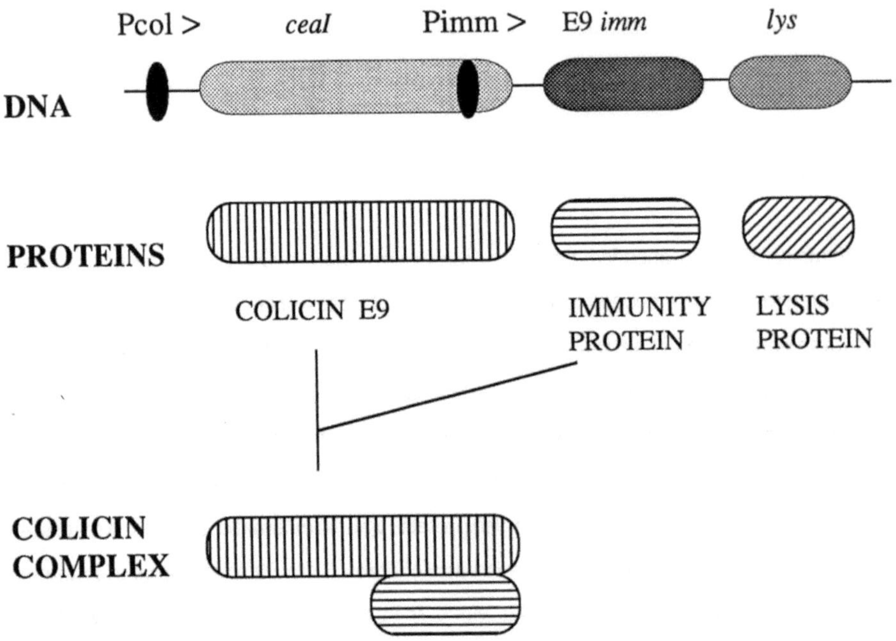

Figure 1. Genetic organization of the Colicin E9 operon. Pcol is an SOS-inducible promoter; Pimm is the immunity gene promoter; *ceal* is the colicin E9 structural gene; E9*imm* is the E9 immunity protein gene; *lys* is the lysis gene which is involved in secretion of the colicin complex from the *E.coli* cell. pColE9 also carries a second immunity gene (E5*imm*) and a second *lys* gene (James *et al.*, 1987, Lau & Condie, 1989).

The colicin E9 immunity gene (E9*imm*) is > 60% homologous to both the E2*imm* and the E8*imm* genes, at both the nucleotide and amino acid level (James *et al.*, 1987). The nucleotide sequences of the 3' ends of the ColE2 (Cole *et al.*, 1985), ColE8 (Uchimara & Lau, 1987; Toba *et al.*, 1988) and ColE9 (Eaton & James, 1989; Lau & Condie, 1989) structural genes have been determined, and show a very high degree of homology. Thus colicins E2, E8 and E9, together with their immunity proteins, constitute a family of closely-related '*in-vivo*' mutant proteins.

We are interested in the specificity of the interaction between the DNAase type E colicins with their cognate immunity proteins. Since the presence of the E9 immunity protein only protects *E.coli* cells against externally added colicin E9, and not against colicin E2 or E8, what is the sequence information within both the immunity protein and the colicin which determines this level of specificity ? This makes the DNAase type E colicins an invaluable system for the study of protein-protein interactions. Information on the specificity determinants will be invaluable in increasing our limited knowledge of the specific interactions between proteins. A particular advantage of this model system is that we can investigate the specificity determinants of both the colicin and the immunity protein, and then use the information obtained to study the two proteins when complexed together.

THE COLICIN E9 PROTEIN

Our strategy to investigate protein-protein interactions between colicin E9 and its cognate immunity protein involved three stages

1) to identify amino-acids in the C-terminal domain of colicin E9 which are candidate specificity determinants for the interaction with the E9 immunity protein.

2) to change the candidate specificity determinants of colicin E9 to those found in the same position in colicin E8, using site-directed mutagenesis.

3) to determine the phenotype encoded by the "mutant" colicin proteins against indicator strains carrying immunity to either colicin E8 or E9.

Identification of candidate specificity determinants

In view of the considerable homology between the E2, E8 and E9 immunity protein sequences (Lau *et al.*, 1984; James *et al.*, 1987), it is perhaps not surprising that there is also considerable sequence homology between the C-terminal domains of the E2, E8 and E9 colicins, to which the immunity proteins specifically bind and thus provide biological protection for *E.coli* cells (Figure 2). Candidate specificity determinants for the interaction between these three colicins and their cognate immunity proteins were identified as those positions where the amino-acid present in each of the colicins was different.

Site-directed mutagenesis of the candidate specificity determinants

Initially, we decided to change each candidate specificity determinant of colicin E9, in turn, to the amino-acid found in the same location in colicin E8. This presents a potential problem in that a "mutant" colicin, which may no longer interact with the E9 immunity protein, could be lethal to the host *E.coli* cell. We therefore constructed plasmid pMC27 (Figure 3), in which a unique *Bam*HI site was introduced into

K	S	F	D	D	F	R	K	A	V	W	E	E	V	S	K	D	
AAA	AGC	TTC	GAC	GAT	TTT	CGG	AAG	GCT	GTA	TGG	GAA	GAG	GTG	TCG	AAA	GAT	**E9**
•••	•A•	•••	•••	•••	•••	•••	•GA	AAA	T•C	•••	•••	••A	•••	•••	•••	•••	**E8**
	N						R	K	F								
•••	•A•	••T	•••	•••	••C	•••	•••	AAA	T•C	•••	•••	••A	•••	••A	•••	•••	**E2**
	N							K	F								

P	E	L	S	K	N	L	N	P	S	N	K	S	S	V	S	K	
CCT	GAG	CTT	AGT	AAA	AAT	TTA	AAC	CCA	AGC	AAT	AAG	TCT	AGT	GTT	TCA	AAA	**E9**
					Q	F			G			K	R	L		Q	
•••	•••	T•A	•••	••G	C•A	••T	••T	•••	G•T	•••	••A	AAA	C•C	T•G	AGC	C••	**E8**
•••	••T	•••	•••	••G	C•A	••T	••A	GGC	••T	•••	•••	ACG	•AC	A••	CA•	••G	**E2**
	D				Q	F	K	G				T	N	I	Q		

G	Y	S	P	F	T	P	K	N	Q	Q	V	G	G	R	K	V	
GGT	TAT	TCT	CCG	TTT	ACT	CCA	AAG	AAT	CAA	CAG	GTC	GGA	GGG	AGA	AAA	GTC	**E9**
	L	A		R	A	R	N	K	D	T					R	S	
••A	•TA	G••	••A	CGG	G••	AGG	••T	••A	G•T	ACT	•••	••T	•••	C•T	CGC	TCA	**E8**
••A	A•A	G•A	••T	••	G•A	AGG	•••	••A	G•C	••A	••A	••T	••T	••G	G••	CG•	**E2**
	K	A			A	R		K	D						E	R	

Y	E	L	H	H	D	K	P	I	S	Q		G	G	E	V	Y	
TAT	GAA	CTT	CAT	CAT	GAC	AAG	CCA	ATT	AGT	CAA	---	GGT	GGT	GAG	GTT	TAT	**E9**
F											D						
•T•	••G	•••	•••	••C	•••	•••	•••	••C	•••	••G	GAT	•••	•••	---	•••	•••	**E8**
•T•	•••	T•A	•••	•••	••T	••A	•••	••C	•••	••G	GAT	•••	•••	---	••C	•••	**E2**
F											D						

D	M	D	N	I	R	V	T	T	P	K	R	H	I	D	I	L	
GAC	ATG	GAT	AAT	ATC	CGA	GTG	ACT	ACA	CCT	AAG	CGA	CAT	ATC	GAT	ATT	CTC	**E9**
				L		I											
••T	•••	••C	••C	C••	••T	A•C	•••	••G	••G	••A	••C	•••	••T	•••	•••	•A•	**E8**
••T	•••	A••	•••	•••	AGT	•••	••C	•••	•••	•••	•••	•••	••T	•••	•••	•AT	**E2**
		N														H	

R	G	K		
CGA	GGT	AAG	TAA	**E9**
		Q		
A••	•••	C••	•••	**E8**
••G	•••	•••	•••	**E2**

Figure 2. Identification of candidate specificity determinants of colicin E9. The nucleotide sequences of the 3' end of the colicin E9 (*ceaI*) gene are aligned with those of the same region of colicin E8 and colicin E2. A dot indicates that the nucleotide sequence of colicin E8 or E2 is identical to that of colicin E9. The amino-acid sequence of colicin E8 and colicin E2 is only shown when it differs from that of colicin E9. Padding spaces are included in the nucleotide sequences in order to maximize the alignment. The presumptive specificity determinants are boxed and for convenience will be identified by the amino-acid residue number, ie, Ser525, Ser526, Val527, Tyr531, Lys545 or Val546.

the 3' end of the colicin E9 structural gene (by changing base 48 in Figure 2 from an A to a G). This change does not effect the amino-acid sequence of the colicin E9 protein. We then constructed an M13mp19 based-clone (pMC26), which contained the 845bp *Bam*HI-*Eco*RI fragment of pMC27. This plasmid encodes a truncated inactive form of the colicin E9 gene and therefore allows mutagenesis and subsequent sequencing to confirm the correct mutagenic change had been made, without the complication of the potential lethality of the "mutant" colicin E9. After mutagenesis of pMC26, with the appropriate mutagenic primer, the "mutant" *Bam*HI-*Eco*RI fragment of colicin E9 was then ligated with the large *Bam*HI-*Eco*RI fragment of plasmid pMC27 (Figure 3), which contains the 5' end of the colicin E9 gene. The reconstructed "mutant" colicin E9 plasmids were then transformed into *E.coli* JM103, or JM103 *btuB*, host containing plasmid pMC28, which carries the E8 *imm* gene on the compatible vector pACYC184.

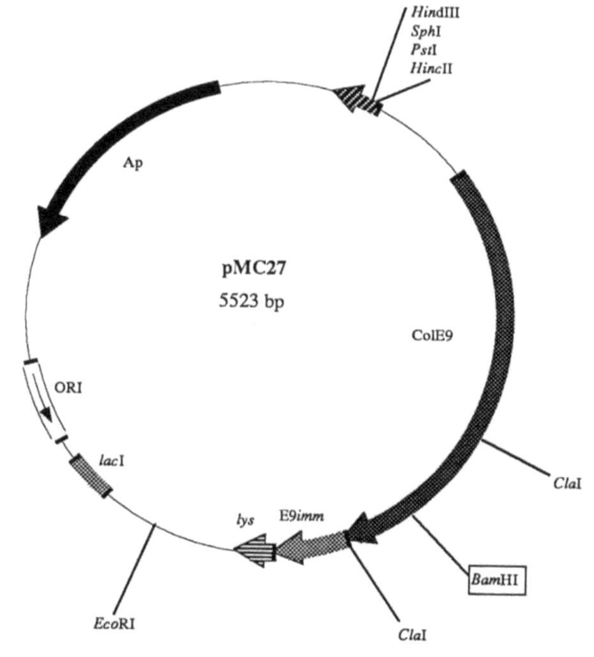

Figure 3. Restriction map of plasmid pMC27. Details of the construction of this plasmid are given in the text.

Phenotypes of site-directed mutants of colicin E9

The phenotypes of the site-directed mutants we have constructed are shown in Table 1 , expressed as colicin titres against three different indicator strains (Curtis & James, 1991). In the initial experiments we changed a single candidate specificity determinant. Since these single mutations did not result in any significant change in the colicin titres, we simultaneously changed four candidate specificity determinants using a single mutagenic primer (D7). The results of the colicin titre tests clearly show that changing the four candidate specificity determinants Ser525, Ser526, Val527 and Tyr531, to the corresponding amino-acids found in colicin E8, altered the phenotype of the resulting "mutant" colicin such that it is active against indicator strains which are immune to both colicin E8 or colicin E9. The additional change of Lys545 and Val546, to the corresponding amino-acids found in colicin E8, further changed the phenotype of the "mutant" colicin closer to that of colicin E8. A single change of either Lys545, or Val546, alone had no effect on the phenotype of the "mutant" colicin; both changes together also had no effect. These results provide conclusive evidence that the primary specificity for the protein-protein interaction between colicin E9 and its immunity protein resides in these six amino-acids.

PLASMID	SPECIFICITY DETERMINANTS[a]						Titres[c]		
							E9*imm*	E8*imm*	pUC18
ColE9	Ser525	Ser526	Val527	Tyr531	Lys545	Val546	$< 5 \times 10^1$	3.5×10^4	2.9×10^3
ColE8	Lys	Arg	Leu	Leu	Arg	Ser	7.9×10^6	$< 5 \times 10^1$	1.1×10^5
Primers[b]									
E32						Ser	$< 5 \times 10^1$	4.5×10^4	2.6×10^3
P24					Arg		$< 5 \times 10^1$	4.6×10^4	1.5×10^3
P37				Leu			$< 5 \times 10^1$	5.4×10^4	3.4×10^3
D87					Arg	Ser	$< 5 \times 10^1$	5.9×10^3	2.0×10^2
P37+E32				Leu		Ser	$< 5 \times 10^1$	6.8×10^3	2.1×10^2
D7	Lys	Arg	Leu	Leu			4.0×10^3	2.6×10^3	2.5×10^3
D7+E32	Lys	Arg	Leu	Leu		Ser	5.9×10^3	5.2×10^3	4.0×10^3
D7+D87	Lys	Arg	Leu	Leu	Arg	Ser	7.0×10^3	3.1×10^2	4.2×10^3

Table 1. Phenotypes of the site-directed mutants of colicin E9
a: The candidate specificity determinants of colicin E9 are shown in the top row of the table, whilst the equivalent amino-acids of colicin E8 are shown in the second row.
b: Using the primers shown in the left hand column, site-directed mutants were prepared in which the amino-acids shown have been changed from those in colicin E9 to those present in colicin E8. Where two primers are shown this indicates two separate site-directed mutagenesis experiments, with the "mutant" ssDNA from the first experiment being used as the template for site-directed mutagenesis using the second primer.
c: The titres of colicin extracts prepared from the site-directed mutants when tested against *E.coli* JM103 containing either the E9*imm*, the E8*imm* gene, or the vector pUC18, are indicated.

Construction of a hybrid E colicin plasmid

The above results show that changing the six candidate specificity determinants of colicin E9 by site-directed mutagenesis, results in a change of phenotype of the "mutant" colicin close to that of colicin E8. It should therefore be possible to replace the E9*imm* gene in this construct with the E8*imm* gene. The restriction map of such a "hybrid" E colicin plasmid, pMC43, is shown in Figure 4 . *E.coli* cells containing plasmid pMC43 grow slowly relative to those containing pMC27. This is presumably because of the residual colicin E9-like activity of the "mutant" colicin. This growth rate differential was abolished by using *E.coli* ***btuB*** as the host strain (data not shown), which suggests that the DNAase activity encoded by this "mutant" colicin is not active inside the producing cell. This is the first experimental evidence to suggest that the DNAase activity of an E colicin is only active after secretion from the producing cell and re-entry via the BtuB receptor protein.

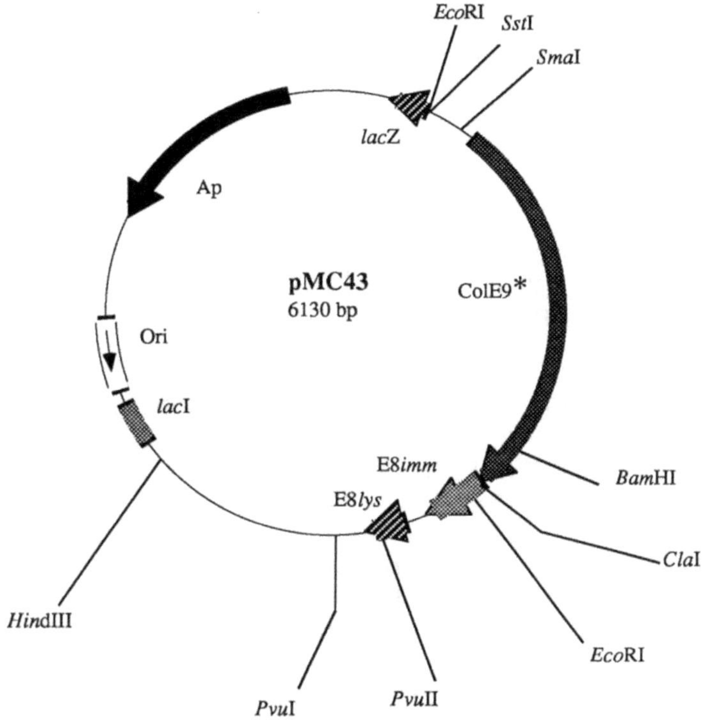

Figure 4. Restriction map of pMC43. This plasmid consists of the "mutant" colicin E9 gene from plasmid pMC26, after mutagenesis with primers D7 and D87, and the E8*imm* and *lys* genes.

THE E9 IMMUNITY PROTEIN

Our initial strategy for the investigation of the specificity determinants in the E9 immunity protein was the same as that which we had successfully used for the colicin E9 protein but also included a gene-fusion approach. Furthermore, we have embarked upon a high field NMR study of this small protein with the aim of deriving its structure in solution; an approach which may provide important clues as to the nature of the specificity determinants.

Identification of candidate specificity determinants

The aligned amino-acid sequences of the E2, E8 and E9 immunity proteins are shown in Figure 5. We have identified 13 candidate specificity determinants in these proteins using the same criteria adopted above for the C-terminal domains of colicin E2, E8 and E9. There are two points to note about the aligned immunity protein sequences, compared to those of the colicins. The first is that there are more differences between the sequences of the immunity proteins than there are for the colicins. This means that the positioning of the padding character in the E8 immunity protein, which is used to optimize the alignment between the three sequences, gives rise to some uncertainty concerning the exact alignment of some of the amino-acids.

```
1         10              20            30
MELKHSISDYTEAEFL|QL|V|TT|ICNADT|SS|EEEL|V|      E9imm
MELKHSISDYTEAEFL|EF|V|KK|ICRAEG|AT|EEDD|N|      E2imm
MELKNSISDYTEAEFK|KI|I|ED|IINCEG|DE|KKQD|D|      E8imm

    40          50           60           70
KLVTHFE|EM|TEHPSGSDLIYYP|KEGD|DDSPSGIVNTV       E9imm
KLVREFE|RL|TEHPDGSDLIYYP|RDDR|EDSPEGIVKEI       E2imm
NLE HFI|SV|TEHPSGSDLIYYP|EGNN|DGSPEAVIKEI       E8imm

       80
KQWRAANGKSGFKQG      E9imm
KEWRAANGKSGFKQG      E2imm
KEWRAANGKSGFKQG      E8imm
```

Figure 5. Identification of the candidate specificity determinants of the E9 immunity protein. The candidate specificity determinants are boxed. One padding space has been introduced into the E8 immunity protein sequence in order to optimize the alignment. The numbering refers to the amino-acids of the E9 immunity protein.

Site-directed mutagenesis of the E9*imm* gene

We cloned the E9*imm* gene on a *Cla*I-*Hpa*II fragment into M13mp19. The 13 candidate specificity determinants have been changed, by site-directed mutagenesis, to the corresponding residues in the E8 immunity protein. Because of the uncertainty in the precise alignment of the three immunity proteins due to the precise positioning of the padding character, some other changes were also made. Both single and multiple mutations have been made, and the phenotypic effect of the mutations has been determined by quantitating the level of immunity conferred by the "mutant" immunity proteins to *E.coli* against extracts of colicin E8 and E9. These immunity tests were performed after sub-cloning the "mutant" *imm* genes back into a pUC vector (Table 2). To our surprise, changing twelve of the candidate specificity determinants (ie. pRW31) did not significantly affect the immunity phenotype conferred by the "mutant" protein. The only residue which led to any change in the phenotype of the "mutant" immunity protein was the change of Val34 to Asp. This change resulted in the appearance of significant immunity to colicin E8, without reducing the immunity to colicin E9. Because the results of this site-directed mutagenesis programme were much less clear cut than with colicin E9, we decided to adopt a gene-fusion approach to identifying important specificity-determining domains of the E9 immunity protein.

Gene-fusions

Plasmid pRW40 is a pUC19 derivative which contains the E9 and E8 *imm* genes in tandem, separated by a tetracycline resistance gene (Figure 6). All three genes are in the same orientation as the *lacZ* gene of the vector and are thus transcribed by the vector promoter. The phenotype conferred by pRW40 to host *E.coli* cells is tetracycline resistance, ampicillin resistance, E9 immunity and E8 immunity. The orientation of the two immunity genes in pRW40, together with the degree of DNA sequence homology between them, allows homologous recombination to take place between the two immunity genes during growth of the host *E.coli* cells. During such homologous recombination the tetracycline resistance gene will be lost. Such recombinant plasmids can be selected for on the basis of size, or by using a unique *Eco*RV restriction site within the tetracycline resistance gene. A number of recombinant plasmids have been obtained by this method. DNA sequencing revealed that most of the recombination events occurred either at the extreme 5" or 3' ends of the immunity genes, however we also obtained some more interesting gene fusions. The phenotypes of the immunity protein encoded by these fusions (Figure 7) indicate the importance of the N-terminal half of the protein in determining the specificity of the interaction of the immunity proteins with their homologous colicins. This region includes the Val34 residue identified as being important by site-directed mutagenesis.

Plasmid	Specificity determinants[a]															Phenotypes[b]	
																E9 extract	E8 extract
ColE9	Glu17	Leu18	Thr20	Thr21	Ser28	Ser29	Glu32	Leu33	Val34	Glu43	Met44	Lys57	Glu58	Gly59	Asp60	> +++++++	
ColE8	Lys	Ile	Glu	Asp	Asp	Glu	Gln	Asp or Gln[e]	Asp	Ser	Val	Glu	Gly	Asn	Asn	+	> +++++++
pRJ320[c]														Asn		> +++++++	
pRJ323												Glu	Gly	Asn	Asn	> +++++++	
pRW24						Glu										> +++++++	
pRW25				Asp												> +++++++	
pRW42	Lys	Ile	Glu	Asp												> +++++++	
pRW49									Asp							> +++++++	+++
pRW53					Asp											> +++++++	
pRW30						Glu		Gln				Glu	Gly	Asn	Asn	> +++++++	
pRW29				Asp		Glu		Gln				Glu	Gly	Asn	Asn	> +++++++	
pRW20				Asp		Glu		Gln		Ser	Val	Glu	Gly	Asn	Asn	> +++++++	
pRW31	Lys	Ile	Glu	Asp		Glu		Gln		Ser	Val	Glu	Gly	Asn	Asn	+++++++[d]	
pRW55	Lys	Ile	Glu	Asp		Glu			Asp	Ser	Val	Glu	Gly	Asn	Asn	++++++[d]	

Table 2. Phenotypes of the E9 immunity protein 'mutants'.
a : The candidate specificity determinants of the colicin E9 gene are shown on the top line, with the amino acids found at the same position in colicin E8 are shown underneath. **b**: The level of immunity conferred to sensitive *E.coli* cells is indicated by the number of + signs. **c**: In the plasmid shown, only the candidate specificity determinants which have been changed, together with the residue to which they have been changed areshown. **d**: In these tests very hazy zones were obtained at all dilutions tested. **e**: The homology between the two proteins is dependent upon the position of the spacing character.

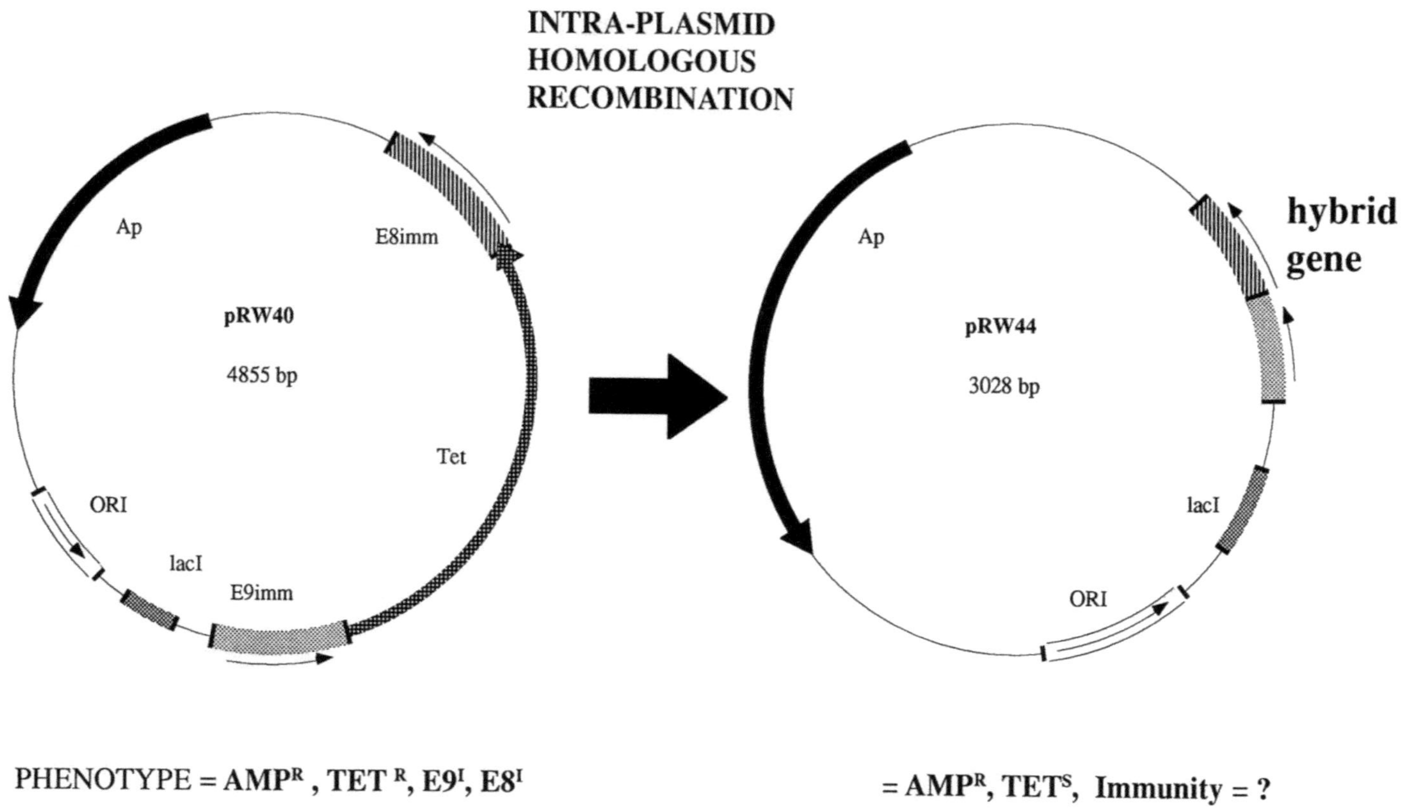

Figure 6. Generation of gene fusions between the E8 and E9 immunity genes by homologous recombination.

```
E9   MELKHSISDYTEAEFLQLVTTICNADTSSEEELVKLVTHFEEMTEHPSGSDLIYYPKEGDDDSPSGIVNTVKQWRAANGKSGFKQG   PHENOTYPE
     **** **********       * *            *  **   *************     * **       * *************
E8   MELKNSISDYTEAEFKKIIEDIINCEGDEKKQDDNLE HFISVTEHPSGSDLIYYPEGNNDGSPEAVIKEIKEWRAANGKSGFKQG
```

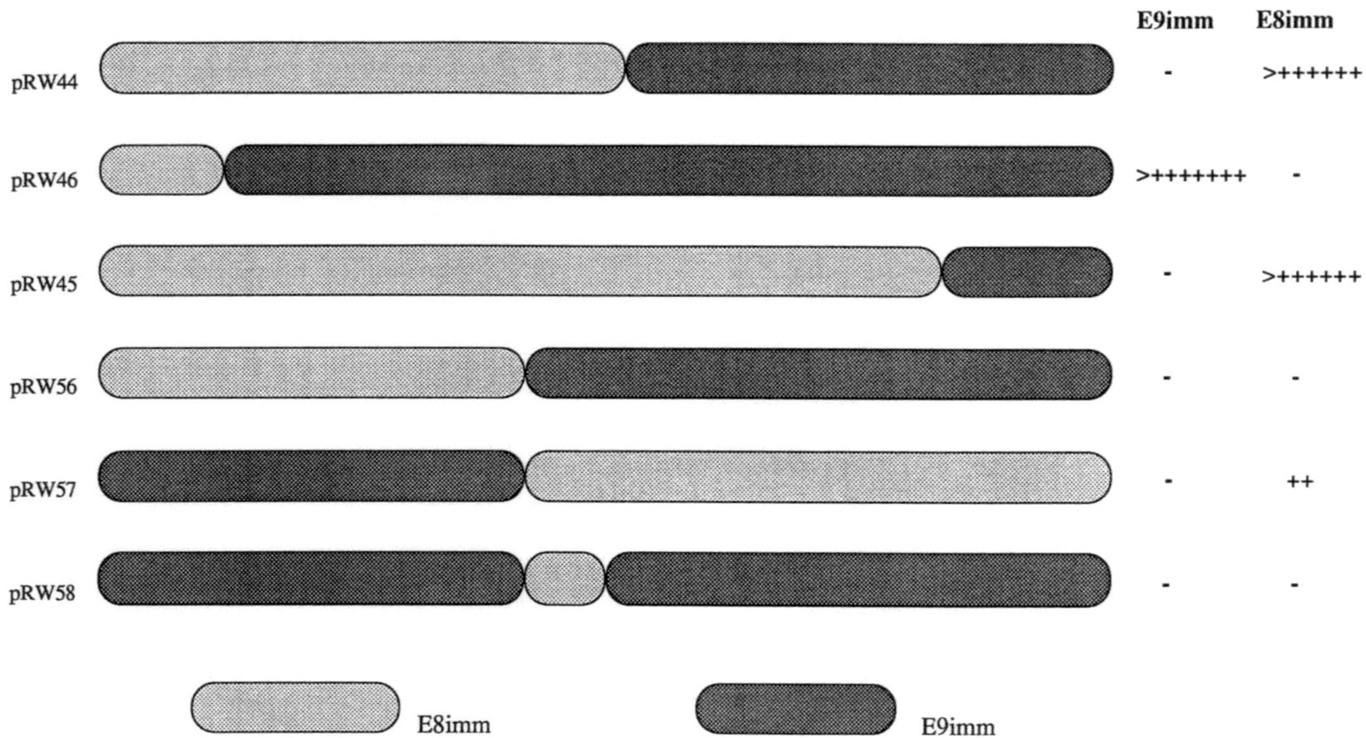

Figure 7. Phenotypes encoded by the immunity gene fusion proteins. The amino-acid sequences of the E9 and E8 immunity proteins are aligned at the top of the Figure. The dots show identical residues in the two proteins. The derivation of each of the fusion proteins is shown, together with the immunity phenotype encoded . The level of immunity was quantitatively determined. - = sensitive; ++ = low level immunity; ++++++ = high level immunity.

Further gene fusions have been created after introducing a unique *Ava*I site within the E9 *imm* gene, in the same position as the unique *Ava*I site in the E8 *imm* gene. The introduction of this *Ava*I site changes Val37 to Glu, and results in the appearance of low level E8 immunity, without affecting the level of E9 immunity. This is perhaps indicative of there being two different, but perhaps partially overlapping, domains within the E8 and E9 immunity proteins. Gene fragments were then exchanged between the two *imm* genes by using the *Ava*I site. The phenotypes of the three resulting fusions (pRW56, pRW57 and pRW58), when compared with that of pRW44, point to the importance of a region of amino-acids between Val34 and Met44 in E colicin immunity.

NMR SPECTROSCOPY OF THE E9 IMMUNITY PROTEIN

Our aim is to determine the three-dimensional structure of the E9 immunity protein in solution and to eventually derive a structure for the colicin/immunity protein complex. Such an investigation requires large quantities of protein. Until recently immunity proteins were purified by the dissociation of the colicin/immunity protein complex, isolated after mitomycin C induction of *E.coli* cells carrying the appropriate Col plasmid. However these procedures require large culture volumes and the yield of immunity protein is poor. We thus sought to overexpress the E9 *imm* gene in order to overproduce large amounts of the E9 immunity protein.

Overexpression of the E9 immunity gene

We have described a method for the selective overproduction of the E9 immunity protein (Wallis *et al.*, 1992), using the expression vector pKK233-2 (Figure 8). Using this system the E9 immunity protein is expressed to approximately 10% of the total cell protein in *E.coli* JM83 (data not shown), and is easily purified to homogeneity in just three steps (ammonium sulphate fractionation, followed by anion exchange and then gel filtration chromatography). Routinely we can now purify >100mg of pure immunity protein from just 5L of cells.

We have extensively characterized the overproduced immunity protein, both *in vitro* and *in vivo*, to show that it is identical to the immunity protein recovered from the colicin E9/immunity protein complex. We have also demonstrated that the purified immunity protein inhibits the DNAase activity of colicin E9 *in vitro*, with complete inhibition of the DNAase activity occurring when the two proteins are present in a 1:1 ratio molar ratio (Wallis *et al.*, 1992). These results have given us confidence in the results of our NMR experiments with this protein.

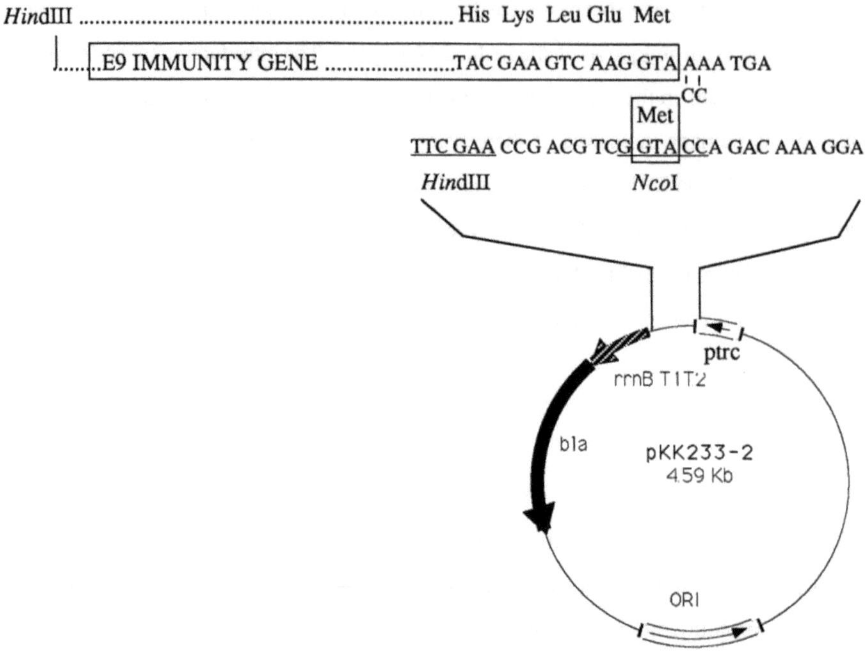

Figure 8. Overexpression of the E9*imm* gene. In the lower part of the figure is the restriction map of plasmid pKK233-2 (Pharmacia/LKB). At the top of the figure is shown the two A - C mutations required to introduce an *Nco*I site in the E9*imm* gene. The *Nco*I site includes the ATG start codon of the E9*imm* gene. Cloning of the 'mutant' E9imm gene as an *Nco*I - *Hin*dIII fragment into pKK233-2 restricted with *Nco*I - *Hin*dIII places the expression of the E9*imm* gene under the control of the very strong ptrc vector promoter.

NMR Spectroscopy

The present objective of the NMR work is to determine the sequence location of the secondary structure elements in the protein as a prelude to determination of its tertiary structure.

Inspection of the ^1H NMR spectrum of Im9 in D_2O facilitates a number of qualitative structural conclusions (Figure 9). Firstly, the resonance line widths (which are determined by the tumbling rate of the protein in solution) are narrow, indicating that the protein is present as the monomer in solution. Secondly, both aliphatic and aromatic resonances are shifted from their "random coil" positions (see the methyls marked 1, 2 and 3 in Figure 9) (Wuthrich, 1986). This suggests that the protein has a well defined tertiary fold.

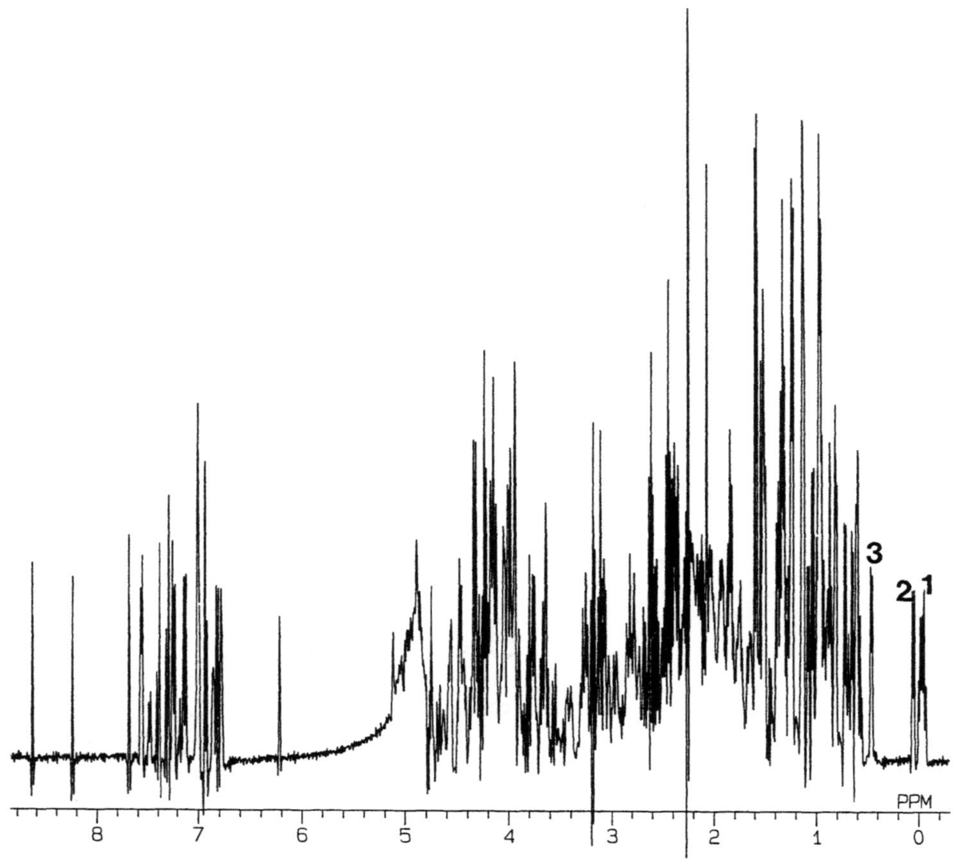

Figure 9. Resolution enhanced one dimensional 400MHz ^{1}H NMR spectrum of Im9 in D_2O (pH 6.8). The peaks marked 1, 2 and 3 correspond to the two Ile δ methyls, an Ile γ methyl and a Val γ methyl resonances respectively.

The first step towards the determination of secondary structure involves the assignment of resonances to specific residues (Wuthrich, 1986).; 2D NMR is employed for this. Typical 400MHz 2D-COSY spectra are shown in Figures 10 and 11. A number of assignments are indicated on the spectra for the side chain (Figure 10) and NH-αCH connectivities (Figure 11).

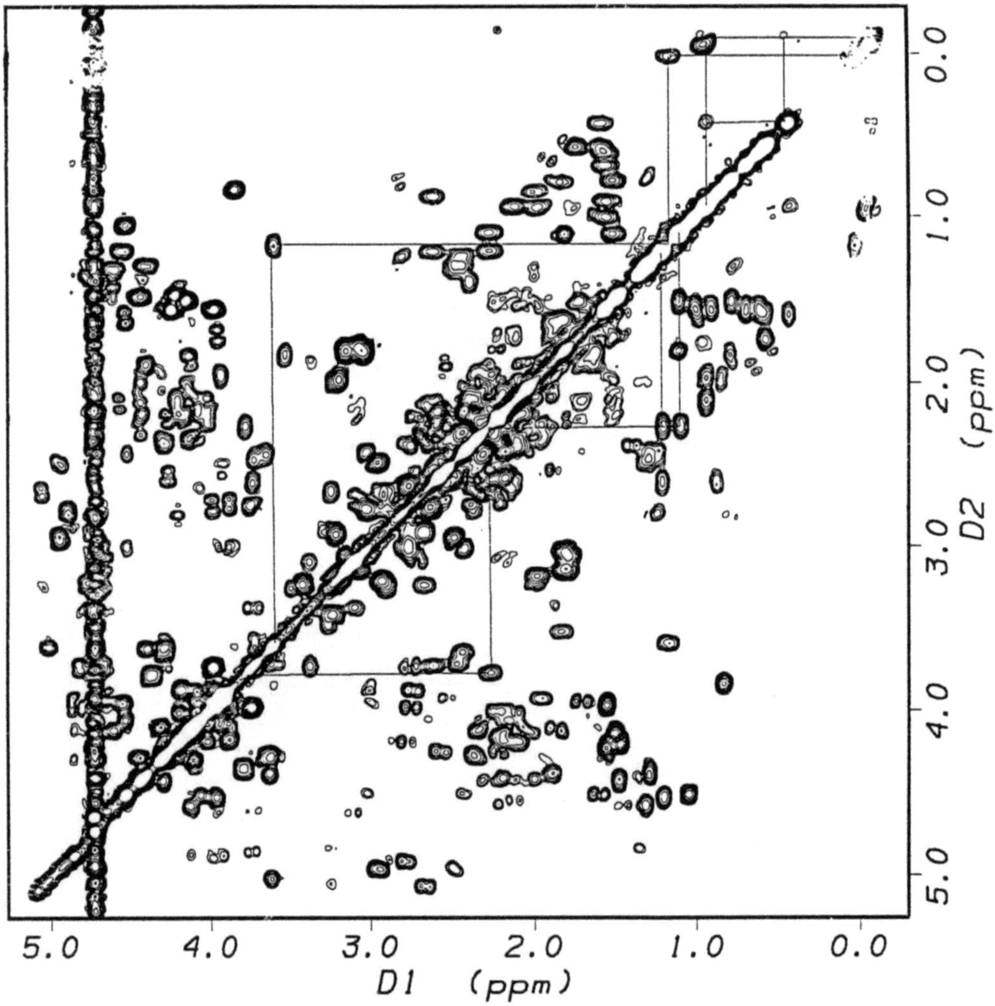

Figure 10. 400MHz 2D COSY spectra of Im9 in D_2O showing the aliphatic region of the spectra. The connectivities for an Ile spin-system and a Val spin system are indicated above and below the diagonal respectively.

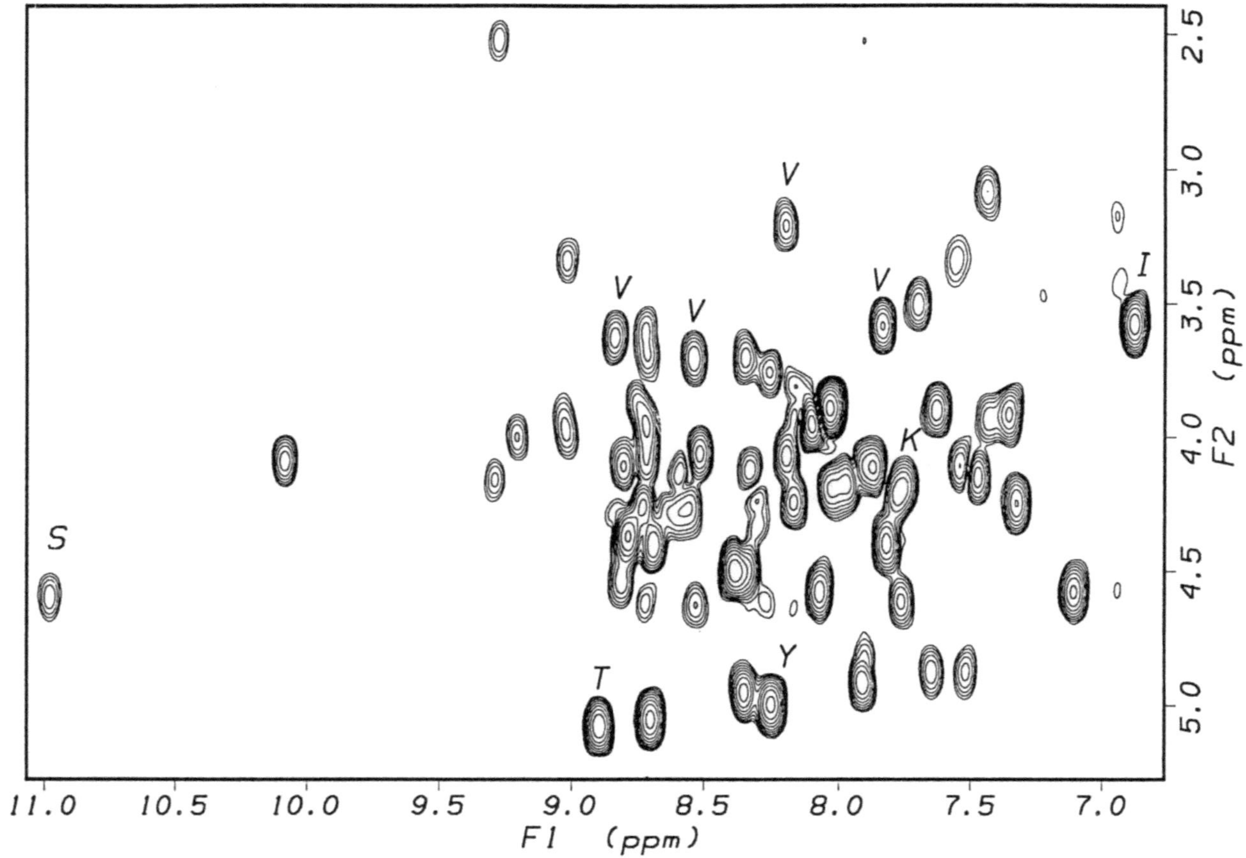

Figure 11. 400MHz 2D COSY spectra of Im9 in 10% D_2O/90% H_2O. The NH (F1)-α (F2) (fingerprints) region is shown, and a number of NH-α connectivities assigned.

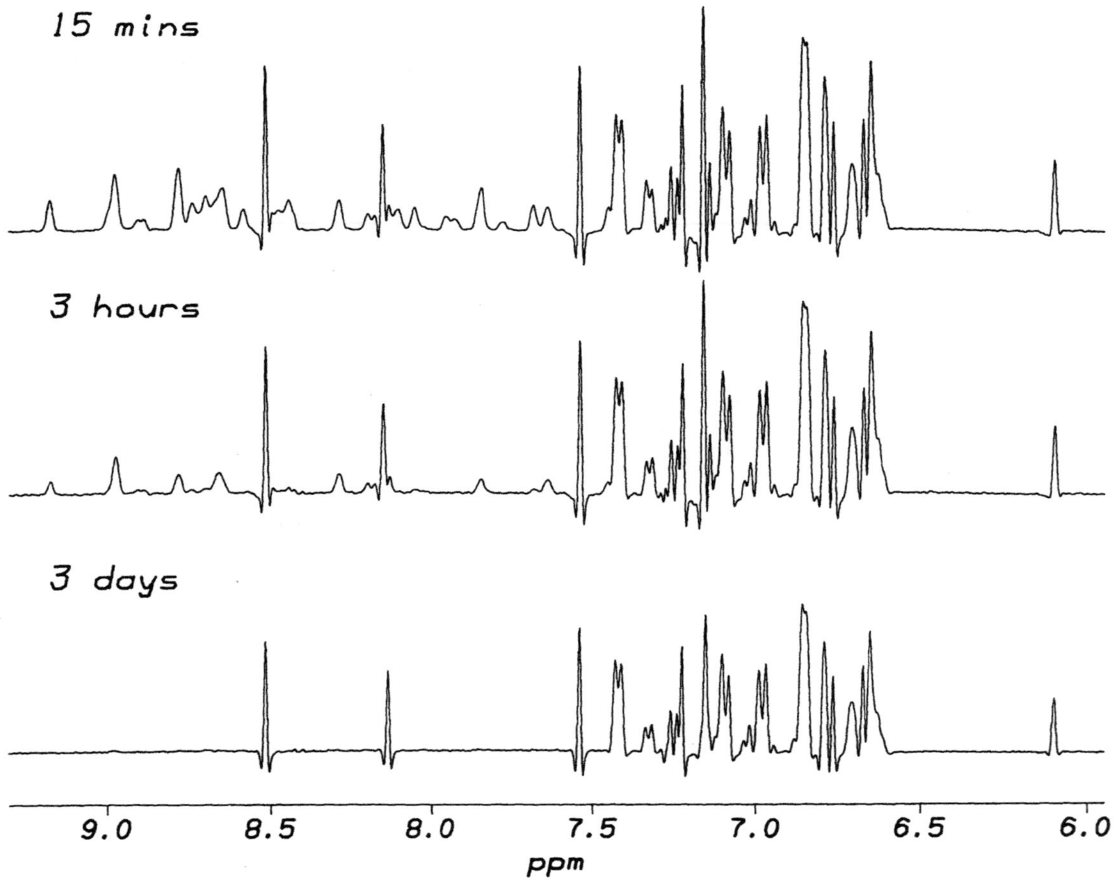

Figure 12. ND/NH exchange of Im9. The fully protonated Im9 protein was dissolved in 100% D_2O at pH 7.0. Spectra were recorded at the time after dissolution indicated in the figure. Full exchange was achieved in 3 days.

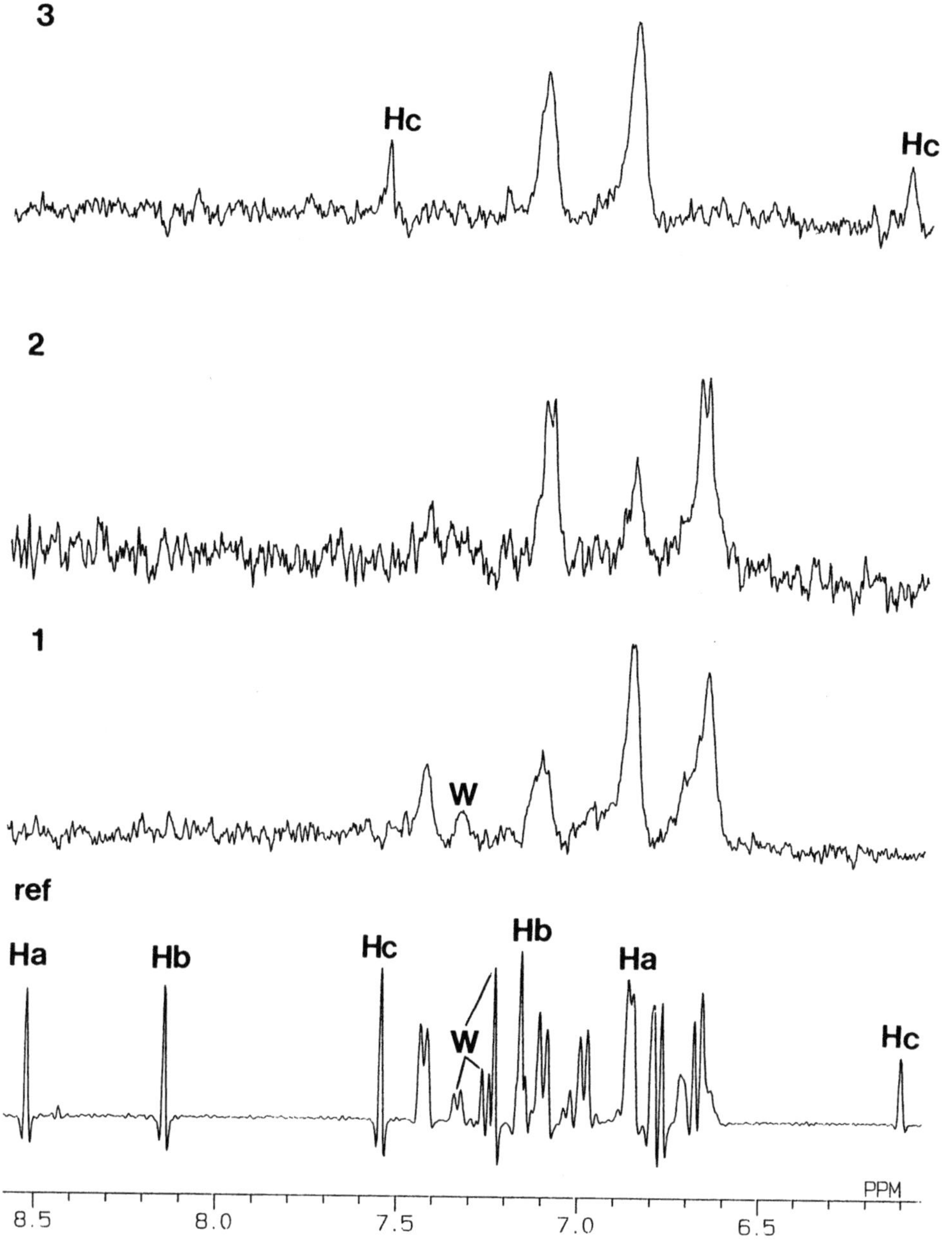

Figure 13. The aromatic region of the 1D difference NOE spectra of Im9 in D_2O. Spectra 1, 2 and 3 are the difference spectra acquired by irradiating the corresponding methyl difference resonances marked in Figure 9.

The chemical shifts of the α protons (F2 axis of Figure 11) have been demonstrated to be strongly dependent on the character and nature of the protein secondary structure (Wishart *et al.*, 1992). Alpha protons experiencing a high frequency shift (compared to random coil values) are believed to form part of a β sheet, conversely a low frequency shift is indicative of a β sheet configuration. Based on this, the shifts of the protons resonances in Figure 11 imply that both β sheet and α helical structural elements are present. However, ND/NH exchange experiments (Figure 12) indicate that there is no large continuous β strand present, as there is in proteins such as bovine pancreatic trypsin inhibitor (Wagner & Wuthrich, 1982). Difference NOE data (Figure 13) show that there is a hydrophobic box present within the protein, comprising of, at least, the sole tryptophan, a histidine, another aromatic residue (Y or F), a valine and isoleucine. The quality of the spectral data is high (approximately 65 of the 82 theoretical NH-α connectivities are resolved at 400MHz), and we are confident that at higher magnetic fields we shall be able to map out the secondary structural elements, and have sufficient longe range NOE's for a tertiary structure determination.

pH titration data have highlighted some unusual titration behaviour of the three histidine residues in the E9 immunity protein. Over the pH range 5.7 - 10.5 only one of the three histidines titrates (and that with a pKa of ~ 7.0) and, surprisingly, one of the non-titrating histidines appears to have a pKa > 10.5, since the chemical shifts of its resonances indicate that it is protonated. The histidine in the hydrophobic box has a pKa <6.0. It appears unlikely that the histidines are specificity determinants but, nevertheless, the unusual properties of at least one histidine require explanation.

Acknowledgements

This work was supported by the Molecular Recognition Inititative of the SERC (UK) and by research grants (to C.K) from the Royal Society, the Society of General Microbiology and the University of East Anglia Research Promotion Fund.

REFERENCES

Bowman CM, Dahlerg JE, Ikemura T, Konisky J, Nomura M (1971) Specific inactivation of 16S ribosomal RNA induced by colicin E3 *in vitro*. Proc Natl Acad Sci USA 68:964-968

Chak KF, James R (1984) Localization and characterization of a gene on the ColE3CA38 plasmid that confers immunity to colicin E8. J Gen Microbiol 130:701-710

Chak KF, James (1986) Characterization of the ColE9-J plasmid and analysis of its genetic organization. J Gen Microbiol 132:61-71

Chak KF, Kuo WS, Lu FM, James R (1991) Cloning and characterization of the ColE7 plasmid. J Gen Microbiol 137:91-100

Cole ST, Saint-Joanis B, Pugsley AP (1985) Molecular characterization of the colicin E2 operon and identification of its products. Mol Gen Genet 198:465-472

Cooper PC, James R (1984) Two new colicins, E8 and E9, produced by a strain of *Escherichia coli* J. J Gen Microbiol 130:209-215

Curtis MD, James R (1991) Investigation of the specificity of the interaction between colicin E9 and its immunity protein by site-directed mutagenesis. Mol Microbiol 5: 2727-2733

DiMasi RD, White J, Schnaitman CA, Bradbeer C (1973) Transport of vitamin B12 in *Escherichia coli:* common receptor sites for vitamin B12 and E colicins on the outer membrane of the cell envelop. J Bacteriol 115:506-573

Eaton T, James R (1989) Complete nucleotide sequence of the colicin E9 *(ceal)* gene. Nucleic Acids Res 17:1761-1761

Herschman HR, Helinski DR (1967) Comparative study of the events associated with colicin induction. J Bacteriol 94: 691-699

Jakes KS, Zinder ND, Boon T (1974) Purification and properties of colicin E3 immunity protein. J Biol Chem 249: 438-444

James R, Jarvis M, Barker DF (1987) Nucleotide sequence of the immunity and lysis regions of the ColE9-J plasmid. J Gen Microbiol 133:1553-1562

Lau PCK, Condie J (1989) Nucleotide sequences from the colicin E5, E6 and E9 operons: Presence of a degenerate transposon-like structure in the ColE9-J plasmid. Mol Gen Genet 217:269-277

Lau PCK, Rowsome RW, Watson RJ, Visentin LP (1984) The immunity genes of colicins E2 and E8 are closely related. Biosci Rep 4:565-572

Masaki H, Toba M, Ohta T (1985) Structure and expression of the ColE2-P9 immunity gene. Nucleic Acids Res 13:1623-1635

Mock M, Pugsley AP (1982) The BtuB group Col plasmids and homology between the colicins they encode. J Bacteriol 150:1069-1076

Schaller K, Nomura M (1976) Colicin E2 is a DNA endonuclease. Proc Natl Acad Sci USA 73:3989-3993

Sidikaro J, Nomura M (1974) E3 immunity substance, a protein from E3-colicinogenic cells that accounts for their immunity to colicin E3. J Biol Chem 249: 445-453

Toba M, Masaki H, Ohta T (1988) Colicin E8, a DNase which indicates an evolutionary relationship between colicins E2 and E3. J Bacteriol 170:3237-3242

Tokuda H, Konisky J (1979) Effect of colicins Ia and E1 on ion permeability of liposomes. Proc Nat Acad Sci USA 76: 6167-6171

Uchimura T, Lau PCK (1987) Nucleotide sequences from the colicin E8 operon: Homology with plasmid ColE2-P9. Mol Gen Genet 209:489-493

Wallis R, Reilly A, Rowe A, Moore GR, James R, Kleanthous C. (1992) *In vivo* and *in vitro* characterization of overproduced colicin E9 immunity protein. Eur J Biochem In press

Watson R, Rowsome W, Tsao J, Visentin LP (1981) Identification and characterization of Col plasmids from classical colicin E-producing strains. J Bacteriol 147:569-577

Wishart DS, Sykes BD, Richards FM (1992) The chemical shift index: A fast and simple method for the assignment of protein secondary structure through NMR spectroscopy. Biochemistry 31:1647-1651

Wuthrich K (1986) NMR of proteins and Nucleic Acids, Wiley, New York

Wagner G, Wuthrich K (1982) Amide proton exchange and surface conformation of the basic pancreatic trypsin inhibitor in solution. Studies with two dimensional NMR. J Mol Biol 160: 343-361

STRUCTURAL STUDIES ON COLICIN E3 AND ITS IMMUNITY PROTEIN

M. Shoham and A. Djebli
Department of Biochemistry
School of Medicine
Case Western Reserve University
10900 Euclid Avenue
Cleveland
Ohio 44106-4935
U. S. A.

Introduction

Colicin E3 is a bacteriocin that penetrates sensitive cells and kills them by an enzymatic mode of action. It acts as a highly specific RNase that cleaves 16S ribosomal RNA at a position 49 bases in from the 3' end (Nomura *et al.*, 1974). Although only a single phosphodiester bond is cleaved and the fragments stay attached to each other following the endonucleolytic attack, protein biosynthesis ceases completely and the infected cell dies. Colicin E3 acts in this specific manner only if all the components necessary for ribosome function are present. Thus, colicin E3 function requires the presence of the 50S subunit in addition to the 30S subunit as well as ribosomal proteins (Jakes & Zinder, 1974; Sidikaro & Nomura, 1974). Knowledge of the three-dimensional structure of colicin E3 will therefore not only provide an understanding of its mode of action but it is expected also to shed light on the structure of the ribosome at large in the vicinity of the cutting site. This region of the ribosome close to the interface between the small and large subunit is obviously critical for protein biosynthesis if a single nick in 16S RNA totally inactivates the ribosome.

The properties of colicin E3 have been reviewed by Karen Jakes (1982). A brief summary and some more recent information are outlined here. *In vivo* colicin E3 acts only on related species of *E. coli* which have the BtuB receptor to which it binds on the surface of the outer membrane of the infected cell. This receptor also binds vitamin B12 as well as bacteriophage BF23 (Sabet & Schnaitman, 1973). *In vitro* colicin E3 will cleave 70S ribosomal particles specifically, but with isolated 16S RNA the result is non-specific cleavage and complete degradation (Ohno-Iwashita & Imahori, 1977; Ohno & Imahori, 1978).

The producing organism is protected from the action of endogenous or exogenous colicin E3 by the presence of an immunity protein. In the case of colicin E3 the immunity protein acts as an inhibitor of the ribonucleolytic function of colicin E3. The 9.8 kDa immunity protein forms a 1:1 complex with the 58 kDa colicin E3. This tight complex can only be dissociated under denaturing conditions (de Graaf & Klaasen-Boor, 1974; Jakes *et al.*, 1974).

NATO ASI Series, Vol. H 65
Bacteriocins, Microcins and Lantibiotics
Edited by R. James, C. Lazdunski and F. Pattus

Colicin E3

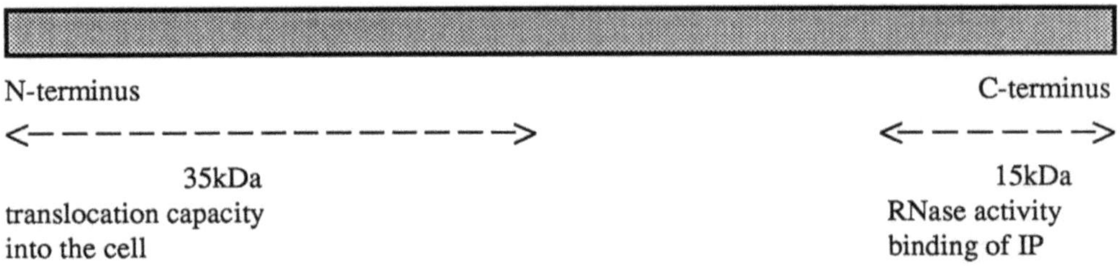

Figure 1. Tryptic Digestion of Colicin E3 yields two stable fragments, the middle portion is completely degraded (Ohno *et al.*, 1977).

Some structural information is available on colicin E3 and its immunity protein. Proteolytic digestion of colicin E3 with trypsin under mild conditions yields two fragments, a 35 kDa amino-terminal fragment and an 15 kDa carboxy-terminal fragment in complex with the immunity protein (Ohno *et al.*, 1977). This suggests a domain structure for colicin E3 (Figure 1). From small angle X-ray scattering experiments the radii of gyration for free immunity protein, free colicin E3, and the complex were measured to be 13.4Å, 75.6Å and 35.5Å, respectively (Levinson *et al.*, 1983). The unusually large radius of gyration of free colicin E3 was interpreted to be the result of dimerization in the absence of immunity protein.

Since the substrate of colicin E3 is a nucleic acid, it is not surprising that colicin E3 is a basic protein with a pI of 9.0 (Levinson *et al.*, 1983). Immunity protein is acidic with a pI of 4.0. When these two proteins interact the charges are largely neutralized to yield a complex with a pI of 6.5. However, the interactions cannot be purely electrostatic in nature since the complex cannot be dissociated by raising the pH to 9.5, a value above the pI for isolated colicin E3.

In addition to being a ribonuclease inhibitor the immunity protein may also facilitate receptor binding or translocation across the inner membrane (Lau & Richards, 1976).

Like other colicins, E3 seems to be composed of three domains, a hydrophobic N-terminal domain responsible for translocation across the inner membrane, a central domain with a binding capacity to the receptor on the outer membrane and a C-terminal "business domain" carrying the catalytic antibiotic-like activity. The colicin E3 molecule thus represents a truly multifunctional protein open to structural studies of such diverse aspects as membrane insertion, protein-nucleic acid interactions, catalytic activity and protein-protein recognition with the receptor and with the immunity protein.

In this paper we present X-ray crystallographic data for free immunity protein as well as the colicin E3-immunity protein complex.

Isolation and Purification of Colicin E3 and its Immunity Protein

The colicin E3-immunity protein complex, as well as free immunity protein, were prepared as described by Frolow and Shoham (1990), following procedures by Levinson *et al.* (1983) and Herschman and Helinski (1967). This preparation of the complex does have some minor contaminants even after the final step of HPLC chromatography (Figure 2).

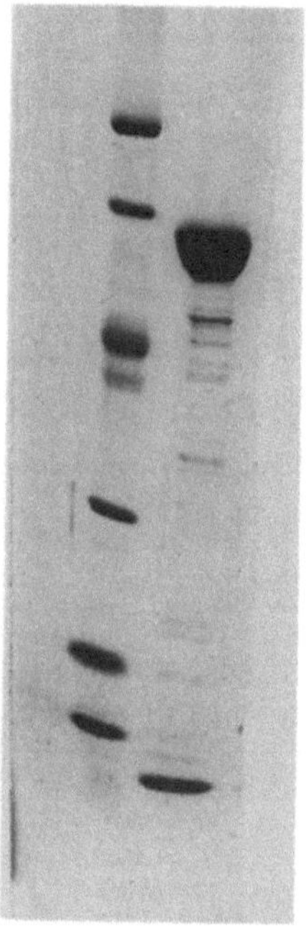

Figure 2. SDS-PAGE of the colicin E3-immunity protein complex after HPLC chromatography (Frolow & Shoham, 1990). The left lane contains molecular weight markers. Even this most pure preparation has several minor bands.

Two Isomers of the Immunity Protein

During the final step of the immunity protein purification on a Biogel P-60 column two peaks were found which contain a protein that migrates on SDS-PAGE with a molecular weight of 10 kDa (Figure 3). These two fractions however, have a different mobility on gel electrophoresis performed under non-denaturing conditions (Figure 4). The two fractions have identical amino acid compositions except that fraction number 1, which migrates faster on the gel, has 1.5 extra Asx residues and 0.5 extra Glx residues. Since both fractions comigrate on SDS-PAGE as well as on a Sephadex G-75 column, they seem to have identical molecular weights. The different migration on a native gel and on a Biogel P-60 column must be due to a charge difference. It seems likely therefore that fraction number 1 has two extra negative charges compared to fraction number 2. It is fraction number 2 that crystallized whereas fraction number 1 never yielded any crystals.

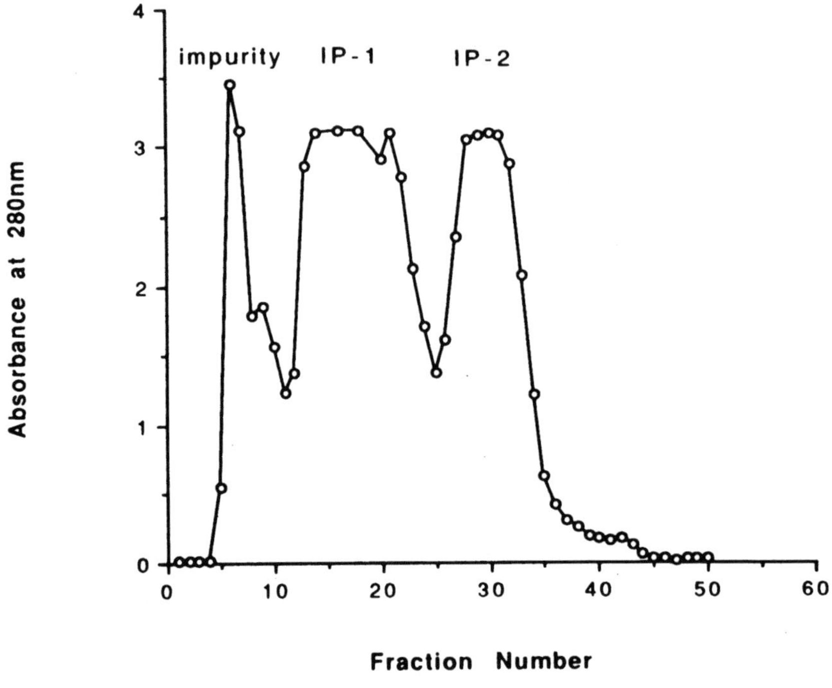

Figure 3. Separation of crude immunity protein on a Biogel P-60 column. The buffer solution consists of 0.1M phosphate, pH 7.0, 10mM NaN_3, 10μM phenylmethylsufonyl fluoride (PMSF). The first peak contains an unspecified yellow impurity. Peaks designated IP-1 and IP-2 each contain a single protein. IP-1 and IP-2 comigrate as a 10 kDa band on SDS-PAGE (data not shown), but they migrate differently on gel electrophoresis under non-denaturing conditions.

Free Immunity Protein forms a Dimer

Both isomers of free immunity protein elute as a dimer of 20 kDa on a Sephadex G-75 column. The asymmetric unit of the crystals also contains a dimer so that it appears that this preparation of immunity protein forms a dimer in solution as well as in the crystalline state.

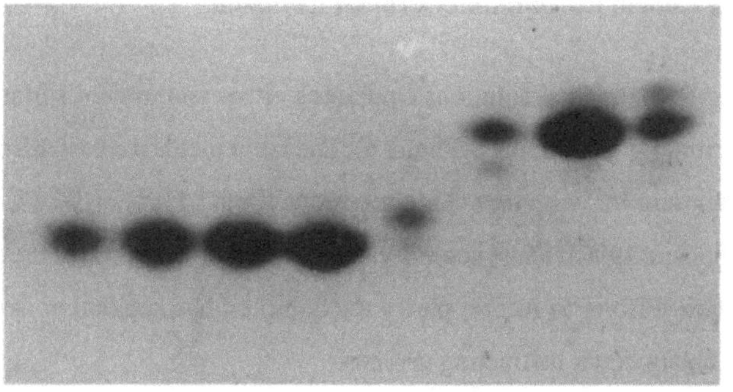

IP-1 IP-2

Figure 4. Polyacrylamide Gel Electrophoresis under Non-denaturing conditions: IP-l and IP-2 migrate differently.

The Immunity Protein is a Highly Specific Inhibitor of RNases

The immunity protein is one of very few known ribonuclease inhibitors. In an effort to find out whether the immunity protein would inhibit other ribonucleases, commercially available microbial RNases were tested for their inhibition by the immunity protein. The assay used was degradation of yeast RNA as followed by absorption at 260nm. The enzymes RNase Tl, RNase C and RNase Ul were used. No inhibition was detected on any of these enzymes at immunity protein concentrations of up to 1000 times higher than the enzyme concentration and up to 40 times higher than the substrate concentration. The conclusion to be drawn from these experiments is that the immunity protein is a highly specific inhibitor RNase. However, in light of some limited sequence homology of the enzymatic domain of colicin E3 and microbial ribonucleases (Figure 5) the potential exists for engineering the immunity protein by site-directed mutagenesis into a more general ribonuclease inhibitor, or at least into an inhibitor of altered specificity.

```
                                   ***
colicin E3   121   NG-GGKRKRWTG-D--KGRKIYEWDSQHGELEGYR-ASDGQHLGSFDPKT
                   |:  .  | ..:.| |     :...|||.    :: :.|. :|.|.. . |:   .
RNAse T1      36   NSYPHKYNNYEGFDFSVSSPYYEWPILSSG-DVYSGGSPGADRVVFN--E

Colicin E3   166   GNQLKG--PDPKRNIKKYL  182
                   .||| |   ...  . ..::
RNAse T1      83   NNQLAGVITHTGASGNNFV  101
```

Figure 5. Sequence homology between the catalytic regions of colicin E3 and ribonuclease Tl. The sequence identity is 25.4%. Similarity is indicated by dots and identity by vertical bars between corresponding residues. The active site sequence tyrosine-glutamic acid-tryptophane of ribonuclease T1 is marked by stars above these residues.

Crystallization of the Colicin E3-Immunity Protein Complex

The complex crystallizes from solutions containing either ammonium sulfate, sodium-potassium phosphate or sodium citrate as a precipitant (Figure 6). The latter yields the best diffracting crystals which are monoclinic, space group $P2_1$ with unit cell dimensions a = 67.71, b = 196.67, c = 85.58 Å and β = 113.67° (Frolow & Shoham, 1990). There is some variability in the resolution of the crystals, the best ones diffract to 3Å resolution. Efforts to further purify the complex by preparative isoelectric focusing are underway in order to obtain better diffracting crystals.

Figure 6. Crystals of the colicin E3 - immunity protein complex obtained from 1.0M citrate, pH 5.6, 10mg/ ml protein. The size of the large cystal is 0.75 x 0.50 mm.

Crystallization of Free Immunity Protein

Free immunity protein crystallizes readily (Shoham *et al.*, 1984). The crystals currently being used for the structure determination are monoclinic, space group C2 with unit cell dimensions a = 80.07, b = 53.72, c = 36.25 Å and β = 93.18° (Figures 7 and 8). There are two molecules per asymmetric unit and the crystals diffract to 1.8Å. The technique of macroseeding was employed to grow crystals as large as 1.5mm in the longest dimension.

Figure 7. A crystal of immunity protein grown by macroseeding in 1.4M $(NH4)_2SO_4$, 30mM NaCl, 30mM MES buffer, pH 6.0, 5mg/ml immunity protein. The size of the crystal is 0.4 x 0.5 mm.

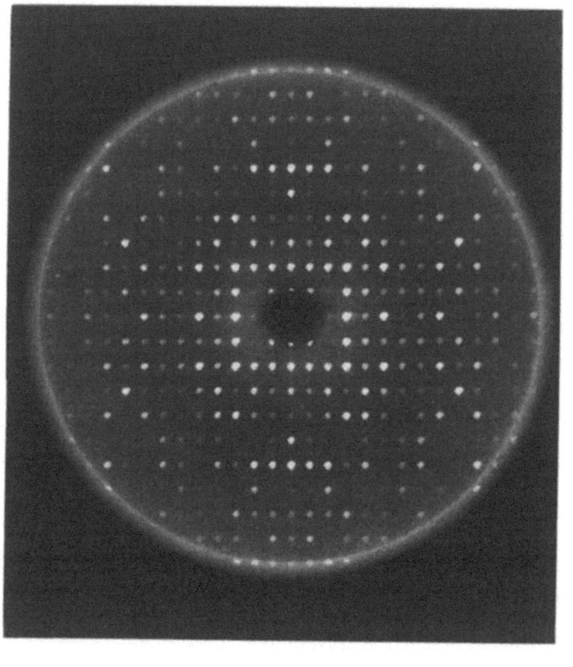

Figure 8. Precession X-ray photograph of an immunity protein crystal: 0KL zone, precession angle 17°.

X-ray Data Collection of Immunity Protein and a Heavy-Atom Derivative

X-ray data have been collected on immunity protein crystals on a Sun Diego Multiwire Area Detector mounted on a Rigaku RU200 rotating anode generator (Table 1).

Table 1

Statistics of Data Collection *

	Resolution (Å)	Completeness (%)	<I/σ>	<I/σ> at highest resolution	Number of reflections reflections	Number of unique	Rsym**
Native	2.0	0.90	20.2	5.4	32,627	9,488	7.3
$NaAuCl_4$***	2.5	0.86	20.8	8.5	32,563	4,681	9.7

* The data were collected on a San Diego Multiwire Dual Area Detector System equipped with a Rigaku AFC6 Four-Circle Goniostat and a Rigaku RU200 Rotating Anode X-ray Generator.

** **Rsym** = $\frac{\Sigma|I - I_{av}}{\Sigma I_{av}}$ **I, intensity; I_{av}, average intensity; σ, standard deviation**

*** Bijvoet pairs were measured for the derivative data set. Data collection speed: 70 sec per frame.

The search for heavy-atom derivatives resulted in a single useful derivative of excellent quality, O.5mM sodium tetrachloroaurate $NaAuC1_4$ (Table 2). There are two gold sites per asymmetric unit, presumably one per molecule (Figure 9). The only other isomorphous derivative found was K_2HgI_4. However, it binds at the exact same locations as the gold compound, and therefore does not yield significant new phase information. Since this derivative is inferior to the gold derivative, it was not used for phasing. No mercury derivative other than K_2HgI_4 was found. This is surprising since the immunity protein contains a single cysteine as well as a single methionine residue (Figure 10). Even a compound as small as mercuric chloride or methyl mercury acetate did not bind to the crystals as evidenced by the lack of changes in the diffraction pattern. This could be interpreted in two ways. Either the sulfhydryl group is buried and inaccessible to solvent or it is engaged in an intermolecular disulfide bridge with the other molecule in the dimer. In order to try and break this alleged intermolecular disulfide bond, a crystal was soaked in 2 mM dithiothreitol (DTT) for 1 day followed by a soak in 5 mM methyl mercury chloride for 1 month. Although the crystal suffered a crack the diffraction pattern was indistinguishable from that of the native protein. This result is inconclusive as far as the oxidation state of the cysteine residue is concerned. This cysteine is inaccessible to mercurials for whatever reason. A likely ligand for the $AuCl_4$-moiety is His64, the single histidine residue in the molecule. $NaAuCl_4$ is usually bound to either histidine or cysteine residues in protein crystals (Blundell & Johnson, 1976). Since the cysteine residue appears to be inaccessible, that leaves His64 as the putative ligand. The two gold sites of the dimer are relatively close to each other, only about 6Å apart. It appears therefore that the two histidines in the dimer are located on the surface with a distance of some 10Å between their imidazole rings.

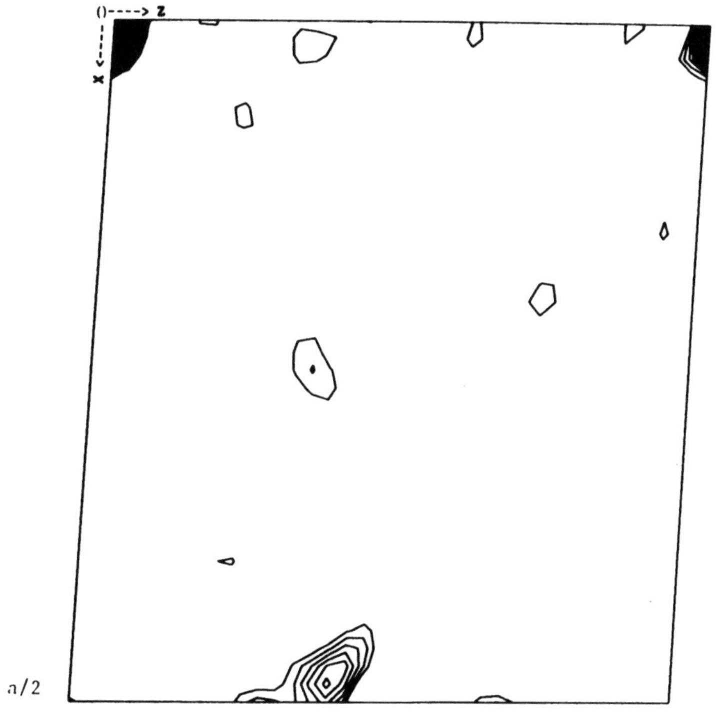

Figure 9. Difference Patterson Map of $NaAuCl_4$ at 2.0Å Resolution: Harker Section v=0. A peak in this map corresponds to a vector between a heavy atom site and its symmetry related position. In addition to the huge peak at the origin, which represents all the self vectors, there are only two significant peaks, both at the bottom of the diagram. The heavy-atom locations derived from these two peaks were subsequently refined (Table 2).

Table 2

Refined Parameters of the $NaAuCl_4$ Derivative

R_{iso}*	x	y	z	Occupancy	B-factor ($Å^2$)	R_{cullis}**	Phasing power***	Figure of merit
.15	.234	.000	.718	.910	16.41	.73	3.0	.61
	.262	.076	.835	.857	13.96			

The CCP4 suite of crystallographic programs (from the Daresbury Laboratory, United Kingdom) was used for all calculations including heavy-atom refinement in programs REFINE, PHASE and PHARE.
* R_{iso} mean discrepancy between derivative and native structure factors.
** $R_{cullis} = (\Sigma|(F_{PH} \pm F_p) - F_{H(calc)}|)/\Sigma|F_{PH}-F_p|)$ for centric reflections.

Non-crystallographic Symmetry in Immunity Protein Crystals

There are two molecules in the asymmetric unit of immunity protein crystals which are related to one another by a local twofold axis (Figure 11). On the assumption that the two $NaAuCl_4$ sites are related to each other by the same molecular dyad the location as well as the inclination of this dyad was uniquely determined.

```
Gly-Leu-Lys-Leu-Asp-Leu-Thr-Trp-Phe-Asp-Lys-Ser-Thr-Glu-Asp-Phe-Lys-Gly-Glu-Glu
1                           10                                              20
Tyr-Ser-Lys-Asp-Phe-Gly-Asp-Asp-Gly-Ser-Val-Met-Glu-Ser-Leu-Gly-Val-Pro-Phe-Lys
21                          30                                              40
Asp-Asn-Val-Asn-Asp-Gly-Cys-Phe-Asp-Val-Ile-Ala-Glu-Trp-Val-Pro-Leu-Leu-Gln-Pro
41                          50                                              60
Tyr-Phe-Asn-His-Gln-Ile-Asp-Ile-Ser-Asp-Asn-Glu-Tyr-Phe-Val-Ser-Phe-Asp-Tyr-Arg
61                          70                                              80
Asp-Gly-Asp-Trp
81          84
```

Figure 10. Amino Acid Sequence of the Immunity Protein (Mochitate *et al.*,1981; Masaki & Ohta, 1985).

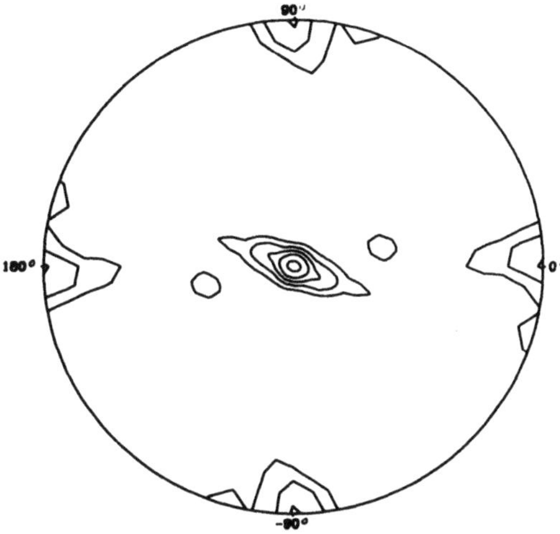

Figure 11. Non-crystallographic Symmetry in Immunity Protein Crystals: Stereographic Projection of a Self-Rotation Search for $\kappa = 180°$. 491 reflections were used in the resolution range 8-5Å. A value of 15Å was used for the radius of integration, in accordance with the measured radius of gyration for the immunity protein (Levinson *et al.*, 1983). Beyond the origin peak in the center which represents the crystallographic twofold axis there is only one other significant peak at $\omega = 90.0°$ and $\phi = 0.8°$ or 89.2°. This peak has a height of 61.3% of the origin. The next highest peak is 35.7% of the origin. The local dyad is therefore in the ac-plane, roughly parallel to either the a-axis or the c-axis. If one assumes the two heavy- atom sites to be related to one another by the local dyad then this local dyad must be roughly parallel to the a-axis.

Status of the Structure Determination of the Immunity Protein

The electron-density map computed with phases from the single heavy-atom derivative including anomalous scattering is partially interpretable. The outline of the dimer is clearly visible but the monomer-monomer interface cannot be delineated at this time. Several β-strands can be identified in the map but the connecting loops cannot be traced unequivocally. Efforts are underway to find additional heavy-atom derivatives. In parallel, non-crystallographic symmetry averaging combined with solvent flattening is being explored to improve the interpretability of the electron-density map.

Acknowledgements

We thank Diane Hughes for skilled technical assistance, James Clifton for a sample of HPLC-purified colicin E3-immunity protein complex, and Peter Proctor for help with the protein purification and for critically reading this manuscript. One of us (A.D) is supported by a postdoctoral fellowship from the

American Heart Association, Northeast Ohio Affiliate. This research is supported by Grant DMB-9018333 from the National Science Foundation. Computing resources were provided by Grant PDS153 from the Ohio Supercomputer Center.

Note added in proof

Since the submission of this manuscript we have found a second useful heavy-atom derivative for immunity protein, namely trimethyl lead acetate.

REFERENCES

de Graaf FK, Klaasen-Boor P (1974) Purification and Characterization of the Cloacin DF13 Immunity Protein. FEBS Lett 40:293-296

Blundell TL, Johnson LN (1976) Protein Crystallography, pp. 224, Academic Press

Frolow F, Shoham M (1990) Crystallization and Preliminary X-ray Investigation of Colicin E3 in Complex with Its Immunity Protein. J Biol Chem 265:10196-10197

Herschman HR, Helinski DR (1967) Purification and Characterization of Colicin E2 and Colicin E3. J Biol Chem 242:5360-5368

Jakes KS (1982) The mechanism of action of colicin E2, colicin E3 and cloacin DF13 in "Molecular action of toxins and viruses" (Cohen, P. and van Heyningen, S., eds), Elsevier Biomedical Press, Amsterdam-New York, Oxford

Jakes KS, Zinder ND (1974) Highly Purified Colicin E3 Contains Immunity Protein Proc Natl Acad Sci USA 71:3380-3384

Jakes KS, Zinder N, Nomura M (1974) Purification and Properties of Colicin E3 Immunity Protein. J.Biol Chem 249:438-444

Lau C, Richards FM (1976) Proteolytic and Chemical Modification of Colicin E3 Activity. Biochemistry 15:3856-3863

Levinson BL, Pickover CA, Richards FM (1983) Dimerization of Colicin E3* in the Absence of Immunity Protein. J Biol Chem 258: 10967- 10972

Masaki H, Ohta T (1985) Colicin E3 and Its Immunity Genes. J Mol Biol 182:217-227

Mochitate K, Suzuki K, Imahori K (1981) Amino Acid Sequence of Immunity Protein (B Subunit) of Colicin E3. J.Biochem (Tokyo) 89:1609-1618

Nomura M, Sidikaro J, Jakes K, Zinder N (1974) In: Nomura M, Tissiers A, Lenzyel P (eds) Ribosomes. Cold Spring Harbor Laboratory, Cold Spring Harbor New York, p 805-814

Ohno-Iwashita Y, Imahori K (1977) Colicin E3 Induced Cleavage of 16S rRNA of Isolated 30S Ribosomal Subunits. J Biochem (Tokyo) 82:919-922

Ohno S, Ohno-Iwashita Y, Suzuki K, Imahori K (1977) Purification and characterization of active component and active fragment of colicin E3. J Biochem (Tokyo) 82:1045-1053

Ohno S, Imahori K (1978) Colicin E3 is an Endonuclease. J Biochem (Tokyo) 84:1637-1640

Sabet SF, Schnaitman CA (1973) Purification and Properties of the Colicin E3 Receptor of *Escherichia Coli*. J Biol Chem 248:1797-1806

Shoham M, Levinson BL, Richards FM (1984) Crystallization and Preliminary X-ray Diffraction Studies of Colicin E3 Immunity Protein. J Mol Biol 177:563-565

Sidikaro J, Nomura M (1974) A Protein from E3-Colicinogenic Cells That Accounts for Their Immunity to Colicin E3. J Biol Chem 249:445-453

STUDY OF THE IMPORT MECHANISMS OF COLICINS THROUGH PROTEIN ENGINEERING AND K+ EFFLUX KINETICS

Hélène Bénédetti, Lucienne Letellier*, Roland Lloubès, Vincent Géli, Daniel Baty and Claude Lazdunski
Centre de Biochimie et de Biologie Moleculaire du C.N.R.S.
31 Chemin Joseph Aiguier
B.P. 71, 13402 Marseille Cedex 9
France

Introduction

Colicin action is initiated by adsorption of the bacteriocin onto a specific receptor protein located in the outer membrane of susceptible cells. The presence or absence of such receptor proteins is a critical factor in defining the activity spectrum of a particular colicin (Graham & Stocker, 1977). The presence of an outer membrane receptor however is not, in itself, sufficient to bring about the lethal events. Following binding, the colicin has to be transported to its ultimate target. Mutant strains, possessing functional receptor, but which are not killed by a particular colicin have been isolated and described. Such mutants are usually denoted as "tolerant".

There are two different pathways for the transport of nutrients across the permeability barrier of the outer membrane. The passive diffusion pathway is used for small hydrophilic molecules (Mr < 700) that can diffuse across porins (Nikaido & Vaara, 1985). Bulky nutrients like iron bearing siderophores, or vitamin B12, that exceed the diffusion limit of the outer membrane are imported through high affinity energy-dependent receptors that allow their transport across the outer membrane. The TonB protein provides the energy for the latter pathway.

Colicins have parasitized both the passive diffusion pathway and the energy-dependent pathway. Group B colicins (B, D, Ia Ib, M...) require in addition to their specific receptors, TonB, ExbB and ExbD (and probably TolQ) (for a review see Postle, 1990; Braun, 1991), while group A colicins (A, El, E2, E3, K, N...) require Tol A, B, Q, R, and TolC for colicin El (for a review see Webster, 1991; Davies & Reeves, 1975a,b). There is no relationship between the lethal activity of the colicin and the translocation pathway which is used.

Consistent with the three steps in the mode of action of colicins; receptor binding, translocation and lethal activity, their polypeptide chains comprise three domains, linearly organized, corresponding to each of these steps. The N-terminal domain is involved in translocation, the central domain is involved in binding to the receptor, and the C-terminal domain carries the lethal activity (Baty *et al.*, 1988; Baty *et al.*, 1990).

* Laboratoire des Biomembranes, U.A. 116, C.N.R.S., Universite Paris-Sud, 91405 Orsay, Cedex, France.

NATO ASI Series, Vol. H 65
Bacteriocins, Microcins and Lantibiotics
Edited by R. James, C. Lazdunski and F. Pattus

The N-terminal domain of colicins specifies the import pathway

While there was ample evidence that the N-terminal domain of colicins is involved in the translocation step, this had never been demonstrated formally. Also, it had not yet been demonstrated that the nature of the N-terminal domain of a given colicin specifies which components of the translocation machinery of a given pathway (passive-diffusion or energy dependent) is used.

To address this question, hybrid colicins between colicin A and colicin El, two pore-forming colicins that belong to group A, were first constructed (Frenette *et al.*, 1991). These colicins both use BtuB as a receptor, however, colicin A requires OmpF (or OmpC with a much lower efficiency) for translocation across the outer membrane while colicin El uses TolC and BtuB (Bénédetti *et al.*, 1989). In addition, the TolA, B, Q, R proteins are required for colicin A to reach its target (inner membrane) while only TolA and TolQ are required for colicin El.

Six different hybrid colicins were constructed by recombining various domains of the two pore-forming colicins A and El (Frenette et *al.*, 1991). These hybrid colicins were purified and their properties were studied. All of them were active against sensitive cells, although to varying degrees (Bénedétti *et al.*, 1991a). From the results one can conclude that: (1) the binding site of OmpF is located in the N-terminal domain of colicin A; (2) the OmpF, TolB, and TolR dependence for translocation is also located in this domain; (3) the TolC dependence for colicin El is located in the N-terminal domain of colicin El; (4) the individual functioning of different domains in various hybrids suggests that domain interactions can be reconstituted in hybrids that are fully active, whereas in others that are much less active, non-proper domain interactions may interfere with translocation.

The same colicin domains can promote transport of enzymatic or ionophoric polypeptides

With respect to their ultimate target in susceptible cells, colicins can be divided into four major groups. The first group includes several bacteriocins, of which colicin D13 and colicin E3 are typical representatives. The primary effect of these bacteriocins is the cleavage of the 16S ribosomal RNA generating a small fragment of 49 nucleotides from the 3'-terminus (De Graaf *et al.*, 1971; Senior & Holland, 1971). The second group consists of bacteriocins which cause degradation of DNA. The best investigated representative of this group is colicin E2 which is an endonuclease (Schaller & Nomura, 1976).

The major group is the third one, the group of pore-forming colicins (A, B, El, N, Ia, Ib, K). Their lethal action is based on the formation of ion channels in the cytoplasmic membrane leading to disruption of the energy potential of sensitive cells (Lazdunski *et al.*, 1988).

The last group is represented by colicin M which inhibits peptidoglycan and lipopolysaccharide biosynthesis through interference with bactoprenyl phosphate recycling (Harkness & Braun, 1989).

Enzymatically active colicins (E2, E3) are released from producing cells as an equimolar complex with their homologous immunity protein. In contrast the immunity proteins directed against pore-forming colicins (Lazdunski *et al.*, 1988), or against colicin M (Ölschäger & Braun, 1987), are localized in the inner membrane of immune cells.

Since chimeric colicins constructs proved to be stable and sometimes fully active, we considered the possibility to transport a colicin with a cytoplasmic target, like colicin E2 or E3, using the N-terminal and central domains of a colicin (like colicin A), targeted to the inner membrane. Two such colicins were constructed (Bénédetti, PhD thesis, 1991). The hybrid AAE2 (N-terminal and central domains of colicin A, C-terminal domain of colicin E2) contained 371 amino acid residues of colicin A and the 255 C-terminal residues of colcin E2 (a total of 626 residues). The hybrid AAE3 (N-terminal and central domains of colicin A, C-terminal domain of colicin E3) contained 371 residues of colicin A and the C-terminal 225 residues of colicin E2 (a total of 596 residues). Both hybrids were correctly expressed and secreted into the extracellular medium by producing cells. They were then purified and assayed *in vivo* and *in vitro*. They had no activity *in vivo*. However, competition assays demonstrated that they could bind to the receptor since they could protect efficiently sensitive cells from colicin A action. A 200 to 400 times higher concentration of colicin was required to kill an equivalent number of cells in the presence of the hybrid colicins.

Despite their lack of *in vivo* activity, hybrid colicins were found to be active *in vitro* after removal of the respective immunity proteins through urea treatment. This data demonstrated that the translocation step was defective with both hybrid colicins, although all structural information was contained within their polypeptide chain.

Such a problem, however, does not mean that the N-terminal and central domains of colicin A are unable to promote transport of the enzymatic C-terminal domains of colicins E2 and E3. Indeed, we have previously reported that every hybrid between colicins A and El, two pore-forming colicins with the same target, is not fully active (Bénédetti *et al.*, 1991a). Some hybrids had a very low activity. This was probably due to non-proper domain interactions that may interfere with translocation. The lack of *in vivo* activity of AAE2 and AAE3 was thus similarly interpreted.

A recent report from Ross *et al.* (1989) demonstrated that the structural genes of colicins are assembled from DNA fragments that encode the domains responsible for colicin activity and uptake, irrespective of the type of activity. We thus devised conditions to directly select active hybrid colicins of the AAE3 type. This technique is based upon exonuclease III and Mung bean nuclease digestion of plasmids carrying colicin genes, or colicin domains cloned in tandem, and homologous recombination. Then, transformants producing active hybrid colicins are selected (Géli *et al.*, manuscript in preparation). Using this technique, active hybrids were isolated. The plasmids were purified and the corresponding gene was sequenced. A typical gene of this type encoded a colicin AAE3, in which the fusion between the N-terminal and central domains of colicin A with the C-terminal domain of colicin E3 occurred at the amino acid residue 426 of colicin E3, which is exactly the limit of the domain endowed with the RNase activity (Ohno-Iwashita & Imahori, 1980; H. Bénédetti, V. Géli and C. Lazdunski, manuscript in preparation).

These results demonstrate that the uptake-determining region of a colicin targeted to the inner membrane (colicin A) can also be used for a colicin which has a cytoplasmic target (16S ribosomal RNA). Moreover, this data also underlines the fact that intradomain folding is probably an important factor for translocation. Proper folding can probably be achieved when the exact limits of the domains are respected. Under different conditions (with inactive hybrid constructs described above) non-proper domain folding and interactions can prevent translocation.

Switch from the passive diffusion pathway to the energy requiring pathway through genetic engineering.

Results previously reported demonstrated that the complete information that specifies interactions with components of translocation machinery is located in the N-terminal domains of colicins (Bénédetti *et al.*, 1991a). It thus seemed possible to promote a switch from the passive diffusion pathway to the energy-requiring pathway simply by substituting N-terminal domains. This was performed with the two pore-forming colicins A and Ib. Colicin A belongs to the group A colicins and is imported through the Tol-pathway with no energy requirement (Bourdineaud *et. al.*, 1990). Colicin Ib is a group B colicin, it is taken up through the energy-dependent TonB pathway.

Through genetic engineering the N-terminal 30 residues of colicin A were replaced by the 85 N-terminal residues of colicin Ib (H. Bénédetti, PhD thesis, 1991). The hybrid protein was expressed but not released into the extracellular medium. It was obtained as a crude fraction from cellular extracts and tested on lawns of an indicator strain.

The hybrid IbAA was inactive against TonB$^-$ and BtuB$^-$ strains, while growth zone inhibition was clearly visible around the drop applied on cell lawns of TolA$^-$, TolB$^-$, TolQ$^-$, TolR$^-$ and OmpF$^-$ strains (Table 1). Thus the deletion of the 30 N-terminal residues of colicin A and their substitution by the N-terminal 55 residues of colicin Ib resulted in a chimeric protein using BtuB as a receptor but no longer using the Tol-dependent pathway. Instead, the IbAA protein is channelled into the TonB pathway, thus demonstrating that the first 85 N-terminal residues of colicin IB are sufficient to redirect a colicin to this pathway. This region contains the consensus "Ton B box" found in both the high affinity receptors and group B colicins (Braun, 1991). While the N-terminal domain (residues 1 to 172) of colicin A contain the information for the interaction with OmpF and TolA, B, Q, R (Bénédetti *et al.*, 1991a), the truncated region 30-172 is no longer sufficient to confer this dependence to the hybrid. Results, previously reported, also provided evidence that a short deletion of 15 residues (residues 15 to 29) in the N-terminal region of colicin A prevented the translocation step (Bourdineaud *el al.*, 1989; Baty et *al.*, 1990). It was also observed that the "Ton B box" in itself, when substituted for amino acid residues 15 to 29 in the colicin A deletion mutant, was not sufficient to promote TonB-dependent import (Baty *et al.*, 1990).

Table I. The hybrid IbAA is imported through the TonB pathway

	OmpF	BtuB	TonB	TolB	TolQ	TolR	TolA
Colicin A	-	-	+	-	-	-	-
Ib AA	+	-	-	+	+	+	+

The tests were carrried out in triplicate on a lawn of *E.coli* mutant cells with 3 µl dilutions of colicin preparations.
+, active ; - , inactive.

Dynamics of colicin import

The transport of colicin through the outer membrane is a dynamic process that probably involves a cascade of events: interaction with a receptor, possible conformational changes, interactions with components of the translocation machinery... etc. To investigate these points, we have studied the kinetics of K^+ efflux from sensitive cells after the addition of colicin A. These kinetics are biphasic, there is first a lag time corresponding to the time needed for translocation of the toxin through the cell envelopeand then K^+ efflux. Denaturing the colicin A with urea, before adding it to the cells, did not affect the properties of the pore but decreased the lag time. After renaturation, the lag time was similar to that of the native colicin (Bénédetti *et al.*, 1991c). This suggests that the unfolding of colicin A accelerates its translocation. Thus, the rate-limiting step in translocation may be at least partial unfolding of the polypeptide chain. This is probably a general feature of colicins, and unfolding could be triggered by their interaction with receptors. Binding of the central domains of colicins (receptor-binding domain) may destabilize interdomain interactions thus initiating unfolding. In the case of colicin A, the N-terminal domain of colicin A would then interact with OmpF (Bourdineaud *et al.*, 1989; Fourel *et al.*, 1990) and then with TolA (Bénédetti *et al.*, 1991b) (Figure 1).

With respect to unfolding, it is striking that none of these colicins, so far sequenced, was found to contain a disulfide bridge despite the fact that they are secreted proteins. It thus seems that disulfide bridges have been counterselected because they may hamper or prevent import.

Like unfolding, the physical interaction of colicins with components of their translocation machinery is also a likely event. Since group B colicins contain a "Ton B-box" like their receptors and, since a physical interaction between their receptors and TonB has been demonstrated (Brewer *et al.*, 1990; Braun, 1991), group B colicins might interact with TonB at some stage during import of the polypeptide chain.

In the case of group A colicins, a strong *in vitro* interaction between TolA and colicins A and E1 has been demonstrated. The C-terminal region of TolA, which is necessary for colicin uptake, was also found to be necessary for colicin A and colicin El binding to occur. Furthermore, only the N-terminal domain of colicin A, which is involved in the translocation step, was found to bind to TolA (Bénédetti *et al.*, 1991b). These results suggest that the interactions might be part of the import process (Figure 1).

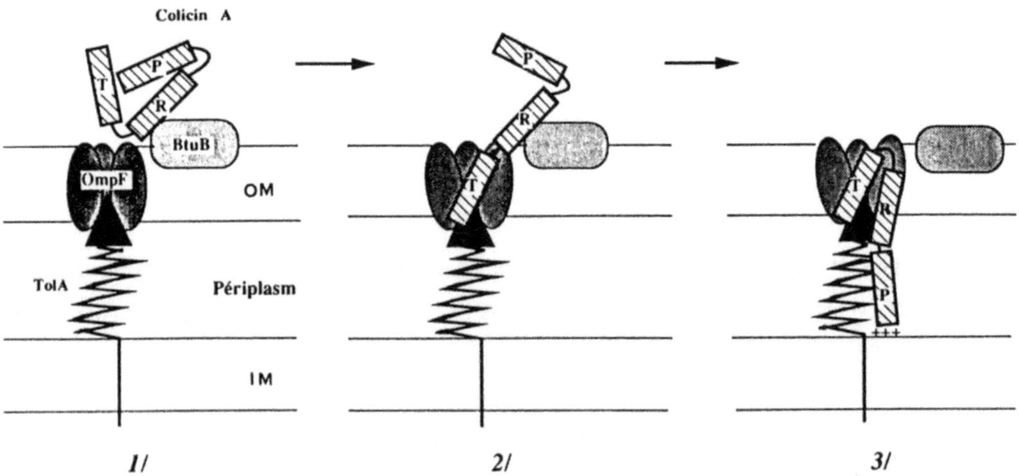

Figure 1. Hypothetical model of translocation of colicin A (Bénédetti, Ph.D thesis).
OM and IM designate the outer and inner membranes of bacteria, respectively. Colicin A is depicted with its 3 domains: **T** is the N-terminal domain; **R** is the receptor-binding domain and **P** is the pore-forming domain. The 3 domains of TolA are represented by: a straight line (N-terminal anchor), a zig-zag line (putative α-helix) and a black triangle (C-terminal domain), respectively. The localization of the latter in the OmpF lumen is hypothetical. The other Tol proteins (B, Q, R) are not represented since the possible interactions with TolA, with colicin A and between themselves have not yet been demonstrated.
Steps 1, 2 and 3 correspond to receptor binding, translocation across the outer membrane and insertion (and pore formation), respectively.

Colicins might span the cell envelope at their translocation sites

It has been reported long ago that there is an intermediate stage in colicin action at which the cells can be rescued through trypsin action ("trypsin rescue state") (Nomura & Nakamura, 1962; Levisohn *et al.*, 1967). Recent results from our laboratory yield new insights about this point (Bénédetti *et al.*, 1991c). The addition of trypsin, under conditions where it had access neither to the periplasmic space, nor to the cytoplasmic membrane, resulted in an immediate arrest of the potassium efflux induced by colicins A and B. The possibility that trypsin may act on a bacterial component required for colicin reception and/or translocation was ruled out. It is thus likely that the arrest of the efflux corresponds to a closing of the pores both with group A and group B colicins. This remote effect of trypsin suggests that part of the polypeptide chain of the colicins may still be in contact with the external medium, even when the pore has formed in the inner membrane (Bénédetti *et al.*, 1991c). It is thus likely that stage 1 is that at which the pore can be closed, and the main transport activities required for the vital functions of the cells can be restored; stage 2 may take place when this closing occurs too long after the addition of colicin, i.e. when the depletion of internal K^+ is total. This depletion, which is accompanied by the depolarization of the inner membrane (Gould & Cramer, 1977; Bourdineaud *et al.*, 1990), and a decrease in the internal ATP (Fields & Luria, 1969), may be responsible for the irreversible damage caused to the cells.

Conclusion

Although considerable progress in our understanding of colicin import has been achieved over recent years, the molecular mechanisms of translocation of their polypeptide chains still remains to be elucidated. Import of precursors into the mitochondria appears to occur at contact sites between the outer and inner membranes (Schleyer & Neupert, 1985). Whether these contact sites pre-exist or are formed during transport remains uncertain. Contact sites have also been identified in *E. coli* cells (Bayer, 1968; Ishidate *et al.*, 1986); the presence at these contact sites of proteins involved in the translocation of phage DNA and colicins have recently been demonstrated (Bourdineaud *et al.*, 1989; Guihard *et al.*, 1991). Further experiments are now being carried out to provide more insights into these mechanisms.

REFERENCES

Baty D, Frenette M, Lloubès R, Géli V, Howard SP, Pattus F, Lazdunski C (1988) Functional domains of colicin A. Mol Microbiol 2:807-811

Baty D, Pattus F, Parker M, Bénédetti H, Frenette M, Bourdineaud JP, Cavard D, Knibiehler M, Lazdunski C (1990) Uptake across the cell envelope and insertion into the inner membrane of channel forming colicins in *E. coli*. Biochimie 72:123-130

Bayer M (1968) Areas of adhesion between wall and membrane of *E. coli*. *J* Gen Microbiol 53:395-404
Bénédetti H (1991) Importation des colicines à travers l'enveloppe d'*Escherichia coli*. PhD Thesis, Universite d'Aix-Marseille I
Bénédetti H, Frenette M, Baty D, Lloubès R, Géli V, Lazdunski C (1989) Comparison of uptake systems for the entry of various BtuB group colicins into *Escherichia coli*. J Gen Microbiol 135:3413-3420
Bénédetti H, Frenette M, Baty D, Knibiehler M, Pattus F, Lazdunski C (1991a) Individual domains of colicins confer specificity in colicin uptake, in pore-properties and in immunity requirements. J Mol Biol 217:429-439
Bénédetti H, Lazdunski C, Lloubès R (1991b) Protein import into *Escherichia coli:* colicins A and El interact with a component of their translocation system. EMBO J 10:1989-1995
Bénédetti H, Lloubès R, Lazdunski C, Letellier L (1991c) Colicin A is translocated through the *Escherichia coli* envelope in an unfolded state and spans the cell envelope. EMBO J: in press
Bourdineaud JP, Howard SP, Lazdunski C (1989) Localization and assembly into the *Escherichia coli* envelope of a protein required for entry of colicin A. J Bacteriol 171:2458-2465
Bourdineaud JP, Boulanger P, Lazdunski C, Letellier L (1990) *In vivo* properties of colicin A: channel activity is voltage dependent but translocation may be voltage independent. Proc Natl Acad Sci USA 87:1037-1041
Braun V, Günter K, Hantke K (1991) Transport of iron across the outer membrane. Bio Metals 4: 1422
Brewer S, Tolley M, Trayer I, Barr G, Dorman C, Hannavy K, Higgins C, Evans J, Levine B, Wormald M (1990) Structure and function of X-Pro dipeptide repeats in the TonB Proteins of *Salmonella typhimuriwn* and *Escherichia coli*. J Mol Biol 216:883-895
Davies JK, Reeves P (1975a) Genetics of resistance to colicins in *Escherichia coli* K12: cross-resistance among colicins of group B. J Bacteriol 123:96-101
Davies JK, Reeves P (1975b) Genetics of resistance to colicins in *Escherichia coli* K12 cross-resistance among colicins of group A. J Bacteriol 123:102-117
De Graaf F, Planta R, Stouthamer A (1971) Effect of a bacteriocin produced by *Enterobacter cloacae* on protein biosynthesis. Biochim Biophys Acta 240:122-136
Fields K, Luria S (1969) Effects of colicins El and K on cellular metabolism. J Bacteriol 97:64- 77
Fourel D, Hikita C, Bolla JM, Mizushima S, Pagès JM (1990) Characterization of OmpF domains involved in *Escherichia coli* K-12 sensitivity to colicins A and N. J Bacteriol 172:3675-3680
Frenette M, Bénédetti H, Bemadac A, Baty D, Lazdunski C (1991) Construction, expression and release of hybrid colicins. J Mol Biol 217:421-428
Gould J, Cramer W (1977) Studies on the depolarization of the *Escherichia coli* cell membrane by colicin El. J Biol Chem 252:5491-5497
Graham A, Stocker B (1977) Genetics of sensitivity of *Salmonella* species to colicin M and bacteriophages T5, Tl and ES18. 1 Bacteriol 130:1214-1253
Guihard G, Boulanger P, Lettelier L (1991) Identification and localization in the contact sites between the outer and inner membrane of *E. coli* of a phage protein involved in the translocation of phage T5 DNA through the bacterial envelope. J Biol Chem: submitted for publication
Harkness R, Braun V (1989) Colicin M inhibits peptidoglycan biosynthesis by interferring with lipid carrier recycling. J Biol Chem 264:6177-6182
Ishidate K, Creeger E, Zuike J, Deb S, Glauner B, Mc Alister T, Rothfield L (1986) Isolation of differentiated membrane domains from *Escherichia coli* and *Salmonella typhimurium*, including a fraction containing attachment sites between the inner and outer membranes and the murein skeleton of the cell envelope. J Biol Chem 261:428-443
Lazdunski C, Baty D, Géli V, Cavard D, Morlon J, Lloubès R, Howard SP, Knibiehler M, Chartier M, Varenne S, Frenette M, Dasseux JL, Pattus F (1988) The membrane channel-forming colicin A: synthesis, secretion structure, action and immunity. Biochim Biophys Acta 947:445-464
Levisohn R, Konisky J, Nomura M (1967) Interaction of colicins with bacterial cells. J Bacteriol 96:811-821
Nikaido H, Vaara M (1985) Molecular basis of bacterial outer membrane permeability. Microbiol Rev 49:1-32

Nomura M, Nakamura M (1962) Reversibility of inhibition of nucleic acids and protein synthesis by colicin K. Biochem Biophys Res Commun 7:306-314

Ohno-Iwashita Y, Imahori K (1980) Assignment of the functional loci in colicin El and E3 molecules by the characterization of their proteolytic fragments. Biochemistry 19: 652-659

Ölschläger T, Braun V (1987) Sequence, expression and localization of the immunity protein to colicin M. J Bacteriol 169:4765-4769

Postle K (1990) TonB and the gram-negative dilemna. Mol Microbiol 4:2019-2025

Ross U, Harkness R, Braun V (1989) Assembly of colicin genes from a few DNA fragments. Nucleotide sequence of colicin D. Mol Microbiol 3:891-902

Schaller K, Nomura M (1976) Colicin E2 is a DNA endonuclease. Proc Natl Acad Sci USA 73, 3989-3993

Schleyer M, Neupert W (1985) Transport of proteins into mitochondria: translocational intermediates spanning contact sites between inner and outer membranes. Cell 43:339-350

Senior B, Holland I (1971) Effect of colicin E3 upon the 30S ribosomal subunit of *Escherichia coli.* Proc Natl Acad Sci USA 68:959-963

Webster R (1991) The *tol* gene products and the import of macromolecules into *Escherichia coli.* Mol Microbiol 5:1005-1011

IMPORT AND EXPORT OF COLICIN M

V. Braun, S. Gaisser, C. Glaser, R. Harkness[1], T. Ölschläger[2] and J. Mende
Mikrobiologie II
Universitat Tübingen
Auf der Morgenstelle 28
D-7400 Tübingen
Germany

Colicin M is unique among the colicins in that it does not belong to the colicins which form pores in the cytoplasmic membrane of target cells, or display nuclease activities (Braun *et al.*, 1974; Harkness & Ölschläger, 1991). Rather, colicin M inhibits murein biosynthesis (Schaller *et al.*, 1982) by interfering with the dephosphorylation of C_{55}-polyisoprenyl pyrophosphate, which leads to cell lysis (Harkness & Braun, 1989a). Colicin M also inhibits lipopolysaccharide O-antigen synthesis since this involves the same step of bactoprenyl phosphate regeneration (Harkness & Braun 1989b). However, prevention of O-antigen synthesis does not kill cells, as viable rough mutants are isolated.

Colicin M is composed of three functional domains

The unique mode of action of colicin M is reflected in its unique structure. The nucleases, for example colicin E3 and cloacin DF13, and the pore forming colicins, for example A, B, and N, show common characteristics as revealed in their hydropathy profile shown in Figure 1. The common uptake route for colicins B and D is also reflected in the very similar N-terminal and central structures of the polypeptide chains, which define the regions of receptor interaction and uptake across the outer membrane (Figure 1). These structural characteristics led us to propose an evolution of colicins in which DNA fragments encoding structural domains were recombined in various combinations leading to the present day colicins (Roos *et al.*, 1989). Colicin M has a unique amino acid sequence, and it is smaller (29,453 Dalton, in the following designated 29 kDa) than colicins usually are (60 to 90 kDa)(Kock *et al.*, 1987). However, it shares, with the other colicins, the same basic design of three structural and functional domains (Figure 2).

The N-terminal domain is required for colicin M uptake, the central domain for binding to the receptor, and the C-terminal domain for the toxic function. This has been demonstrated with a fragment obtained by partial cleavage with protease K in which a 3kDa fragment at the N-terminus has been deleted (Dreher *et al.*, 1985). The fragment was inactive against whole cells but was as active as the complete colicin when the normal transport through the outer membrane (see later) was bypassed by osmotic shock treatment of the cells. The fragment prevented killing by complete colicin M and it inhibited infection by

---

[1] Connaught Center of Biotechnology, Willowdale, Ontario, Canada. [2] Walter Reed Army Medical Center, Washington, D.C. U.S.A

NATO ASI Series, Vol. H 65
Bacteriocins, Microcins and Lantibiotics
Edited by R. James, C. Lazdunski and F. Pattus

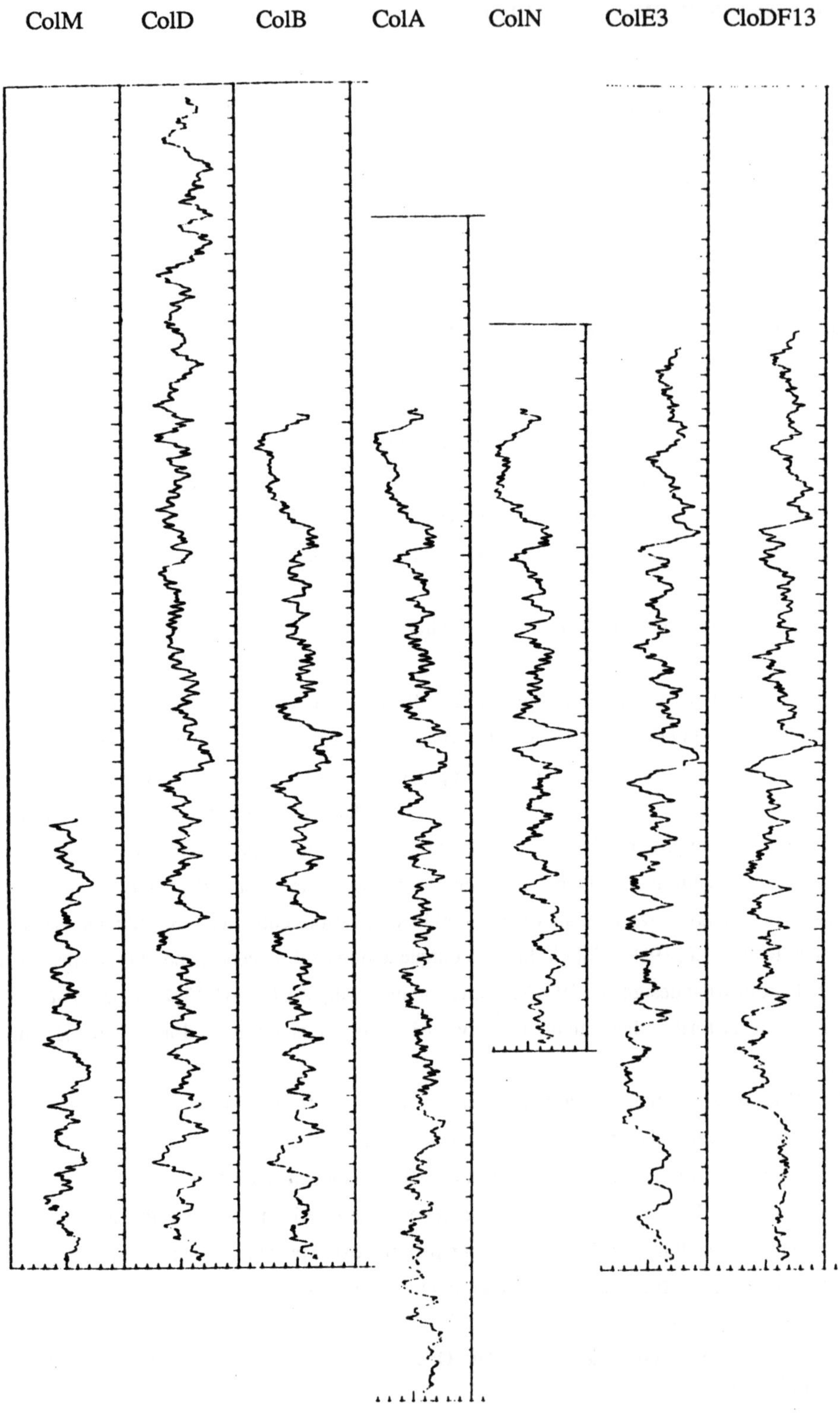
ColM
ColD
ColB
ColA
ColN
ColE3
CloDF13

Figure 1. Kyte-Doolittle hydropathy profile of (from above) colicins M, D, B, A, N, E3, and cloacin DF13. Colicin M does not fit any of these sequences. Colicins D and B are very similar in the N-terminal half reflecting the same uptake requirement. Colicin B differs entirely from D in the C-terminal region since it forms pores in the cytoplasmic membrane, in contrast to D which inhibits protein biosynthesis by an unknown mechanism. The pore-forming colicins B, A and N are very similar in the C-terminal domain which determines the pore-forming activity. Colicin E3 and cloacin DF13 are both ribosomal RNA nucleases and are very similar in their C-terminal structure. Although they bind to different receptors, their uptake seems to follow a similar mechanism because the N-terminal half of both colicins is also very similar (see Roos *et al.*, 1989 for further discussion).

phage T5 (which uses the same FhuA receptor). This demonstrates that the receptor binding domain of the truncated colicin was still intact (Figure 2). Specific release of the C-terminal lysine and arginine residues by carboxypeptidase B abolished colicin M killing activity. Localization of the activity domain to the C-terminal region was further shown by isolating point mutants in this region with strongly reduced (10^2 to 10^4 fold) colicin activities. These derivatives were still able to prevent the lethal action of wild type colicin M, probably by competition for the FhuA receptor.

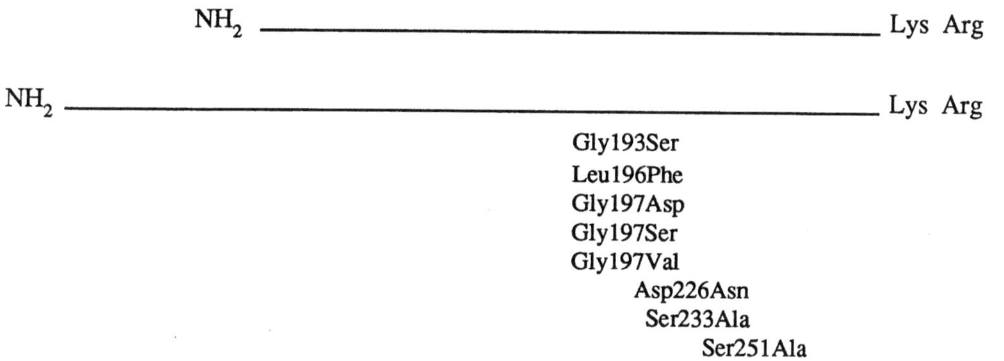

Figure 2. Comparison of colicin M and of an N-terminally truncated derivative emphasizing the functionally important N-terminal region for uptake, and the C-terminal Lys Arg residues for activity. The amino acid replacements which reduce colicin M activity 10^2 to 10^4 -fold were the result of a random chemical mutagenesis using hydroxylamine, with the exception of the replacements of the serine residues at positions 233 and 251 of the amino acid sequence by alanine, which have been done by site-directed mutagenesis.

Import of colicin M

The target site of colicin M resides in the cytoplasmic membrane since the carrier lipid is located there, and also the pyrophosphatase activity has been found in the membrane fraction (Siewert & Strominger, 1967). To gain access to its target, colicin M has to overcome the outer membrane barrier and enter the cytoplasmic membrane. The question arises how a protein is taken up across the outer membrane and into the cytoplasmic membrane. Both membranes do not allow the free passage of proteins. The outer membrane only becomes permeable for proteins, for example to hen egg white lysozyme and bovine trypsin, by treatment with EDTA, which releases divalent cations, mostly Mg^{2+}. Additional destruction of the murein layer strongly increases outer membrane permeability. Another method is the above mentioned osmotic shock treatment which renders the outer membrane temporarily permeable to proteins. For the uptake of colicin M into untreated cells, a special uptake mechanism must exist.

1. Time-dependence and reversibilitv of colicin M translocation through the outer membrane

Binding of colicin M to the FhuA outer membrane protein is required for colicin M uptake. A substantial amount of radioactively labeled colicin M was bound to mutants lacking FhuA, without affecting cells. The time colicin M remains at the receptor was determined through competition experiments with ferrichrome, an iron siderophore which is also transported via the FhuA receptor. Addition of ferrichrome 2 min after colicin M protected most of the cells. Half of the cells died when ferrichrome was added 4 min after colicin M, and 6 min later protection was no longer achieved, suggesting that during this period, colicin M has been translocated from the FhuA receptor into the interior of the cell. Similar results were obtained when colicin M was inactivated with trypsin, sodium dodecyl sulfate (SDS) or colicin M specific antibodies, all of which do not penetrate the outer membrane (Schaller *et al.*, 1981a). These agents inactivate colicin M only as long as it is bound to the receptor at the cell surface. The experimental conditions were chosen such that cells lysed 20 min after colicin M addition. Subsequent addition of trypsin, SDS and antibodies during the first 4 minutes of colicin M treatment rescued most of the cells from being killed. At later times, progressively more cells were killed indicating that after 4 minutes more and more colicin M has been taken up. Interestingly, colicin M in solution is largely resistant to trypsin, but when bound to the receptor it is trypsin sensitive, suggesting a conformational change of colicin M at the receptor.

In contrast to SDS, trypsin and antibodies, addition of 0.2 mM EGTA (glycol-bis(β-amino-ethyl ether)-N,N'-tetraacetate) rescued cells up to the point where they started to lyse. Inhibition by EGTA was abolished by adding 0.2 mM of Ca^{2+} ions (Figure 3). EGTA at the concentration used did not affect the growth rate, nor did it prevent binding of colicin M to the cells, or remove colicin M from the receptor. Under the influence of EGTA, colicin M again became susceptible to degradation by trypsin since trypsin prevented cell lysis when added before EGTA was removed, or was titrated out by adding excess Ca^{2+}. Apparently, Ca^{2+} was required to maintain colicin M at its target site in the cytoplasmic membrane, either by contributing to the binding of colicin M at the target site, or by being involved in the uptake of colicin M. The latter function would imply that colicin M is transported continously during its action. Otherwise it would leak out of the cytoplasmic membrane and reappear at the cell surface receptor where it again becomes susceptible to inactivation.

The reversible translocation of colicin M across the outer membrane by the antagonists EGTA and Ca^{2+} could also be shown with a temperature sensitive colicin M tolerant mutant. This rules out the

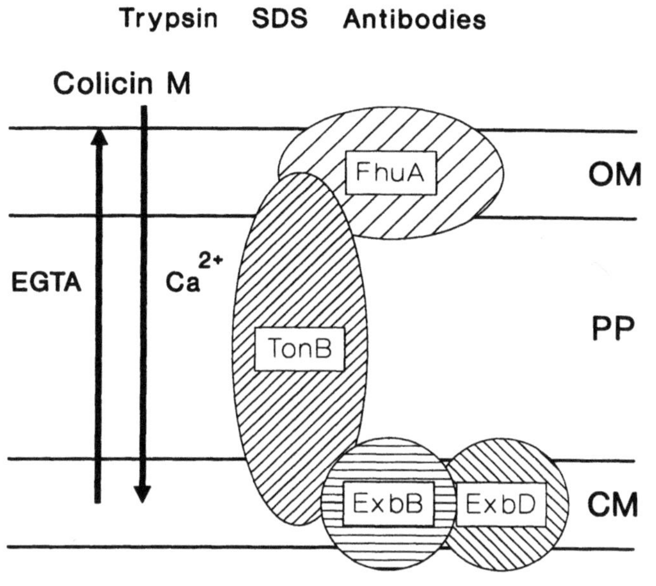

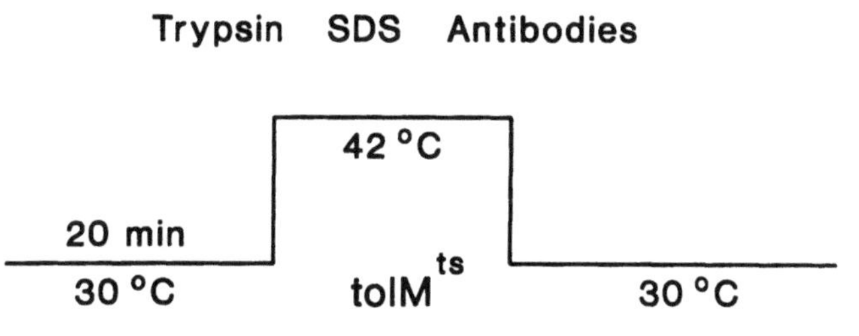

Figure 3. Reversible translocation of colicin M across the outer membrane of *E.coli*. Upper panel: Colicin M is first bound to the FhuA receptor from where it is taken up across the outer membrane (OM) in a process that requires the activity of the TonB and the ExbB and ExbD protein, of which regions have been localized in the cytoplasmic membrane (CM), and the periplasm (PP). The model suggests an interaction of the 4 proteins as revealed by stabilization of TonB by FhuA and ExbB, and of ExbD by ExbB. Ca^{2+} is required for uptake. Addition of EGTA renders transported colicin M again susceptible to inactivation by trypsin, SDS and colicin M-specific antibodies. Lower panel: description of the experiment in which colicin M was added to a temperature sensitive tolerant mutant (*tolM*^{ts}) at 30°C. After 30 min the cells started to lyse. If the temperature was raised to 42°C after 20 min, cells were not killed as long as the higher temperature was maintained. Lowering of the temperature to 30°C rendered cells sensitive to colicin M. At 42°C, but not at 30°C, colicin M was susceptible to inactivation by trypsin, SDS and antibodies.

possibility that EGTA, like EDTA, rendered the outer membrane permeable to trypsin, SDS and antibodies. The mutant was specifically resistant to colicin M at 42°C, while at 30°C it showed a 10 fold lower sensitivity than the parent strain (Schaller *et al.*, 1981b). Shifting cells from 30°C to 42°C in the presence of colicin M prevented lysis. Protection by temperature shift could be achieved during the entire incubation period up to the point where cells started to lyse. After shift-down to 30°C, cells lysed after a time which was somewhat shorter than the time required to obtain lysis during continous incubation at 30°C. During this procedure colicin M stayed cell-associated. After the shift-up experiment colicin M became susceptible to trypsin, SDS and antibodies at 42°C (Figure 3). In this experiment the temperature change had the same effect as EGTA in the previous experiments. It was concluded that under both conditions colicin M was reversibly translocated across the outer membrane.

2. Energy requirement of colicin M uptake

Colicin M adsorbed to energy-starved cells (treated with dinitrophenol) remained at the FhuA receptor, since trypsin treatment of colicin M treated cells prevented their killing as long as the cells were not energized. Progressively more cells were killed (20%, 99%, 99.9%) when trypsin was added 2, 4 and 8 min after reenergization of cells with glucose in the presence of oxygen (Braun *et al.*, 1980).

Another means of studying the energy dependence of colicin M uptake was through inhibition of phage T5 adsorption. Colicin M bound to the FhuA receptor inhibited adsorption of phage T5. In energy-starved cells, T5 inhibition was much stronger than in energized cells (Braun *et al.*, 1980). Since adsorption of phage T5 occurs independent on the energy state of the cells, the energy dependence must reflect the amount of colicin M at the receptor, which is higher in the unenergized cells than in the energized cells in which colicin M is transported through the outer membrane. Killing of cells is by suicide since the cells actively contribute to the uptake of the colicin.

3. TonB, ExbB and ExbD dependence of colicin M uptake across the outer membrane

Mutants in the *tonB* gene were completely resistant to a highly active colicin M solution with a dilution titer of 10^4 (Table 1).Colicin M inhibited adsorption of phage T5 which was most pronounced in *ton*B mutants, indicating that in *tonB* mutants colicin M remained at the FhuA receptor. T5 adsorption was not affected by TonB. Mutants in the *exb* locus, consisting of the *exbB* and *exbD* genes (Eick-Helmerich & Braun, 1989), showed a strongly decreased sensitivity (Table 1) in that a turbid growth inhibition zone was obtained. The requirement for TonB, and for ExbBD, could be bypassed by osmotic shock treatment (Braun *et al.*, 1980; Eick-Helmerich & Braun, 1989). These results imply that TonB and ExbBD are involved in outer membrane transport, and that they are not required for the insertion of colicin M into the cytoplasmic membrane. Since the target site of colicin M has not been localized within the cytoplasmic membrane, surface association of colicin M with the cytoplasmic membrane may suffice for inhibition of dephosphorylation.

Table 1. Colicin M sensitivity of *E.coli* K-12 and derivatives

Strain	Endpoint dilution of colicin M
GMl wild type	3 (4)
HEl *exbB(D)*	(2)
HEl pKE7 *exbB⁺D⁺*	3 (4)
HEl pHH5 *tolQ⁺R⁺*	2 (3)
HEl pCHl *tolQ⁺*	(2)
HEl pCH2 *tolR⁺*	r
HEl pCH3 *tolR⁺*	(2)
TPS13 *tolQ(R)*	3 (4)
HE2 *tolQ(R) exbB(D)*	r
HE2 pHH5 *tolQ⁺tolR⁺*	(3)
HE2 pCHl *tolQ⁺*	r
HE2 pCH2 *tolR⁺*	r
HE2 pCH3 *tolR⁺*	r
2259 *tolR*	2 (4)
HE10 *tolR exbB(D)*	r
HE10 pKE7 *exbB⁺D⁺*	3 (4)
HE10 pCHl *tolQ⁺*	r
HE10 pCH3 *tolR⁺*	r
HE5 *tolA*	3 (4)
HE7 *tolA exbB(D)*	3 (4)
HE6 *tolB*	3 (4)
HE8 *tolB exbB(D)*	2 (3)

HEl, HE2 and TPS13 are derivatives of GMl, HE10 is a derivative of 2259. Nomenclature: HEl is an *exb*B mutant and probably also an *exbD* mutant (therefore designated *exbB(D)*) of GMl. HEl pKE7 carries the chromosomal *exbB(D)* mutations and contains plasmid pKE7 that encodes wild type *exbBD* genes (see also Braun, 1989; Braun *et al.*, 1991). The last of 10-fold dilutions which resulted in a clear (for numbers in brackets, turbid) zones of growth inhibition are listed. For example 2 indicates that the colicin stock solution could be diluted 10^2 - fold to yield a clear zone on a lawn of bacteria seeded on tryptone-yeast agar plates. r, cells were totally resistant.

4. Amino acid sequences involved in colicin M uptake. The Tonb box of TonB dependent receptors and colicins

Colicin M contains the pentapeptide Glu Thr Leu Thr Val sequence from residue 2 to 6, and FhuA contains the very similar sequence Asp Thr Ile Thr Val at positions 7 to 11 of the mature polypeptide. Homologous sequences have been found at the N-terminus of all the other colicins which require TonB for uptake (colicins B, D, Ia, Ib), and at the N-terminal end of all TonB dependent receptors (FecA, FhuE,

Cir, FepA, IutA). This consensus sequence was designated TonB box to indicate its likely participation in the TonB dependent transport across the outer membrane. Evidence for a direct involvement of the TonB box came from point mutations in the TonB box of FhuA. The mutants were still fully active in the TonB independent infection by phage T5, but were impaired to various degrees, correlated with the type of mutation, in the TonB dependent uptake of colicin M, ferrichrome, albomycin, and in their sensitivity to the TonB dependent phages Tl and ϕ80 (Schöffler & Braun, 1989). The missing or reduced activities of the FhuA TonB box mutations could partially be restored by point mutations in the *tonB* gene, which were all located in the same triplet, causing the replacement of the glutamine residue number 160 by either leucine or lysine. Intensive searching for more *tonB* mutants suppressing FhuA TonB box mutations resulted in the isolation of only one additional *tonB* suppressor mutation (arginine 158 replaced by leucine), which however compensated only weakly one of the FhuA TonB box mutations (Günter & Braun, 1990). Suppression of a mutation in one gene by a mutation in another gene is taken as genetic evidence for interaction of the two gene products.

In addition, we have isolated TonB box mutations in colicin B and demonstrated that the mutants were unchanged in their binding to the FepA outer membrane receptor but deficient in uptake. Some mutations were partially suppressed by the *tonB* mutations described, indicating that the TonB box of colicin B is required for its uptake, and that it is involved in the TonB dependent step (Mende & Braun, 1990).

We have sequenced the *tonB* gene of *Serratia marcescens* in the expectation that comparison with the *E.coli tonB* sequence may reveal important regions of TonB activity. The sequence around the glutamine residue number 160 of *E.coli* (S R N Q P Q) differs somewhat in *S.marcescens* (S R A N P L). Accordingly, the cloned *S.marcescens tonB* gene did not fully complement all *E.coli* TonB dependent functions in an *E.coli tonB* mutant. Interestingly, complementation was weakest with those colicins in which the TonB box sequences showed the largest deviations from the TonB box consensus sequence (Gaisser & Braun, 1991). Transformants of *E.coli* carrying the *S. marcescens tonB* gene were resistant to colicins Ia and Ib (TonB box sequence E I M A V), and to two colicin B TonB box derivatives (wild type D T M V V, mutants D T M V A and D T M V G), and were 10 times less sensitive to colicin D (H S M V V). We take this as further evidence that the TonB box domains of the colicins and the receptors physically interact with the region defined by glutamine residue number 160 in the *E.coli* TonB protein.

5. Physical interactions between the TonB, FhuA and ExbB,ExbD proteins

Colicin resistance caused by mutations in the *fhuA*, *tonB*, *exbB* and *exbD* genes implied their involvement in the uptake of colicin M. To gain insight into the uptake mechanism, it was investigated whether these proteins physically interact. The concentration of the TonB, ExbB and ExbD proteins is too low, in the order of a few hundred molecules per cell, to study their interaction *in vitro* with the isolated proteins, or by *in situ* cross-linking. The TonB protein was shown to be functionally (Kadner & McElhaney, 1978) and physically unstable, exhibiting a half life of about 10 min at 42°C (Postle & Skare,

1988). We reasoned that interaction of the FhuA, ExbB and ExbD proteins with the TonB protein could delay TonB degradation and thus provide a tool to demonstrate such an interaction *in vivo*. This indeed has been found (Figure 4).

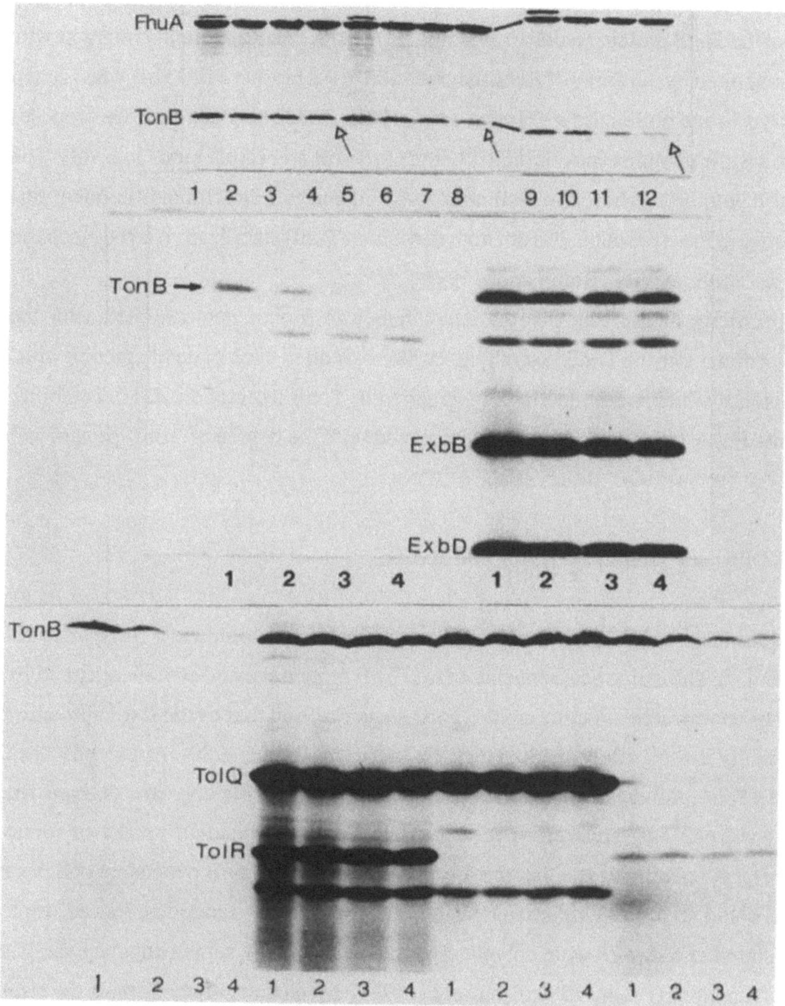

Figure 4. Stabilization of the TonB protein by the FhuA protein (upper panel), the ExbB protein (middle panel), and the TolQ protein (lower panel). The *tonB* gene was contained, together with either the *exbB*, *exbD*, or the *tolQ*, *tolR* genes, on the same plasmid. The genes were transcribed by the phage T7 RNA polymerase. The figure shows autoradiographs of (^{35}S)methionine labeled wild type TonB protein co-expressed with wild type FhuA protein (lanes 1 to 4), with a FhuA mutant carrying an isoleucine to proline substitution (I9P) at residue 9 in the TonB box of the mature protein (lanes 5 to 8), and with FhuA (V11D) (lanes 9 to 12). Cells were pulse-labeled and then chased for 1, 15, 30 and 60 min (lanes 1, 2, 3, 4) with a surplus of unlabeled methionine. The arrows point to the 60 min chase values. Middle and lower panel: in the left 4 lanes cells only expressed TonB which was degraded during the chase period. Lower panel: in the right 4 lanes TolR was weakly expressed, which we always observed in the absence of TolQ. TolR alone displayed no, or only a weak TonB stabilization.

The TonB protein was stabilized by the FhuA and by the ExbB protein. The TonB protein had to be overexpressed to be identified on autoradiographs. To observe a delay in TonB degradation, the FhuA and ExbB protein also had to be overexpressed. This was achieved by cloning these genes on a plasmid under the control of a phage T7 promoter, and by exclusive transcription via the very active T7 RNA polymerase. The ExbD protein itself was unstable in the absence of the ExbB protein, so that no stabilization of the TonB protein could be measured. However, ExbB apparently interacts with ExbD. This finding was supported by the delay of ExbD degradation in the presence of ExbB when spheroplasts were treated with trypsin and proteinase K (Fischer *et al.*, 1989). Based on these data, we propose for an active TonB protein a triple complex in which ExbB binds to TonB and ExbD binds to ExbB. The fact that all three proteins had to be expressed in similar amounts indicates a stoichiometric rather than a catalytic interaction between these proteins. It is not known whether TonB stabilization is the mechanism by which ExbB increases TonB activity (Braun *et al.*, 1991) .

The specificity of the FhuA-TonB interaction was further demonstrated with the previously described point mutants in the TonB box of FhuA which exhibited weak or no interaction with TonB. FhuA TonB box mutants, which were functionally impaired in TonB interaction, did not stabilize TonB. The extent of TonB-FhuA functional interaction correlated with the degree of TonB protein stabilization by FhuA (Günter & Braun, 1990; Braun *et al.*, 1991).

6. TolO and TolR can substitute ExbB and ExbD

Mutations in the *exbBD* genes render cells partially but not completely resistant to colicin M (and to colicins B and D). The amino acid sequence of the ExbB protein showed 26% identity and 79% similarity (amino acids of comparable structure) to the TolQ sequence, and that of the ExbD protein 25% identity and 70% similarity to the TolR sequence (Eick-Helmerich & Braun, 1989). Apparently, the ExbB/TolQ, and the ExbD/TolR proteins are evolutionarily related and presumably are derived from common ancestors. Since TolQ and TolR are involved in the TonB independent uptake of group A colicins (Webster, 1991), we examined whether the residual uptake of the TonB dependent colicins by *exbB* and *exbD* mutants could arise from a partial substitution by the TolQR functions. Indeed, triple mutants in *exbBD tolQ*, and in *exbBD tolR* were completely resistant (Table 1), supporting the hypothesis of partial substitution of functions (Braun, 1989; Braun *et al.*, 1991). Much more dramatic were the results of a study using the TonB dependent phages Tl and ϕ80. Irreversible binding of these phages to the FhuA receptor, leading to infection, requires an energized, TonB active cell. Mutants in *exbBD* or *tolQR* hardly reduced the plaque-forming units, but *exbBD tolQ* and *exbBD tolR* triple mutants were completely resistant. These mutants became sensitive to phages Tl and ϕ80 when they overproduced the TonB protein. This was taken as evidence that ExbBD and TolQR acted via TonB. Apparently, in the overproducing strain, some TonB was in an active conformation even in the absence of ExbBD. This low TonB activity restored only Tl and ϕ80 sensitivity, for which in principle one phage particle suffices to infect a cell. An *exbB(D)* mutant (HEl in Table l) became colicin M sensitive when the TolQR proteins were overproduced in a transformant carrying the *tolQR* genes on a multicopy plasmid (pHH5). This result is particularly interesting in regard to substitution of the ExbBD functions by the TolQR functions. Restoration of colicin sensitivity was less pronounced in the HE2 *tolQR exbBD* pHH5 transformant (Table 1). Transformation of the HEl and HE2 strains with plasmids encoding single *tolQ* or *tolR* genes did not increase colicin M sensitivity. Rather, the strains became resistant. Resistance apparently depended on the amounts of TolQ and TolR proteins present since pCH3 with a low TolR expression (data not shown) did not alter the colicin M sensitivity of HEl (Table 1). We assume that a balanced relationship between the amounts of TonB, ExbBD and TolQR proteins must exist - overproduction of only one of these proteins distorts the functional complex.

Physical interaction between the TolQ and the TonB proteins was demonstrated by stabilization of the TonB protein in the presence of overproduced TolQ (Figure 4). The TolR protein was more weakly expressed than the TolQ protein. TolR was barely detectable on autoradiographs in the absence of TolQ (Figure 4), suggesting that TolR may be stabilized by TolQ, like ExbD by ExbB. The low amounts of TolR, like ExbD, did not prevent degradation of TonB. These results clearly demonstrate the structural and functional similarity of the ExbB/TolQ and the ExbD/TolR proteins, respectively.

Although the TolA and TolB proteins encoded by the *tol* region do not seem to play a direct part in the TonB dependent uptake across the outer membrane of *exbB*$^+$*D*$^+$ cells, they influence TonB activity in certain genetic constructions. For example, colicin M sensitivity of an *exbB* (*D*) mutant was nearly normal when the mutant carried an additional mutation in *tolA* or *tolB* (strains HE7 and HE8, Table 1). These unexpected phenotypes further indicate the existence a functional as well as structural relationship between the Tol and the ExbBD proteins.

7. Molecular model of colicin M uptake

From the preceding results the following model of colicin M uptake is proposed. Colicin M first binds to the FhuA outer membrane receptor protein. It is then taken up in an energy-coupled process through the action of the TonB protein. The TonB protein is anchored via its N-terminal end in the cytoplasmic membrane and the rest of the molecule is exposed in the periplasmic space (Hannavy *et al.*, 1990; Postle & Skare, 1988). TonB binds to the FhuA protein and induces the release conformation of FhuA, which in turn results in the vectorial translocation of colicin M across the outer membrane. TonB itself assumes two conformations, an energized and an unenergized one. Energization of TonB takes place in the cytoplasmic membrane. Induction of the receptor release conformation consumes energy so that TonB dissociates from the FhuA receptor and has to be reenergized to induce the next round of colicin M uptake. Since both, FhuA and colicin M, contain a TonB box, the TonB protein most likely interacts with both, either sequentially (FhuA first) or simultaneously. TonB serves as a coupling device between the outer and the cytoplasmic membrane acting as an energy transducer from the cytoplasmic membrane to the outer membrane receptors. The ExbB protein, and in its absence the TolQ protein (with less efficiency than ExbB), binds to the TonB protein. They are either involved in the energy transduction from the cytoplasmic membrane to TonB, or in stabilization of TonB. The activity of the ExbB and TolQ proteins depends on the ExbD and the TolR proteins, respectively. Once transferred through the outer membrane, colicin M spontaneously interacts with its target. It is unknown how far it has to enter the cytoplasmic membrane in order to reach the target. The immunity protein (14 kDa), which renders cells immune to the lethal action of colicin M, has been localized in the cytoplasmic membrane. Degradation by trypsin left a 6.5 kDa fragment in spheroplasts suggesting exposure of a substantial portion of the immunity protein to the periplasmic space (Ölschläger *et al.*, 1991). Proteolytic degradation of the immunity protein was inhibited by addition of colicin M, which makes it likely that immunity is the result of colicin M binding to the immunity protein. The location of the immunity protein suggests that this occurs at the periplasmic side of the cytoplasmic membrane. It is inferred that colicin M does not deeply penetrate into the cytoplasmic membrane. In addition, it is hard to imagine how colicin M can reappear at the cell surface upon treatment with EGTA when most of the molecule has been inserted in the cytoplasmic membrane.

Specific export of colicin M

Colicin M is usually encoded together with colicin B on large self-transmissible plasmids. In the plasmids studied to some detail (pColBM-C1139 and pColBM-F166) transcription of the colicin M activity gene (*cma*) starts upstream of the colicin B activity gene (*cba*) (Figure 5). The immunity genes (*cmi* and *cbi*) are transcribed with opposite polarity (Figure 5). No lysis protein was encoded by the sequenced 6116 base pairs DNA fragment which includes large flanking regions in addition to the *cma cmi cba* and *cbi* genes (Ölschläger & Braun, 1987; Schramm *et al.*, 1987; Schramm *et al.*, 1988; Thumm *et al.*, 1988). The open reading frame of a small fragment (designated "cbl" in Figure 5) showed some resemblance to lysis proteins, but contained, for example, a tyrosine residue in place of a cysteine residue, which in all the other lysis lipoproteins forms the site of signal peptide cleavage and lipid modification.

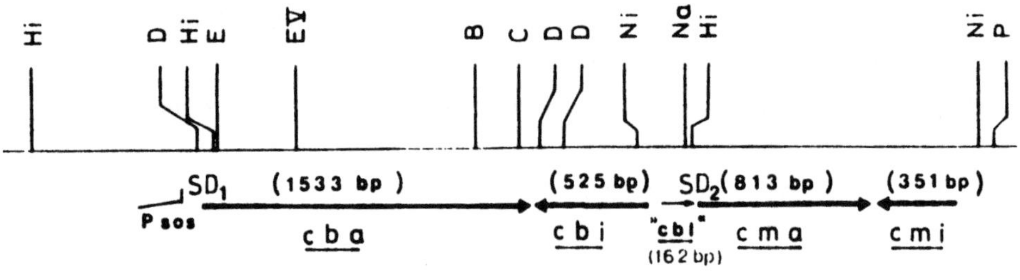

Figure 5. Arrangement of the genes encoding colicins B and M on large plasmids. P_{SOS} indicates the mitomycin-inducible SOS promoter; SDl and SD2 the ribosome binding sites; bp base pairs; Hi the cleavage site of the restriction enzyme *Hinc*II, D *Dra*I, E *Eco*RI, EV *Eco*RV, B *Bst*EII, C *Cla*I, Ni *Nsi*I, P *Pst*I; the arrows indicate the transcription polarity.

A Tn*1000* transposon insertion in "*cbl*" did not reduce secretion of colicin B (no colicin M was synthesized due to the strong polar effect on *cma* expression), showing that this region did not encode a release or lysis protein. Strains containing the cloned *cba cbi cma cmi* genes released only 1 to 10 % of the colicins into the culture medium (Figure 6).

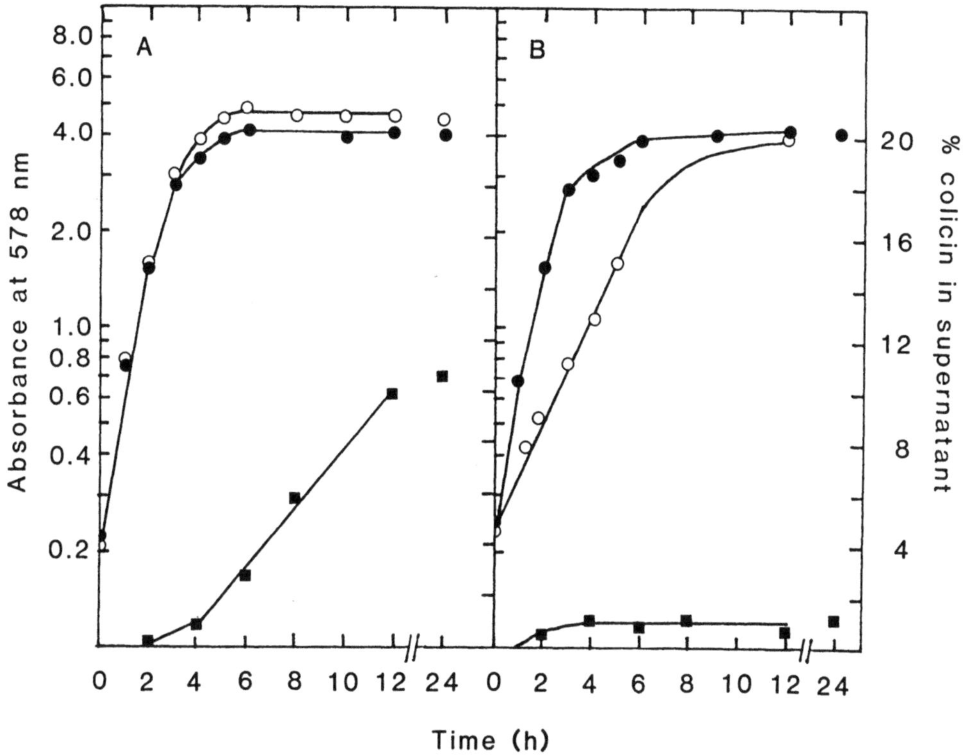

Figure 6. Growth and colicin release by *E.coli* strains pop2135 carrying the natural plasmid pColBM-C1139 (panel A) and pop2135 carrying cloned *cma* on pRTl and cloned *cmi* on pT03 (panel B). Colicin M synthesis was induced with mitomycin C (200 ng/ml), or by temperature upshift for the cloned genes which were under phage lambda cI_{857} control. Optical density of non-induced cultures (○); and induced cultures (●), extracellular colicin M (■) expressed as a percentage of total colicin.

The strain we have originally chosen for studying both colicins (Schaller *et al.*, 1981a; Ölschläger *et al.*, 1984) released 50 % of the intracellular colicins. It turned out that this unusually high secretion was the property of the *E.coli* strain C1139, and was not determined by the pColBM-C1139 plasmid. Regardless whether strains carried natural single copy pColBM plasmids, or the colicin genes cloned on small multicopy plasmids, the percentage of colicins released was between 1 and 10 %. Higher values were obtained with 0.1% Triton X-100 in the culture medium, which not only stabilized secreted colicin M (Schaller *et al.*, 1981a) but apparently also facilitated its release, as the concentration of colicin B in the culture supernatant also increased.

If there is no lysis gene how are these colicins released? Upon induction of colicin synthesis by mitomycin in cells containing natural plasmids, or by IPTG, or temperature upshift with plasmids carrying subcloned colicin genes, there was not much change in culture growth rate (Figure 6). Colicin M was the only, or predominant protein in the culture supernatant (Harkness & Braun, 1990), indicating that the cells did not become generally leaky for cytoplasmic proteins.

Study of the sensitivity of cells carrying cloned *cma* in the absence of *cmi* shed further light on colicin M release. Transformation of cells with *cma* resulted in only a few transformants. These turned out to be either mutated in *fhuA* or in *tonB*. In contrast, many transformants were obtained if *fhuA* or *tonB* mutants were transformed with *cma* plasmids. *cma* cloned downstream of the lambda PL promoter under the control of the temperature-sensitive lambda cI_{857} repressor could be transferred into *fhuA*$^+$ *tonB*$^+$ cells at 30°C. Cells lysed when *cma* expression was induced at 42°C (Harkness & Braun, 1990). However, they continued to grow at the same rate when trypsin was added prior to *cma* induction. These experiments demonstrated that colicin M did not kill cells as long as released colicin M did not enter the cells (in the transport negative *fhuA* and *tonB* mutants, or when it was destroyed by trypsin). Since all cells produced large amounts of colicin M at 42°C, these results cannot be explained by the assumption that only a few cells produced colicin M and lysed, whereas most of the cells were only killed when they took up the colicin. Apparently, only imported colicin M had access to the target in the cytoplasmic membrane while colicin M during export, where it also had to cross the cytoplasmic membrane, did not inhibit murein biosynthesis. These data also imply that exported colicin M did not appear in the periplasm, from where it could enter the cytoplasmic membrane from the same side as in normal uptake. We therefore have to postulate that the export and import of colicin M follow different routes. Export could occur through adhesion sites between the outer and the cytoplasmic membrane, in this way avoiding the periplasm. Since colicin M kills cells when brought into the periplasm by osmotic shock treatment, this rules out a specific FhuA and TonB requirement for insertion into the cytoplasmic membrane. The concentration of colicin M required to kill cells by osmotic shock is much higher than that required *in vivo*. This most probably reflects the low efficiency of colicin M transfer by osmotic shock compared to the energy-coupled transport, and does not point to an involvement of FhuA/TonB in the cytoplasmic membrane interaction. However, the possibility is not excluded that colicin M remains fixed with a small portion of the polypeptide chain to FhuA and by this means is oriented to correctly react with its cytoplasmic membrane target. This would occur in transport active cells.

Disturbance of the colicin M specific export route

Colicin M was fused with the signal sequence of the murein lipoprotein (Lpp) to examine whether colicin M is transported across the cytoplasmic membrane via the signal sequence-dependent general protein export pathway, and whether the colicin M polypeptide chain contained intrinsic information to subsequently cross the outer membrane (Ölschläger, 1991). The *lpp '-cma* gene fusion was preceded by the *lpp lac* tandem promoter, the lipoprotein signal sequence and a linker encoding 9 amino acids (Figure 7). The cysteine cleavage and modification site of the lipoprotein was converted to a lysine residue, which is recognized by the Lep signal peptidase. The *lpp '-cma* fusion contained two potential Shine-Dalgarno ribosome binding sites (marked SD in Figure 7). The colicin M immunity gene was cloned onto the same plasmid. After IPTG induction of *lpp '-cma* expression, 50% of the colicin M activity was in the culture supernatant. However, the high amount of colicin M activity in the supernatant was not the result of colicin M secretion, rather it was the result of cell lysis. Lysis depended on the activity of colicin M since a colicin M tolerant mutant did not lyse, and released only 1% of the total colicin M activity. Apparently, the Lpp'-Cma fusion protein disturbed the separate colicin M export pathway. Lpp'-Cma and Cma, or either one of them (two protein bands are seen on SDS gels) got access to the colicin M target site, inhibited murein biosynthesis and caused cell lysis. The fusion redirected the protein, taking Cma directly to, or in close proximity of, its target. Both the export and the import pathways could have been bypassed. Although the cells expressed the immunity gene, the amount of colicin M at the target site was so high that immunity broke down. This assumption is supported by the immunity breakdown which was achieved with high concentrations of colicin M when cells overexpressed the FhuA, TonB and ExbBD uptake components (Ölschläger *et al.*, 1991). Lysis of cells forming the Lpp'-Cma fusion protein further supports the existence of a colicin M export route which differs from the import route.

```
                                                          SD
                                                  TCTAGAGGGTATTAATA

ATGAAAGCTACTAAACTGGTACTGGGCATCCTGGGTTCTACTCTGCTGGCTGGGAAGGAATTAACT
MetLysAlaThrLysLeuValLeuGlyIleLeuGlySerThrLeuLeuAlaGlyLysGluLeuThr
   SD                                                 ↑
TATAAGGAGTTATGTATGGAAACCTTAACTCATGCACCATCACCATCAACT
TyrLysGluLeuCysMetGluThrLeuThrHisAlaProSerProSerThr
```

Figure 7. N-terminal nucleotide and amino acid sequence of the lipoprotein-colicin M (Lpp'-Cma) fusion protein. The underlined Met indicates the translation initiation site of Cma, the arrow the potential signal sequence cleavage site. SD, Shine-Dalgarno ribosome binding site.

Summary

The unique action of colicin M differentiates it from the other colicins (most, if not all, are pore formers or nucleases). In common with the other colicins, colicin M displays the three domain structure arrangement; one domain for receptor interaction, one for uptake, and one for colicin-specific activity (lethal function). This frequently found arrangement suggests that during their evolution, recombination events occurred which gave rise to common characteristicts among the colicins, sharing receptors, uptake routes, or having common modes of action. The binding of colicin M to its outer membrane receptor, FhuA, and subsequent cellular uptake were found to be a time dependent process requiring about 4 min. Differential protease sensitivity indicates that upon interacting with FhuA, colicin M undergoes a conformational change. Transport across the outer membrane is a reversible process. Colicin M uptake is energy dependent and requires a functional TonB protein. The TonB box consensus sequence found in FhuA and colicin M (as well as in all the other TonB dependent receptors and colicins) suggests a physical interaction between these proteins occurs. This has been demonstrated by TonB box mutations in FhuA which are compensated by mutations in TonB, and by the stabilization of TonB by FhuA. TonB activity requires the ExbB and ExbD proteins which reside, like TonB, in the cytoplasmic membrane. A triple complex between TonB, ExbB and ExbD is suggested by the stabilization of TonB and ExbD by ExbB. The function of ExbB and ExbD can partially be replaced by the TolQ and TolR proteins, which share sequence homologies with ExbB and ExbD, respectively. How colicin M leaves the producing cell is not known. Release is an inefficient process in which only about 10 % of the intracellular colicin finds its way to the external medium. The observation that cells releasing colicin M are not killed by self-produced colicin when they are lacking the immunity protein indicates the uptake route to be independent of and separated from the export route. The study of colicin M has led to a better understanding of the functions and interactions of proteins involved in the uptake of biopolymers across two membranes.

Acknowledgements

This work was supported by the Deutsche Forschungsgemeinschaft (SFB 76, Br330/8-2, Br330/9-2) and the Fonds der Chemischen Industrie.

References

Braun V (1989) The structurally related *exbB* and *tolQ* genes are interchangeable in conferring TonB-dependent colicin, bacteriophage, and albomycin sensitivity. J Bacteriol 171:6387-6390

Braun V, Frenz J, Hantke K, Schaller K (1980) Penetration of colicin M into cells of *Escherichia coli*. J Bacteriol 142:162-168

Braun V, Günter K, Hantke K (1991) Transport of iron across the outer membrane. Biol Metals 4:14-22

Braun V, Schaller K, Wabl MR (1974) Isolation, characterization, and action of colicin M. Antimicrob Agents Chemother 5:520-533

Dreher R, Braun V, Wittmann-Liebold B (1985) Functional domains of colicin M. Arch Microbiol 140: 343-346

Eick-Helmerich K, Braun V (1989) Import of biopolymers into *Escherichia coli:* nucleotide sequences of the *exbB* and *exbD* genes are homologous to those of the *tolQ* and *tolR* genes, respectively. J Bacteriol 171:5117-5126

Fischer E, Günter K, Braun V (1989) Involvement of *exbB* and *tonB* in transport across the outer membrane of *Escherichia coli*: phenotypic complementation of *exb* mutants by overexpressed *tonB* and physical stabilization of TonB by ExbB. J Bacteriol 171:5127-5134

Gaisser S, Braun V (1991) The *tonB* gene of *Serratia marcescens*: sequence, activity and partial complementation of *Escherichia coli tonB* mutants. Mol Microbiol 5:2777-2787

Günter K, Braun V (*1990) In vivo* evidence for FhuA outer membrane receptor interaction with the TonB inner membrane protein of *Escherichia coli*. FEBS Lett 274:85-88

Hannavy K, Barr GC, Dorman CJ, Adamson J, Mazengera LR, Gallagher MP, Evans JS, Levine BA, Trayer IP, Higgins CF (1990) TonB protein of *Salmonella typhimurium*, a model for signal transduction between membranes. J Mol Biol 216:897-910

Harkness RE, Braun V (1989a) Colicin M inhibits peptidoglycan biosynthesis by interfering with lipid carrier recycling. J Biol Chem 264:6177-6182

Harkness RE, Braun V (1989b) Inhibition of lipopolysaccharide O-antigen synthesis by colicin M. J Biol Chem 264:14716-14722

Harkness RE, Braun V (1990) Colicin M is only bactericidal when provided from outside the cell. Mol Gen Genet 222:37-40

Harkness RE, Ölschläger T (1991) The biology of colicin M. FEMS Microbiol Rev 88:27-42

Kadner RJ, McElhaney G (1978) Outer membrane-dependent transport systems in *Escherichia coli*: turnover of TonB function . J Bacteriol 134:1020-1029

Kock J, Ölschläger T, Kamp RM, Braun V (1987) Primary structure of colicin M, an inhibitor of murein biosynthesis. J Bacteriol 169:3358-3361

Mende J, Braun V (1990) Import-defective colicin B derivatives mutated in the TonB box. Mol Microbiol 4:1523-1533

Ölschläger T (1991) A colicin M derivative containing the lipoprotein signal sequence is secreted and renders the colicin M target accessible from inside the cells. Arch Microbiol 156:449-454

Ölschläger T, Braun V (1987) Sequence, expression, and localization of the immunity protein for colicin M. J Bacteriol 169:4765-4769

Ölschläger T, Schramm E, Braun V (1984) Cloning and expression of the activity and immunity genes of colicin B and M on ColBM plasmids. Mol Gen Genet 196:482-487

Ölschläger T, Turba A, Braun V (1991) Binding of the immunity protein inactivates colicin M. Mol Microbiol 5:1105-1111

Postle K, Skare JT (1988) *Escherichia coli* TonB protein is exported from the cytoplasm without proteolytic cleavage of its amino terminus. J Biol Chem 263:11000-11007

Roos U, Harkness RE, Braun V (1989) Assembly of colicin genes from a few DNA fragments. Nucleotide sequence of colicin D. Mol Microbiol 3:891-902

Schaller K, Dreher R, Braun V (1981a) Structural and functional properties of colicin M. J Bacteriol 146: 54-63

Schaller K, Krauel A, Braun V (1981b) Temperature-sensitive, colicin M-tolerant mutant of *Escherichia coli* . J Bacteriol 147: 135-139

Schaller K, Höltje JV, Braun V (1982) Colicin M is an inhibitor of murein biosynthesis. J Bacteriol 152: 994-1000

Schöffler H, Braun V (1989) Transport across the outer membrane of *Escherichia coli* K12 via the FhuA receptor is regulated by the TonB protein of the cytoplasmic membrane. Mol Gen Genet 217: 378-383

Schramm E, Mende J, Braun V, Kamp RM (1987) Nucleotide sequence of the colicin B activity gene *cba*: consensus pentapeptide among TonB-dependent colicins and receptors. J Bacteriol 169:3350-3357

Schramm E, Ölschläger T, Troger W, Braun V (1988) Sequence, expression and localization of the immunity protein for colicin B. Mol Gen Genet 211:176-182

Siewert G, Strominger JL (1967) Bacitracin: an inhibitor of the dephosphorylation of lipid pyrophosphate, an intermediate in biosynthesis of the peptidoglycan of bacterial cell walls. Proc Natl Acad Sci USA 57:767-773

Thumm G, Ölschläger T, Braun V (1988) Plasmid pColBM-C1139 does not encode a colicin lysis protein but contains sequences highly homologous to the D protein (resolvase) and the oriV region of the miniF plasmid. Plasmid 20:75-82

Webster RE (1991) The *tol* gene products and the import of macromolecules into *Escherichia coli* . Mol Microbiol 5:1005-1011

TolA: STRUCTURE, LOCATION AND ROLE IN THE UPTAKE OF COLICINS

Robert E. Webster and Sharyn K. Levengood
Department of Biochemistry
Duke University Medical Center
Durham, NC 27710
USA

Introduction

Colicins are bacterial protein toxins which are antagonistic to *Escherichia coli* and other related species. Their mode of action varies, ranging from the ability to form ion sensitive channels in bacteria to endonucleolytic cleavage of DNA and RNA or the lysis of susceptible bacteria. Genetic analysis has shown that the mechanism by which the colicins are able to reach their target of action involves a minimum of two steps. The toxin first must recognize a specific outer membrane receptor. Bacteria which lack a functional receptor cannot bind the colicin and have been called resistant. Other mutant bacteria have been isolated which are not affected by the colicin molecules but have a functional receptor to which the colicin can bind normally. Presumably these mutants, which have been defined as tolerant, are defective in a second step required for the toxin to be translocated to its site of action.

Analysis of the genetics of colicin tolerance in *Escherichia coli* (Davies & Reeves, 1975a,b) indicated that there are two pathways used for the translocation of the colicins into the bacteria subsequent to its binding to a specific outer membrane receptor. The group A colicins (including A, El, E2, E3, K, L, N and S4) appears to require the products of *tolA* and *B* for their activity. Other studies showed that the products of the *tolQ* and *R* also are necessary for the uptake of these colicins (reviewed in Webster, 1991). The role that the *tol* gene products play in the normal physiology of the bacteria is presently unknowm. It has been suggested that the function of the Tol proteins might be to maintain the integrity of the outer membrane since mutations in the *tol* genes result in increased sensitivity of bacteria to detergents and dyes (Nagel de Zwaig & Luria, 1967; Nomura & Witten, 1967; Sun & Webster, 1986), and increased release of periplasmic proteins into the medium (Lazzaroni *et al.* 1989).

A separate group of colicins, the group B colicins (including B, D, G, H, Ia, Ib, M, Q and V) was found to require the products of TonB for entry into the bacteria. TonB is part of a high-affinity active transport system required for the uptake of vitamin B_{12}, iron-siderophore complexes and the group B colicins (Kadner, 1990; Postle, 1990). Uptake of the group B colicins need the products of the *exbB* and *exbD* genes as well as TonB (Fischer *et al.*, 1989; Braun *et al.*, 1991). Mutations in the TonB system do not appear to affect the uptake of the group A colicins nor do they exhibit any change in their membrane properties.

NATO ASI Series, Vol. H 65
Bacteriocins, Microcins and Lantibiotics
Edited by R. James, C. Lazdunski and F. Pattus

Sun and Webster (1986) isolated *E.coli* mutations which are tolerant to infection by the filamentous single-stranded DNA bacteriophage. Most of these bacteriophage use the tip of a specific conjugative pilus as a receptor. These tolerant mutations mapped in or near *tolA* and had an increased sensitivity to various detergents. Using this phenotype, a 4.7 kilobase fragment of DNA was cloned into a multicopy plasmid which could complement both the bacteriophage tolerent mutations and the peviously described group A colicin tolerant *tolA* and *tolB* mutations. Sequence analysis showed this region to code for five open reading frames, two of which encoded the TolA and TolB proteins (Sun & Webster, 1987; Levengood & Webster, 1989). Genetic analysis showed that the products of two other open reading frame encoding regions, *tolQ* and *tolR*, were also needed for filamentous phage infection as well as import of the group A colicins. Analysis by a number of groups has shown that the products of *tolQR* and *A* are required for colicin El, E2, E3, A and K. These three gene products also are required for the DNA of filamentous single stranded DNA phage, that initiate infection by binding to the tip of a specific conjugative pilus, to be translocated into the bacteria (Bradley & Whelan, 1989; Webster, 1991). The TolB protein appears to be required for the action of all the group A colicins, except El, and is not needed for filamentous phage infection.

The *tolQRAB* cluster and its products

The *tolQRAB* genes are contiguous with each other at 17 min on the *Escherichia coli* map, having the order *tolQ*, *R*, *A* and *B* as shown in Figure 1 (Sun & Webster, 1986; 1987). The pattern of expression of these genes is not clear. Examination of the DNA sequence would suggest that potential promoters might be present upstream of *ORF1 tolA* and *tolB* leading to the transcription pattern depicted in Figure 1. This prediction would suggest the *ORF1*, *tolQ* and *tolR* might well constitute an operon. There are genetic experiments using bacteriophage Mu which suggest that *tolA* might also be part of this operon (Bernstein, 1973). It was observed that a number of Mu insertions in a region thought to contain *tolQR* (formerly named *tolP*) could not be complemented by an F'*tolQRAB* episome carrying a mutation in *tolA*, but could be complemented if the mutation was in *tolB*. Other experiments are inconsistant with these results (Sun & Webster, 1986; 1987). In this case, the effects of Tn*5* insertions on *tolQR* and *A* expression were analyzed when this gene cluster was present in a multicopy plasmid. None of the insertions appeared to affect the ability of distal genes to complement mutations in their respective chromosomal loci. It is possible that there is a weak secondary promoter for *tolA* which would allow enough TolA protein to be expressed for complementation to occur because of the multicopy nature of the plasmid. This would negate any polar effect of the upstream insertion mutation. Certainly more experimental evidence will be necessary to determine if the transcription pattern predicted by the sequence (Figure 1) accounts for the *in vivo* expression of these genes.

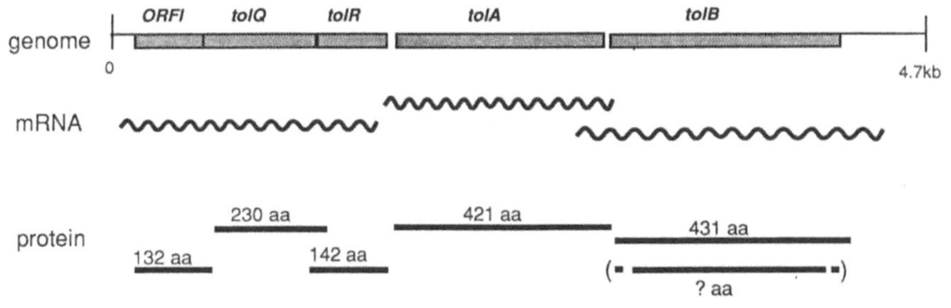

Figure 1. Schematic diagram of the *tolQRAB* locus and its products. Genome refers to the 4.7 kb fragment of DNA containing the *tolQRAB* gene cluster plus an additional sequenced open reading frame *(ORFI)*. The mRNA shown assume the presence of three potential promoters predicted from the DNA sequence. The proteins shown are those predicted from the DNA sequence and have been identified as radiolabeled products in maxi or mini cell systems (Sun & Webster, 1987; Levengood & Webster, 1989).

The protein products of the *tolQRAB* genes and *ORF1* have been identified and they correspond to the sizes predicted from the DNA sequence (Figure 1). TolQ is a 25.5 kDa membrane protein containing three potential membrane spanning regions (Sun & Webster, 1987). It appears to be associated with the inner membrane, or the attachment sites between the inner and outer membrane (Bourdineaud *et al.*, 1989). TolR is a 15.5 kDa protein of 142 amino acids and has one potential membrane spanning region (Sun & Webster, 1987). Preliminary results using anti-TolR antibody have demonstrated that it is tightly associated with the membrane fraction (Muller & Webster, unpublished results). TolA is a 44.2 kDa protein which is associated with the inner membrane and has the orientation described below (Levengood & Webster, 1989). Approximately 800 molecules of this protein are present in a normal growing *Escherichia coli* bacterium (Levengood *et al.*, 1991).

TolB encodes two proteins, although the sequence predicts only one open reading frame in this region (Levengood & Webster, 1989). The large 47.5 kDa molecule, TolB, contains 431 amino acids as predicted by the DNA sequence and is found tightly associated with the cytoplasmic membrane. The smaller protein, TolB*, has an Mr of 43 kDa and is located exclusively in the periplasm. Since there is only one open reading frame encoded in the region of the *tolB* gene, the TolB* would appear to be the result of post-translational processing, or an in-frame initiation within the same open reading frame.

The effect of mutations in each of these four *tol* genes on the action of various group A colicins has been tested in numerous laboratories (Davies & Reeves, 1975; Sun & Webster, 1987; Baty *el al.*, 1990; Braun *et al.*, 1991). Table I is an example of the results from one such test. A single amber or insertion mutation in *tolQ*, *R* or *A* results in tolerance against the effects of those colicins in group A (El, E2 and E3) but not for colicin D, a member of group B. The mutant in *tolB*, the nature of which is unknown, confers tolerance to E2 and E3 but not El. Tolerance exhibited by the *tolQR* or *B* mutations can be overcome by treating the bacteria with 10^3 to 10^4 times more colicin. Therefore, these mutants can be considered partially tolerant. However, *tolA* mutant bacteria are extremely tolerant, and are not affected by colicin solutions 10^6 more concentrated than the minimal colicin concentration needed to kill wild type bacteria. This strong dependence on the presence of TolA for translocation of colicins into the bacteria suggested that TolA was one of the major proteins required for the Tol import system. In order to better understand its function, our laboratory proceeded to further examine its structure and location within the bacterium.

Table I. Properties of *tolQRAB* mutants

	Bacterial Strain			
	tolQ$_{am}$	*tolR*::Cm	*tolA*::Tn*l0*	*tolB*593
Colicin El	T_p	T_p	T	S
Colicin E2	T_p	T_p	T	T_p
Colicin E3	T_p	T_p	T	T_p
Colicin D	S	S	S	S
fl phage	T	T	T	S

Tolerance to a specific colicin was determined by observing the zone of inhibition of growth caused by a 2 μl sample of varying concentrations of the colicin in question. **S** refers to sensitivity, **T** refers to tolerance against a $>10^{5\text{-}6}$ greater concentration of colicin than required to inhibit growth of the sensitive strain. **Tp** refers to partial tolerance phenotype. In this case only a $10^{3\text{-}4}$ greater concentration of colicin is needed for inhibition of growth.

Structure and location of *tolA* in the bacterium

Computer analysis of the amino acid sequence of TolA predicted that this protein might contain three domains (Levengood *et al.*, 1991). As depicted in Figure 2, the amino terminal 47 residues contain a 21 amino acid hydrophobic region capable of forming a membrane spanning helix. This is followed by a second domain of approximately 260 amino acids, of which 220 residues are predicted to form a continuous α-helix. It contains a high percentage of alanine, lysine and glutamic or aspartic acid residues, many of which form a repeat of the sequence Lys-Ala-Ala-Ala-(Glu/Asp). This helical domain is connected to domain 1 via a short sequence of 5 glycine residues. The remaining carboxy- terminal 120 amino acids are predicted to form the third domain. It is considered to be required for TolA function, as it contains the site of the non-functional *tolA*592 mutation (Webster & Levengood, 1989).

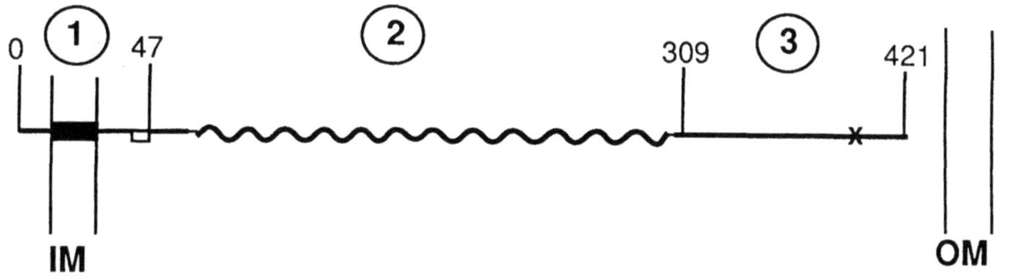

Figure 2. Schematic diagram of the domain structure and cellular localization of TolA. The amino acids delineating the proposed domains 1, 2 and 3 are shown. **IM** and **OM** refer to inner and outer membranes., respectively. The black box shown within the inner membrane delineates the 21 amino acid membrane spanning region. The small open box adjacent to amino acid 47 refers to the polyglycine region (see text). The **X** shown in the carboxyl terminal region of TolA is the position of the 4bp deletion found in *tolA*592.

In an attempt to test these predictions, a series of plasmids were constructed containing the regions encoding domain 2 (TolA-2), domains 2-3 (TolA-2,3) and domain 3 (TolA-3) under the control of an inducible promoter (Levengood *et al.*, 1991). These plasmids were introduced into bacteria and the respective proteins overproduced and purified. Circular dichroism analysis of each of these proteins showed the degree of helical structure to be greater than 90% in domain 2 as predicted. Domain 3 was estimated to contain approximately 10% helical structure by a similar analysis. TolA-2,3, containing both regions, had the expected intermediate helical content.

Antibodies were obtained to the purified TolA-2,3 protein and used to probe the location and topology of the intact TolA protein in wild type bacteria (Levengood *et al.*, 1991). Preliminary experiments showed that all of the protein was tightly associated with the membrane fraction. The membranes were separated into inner and outer membrane fractions and analyzed for the presence of TolA using the antibody to TolA-2,3 in the Western blot analysis shown in Figure 3. All of the TolA protein fractionated with the inner membrane, showing that TolA is an integral cytoplasmic membrane protein. The same experiment was repeated in a *tolA*::Tn*10* strain containing the plasmid encoding only the TolA-2,3 protein.

None of this protein, which contained only domains 2 and 3, was associated with the inner membrane. Rather, it was all found in the cytoplasmic fraction, strongly indicating that the signal for inserting and anchoring TolA in the membrane resides in domain 1. Presumably, TolA is anchored in the cytoplasmic membrane by the membrane spanning 21 residue region (amino acid 14-34) located in domain 1 as depicted in Figure 2.

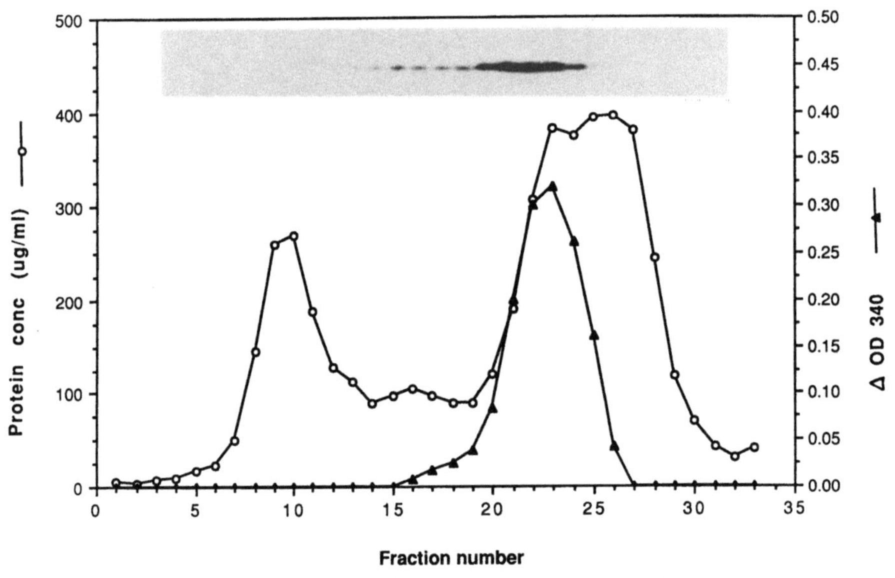

Figure 3. Localization of TolA protein to the cytoplasmic membrane of *Escherichia coli*. Cytoplasmic and outer membranes were isolated as described in Webster and Levengood (1987). The protein profile (0-0) of the final sucrose equilibrium gradient is shown. The NADH oxidase activity (▲-▲) marks the position of the cytoplasmic membrane. The outer membrane is located between samples 5 and 12. A sample of each fraction was subjected to SDS polyacrylamide gel electrophoresis followed by Western blot analysis probed with anti-TolA and I^{125} protein A. The resulting radioautogram is shown above the protein profile.

Proteolysis experiments were used to analyze the topology of TolA in the membrane (Levengood *et al.*, 1991). Bacteria which had been treated with EDTA to permeabilize the outer membrane, or bacteria which had been converted to spheroplasts by the action of lysozyme, were treated with low levels of trypsin. The proteins remaining after such treatment were separated by polyacrylamide gel electrophoresis and subjected to Western blot analysis using the antibody to TolA-2,3 and [I^{125}] protein A as a probe. No antibody reactive material was detected in the bacteria or spheroplasts following treatment with trypsin.

Control experiments showed that cytoplasmic proteins were unaffected by trypsin under these conditions (Levengood *et al.*, 1991). These data are consistent with domains 2 and 3 of TolA being exposed on the periplasmic side of the inner membrane, as depicted in Figure 2.

This localization of the TolA molecule would place the carboxyl-terminal domain 3 near the inner face of the outer membrane, where it could presumably interact with some of the membrane components. In this configuration, domain 3 would be connected to the membrane associated domain 1 by the long helical tether provided by domain 2. One function of TolA might be to connect the inner and outer membrane by an interaction of its carboxyl-terminal domain with the receptor-colicin complex in the outer membrane. This would imply some deformability in the helical structure as an uninterrupted helix of 220 amino acids would be approximately 32 nm long, a length which might be greater than the distance between the cytoplasmic and outer membrane (Kellenberger, 1990; Van Wielink & Duine, 1990). Perhaps the polyglycine region at the end of domain 1 (Figure 2) acts as a swivel for the domain 2 helical tether to allow part of domain 3 to interact with a specific colicin-receptor complex.

Such a hypothesis for the function of TolA implies that domain 3 might have a region which recognizes the colicin molecule. Recently, Bénédetti *et al.* (1991) have obtained *in vitro* evidence for such an interaction. They reported that colicins El and A were able to bind to TolA which had been immobilized to nitrocellulose by the standard Western blot procedure. This binding appeared to be specific for group A colicins since colicin B, a group B colicin, was not able to interact with TolA molecule. When a TolA molecule missing the carboxyl-terminal 44 amino acids was used in the same experiment, it was unable to bind the El or A colicins, indicating that domain 3 is required for this interaction.

If the *in vivo* function of TolA does require domain 3 to recognize the colicin, or at least the colicin-receptor complex, then it should be possible to inhibit this interaction by supplying an excess amount of soluble domain 3 to the periplasm. To test this, we have constructed a plasmid encoding a fusion protein composed of the signal sequence of the ribose binding protein (RBP) attached to the carboxyl-terminal 131 amino acids of TolA (Figure 4). Expression of this fusion gene results in the secretion of a 135 residue protein into the periplasm. This protein contains all of domain 3 as shown in Figure 3. Using this technique, it is possible to export up to 30,000 of these molecules into the periplasm of a wild type bacterial cell without interfering with cell growth. Preliminary evidence suggests that wild type bacteria containing this soluble domain 3 protein in the periplasm leak periplasmic proteins into the media and are sensitive only to higher levels of colicin than normal bacteria. It is possible that this phenotype is the result of the soluble domain 3 competing with the wild type TolA for recognition of outer membrane proteins and receptor colicin complexes. Although more studies must be done, these results are consistant with the above hypothesis for the role of domain 3 in TolA function.

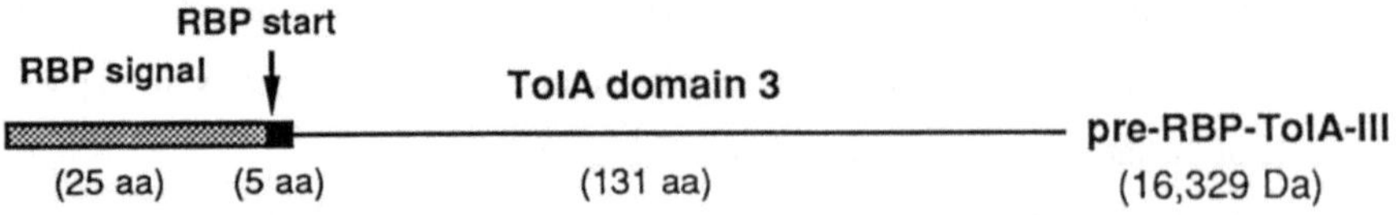

Figure 4. Schematic diagram of fusion protein between the signal sequence of the ribose binding protein (RBP) and TolA domain 3. Pre-RBP-TolA-III is composed of the 25 amino acid signal sequence of RBP, the first 5 amino acids of the mature RBP and the 131 carboxyl terminal amino acids of TolA.

The *tol* versus the *ton* system

The uptake of the group A colicins is totally independent of the TonB function suggesting a complete separation of the *tol* and *tonB* system. Yet there are several similarities in the components of both systems. As mentioned above, the products of *exbB* and *exbD* are necessary for TonB to be fully active in the uptake of the group A colicins. Sequence analysis of the *exbBD* genes show that the ExbB and D proteins are highly homologous in size and amino acid sequence to the TolQ and TolR proteins of the *tol* system (Eick-Helmerich & Braun, 1989). This similarity extends to the function of these genes. Mutations at the *exbBD* locus are partially tolerant to colicin B, D and M but become fully tolerant to these colicins in *exbBD, tolQ* double mutants. This suggests that TolQ can substitute to a certain extent for *exbB* function in TonB dependent colicin uptake (Braun, 1989). Similarly, the partial tolerance of *tolQ* and *R* mutations to the group A colicins (Table I) can overcome by the addition of an *exbB* mutation, again indicating an overlap in the function of ExbB and TolQ in both systems (Braun *et al.*, 1991).

There also are structural and topographical similarities between the TolA and TonB proteins. TonB protein is anchored in the cytoplasmic membrane probably via the hydrophobic amino terminus (Postle, 1990). The remainder of the molecule extends into the periplasmic space and contains a proline rich central region proposed to form an extended constrained conformation. The region of TonB in the periplasm is proposed to interact with specific outer membrane receptor-ligand (colicin) complexes causing a conformational change allowing the translocation of the ligand (colicin) across the outer membrane into the periplasm. Since the translocation of the colicin across the outer membrane requires energy, TonB is thought to be the energy transducer, coupling cytoplasmic membrane energy to the outer membrane translocation process. Here the analogy ends between TolA and TonB. One can detect no protein sequence homologies between the proteins, nor can any significant cross-reactivity be detected between TonB and

TolA and their respective antibodies (see Figure 5). These differences, coupled with the observation that mutations in the *tolA or tolB* genes confer the greatest amount of tolerance in their respective systems, suggest that TolA and TonB are the proteins which determine the specificity of the *tol* and *ton* systems. It is this specificity that explains why the colicins are divided so clearly into groups A and B.

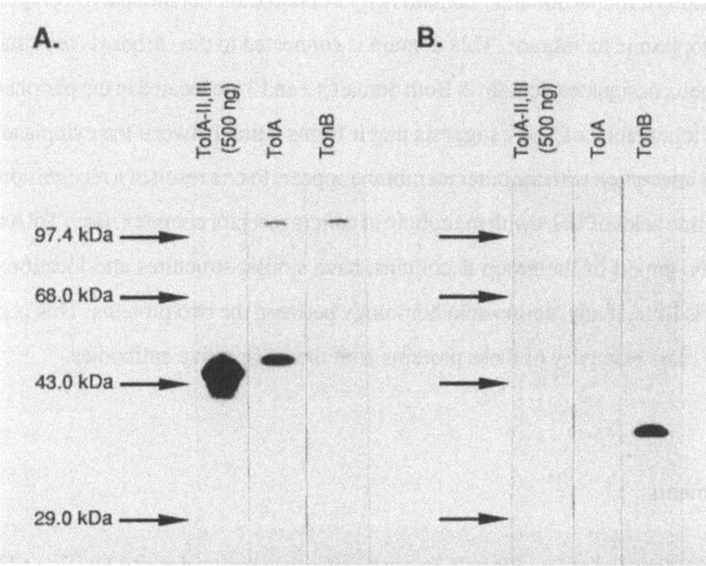

Figure 5. Test for cross reactivity between anti-TolA and TonB. Bacteria containing multicopy plasmids encoding either TolA or TonB were subjected to SDS-polyacrylamide gel electrophoresis and Western blotting analysis using $[I^{125}]$ protein A as a probe. A. Western blot using antibody to TolA-2,3. The lane designated TolA-II,III contains only the TolA-2,3 protein; the second lane is protein from bacteria containing the *tolA* plasmid; the third lane uses bacteria containing the *tonB* plasmid. B. This blot is the same as in A. except antibody to TonB was used.

The proteins of both the *tol* and the *ton* systems have been utilized by specific groups of colicins to gain entrance into the bacterium. Although the *ton* system can be assigned a function in the normal physiology of the bacterium, it has not been possible to do this for the proteins of the *tolQRAB* gene cluster. Determining the mechanism by which these Tol proteins are able to facilitate the tranlocation of the colicins, or of filamentous phage DNA, into the cell may help decipher the normal role of these proteins in the bacteria. Also, such knowledge can give a clearer picture of how macromolecules can be transported across biological membranes into a cell

Summary

The products of the *tolQRAB* cluster are required for efficient translocation of the group A colicins to their respective targets of action. The inner membrane location of the Tol proteins is consistent with their function, being required subsequent to the interaction of the colicin with its specific outer membrane receptor. TolA plays the major role in the translocation process and appears to be composed of three domains. The amino terminal domain 1 contains a hydrophobic membrane spanning region which anchors TolA to the cytoplasmic membrane. This domain is connected to the carboxyl- terminal domain 3 via a 220 residue α-helix designated domain 2. Both domains 2 and 3 are located in the periplasm. The structure and membrane topography of TolA suggests that it forms a link between the cytoplasmic and the outer membrane. The interaction with the outer membrane appears to be a result of a recognition of the carboxyl-terminal 120 amino acids of TolA with the colicin or colicin receptor complex. Both TolA and TonB, which is needed for the import of the group B colicins, have similar structures and locations in the bacteria. However there is little, if any, amino acid homology between the two proteins. This is confirmed by the absence of any cross reactivity of these proteins with their respective antibodies.

Acknowledgements

We thank Mary Jo Outlaw for help in processing the manuscript and Dr. Eva Clark and Michelle Muller for sharing some experimental results. Some of the work reported was supported by Public Health Service Grant GM-18305 from the National Institute of General Medical Sciences, U.S.A.

REFERENCES

Baty D, Pattus F, Parker M, Bénédetti H, Frenette M, Bourdineaud JP, Cavard D, Knibiehler M, Ladunski C (1990) Uptake across the cell envelope and insertions into the inner membrane of ion channel-forming colicins of *E. coli*. Biochimie 72:123-130

Bénédetti H, Lazdunski C, Lloubès R (1991) Protein import into *Escherichia coli:* colicins A and El interact with a component of their translocation system. EMBO J 10:1989-1995

Bernstein A (1973) The *E. coli* cell surface: on the genetic control of the *tolPAB* cluster. Mol Gen Genet 123:111-121

Bourdineaud JP, Howard SP, Lazdunski C (1989) Localization and assembly into the *Escherichia coli* envelope of a protein required for entry of colicin A. J Bacteriol 171:2458-2465

Bradley DE, Whelan J (1989) *Escherichia coli tolQ* mutants are resistant to filamentous bacteriophages that adsorb to the tips, not the shafts, of conjugative pili. J Gen Microbiol 135:1857-1863

Braun V (1989) The structurally related *exbB* and *tolQ* genes are interchangable in conferring *tol*B-dependent colicin, bacteriophage and albomycin sensitivity. J Bacteriol 171:6387-6390

Braun V, Günter K, Hantke K (1991) Transport of iron across the outer membrane. Biol Metals 4:14-22

Davies JK, Reeves P (1975a) Genetics of resistance to colicins in *Escherichia coli* K-12: cross-resistance among colicins of group B. J Bacteriol 123:96-101

Davies JK, Reeves P (1975b) Genetics of resistance to colicins in *Escherichia coli* K-12: cross-resistance among colicins of group A. J Bacteriol 123:102-117

Eick-Helmerich K, Braun V (1989) Import of biopolymers in *Escherichia coli:* nucleotide sequences of the *exbB* and *exbD* genes are homologous to those of *tolQ* and *tolR* genes respectively. J Bacteriol 171:5117-5126

Fischer E, Günter K, Braun V (1989) Involvement of ExbB and TonB in transport across the outer membrane of *Escherichia coli:* phenotype complementation of *exbB* mutants by overexpressed *tonB* and physical stabilization of TonB by ExbB. J Bacteriol 171:5127-5134

Kadner RJ (1990) Vitamin B_{12} transport in *Escherichia coli:* energy coupling between membranes. Mol Microbiol 4:2027-2033

Kellenberger E (1990) The "Bayer bridges" confronted with results from improved electron microscopy methods. Mol Microbiol 4:697-705

Lazzaroni J-C, Fognini-Lefebvre N, Portalier R (1989) Cloning of the *excC* and *excD* genes involved in the release of periplasmic proteins by *Escherichia coli* K12. Mol Gen Genet 218:460-464

Levengood SK, Beyer Jr WF, Webster RE (1991) TolA; a membrane protein involved in colicin uptake contains an extended helical region. Proc Natl Acad Sci USA 88:59395943

Levengood SK, Webster RE (1989) Nucleotide sequence of the *tolA* and *tolB* genes and localization of their products, components of a multistep translocation system in *Escherichia coli.* J Bacteriol 171:6600-6609

Nagel de Zuaig R, Luria S (1967) Genetics and physiology of colicin-tolerant mutants of *Escherichia coli.* J Bacteriol 94:1112-1123

Nomura M, Witten C (1967) Interactions of colicins with bacterial cells. III. Colicin-tolerant mutations in *Escherichia coli.* J Bacteriol 94:1093-1111

Postle K (1990) TonB and the gram-negative dilemna. Mol Microbiol 4:2019-2026

Sun T-P, Webster RE (1986) *fii,* a bacterial locus required for filamentous phage infection and its relation to colicin-tolerant *tolA* and *tolB*. J Bacteriol 165:107115

Sun T-P, Webster RE (1987) Nucleotide sequence of a gene cluster involved in entry of E colicins and single-stranded DNA of infecting filamentous phage in *Escherichia coli.* J Bacteriol 169:2667-2674

Van Wielink JE, Duine JA (1990) How big is the periplasmic space? Trends Biochem Sci 15:136-137

Webster RE (1991) The *tol* gene products and the import of macromolecules into *Escherichia coli.* Mol Microbiol 5:1005-1011

DOMAINS OF THE ESCHERICHIA COLI BtuB PROTEIN INVOLVED IN OUTER MEMBRANE ASSOCIATION AND INTERACTION WITH COLICIN TRANSLOCATION COMPONENTS

Robert J. Kadner, Bei-Yang Wei and Wolfgang Köster
Department of Microbiology, School of Medicine,
University of Virginia,
Charlottesville,
Virginia 22908
USA

The lethal action of colicins, which are the bacteriocins active on cells of *Escherichia coli*, can occur through several different mechanisms. Some colicins form a pore or channel in the bacterial cytoplasmic membrane, thereby leading to dissipation of the proton motive force and leakage of cytoplasmic ions and metabolites. Other colicins act in the cytoplasm as nucleases, leading to disruption of the genome or inactivation of ribosomes. Whatever their mechanism, all colicins require specific uptake systems to mediate their attachment to susceptible cells and their efficient translocation to their appropriate sites of action. The multiphasic uptake process is comprised of multiple gene products and provides an experimentally accessible model for a process of profound biological significance, namely, the passage of macromolecules across cellular membranes.

The colicin uptake process can be divided into two basic steps, adsorption to a specific receptor on the cell surface and translocation from the receptor across the cell membrane(s) to the target. Colicin receptors are responsible for the host range specificity of most bacteriocins, although there are examples where the host range is also determined by components of the translocation step. Receptor proteins have been genetically defined by the isolation of colicin resistant mutants, which are defective in the adsorption of the colicin to the cell. Mutations affecting a receptor protein confer resistance to only those colicins which use that receptor. Many of these receptors have been found to have beneficial physiological roles, usually as a component of a nutrient transport system, thereby accounting for the paradoxical presence of evolutionarily detrimental routes for entry of lethal agents like the bacteriocins or bacteriophages. For example, all of the E colicins use the BtuB protein in the outer membrane as receptor; this protein is an essential component of the high-affinity transport system for vitamin B_{12}. Colicins B and D adsorb to susceptible cells by means of the FepA protein, which normally allows the entry of ferric iron chelated to the siderophore produced by *E. coli* strains, enterobactin. Enterobactin transport is necessary for growth of the bacteria under aerobic conditions and neutral or alkaline pH.

The subsequent translocation steps involved in the transmission of a colicin from its receptor to its target have been defined genetically by the isolation of tolerant mutants. Tolerant mutants are defined as

NATO ASI Series, Vol. H 65
Bacteriocins, Microcins and Lantibiotics
Edited by R. James, C. Lazdunski and F. Pattus

those in which adsorption of the colicin is normal, but the colicin remains bound to its receptor and is unable to be transmitted to its lethal target. Davies & Reeves (1975a;1975b) showed that colicins could be divided into two separate groups, A and B, depending on whether they require the *tolA* or the *tonB* gene products. Killing by the group A colicins (including A, El, E2, E3, K, L, and N) is dependent on the TolA, B, C, Q, and R proteins. Only TolA and TolQ are required by all of the group A colicins, and the other colicins require different combinations of the other *tol* gene products. Many of the Tol proteins play some role in the integrity of the cell envelope, since many *tol* mutants exhibit altered membrane permeability, susceptibility to dyes and detergents, and release of periplasmic constituents. The group B colicins (which include B, D, G, H, I, M, and V) are Tol-independent, but require the TonB, ExbB, and ExbD proteins.

A major question in bacteriocin research concerns the mechanism of colicin translocation and the identification, localization, and role of the various gene products involved in that process. For example, is there a direct interaction between the translocation proteins and the receptor proteins or between them and the colicins? Evidence for some interaction has been obtained for TonB and some of the TonB-dependent receptors (Bell *et al.*, 1990). This possibility remains to be detected for the Tol system, although there is evidence for an interaction of the colicin and translocation components (Bénédetti *et al.*, 1991). The TolA protein has recently been shown to have a very unusual structure that might allow it to interact with other proteins in the periplasm or outer membrane (Levengood *et al.*, 1991).

The third ingredient for colicin uptake is metabolic energy, needed for the transmembrane movement of the colicin molecules. How this energy is coupled to protein movement is yet to be defined.

The BtuB system

Our research has centered on the BtuB protein and was recently reviewed (Kadner, 1990). This protein is a minor constituent of the outer membrane, present in about 200 copies per cell (Heller *et al.*, 1985). The BtuB polypeptide is synthesized with a 20-amino acid amino-terminal signal sequence which is cleaved off during passage of the protein across the cytoplasmic membrane, on its way to the outer membrane (Heller & Kadner, 1985). The mature portion of this protein is 594 amino acids long and, like most other outer membrane proteins, is likely to be very rich in β-sheet conformations. Its involvement in several different transport processes makes it a very interesting model for studies of nutrient and macromolecular transport across the outer membrane. These transport processes and the other components that participate are presented in very schematic manner in Figure 1. Our current goals are to define the domain structure and protein motifs of BtuB that contribute to ligand binding, energy coupling, outer membrane localization, and interaction with other outer membrane components.

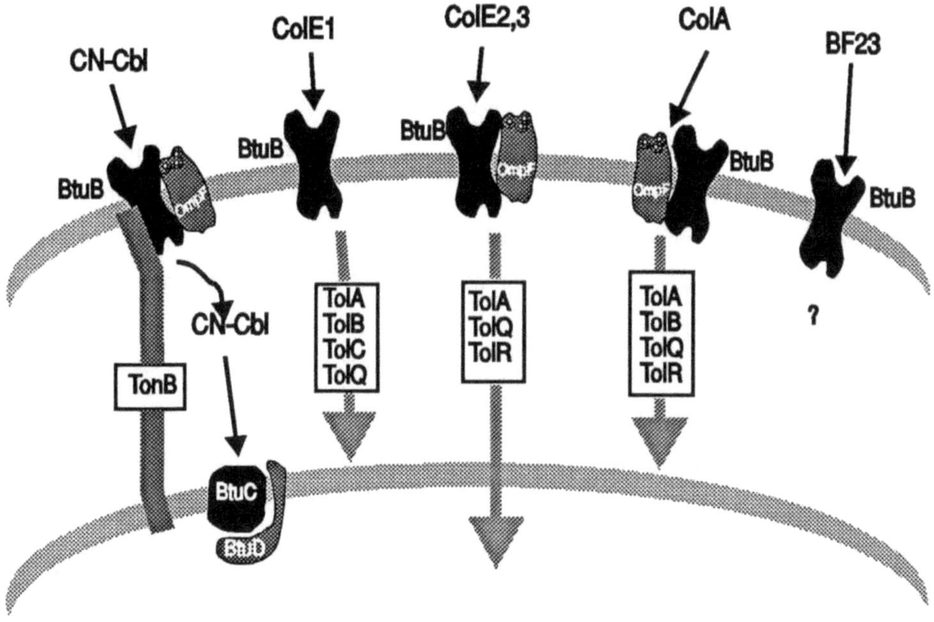

Figure 1. Schematic diagram of the BtuB-dependent transport activities. The outer and cytoplasmic membranes are shown with the protein components needed for the uptake of the various transport substrates indicated. The cellular location of the Tol proteins and the existence of defined complexes is not intended.

The physiological substrate for BtuB transport function is vitamin B_{12} and other corrinoid compounds. These are dispensable micronutrients that are required for growth only under certain conditions (reviewed in Jeter *et al.*, 1987). Adenosyl-cobalamin is the cofactor for the enzyme, ethanolamine ammonia lyase, whose action is needed for utilization of ethanolamine as carbon or nitrogen source. Adenosyl-cobalamin also appears to be a cofactor in the terminal step of synthesis of the nucleotide, queuosine, present in the anticodon loop of certain tRNAs. Methylcobalamin is a cofactor for the MetH isoenzyme which carries out the terminal step in methionine biosynthesis. Thus, *metE* mutants, which are defective in the cobalamin-independent isoenzyme for this step, require either methionine or a source of cobalamin. A *metE* strain of *E. coli* grows well in the absence of methionine with vitamin B_{12} concentrations as low as 1 nM. A *btuB* mutation that eliminates the BtuB protein in the outer membrane reduces cobalamin transport drastically, and maximal growth of these mutants requires the presence of 1 to 5 μM vitamin B_{12} (Bassford & Kadner, 1977). This strong dependence on the presence of an outer membrane protein arises because cobalamins are too large and are normally too scarce in the environment to be able to enter efficiently through the OmpF/C porin diffusion channels across the outer membrane.

BtuB is also the receptor for the E colicins (El to E9), which are nine different proteins having different mechanisms of cell killing. BtuB also participates with porin OmpF to form the high-affinity receptor for colicin A (Chai *et al.*, 1982; Bénédetti *et al.*, 1991). Finally, BtuB is the receptor for adsorption of bacteriophage BF23, a close relative of T5.

The entry of each of these classes of transport substrates differs with respect to the requirement for the translocation proteins involved in the subsequent uptake steps. Cobalamin transport is dependent on the *tonB* gene product; the *exbBD* gene products are somewhat required for transport. Entry of the colicins is independent of TonB, but does require the action of the *tolABCQR* and *ompF* gene products in varying combinations depending on the identity of the colicin (Davies & Reeves, 1975; Sun & Webster, 1987; Bénédetti et al., 1989). Even though TonB is not required for entry of the colicins that use BtuB, the fact that cobalamin entry requires TonB suggests that BtuB may provide a model for TonB action in uptake of the group B colicins.

Interaction of BtuB and TonB

Genetic study of the interaction of BtuB and TonB was stimulated by comparison of the deduced amino acid sequences of several TonB-dependent outer-membrane transport proteins, including BtuB (Heller & Kadner, 1985), FhuA [receptor for ferrichrome, colicin M, and phages Tl, T5 and ϕ80](Coulton *et al.*, 1986), FepA [receptor for ferric-enterobactin and colicins B and D](Lundrigan & Kadner, 1986), IutA [receptor for aerobactin and cloacin DF13](Krone *et al.*, 1985), Cir [receptor for ferric-2,3-dihydroxybenzoate and colicins I] (Nau & Konisky, 1989), and FecA [receptor for ferric citrate](Pressler *et al.*, 1988). All of these proteins share segments of reasonably conserved sequence, interspersed with much larger variable regions. It seems likely that some of these conserved regions might be involved in interaction with TonB and the variable regions might contribute to substrate binding.

The first indication for this possibility came when Knut Heller showed that a mutation in the first of these conserved segments in BtuB, very near the amino terminus of the mature protein, conferred a phenotype like that expected for an inability to interact with TonB (Heller & Kadner, 1985; Bassford & Kadner, 1977). This mutant, termed L8P because it was a substitution of a Pro for the Leu normally present at position 8 of the mature polypeptide, showed normal binding of all of its ligands and full uptake of the phage and colicins, but was completely defective in active transport of vitamin B_{12}. This is exactly the

Table 1. Mutations in the TonB box of BtuB affect utilization of vitamin B_{12}

Sequence								Growth on 5 nM CN-Cbl	Transport (% WT)
6	7	8	9	10	11	12	13		
asp	thr	leu	val	val	thr	ala	asn	+	100
GLU	thr	leu	val	val	thr	ala	asn	+	94
GLY	thr	leu	val	val	thr	ala	asn	+	100
asp	ALA	leu	val	val	thr	ala	asn	+	75
asp	ASN	leu	val	val	thr	ala	asn	+	63
asp	LEU	leu	val	val	thr	ala	asn	+	88
asp	PRO	leu	val	val	thr	ala	asn	+	110
asp	SER	leu	val	val	thr	ala	asn	+	127
asp	thr	ARG	val	val	thr	ala	asn	+	62
asp	thr	GLN	val	val	thr	ala	asn	+	98
asp	thr	GLU	val	val	thr	ala	asn	+	106
asp	thr	ILE	val	val	thr	ala	asn	+	122
asp	thr	PRO	val	val	thr	ala	asn	-	<1
asp	thr	leu	ALA	val	thr	ala	asn	+	144
asp	thr	leu	CYS	val	thr	ala	asn	+	54
asp	thr	leu	GLY	val	thr	ala	asn	+/-	23
asp	thr	leu	ILE	val	thr	ala	asn	+	81
asp	thr	leu	LEU	val	thr	ala	asn	+	82
asp	thr	leu	PRO	val	thr	ala	asn	+/-	30
asp	thr	leu	THR	val	thr	ala	asn	+	104
asp	thr	leu	val	ALA	thr	ala	asn	+	91
asp	thr	leu	val	GLY	thr	ala	asn	-	<1
asp	thr	leu	val	ILE	thr	ala	asn	+	98
asp	thr	leu	val	PRO	thr	ala	asn	-	
asp	thr	leu	val	ASP	thr	ala	asn	+	113
asp	thr	leu	val	val	ASN	ala	asn	+	48
asp	thr	leu	val	val	ILE	ala	asn	+	102
asp	thr	leu	val	val	thr	ASP	asn	+	89
asp	thr	leu	val	val	thr	GLY	asn	+	88
asp	thr	leu	val	val	thr	THR	asn	+	101
asp	thr	leu	val	val	thr	VAL	asn	+	104
asp	thr	leu	val	val	thr	ala	THR	+	55
asp	thr	leu	val	val	thr	ala	TYR	+	82

The wild-type amino acid residues of the mature portion of BtuB from 6 to 13 are shown in lower case, and the single amino acid substitutions in upper case. The growth with vitamin B_{12} in place of methionine and the relative rate of vitamin B_{12} transport of strains carrying the mutations in single copy are indicated in the two columns.

phenotype of a *tonB* mutant without the effects on the iron transport proteins. Heller subsequently found that this deficiency could be suppressed by mutations in which the Gln residue at position 165 in *tonB* was changed to Leu or Lys (Heller *et al.*, 1988).

Subsequently, that conserved region, which was termed the TonB box by Volkmar Braun, was subjected to site-directed mutagenesis to expand our understanding of the role of individual amino acid residues in that region in the function of the respective proteins. Over forty separate amino acid substitutions have been isolated in this region in the BtuB (Gudmundsdottir *et al.*, 1989), FhuA (Schoffler & Braun, 1989), and Cir (Bell *et al.*, 1990) proteins. Table 1 summarizes results obtained with the mutations affecting BtuB (taken from Gudmundsdottir *et al.*, 1989). Each residue could be changed to another residue of quite different chemical type without serious effect on vitamin B12 transport activity. Only mutations that substituted a Gly or Pro at residues 8 or 10 had a major effect on function, although some other substitutions affected activity to a lesser degree. We concluded that specific amino acid side chains were not recognized in this region of the protein, although the conformation of the polypeptide backbone was important.

Insights into protein function can often be achieved by genetic analysis of revertant or suppressor mutations that restore function to a mutated form of the protein. Suppressor mutations in *btuB* were obtained, in which the presence of the Pro at position 8 or the Gly at position 10 could be tolerated if Ala at 12 was changed to Val or Thr-7 was changed to Ile, respectively (Gudmundsdottir *et al.*, 1989). Although the structural consequences of these changes are not apparent, it is clear that the distortions induced by the first mutations could be compensated by certain specific additional changes in the same region of the protein.

Suppressor mutations that restored some function to the TonB box mutants were also found which had changes unlinked to the *btu*B gene. These were found to occur in plasmid-borne copies of the *tonB* gene. Consistent with the earlier work of Heller *et al*, every one of these mutations affected the same codon in *tonB*, and changes of Gln 165 to Leu, Lys, or Pro were obtained (Bell *et al.*, 1990). The isolation of such second-site suppressors may indicate that the two gene products physically interact during their function. One important test for an interaction is whether suppression displays allele specificity, i.e., whether different amino acid substitutions in one protein respond preferentially to different amino acid substitutions in the other protein. This was found to be the case. As indicated in Table 2, the three *btuB* mutations, L8P, V10G and V10P, responded very differently when combined with the various *ton*B alleles, Q165L Q165K or Q165P.

A very interesting result arose from comparison of the effects of the *ton*B suppressor mutations on the V10G mutation in *btuB* with the same V10G substitution in the *cir* gene for the colicin I receptor. This mutation was isolated by Cynthia Nau and Jordan Konisky. The pattern of suppression was markedly

Table 2. Allele-specific suppression on TonB box mutations by mutations in *tonB*.

tonB allele	Growth on CN-Cbl *btuB* allele			Response to Col Ia *cir* allele
	L8P	V10G	V10P	V10G
WT	-	-	-	R
Q165L	++	++	++	S
Q165K	++	++	-	R
Q165P	+/-	++	-	R

different for the two genes (Bell *et al.*, 1990). We concluded that the interaction of the TonB protein was not with the TonB box region of the receptor by itself, but with that portion in combination with another portion(s). This possibility makes some sense in light of the strong sequence conservation of this region but the lack of requirement for any particular amino acid side chain.

This genetic evidence for a direct interaction between the TonB-dependent transporters and TonB has been supported by biochemical studies from Higgins' group and collaborators, which indicated that a portion of the TonB protein binds to the FhuA protein (Hannavy *et al.*, 1990). It remains to be resolved which parts of TonB are involved in these processes, since the Gln-165 site identified in the genetic studies was not included in the fragment used for the biochemical studies. Nonetheless, there is growing evidence for a direct interaction of TonB protein with its cognate transporters.

BtuB-PhoA Fusions

Information about the cellular localization and topological orientation of secreted and membrane proteins has been obtained through analysis of hybrid proteins in which the mature portion of alkaline phosphatase, PhoA, is linked to various lengths of the amino terminal portion of a target protein. PhoA fusions to periplasmic proteins all have similar levels of alkaline phosphatase activity and are located in the periplasmic space, as long as the signal sequence of the target protein was included to provide the

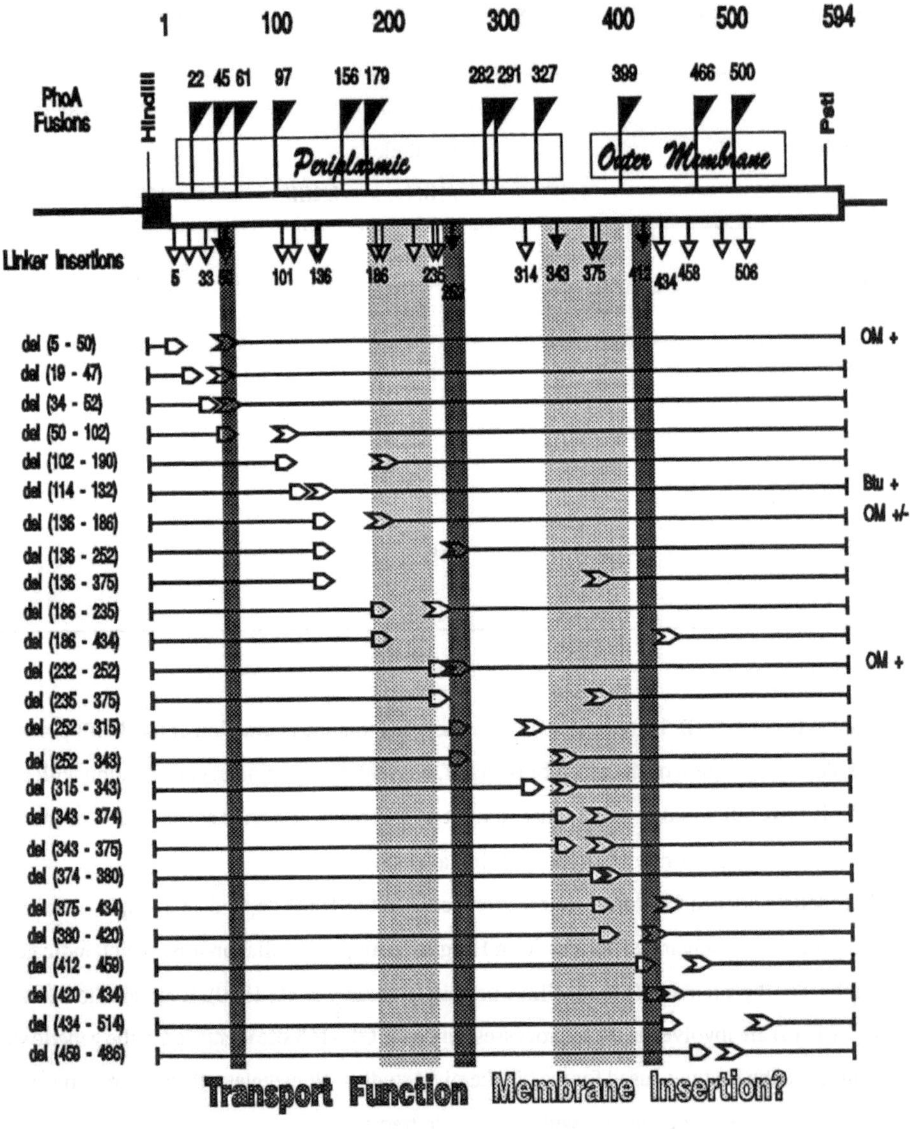

Figure 2. Summary of mutations in the *btuB* gene. The cordinates along the top indicate the amino acid residues in the mature portion of the polypeptide. PhoA fusions are indicated along the top of the linear representation of BtuB; the position of linker insertions are shown below. The genetic content of the *btuB* deletion mutations are shown. The same endpoints were used for construction of duplications of the regions deleted in this figure. The dark vertical stripes represent areas necessary for BtuB transport function, while the lighter stripes represent the segments proposed to be involved for outer membrane localization. All of the deletions that have lost transport function affect one or more of these areas, except for those near the C-terminus.

information needed for secretion of the PhoA moiety across the cytoplasmic membrane. PhoA fusions to cytoplasmic membrane proteins exhibit a characteristic periodicity, in which fusions having high enzymatic activity alternate with fusions of low activity. The high-activity fusions are likely to be those in which the amino acid residues around the fusion junction are located in the intact molecule on the periplasmic side of the cytoplasmic membrane, whereas in the low-activity fusions, the fusion junction, and hence the PhoA moiety, is in the cytoplasmic side of the membrane.

PhoA fusions to the outer membrane proteins FhuA (Coulton *et al.*, 1988, Günter & Braun, 1988) and FepA (Murphy & Klebba, 1989) have been described to have the following interesting properties. As long as the signal sequence is included, all hybrid proteins are secreted across the cytoplasmic membrane. Fusions with a relatively short contribution of the outer membrane protein, up to 88 amino acids of FhuA or 178 amino acids of FepA, were localized in the periplasmic space. Longer fusions, containing at least 180 amino acids of FhuA or 227 amino acids of FepA, were associated with the outer membrane.

Similar behavior was seen for BtuB-PhoA fusion proteins. A series of *btuB::phoA* fusion genes was isolated using the transposon Tn*phoA* of Manoil and Beckwith (1986). The fusion junctions were distributed throughout the mature portion of the BtuB polypeptide from amino acid 22 to 500 (Figure 2). All BtuB-PhoA fusions containing 22 to 327 amino acid residues of BtuB displayed similar levels of alkaline phosphatase activity. They appeared to be located in the periplasmic space, as shown by the absence of BtuB-related polypeptides or alkaline phosphatase activity from the outer membrane fraction and the fact that osmotic shock released comparable proportions of the total alkaline phosphatase and β-lactamase activities.

In contrast, the BtuB-PhoA fusions containing more than 399 amino acid residues of BtuB appeared to be associated with the outer membrane. Transformants carrying these genes on a plasmid exhibited a severe growth deficiency. These plasmids could not be stably maintained in most laboratory strains, with the exceptions of CC118, in which the mutations were initially isolated, and strain M16, which is resistant to the lethal incorporation of hybrid proteins in which export signals from a secreted protein are fused upstream of β-galactosidase. Strain CC118 carrying these fusion genes had substantially reduced (ca. 10-fold) levels of alkaline phosphatase activity. These also released a large proportion of the cellular β-lactamase activity into the medium, indicating some deficiency in the outer membrane barrier properties. Outer membranes of these strains contained alkaline phosphatase activity and a CN-Cbl-repressible polypeptide of the size expected for the full-length BtuB-PhoA hybrid protein. Thus, association of BtuB-PhoA hybrid proteins with the outer membrane appears to require the presence of amino acid sequences between residue 327 and 399 in BtuB. Thus, a substantially greater amount of BtuB appears to be required for outer membrane association than was seen with the related FhuA and FepA proteins.

Linker insertions in *btuB*

Isolation of mutations in a gene can often provide important information about the function of specific amino acid residues or domains in the functions of that proteins. We are very interested in identifying the portions of BtuB that are responsible for colicin binding. However, we have thus far been unsuccessful in the isolation of mutants that are insensitive to the E colicins but still capable of vitamin B_{12} transport, namely, those that might define the amino acid residues specifically involved in colicin binding.

As an alternative approach, we have described the construction, by *in vitro* techniques, of mutations in which a 6-bp linker was inserted into each of 10 sites within the *btuB* gene (Gudmundsdottir *et al.*, 1988); we have recently increased this collection to 26 sites. The majority of the insertions had no obvious effect on BtuB function, as assayed by growth response and transport of vitamin B_{12} or susceptibility to colicins El and E3 or to phage BF23. However, BtuB function was completely inactivated by insertions after BtuB residues 47, 50, 252, and 412 [as with the PhoA fusions, all residue coordinates are numbered from the start of the mature portion of the polypeptide]. Insertion after residue 343 caused substantial impairment. Thus, these insertions helped identify several regions of the protein that are required for its function, and also showed that 2 amino acids could be inserted at numerous other sites without effect.

Deletions in *btuB*

A significant advantage coming from the construction of these insertion mutations is that they allowed the simple construction of deletion mutations in which defined portions of the *btuB* gene were removed. A series of 25 deletions within *btuB* were prepared by deleting the region between appropriate *Bam*HI linkers. These deletions were chosen to maintain the normal reading frame of the BtuB protein, so the wild-type BtuB amino acid sequences are present beyond the region of the deletion. These deletions spanned the *btu*B coding region in a continuous manner from amino acid residue 5 to 506 (Figure 2). The deletion mutations were tested for their ability to confer the $BtuB^+$ phenotype when they were present on multicopy plasmids or when they were present on the chromosome in single gene copy. With the exception of del(114-132), all deletion mutations were unable to confer a $BtuB^+$ phenotype in a *btuB* host.

Each deletion that interfered with proper BtuB function could have lost a protein segment necessary for transport activity, or for export and insertion into the outer membrane, or both. To examine these possibilities, the outer membrane polypeptides of strains carrying some of the internal *btuB* deletion mutations on a multicopy plasmid were resolved by SDS-polyacrylamide gel electrophoresis. As shown in Figure 3, most of these mutant proteins were not detectable in the outer membrane, whereas the full-length BtuB polypeptide represented about 25% of the outer membrane proteins.

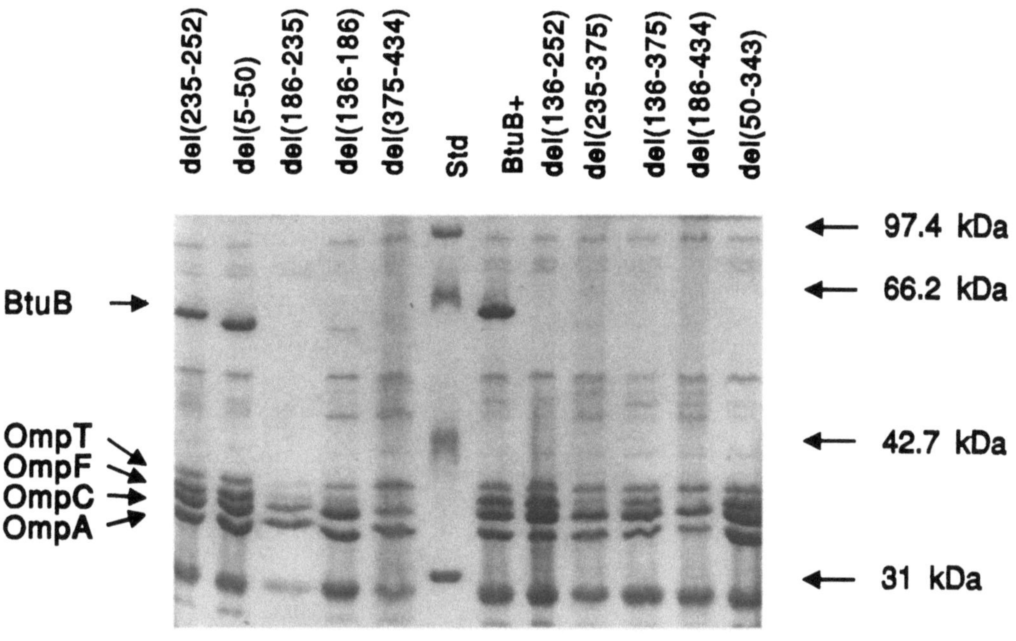

Figure 3. Outer membrane proteins in cells carrying plasmids with internal deletion mutations in *btuB*. The location of wild-type BtuB is indicated. Gels are stained with Coomassie blue. This figure is reproduced from Koster *et al.* (1991), with permission of the American Society for Microbiology.

Deletion polypeptides del(5-50) and del(235-252) were found in the outer membrane in roughly normal amounts, and del (136-186) polypeptide was present in reduced amount. The lack of transport activity of the first two mutants is expected because the linker insertion at one end of each of these deletions (those at amino acid 50 and 252) strongly depressed transport activity by themselves.

Expression of some of the deletion proteins appeared to have a substantial effect on the cell, even though these proteins did not accumulate in the outer membrane, as shown by the substantially reduced levels of the OmpF porin protein in many of the transformant strains. This finding is likely to be significant because previous work has shown that BtuB and OmpF form complexes, which are stable through protein purification (Imajoh *et al.*, 1982) and which appear to comprise the high-affinity receptor for colicin A (Chai & Foulds, 1982; Bénédetti *et al.*, 1989).

Duplications within *btuB*

Availability of the linker insertions also made it possible to construct partial duplications in the BtuB polypeptide covering the same regions as had been removed in the deletions described above. In this way, duplications were prepared that spanned almost the entire length of the polypeptide. In light of the lack of BtuB function of the deletion derivatives, it was surprising that all of the duplication constructs did confer a $BtuB^{+}$ phenotype, although not all were fully active.

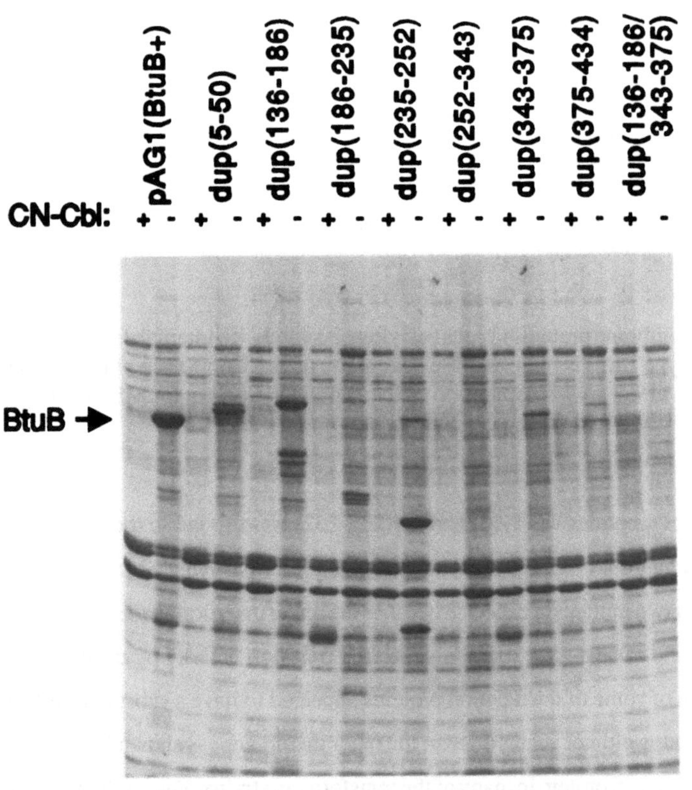

Figure 4. Outer membrane proteins in cells carrying plasmids with duplications in *btuB*. Cells were grown in the absence or presence of repressing levels of vitamin B_{12} (5 uM). This figure is reproduced from Koster *et al.* (1991), with permission of the American Society for Microbiology.

Analysis of the outer membrane polypeptide content revealed that all of the duplication polypeptides were inserted in the outer membrane, although not all were present in the same amount as the full-length BtuB protein. The level of function correlated with the amount of the polypeptide in the outer membrane. Polypeptides of the size expected for each full-length duplication product were observed, but all of the duplication polypeptides were found to be subject to cleavage (Figure 4). Comparison of the protein content of isolated outer membranes, with that of intact cells transferred directly to sample buffer, indicated that some of the proteins were cleaved in the intact cells, whereas others appeared to be cleaved only following cell disruption, during preparation of the outer membrane fraction (Sarkosyl solubilization). The sizes of the BtuB fragments that were formed were consistent with the occurrence of a cleavage event within the duplication region of the protein.

Conclusions

The different localization of BtuB-PhoA hybrid proteins depending on the length of BtuB sequence present suggests that a segment between amino acids 327 and 399 is necessary for outer membrane insertion. Similar behavior was seen by Coulton *et al.* (1988) in the case of FhuA, and by Murphy and Klebba (1989)in the case of FepA, except that the corresponding region needed for outer membrane association occurred earlier in the polypeptide, between residues 180 and 220. This behavior could mean that there is a specific sequence or sequences that must be present for insertion into the outer membrane. There must then be some other component of the outer membrane that recognizes this sequence and forms a complex with the protein that is necessary for its proper localization. Nikaido's group has shown that nascent porin chains must bind to lipopolysaccharide to allow them to undergo the conformation change that leads to trimerization of the porin subunits, which itself appears to be important for stable insertion into the outer membrane (Sen & Nikaido, 1990). It is likely that all outer membrane proteins form complexes with lipopolysaccharide and this may require specific polypeptide sequences. Alternatively, it could be that a certain minimum length of the outer membrane protein must be present to allow the protein to assume the appropriate conformation that allows its stable insertion in the outer membrane. It would be hard to see why such similar proteins would display such different minimum lengths.

Deletions that removed segments over most of the length of BtuB interfered with outer membrane insertion, whereas the duplication of these same segments did not interfere. We conclude from this that the proper progression of inwardly and outwardly directed transmembrane segments is not required for insertion into the outer membrane, since the duplications must have disrupted this progression if the deletions did. The ability of the duplication polypeptides to be inserted into the outer membrane suggests that this insertion process possesses considerable plasticity in its ability to tolerate additional sequences. The fact that all duplications were active in transport despite the fact that some carried insertions in regions that are critical for function means that the spacing between these critical regions is not essential and that a wild-type segment can compensate for a disrupted segment within the same polypeptide chain. This is a new form of internal complementation. Although the additional sequences did not interfere with localization, they must not have been properly folded into the final structure, since they were the sites of proteolysis. Despite that, the intact protein, at least, functions in transport. It is not known whether the products of proteolysis are active in transport.

Our results so far can be explained by postulating that there are two regions in BtuB that must be present for outer membrane association: one in the same place as in FhuA and FepA (180 to 220), the other between 327 and 400. Both must be present for BtuB. There is some sequence relatedness among these various regions proposed to be needed for export. This postulation of two essential segments is reasonable since BtuB might have to interact, not only with LPS, but also with OmpF.

The phenotype of most of the deletions in BtuB can be explained by assuming that three regions of the protein, defined by insertions at 47-50, 252, and 412-434, are essential for function, and that two regions (180-220 and 327-400) are essential for incorporation into the outer membrane. More deletions are being prepared to test this hypothesis.

REFERENCES

Bassford PJ, Kadner RJ (1977) Genetic analysis of components involved in vitamin B_{12} uptake in *Escherichia coli*. J Bacteriol 132:796-805

Bell PE, Nau CD, Brown JT, Konisky J, Kadner RJ (1990) Genetic suppression demonstrates interaction of TonB protein with outer membrane transport proteins in *Escherichia coli*. J Bacteriol 172:382-3829

Bénédetti H, Frenette M, Baty D, Knibiehler M, Pattus F, Lazdunski C (1991) Individual domains of colicins confer specificity in colicin uptake, in pore properties and in immunity requirement. J Mol Biol 217:429-439

Bénédetti H, Frenette M, Baty D, Lloubès R, Géli V, Lazdunski C (1989) Comparison of the uptake systems for the entry of various BtuB group colicins into *Escherichia coli*. J Gen Microbiol 135:3413-3420

Bénédetti H, Lazdunski C, Lloubès R (1991) Protein import into *Escherichia coli*: colicins A and El interact with a component of their translocation system. EMBO J 10:1989-1995

Chai T-J, Wu V, Foulds J (1982) Colicin A receptor: role of two *Escherichia coli* outer membrane proteins (OmpF protein and *btuB* gene product) and lipopolysaccharide. J Bacteriol 151:983-988

Coulton JW, Mason P, Cameron DR, Carmel G, Jean R, Rode HN (1986) Protein fusions of β-galactosidase to the ferrichrome-iron receptor of *Escherichia coli*. J Bacteriol 165:181-192

Coulton JW, Reid GK, Campana A (1988) Export of hybrid proteins FhuA'- 'LacZ and FhuA'-'PhoA to the cell envelope of *Escherichia coli* K-12. J Bacteriol 170:2267-2275

Davies JK, Reeves P (1975a) Genetics of resistance to colicins in *Escherichia coli* K-12: cross-resistance among colicins of group B. J Bacteriol 123:96-101

Davies JK, Reeves P (1975b) Genetics of resistance to colicins in *Escherichia coli* K-12: cross-resistance among colicins of group A. J Bacteriol 123:102-117

Gudmundsdottir A, Bell PE, Lundrigan MD, Bradbeer C, Kadner RJ (1989) Point mutations in a conserved region (TonB box) of *Escherichia coli* outer membrane protein BtuB affect vitamin B_{12} transport. J Bacteriol 171:6526-6533

Gudmundsdottir A, Bradbeer C, Kadner RJ (1988) Altered binding and transport of vitamin B_{12} resulting from insertion mutations in the *Escherichia coli btuB* gene. J Biol Chem 263:14224-14230

Günter K, Braun V (1988) Probing FhuA'-'PhoA fusion proteins for the study of FhuA export into the cell envelope of *Escherichia coli* K12. Mol Gen Genet 215:69-75

Hannavy K, Barr GC, Dorman CJ, Adamson J, Mazengera LR, Gallagher MP, Evans JS, Levine BS, Trayer IP, Higgins CF (1990) TonB protein of *Salmonella typhimurium*. A model for signal transduction between membranes. J Mol Biol 216:897-910

Heller K, Kadner RJ (1985) Nucleotide sequence of the gene for the vitamin B_{12} receptor protein in the outer membrane of *Escherichia coli*. J Bacteriol 161:904-908

Heller K, Kadner RJ, Gunther K (1988) Suppression of the *btuB451* mutation by mutations in the *tonB* gene suggests a direct interaction between TonB and TonB-dependent receptor proteins in the outer membrane of *Escherichia coli*. Gene 64:147-153

Heller K, Mann BJ, Kadner RJ (1985) Cloning and expression of the gene for the vitamin B_{12} receptor protein in the outer membrane of *Escherichia coli*. J Bacteriol 161:896-903

Imajoh S, Ohno-Iwashita Y, Imahori K (1982) The receptor for colicin E3. Isolation and some properties. J Biol Chem 257:6481-6487

Jeter R, Escalente-Semerena JC, Roof D, Olivera B, Roth J. In: Neidhardt F (ed) *Escherichia coli* and *Salmonella typhimurium*. ASM, Washington DC, 1987, p. 551-556.

Kadner RJ (1990) Vitamin B12 transport in *Escherichia coli*: energy coupling between membranes. Molec Microbiol 4:2027-2033

Köster W, Gudmundsdottir A, Lundrigan MD, Seiffert A, Kadner RJ (1991) Deletions or duplications in the BtuB protein affect its level in the outer membrane of Escherichia coli. J Bacteriol 173:5639-5647

Krone WJA., Stegehuis F, Koningstein G, Doorn CV, Roosendaal B, de Graaf FK, Oudega B (1985) Characterization of the pColV-K30 encoded cloacin DF13/aerobactin outer membrane receptor protein of *Escherichia coli*: isolation and purification of the protein and analysis of its nucleotide sequence and primary structure. FEMS Microbiol Lett 26:153-161

Levengood SK, Beyer WF, Webster RE (1991) TolA: a membrane protein involved in colicin uptake contains an extended helical region. Proc Natl Acad Sci USA 88:5939-3943

Lundrigan MD, Kadner RJ (1986) Nucleotide sequence of the gene for the ferrienterochelin receptor FepA in *Escherichia coli*. J Biol Chem 261:10797-10801

Manoil C, Beckwith J (1986) A genetic approach to analyzing membrane protein topology. Science 233: 1403-1408.

Murphy CK, Klebba PE (1989) Export of FepA::PhoA fusion proteins to the outer membrane of *Escherichia coli* K-12. J Bacteriol 171:5894-5900

Nau CD, Konisky J (1989) Evolutionary relationship between the TonB-dependent outer membrane transport proteins: nucleotide and amino acid sequences of the *Escherichia coli* colicin I receptor gene. J Bacteriol 171:1041-1047

Pressler U, Staudenmaier H, Zimmermann L, Braun V (1988) Genetics of the iron dicitrate transport system of *Escherichia coli*. J Bacteriol 170:2716-2724

Schoffler H, Braun V (1989) Transport across the outer membrane of *Escherichia coli* K12 via the FhuA receptor is regulated by the TonB protein of the cytoplasmic membrane. Mol Gen Genet 217:378-383.

Sen K, Nikaido H (1990) *In vitro* trimerization of OmpF porin secreted by spheroplasts of *Escherichia coli*. Proc Natl Acad Sci USA 87:743-747

Sun T-P, Webster RE (1987) Nucleotide sequence of a gene cluster involved in entry of E colicins and single-stranded DNA of infectmg filamentous bacteriophages into *Escherichia coli*. J Bacteriol 169:2667-2674

A STRUCTURE-FUNCTION ANALYSIS OF BtuB, THE *E.COLI* VITAMIN B_{12} OUTER MEMBRANE TRANSPORT PROTEIN.

R.J.Ward, S.E.Hufton, N.A.C.Bunce, A.J.P.Fletcher and R.E.Glass*
Department of Biochemistry
University of Nottingham
Medical School
Queens medical Centre
Nottingham, NG7 2UH
England

INTRODUCTION

Vitamin B_{12}, Structure, Occurrence and Synthesis

Vitamin B_{12} or cobalamin is a member of a group of compounds known as the corrinoids which are the most complex, nonpolymeric molecules so far found in nature. They exist in trace amounts throughout the natural world, usually within the concentration range 1×10^{-9} to 1×10^{-15}M (Bradbeer, 1982). Vitamin B_{12} is found in virtually all animal tissues to a greater or lesser degree, its source being ingested tissue or synthesis by gut flora.

The structure of cobalalmin (see Figure 1), was elucidated during the 1950's by a combination of chemical degradation techniques (Folkers & Wolfe, 1954; Bonnett *et al.*, 1957) and the X-ray crystallographic approach of Hodgkin *et al.* (1956). This work revealed cobalamin to be a large molecule (1356 Daltons), based upon a corrin ring with a cobalt atom at its centre. Above and below the plane of the corrin ring, α and β substituents are linked to the cobalt atom, the α substituent being an invariant dimethylimidazole group. The β substituent can, in contrast, vary, producing the range of cobalamins shown in Figure 1. Cyanocobalamin, has a cyanide group as its β substituent and hence its structure contains a carbon cobalt bond, the only example so far found in nature (Sennett *et al.*, 1981).

The synthesis of vitamin B_{12} is carried out exclusively by microorganisms (Beck, 1982) and is a complex multistep pathway. Many species can only catalyse a fraction of the steps involved. *E.coli*, for instance, is unable to catalyse the formation of the corrin ring (Friedman & Cagen, 1970) but is able to convert cobinamides to corrinoids (Ford *et al.*, 1955) and introduce the 5' deoxyadenosyl group (Volcani *et al.*, 1961). The related organism *Salmonella typhimurium* however is known to be able to synthesize cyanocobalamin *de novo*, but only in the absence of oxygen (Jeter *et al.*, 1984).

* Author to whom correspondence should be addressed

NATO ASI Series, Vol. H 65
Bacteriocins, Microcins and Lantibiotics
Edited by R. James, C. Lazdunski and F. Pattus

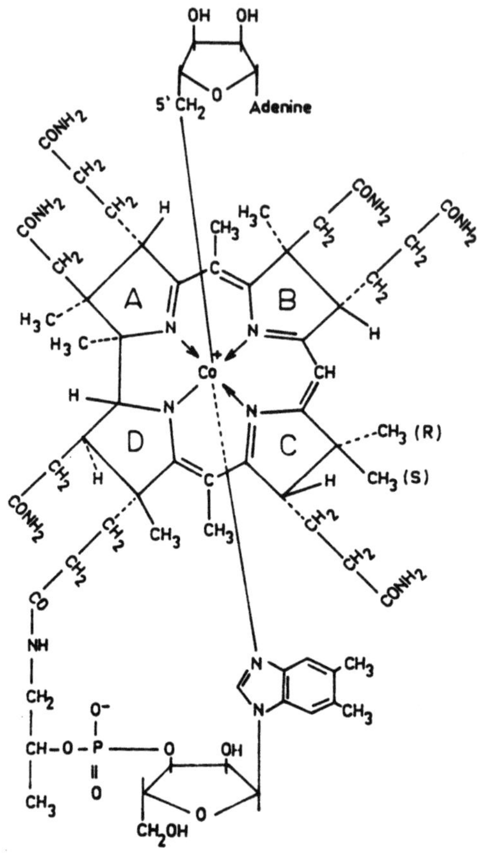

Figure 1. The structure of cobalamin (Schneider & Stroinski, 1987).

The exact sequence of biosynthetic reactions involved in the production of vitamin B_{12} is unclear and indeed it has been estimated that as many as 30 enzymes may be involved (Jeter *et al.*, 1987). The pathway is known to consist of three major branches termed Cob1, Cob11 and Cob111, based upon the nutritional requirements of the *cob* mutants (Jeter *et al.*, 1987).

Role of Vitamin B_{12} in *E.coli*

Wild-type *E.coli* has no absolute requirement for vitamin B_{12} although it is known to function as a cofactor in at least three and possibly four enzyme systems. These are listed in Figure 2. *E.coli* is able to bypass these reactions either by using different enzyme systems to achieve the same end or by uptake of the end product of the vitamin B_{12}-dependent reaction. This can be illustrated by considering the enzyme system b), the vitamin B_{12}-dependent homocysteine transmethylase. The enzyme encoded by the *metH* gene is one of two able to catalyse the final step in methionine biosynthesis, the conversion of homocysteine to methionine (See Figure 3). The alternative enzyme, encoded by the *metE* gene is a much less efficient catalyst of the reaction and is required at a far higher concentration than the vitamin B_{12} dependent enzyme.

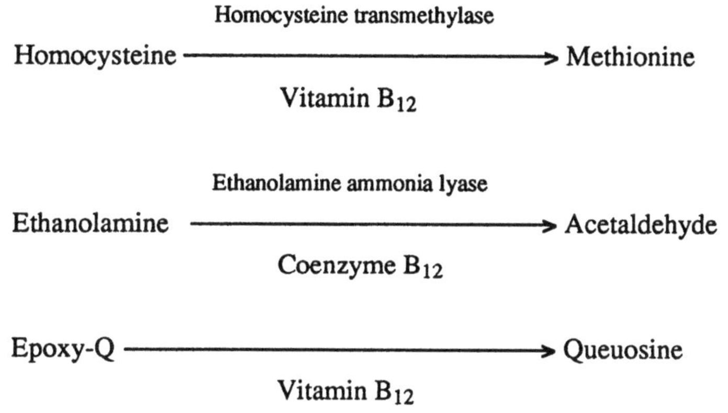

Figure 2. Vitamin B_{12}-dependent reactions in *E.coli.*
(Propanediol dehydratase, which enables propanediol to be used as a sole carbon source, is reported to be vitamin B_{12}-dependent in *Salmonella typhimurium*; Roof & Roth, 1989).

It is the exploitation of this system which helps to make the study of the vitamin B_{12} transport system particularly attractive since *metE* mutants, defective in the vitamin B_{12} independent transmethylase, require either vitamin B_{12} or methionine to grow. Consequently the uptake of vitamin B_{12} can be monitored directly and with great sensitivity.

Vitamin B_{12} Transport in *E.coli.*

The Gram-negative cell envelope such as that possessed by *E.coli* is extensively adapted to protect the cell against extracellular insult. The result of this is that the envelope constitutes a considerable barrier to the uptake of nutrients into the cell cytoplasm. This envelope can be considered to be composed of three distinct regions, the outer membrane, the periplasm (including the murein sacculus) and the cytoplasmic or inner membrane. To overcome this problem of nutrient , *E.coli* has evolved a wide variety of transport systems. The ability of *E.coli* to take up vitamin B_{12} from its growth media was first demonstrated by Oginsky in 1952 using Co^{60} labelled vitamin B_{12}. The uptake of the vitamin was shown to be a two-step process (DiGirolamo & Bradbeer, 1971). The first step is rapid, completed in under one minute and consists of the energy independent binding of vitamin B_{12} to the cell surface. The second part of the process is a slower rate determining step which involves the energy dependent transfer of the vitamin into the interior of the cell (Bradbeer, 1982).

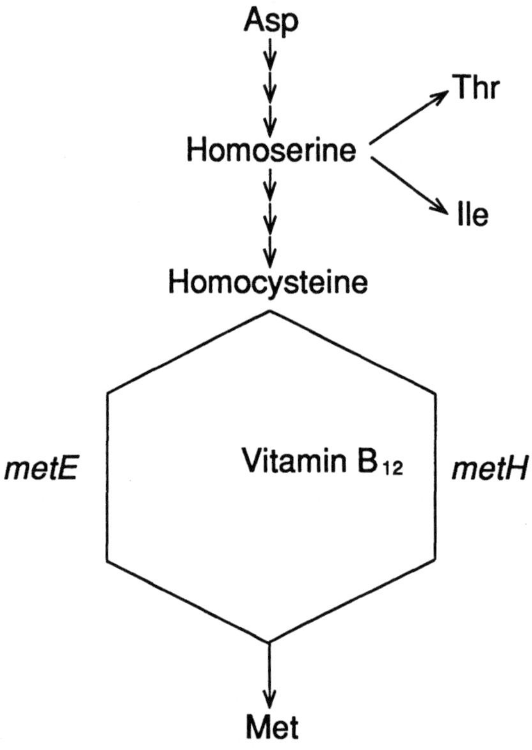

Figure 3. The terminal step of Methionine Biosynthesis in *E.coli*.

The initial, cell-surface binding site for vitamin B_{12}, its receptor protein, is firmly located within the structure of the outer membrane. This receptor is the product of the *btuB* gene and its function is to adsorb vitamin B_{12} and then transport it across the outer membrane into the periplasm where other components of the transport system complete the uptake process into the cytoplasm. The transport of vitamin B_{12} requires other gene products in addition to the BtuB protein. These include TonB, ExbB and possibly ExbD and two of the products of the *btuCED* operon. In addition, the correct function of the BtuB protein is reported to require lipopolysaccharide and probably Ca^{2+} ions (Bradbeer *et al.*, 1986).

The BtuB Protein

The BtuB protein has a molecular weight of 66000 daltons (Armstrong, 1988; Heller *et al.*, 1985). Its quaternary structure is unknown; the work of Nau and Konisky, 1989, showed that no homology exists with the porins which are known to form trimers in the outer membrane.

The *btuB* gene, which is located at 89 minutes on the *E.coli* chromosome, has been sequenced (Armstrong, 1988; Heller *et al.*, 1985) and is predicted to encode a peptide of 614 amino acids. The first 20 of which comprise a typical leader sequence found in the precursors of periplasmic or outer membrane proteins. (Michaelis & Beckwith, 1982). The deduced primary structure contains a high proportion of charged amino acids (22%). There are few non-polar helix-forming regions and so the membrane spanning domains are likely to be β-sheet structures.

The BtuB protein is present at around 200-250 copies per cell under conditions of low exogenous vitamin B_{12} concentration, that is, less then 1×10^{-9}M (Moir *et al.*, 1987). In the presence of concentrations greater then 1×10^{-7}M, the synthesis of the receptor is repressed and the copy number may be as low as 10-20 per cell (Kadner, 1978). Strains harbouring multiple copies of the *btuB* gene also show this repression, suggesting that there is no titration of regulatory molecules (Heller *et al.*, 1985; Moir *et al.*, 1987). Although these strains may contain BtuB at up to 50 times the wild type copy number, the number of functional receptors is only increased by 5-7 times (Moir *et al.*, 1987).

Mutants altered in *btuB* have been isolated and called *btuR* (Lundrigan *et al.*, 1987). This locus has since been shown to be involved in the synthesis of adenosylcobalamin, which may in turn mediate *btuB* repression, in a *btuR* $^{+}$ background (Lundrigan & Kadner, 1989). The work of Lundrigan and Kadner (1988) demonstrated that the *btuB* mRNA transcript is initiated some 250 nucleotides upstream of the start of translation. Numerous point mutations in the 239 nucleotide untranslated leader region caused constitutive expression of a *btuB-lacZ* fusion gene. This lead the authors to suggest a role for mRNA secondary structure in the regulation of expression of the *btuB* gene. Lundrigan *et al.* (1991) identified several regions of dyad symmetry in the leader region and also early in the transcribed region of the mRNA which may have a role. They concluded that a model for *btuB* regulation may involve secondary structures in the transcript able to affect both transcription and translation of the gene.

The Multivalent Nature of the BtuB Protein

In addition to its role as the outer membrane cobalamin receptor, the BtuB protein has also been shown to act as a receptor for the nine E colicins (colicins E1-E9), colicin A and the bacteriophage BF23. Thus, BtuB functions as a receptor for twelve different ligands, all of which are able to demonstrate competitive binding kinetics.

It has been shown that while all the BtuB copies in the outer membrane are able to transport vitamin B_{12} and mediate the lethal effects of BF23, only a proportion are able to act to allow the lethal effect of the colicins (Bassford *et al.*, 1977). Bradbeer, (1982) suggested that the receptors can only mediate colicin action when newly inserted into the membrane, possibly while associated with areas of membrane adhesion or fusion, however the existence of such zones is now the subject of some debate (Kellenberger, 1990).

Despite the competitive binding kinetics of the twelve ligands, they do not all appear to use exactly the same binding site and some require the presence of additional components in order to bind. Bradbeer, (1982) showed that purified BtuB can bind colicin E3, but not vitamin B_{12} or colicin E1. The binding of vitamin B_{12} has been reported to require the presence of lipopolysaccharide (Bradbeer, 1982) and the importance of Ca^{2+} ions was demonstrated by White *et al.*, (1973). It has since been suggested that Ca^{2+} ions may stabilise the BtuB-LPS interactions (Bradbeer *et al.*, 1986).

Corrinoid specificity of BtuB

E.coli BtuB has no specificity for the β-substituent of corrinoids and will transport adenosylcobalamin, cyanocobalamin, methylcobalamin and aquacobalamin with equal efficiency (Kenley *et al.*, 1978). Cobinamide was also bound by the receptor with the same degree of tenacity, indicating that the phosphate, ribose and benzimidazole moieties do not participate in binding (Figure 1).

The work of Bradbeer *et al.* (1978), and Kenley *et al.* (1978), has shown that the high affinity of the receptor for cobalamins results from highly specific interactions with the b-proprionamide groups combined with multiple, less specific interactions with other parts of the molecule.

The BtuB group colicins

Colicins are bacteriocins, protein antibiotics produced by coliform bacteria such as *E.coli*, which are able to kill susceptible bacteria of the same species as those producing the colicin. The colicins utilising BtuB as their receptor are listed in Table 1, and are known as the BtuB group colicins. The BtuB group colicins can be divided into three groups upon the basis of their modes of action against sensitive bacteria. The first group act as ionophores, dissipating the cytoplasmic membrane potential, for example colicin E1. The second and third groups both act as enzymes in the cytoplasm, one group as DNAses, for instance E2 and the other as RNAses such as E3.

The genes for colicin production are carried on plasmids which also encode a gene for the specific immunity protein to the colicin and a gene specifying the release of the colicin by cell lysis (Luria & Suit, 1987). All colicinogenic bacteria produce a specific immunity protein to the colicin that they manufacture. [In some cases immunity proteins are also produced to other colicins, Pugsley & Oudega, 1987]. In the case of E2 and E3 the immunity protein reacts with the carboxy-terminal, enzymatic part, of the protein. The immunity proteins do not cross-react however (Luria & Suit, 1987). The immunity protein for colicin E1 has been shown to reside in the cytoplasmic membrane (Geli *et al.*, 1989), though its mode of action is uncertain. Colicin molecules are organised into three different domains along the polypeptide chain. The central domain appears to be involved in receptor binding, the amino terminal seems to be required for translocation across the outer membrane and the carboxy terminal domain carries the lethal, enzymatic or ionophoric activity (Baty *et al.*, 1988).

The killing of cells by colicins follows single hit inactivation kinetics (Neville & Hudson, 1986). Exposure to colicins results in the number of viable bacteria declining in a log linear fashion with increasing exposure (Nomura, 1964). The implication of this is that a single molecule of colicin is able to cause cell death. However, the number of molecules actually required may be up to 10^3 times the number of cells. This could be explained by the inability of some of the receptors on a sensitive cell to mediate the lethal effects of the colicins (Luria & Suit, 1987). Presumably most of the colicin molecules adsorbed by the receptors are destroyed by proteases before they have a chance to act (Brey, 1982).

Table 1. The BtuB group colicins

Colicin	Molecular weight (Kd)	Plasmid	Colicin Activity
A	63	pCol A-ca31	Ionophore
E1	57	pCol E1-K53	Ionophore
E2	62	pCol E2-P9	DNAse
E3	58	pCol E3-CA38	16S RNAse.
E4	64	pCol E4-CT7	NK
E5	66	pCol E5-O99	RNAse*
E6	66	pCol E6-CT14	RNAse*
E7	64	pCol E7-K317	DNAse
E8	66	pCol E8-J	DNAse
E9	66	pCol E9-J	DNAse

NK, Not known. *, By sequence homology. Table based upon Pugsley, 1985.

Bacteriophage BF23

BF23 is closely related to bacteriophage T5 such that they can undergo phenotypic mixing, genetic recombination and substitution of gene products (Heller, 1984). Their physical characteristics are similar, they are of almost identical molecular weight, and their genomes show a high degree of similarity (McCorquedale, 1978). The adsorption to the cell surface occurs in two stages, reversible binding to the surface followed by irreversible binding to a specific receptor. The first step is nonspecific and may occur with any part of the phage, the second is mediated by a defined part of the phage tail (Schwartz, 1976). The injection of the phage DNA of this type is unusual in that it occurs in two steps. Part of the genome is first transferred, the injection then stops and only after the genes in this region are expressed is the injection of the DNA completed (McCorquedale, 1978).

Other Genes Involved in Vitamin B_{12} and lethal ligand Uptake

A number of other gene products are thought to be directly involved in the transport of vitamin B_{12} into *E.coli* in addition to the BtuB protein. A possible scheme for their locations and interactions is shown in Figure 4. The potential functions of these proteins will be discussed in detail in this section.

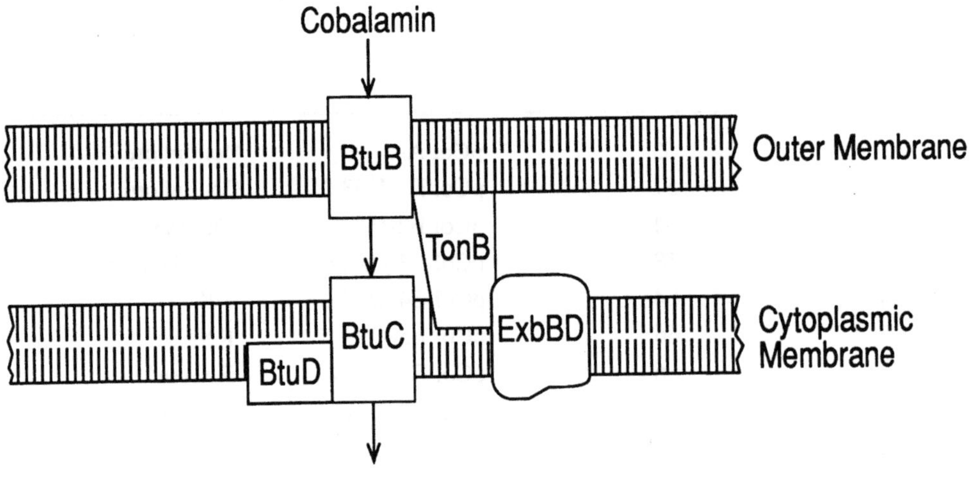

Figure 4. A Possible Scheme for the energy-Dependent Transport of Vitamin B_{12} into *E.coli*.

btuC* and *D

The *btuCED* operon is located at 38 minutes on the *E.coli* genetic map. It was originally thought to encode a periplasmic binding protein-dependent transport system for the energy-dependent passage of cobalamin across the cytoplasmic membrane. The *btuE* gene was thought to encode the periplasmic binding protein, whose existence was reported by Bradbeer *et al.* (1978). Present evidence however would appear to indicate that this is not the complete story as the *btuE* gene appears to have been shown to have no essential role in vitamin B_{12} transport despite its location in the *btuCED* operon (Rioux & Kadner, 1989; Bradbeer, 1991).

btuC mutants are only slightly disabled in vitamin B_{12} uptake in a *btuB*$^+$, *tonB*$^+$ background (Bassford & Kadner, 1977). If either gene is defective, then growth can only occur on elevated levels of cobalamin ($5x10^{-6}$M) and the *btuC* gene is essential (Kadner, 1978). Thus *btuC*$^-$ cells can accumulate vitamin B_{12} in an energy dependent manner, but, in contrast to the wild type, labelled cobalamin can be released by treatments which disrupt the outer membrane. This would suggest that *btuC* encodes a component of a cytoplasmic membrane vitamin B_{12} transport system (Reynolds *et al.*, 1980). *btuD* mutants exhibit a similar phenotype to those in *btuC*. The products of the *btuC* and *D* genes have been shown to be integral and peripheral cytoplasmic membrane proteins respectively, while *btuE* is periplasmic (DeVeaux *et al.*, 1986).

The *btuCED* has been sequenced (Friedrich *et al.*, 1986) and areas of homology found between *the products of btuC* and *btuD* and other cytoplasmic membrane proteins, particularly to integral membrane proteins from two iron transport systems in the case of *btuC* (Staudenmaier *et al.*, 1989). BtuD has homology with the conserved regions of peripheral membrane components of various binding protein dependent transport systems (Friederich *et al.*, 1986). These regions have been proposed as nucleotide binding domains since they are also present in some ATP binding proteins (Higgins *et al.*, 1985). Areas of weak homology have also been found between *btuB* and *btuD* hinting at possible vitamin B_{12} binding domains (Friederich *et al.*, 1986).

tonB

The *tonB* gene is located at 28 minutes on the *E.coli* genetic map. Mutants at this locus are disabled in siderophore-mediated iron transport (Williams, 1979) and vitamin B_{12} transport, they are tolerant to bacteriophages T1 and ϕ80 and also tolerant to the B group colicins, that is, colicins B,D,G,H,Ia,Ib,M and V (Davis& Reeves, 1975). The *tonB* gene has been cloned and sequenced (Postle & Good, 1983). The deduced amino acid sequence predicts a protein of 26000 daltons, whose hydropathy profile suggests that hydrophobic regions exist at the amino- and carboxy-termini separated by a central hydrophilic region (Postle & Good, 1983). The amino-terminus forms an export signal which remains intact after expulsion from the cytoplasm (Postle & Skare, 1988). If TonB behaves in a similar way to other proteins with an uncleaved export signal then the amino-terminus may serve to anchor it in the cytoplasmic membrane. The hydrophilic region is probably located within the periplasmic space (Postle & Skare, 1988) while the location of the carboxy terminus is unclear. Analysis of fusion proteins suggests a periplasmic location for the carboxy terminal, but contact with the outer membrane cannot be ruled out (Postle, 1990).

A wide variety of functions have been ascribed to the TonB protein. At present, however, the most likely role appears to be that of energy transducer between the inner and outer membranes. The outer membrane is essentially non-energised and so energy-dependent transport across it must be driven by energy obtained from some other part of the cell. If the cytoplasmic membrane is treated with ionophores to dissipate its membrane potential, then energy-dependent outer membrane transport is also abolished indicating that the energy comes from the cytoplasmic membrane. *tonB* mutants behave in a similar way to cells treated with ionophores which provides some evidence that TonB is acting to transduce energy from the cytoplasmic membrane (Postle, 1990).

The TonB protein is thought to interact directly with the outer membrane receptors for iron siderophore complexes and vitamin B_{12} through a highly conserved region found close to their amino termini called the 'TonB box'. The suppression of mutations in the TonB box by mutations in the TonB protein has been demonstrated (Bell *et al.*, 1990).

ExbBC and D and Tol proteins

Mutations in the unlinked *exbB* gene have been shown to modify the effect of TonB. *exbB* mutants occur at 65 minutes on the *E.coli* genetic map and exhibit much reduced levels of vitamin B_{12} and iron siderophore transport (Hankte & Zimmerman, 1981). They are also insensitive to the B group colicins, but to a lesser extent than *tonB* mutants. It has been suggested that ExbB has a role in stabilising the TonB protein, based upon measurements of its half life in the presence and absence of ExbB (Postle, 1990). Subsequent work, however, has shown that chromosomally encoded TonB is structurally stable but functionally unstable (Skare & Postle, 1991). The TolQ protein which was thought to have a similar structure to TonB (Braun, 1989) has been shown, by the same work, to play a minimal role in TonB-dependent transport.

The ExbC and D proteins may also have an effect upon the TonB protein. The *exbC* locus is little known other than that mutants in the gene hyper-excrete enterochelin in the same way as *tonB* mutants suggesting a connection. *exbD* was discovered during the sequencing of the *exbB* gene (Eick-Helmerick

& Braun, 1989), and appears to be part of the same operon. Its amino acid sequence shows significant similarity to the TolR protein (found in the same operon as TolQ) but it does not appear to stabilise TonB (Fischer *et al.*, 1989).

Nonsense Suppression.

Nonsense suppression is part of the larger aspect of informational suppression, which includes missense and frameshift suppression. Informational suppression is the term used to describe the process by which partial or total restoration of the wild-type phenotype is achieved by the alteration of the translational apparatus (Glass *et al.*, 1982). Nonsense suppression is made possible by the existence of altered tRNA species, generally carrying mutant anticodons capable of base pairing with translational termination codons. Thus, these prevent premature chain termination which would otherwise occur in a nonsense mutant in the absence of a suppressor.

CAG Gln	UGG Trp	UAA Stop
AAG Lys	UUG Leu	UAU Tyr
GAG Glu	UCG Ser	UAC Tyr

UAG Amber

Figure 5. Nonsense mutations.

There are nine sense codons able to mutate to an amber triplet in a single base change (See Figure 5). All of the seven amino acids encoded at potential amber sites can now be inserted by amber suppression. It is thus possible to recreate the wild-type polypeptide and also protein variants which differ only by one amino acid. The characteristics of the suppressors used are shown in Table 2. The function of the protein produced by nonsense suppression is dependent upon two factors, suppressor efficiency and the suitability of the particular amino acid inserted at the site. Thus, the amino acid must be compatible and inserted at a high enough frequency to allow sufficient copies of the protein to be made. An additional constraint is the nucleotide sequence around the mutation, or the genetic context. For instance, the efficiency of suppression varies depending upon the nature of the base adjacent to the 3' side of the nonsense triplet (Glass *et al.*, 1982).

Table 2. Nonsense suppressors used in this study.

Suppressor	Mutant allele	Codon	Efficiency [a]	Amino acid inserted
Su1	*supD*	UAG	50	Serine
Su2	*supE*	UAG	22	Glutamine
Su3	*supF*	UAG	68	Tyrosine
Su5	*supG*	UAA/G	1[b]	Lysine
Su6	*supP*	UAG	55	Leucine

[a] Mean absolute suppressor efficiency as determined at two sites in *rpoB*, expressed as a percentage (Glass *et al.*, 1982). [b] Estimated value. Note: Suppressor efficiency is context dependent.

An extensive collection of *btuB* nonsense mutants have been isolated in this laboratory. They were isolated by selecting for colicin E3 resistance and then screening for growth on vitamin B_{12} plates after the introduction of nonsense suppressors. The basis of the present work is the characterisation of this group of nonsense mutants and their suppressed derivatives (See Table 2).

MATERIALS AND METHODS.

Bacteria and Phage

The bacteria and bacteriophage used were as follows:

KLR320: *his29*(Am), *metE*, *proB*, *trpA605*(Am), *lacI3*, *lacZ118*(Oc), *recA::Tn10*, *gyrA*, *rpsL*.

Nonsense mutants: KLR320, *btuB*(Non).

AJ5001: *argA*, *metB*, *lacZ53*(Am), *rpsL*, *recA1*, F196 *supD32*.

AJ5002: as AJ5001 except F100-13 *supE*.

AJ5003: as AJ5001 except F123 *supF*.

AJ5005: as AJ5001 except F125 *supG*.

AJ5006: as AJ5001 except F119 *supP*.

BZB2104: as W3110 except pColE1-K53.

BZB2106: as W3110 except pColE3-CA38.

W3110: Wild type but F^-lambda$^-$ IN(*rrnD-rrnE*).

Phage BF23: Wild type.

Media

Cells were grown on either LB rich media or M9 minimal media in liquid form or solidified by the addition of 1.5% Difco agar. LB rich media was made with 10g Difco bacto tryptone, 5g bacto yeast extract and 10g NaCl per litre. M9 minimal media was made up as follows: 42mM Na_2HPO_4, 22mM KH_2PO_4, 19mM NH_4Cl, 8mM NaCl, 2mM $MgSO_4$, 0.1mM $CaCl_2$, and 0.01mM vitamin B_1 (thiamine HCl). This was supplemented with 0.2% sugar (glucose or lactose), and amino acids as required at $20mgml^{-1}$.

Construction of Suppressed Derivatives

5ml LB overnight cultures of the donor and recipient strains were grown up and washed three times in cell buffer, (22mM KH_2PO_4, 49mM Na_2HPO_4, 68mM NaCl, 68mM $MgSO_4$, pH 7.1), and then resuspended in 1ml of the same solution before cross streaking onto minimal media. Derivatives were selected by suppression of the *his* and *trp* amber mutations.

Growth on Vitamin B_{12}

To test for growth on vitamin B_{12}, strains were streaked out onto minimal media containing appropriate supplements, (no methionine), and vitamin B_{12} at 10^{-8} to 10^{-12}M.

Colicin and BF23 Sensitivity Assays

Minimal plates were overlayed with with 5ml minimal top agarose (0.35%) inoculated with 200ml minimal overnight culture. After setting 5 or 10ml drops of a serial dilution of colicin or phage preparations were applied to the plates and allowed to dry. Incubation was carried out at 30°C to avoid temperature sensitivity effects. Crude colicin preparations were obtained according to the method of Pugsley and Oudega, 1987. BF23 was prepared by the method of Hufton, 1991.

Vitamin B_{12} Binding and Uptake

Vitamin B_{12} binding and uptake assays were performed according to the method of Gudmunsdottir *et al.*, 1988.

Cloning and Sequencing

The mutations were localised by marker-rescue and then amplified by the PCR reaction using a kit supplied by the Cetus Corporation. PCR fragments were digested with *Eco*R1 and *Bam*H1 before cloning into these sites in the pTZ18R polylinker. Sequencing was carried out using the Pharmacia T7 sequencing kit. All methods were performed according to Sambrook, 1989.

RESULTS.

The work carried out consists of four main stages, the isolation of nonsense mutants, the phenotypic analysis of their suppressed derivatives, mapping and structure-function correlation (Figure 6). The results of the phenotypic analysis and mapping of the 37 suppressed derivatives of the 13 mutants are shown in Table 3. These will be discussed in the next section.

Table 3. Phenotypic analysis of the BtuB variants.

Amino acid changes.	Sensitivity to lethal agents.			Growth on vitamin B12	Vitamin B12 binding	Uptake of vitamin B12
	Colicin E1	Colicin E3	BF23			
Wild-type	100	100	100	+	100 [0]	100 [0]
btuB deletion	0.01	0	0	–	0 [0]	0 [0]
Ser60-Lys	0.03	3	0.1	+	0 [0]	1.2 [0.6]
Gln62-Ser	100	100	10	+	23 [17]	40 [23]
-Gln	100	100	1	+	8 [1.5]	10 [4]
-Tyr	100	100	10	+	54 [31]	10 [3.8]
-Lys	**100**	**100**	**0.00001**	**+**	**14 [8]**	**10 [3]**
-Leu	100	100	0.1	+	3 [1]	9 [2]
Gln150-Ser	100	100	10	+	7 [4]	20 [12]
-Gln	100	100	10	+	25 [15]	30 [18]
-Tyr	100	100	10	+	45 [23]	40 [9]
-Lys	**100**	**100**	**0.1**	**+**	**32 [14]**	**5 [2.3]**
-Leu	100	100	10	+	9 [3]	20 [8]
Gln158-Lys	<0.01	0.03	0.001	+	0 [0]	2.6 [0.6]
Trp214-Ser	100	100	1	+	82.6 [14]	79.2 [8]
-Gln	100	100	10	+	49.2 [21.6]	84 [45.5]
-Tyr	**100**	**100**	**100**	**+**	**372 [222]**	**240 [58]**
-Lys	**3**	**10**	**0.0001**	**(+)**	**5.8 [4.3]**	**4.3 [4.2]**
-Leu	100	100	0.01	+	30.3 [34.2]	29.4 [11]

Amino acid changes.	Sensitivity to lethal agents			Growth on vitamin B12	Vitamin B12 binding	Uptake of vitamin B12
	Colicin E1	Colicin E3	BF23			
Gln299-Ser	100	100	10	+	14 [9]	10 [7]
-Gln	100	100	1	+	4.5 [2.1]	3 [2]
-Tyr	100	100	1	+	86 [27]	50 [31]
-Lys	**100**	**100**	**<0.000001**	**+**	**18 [12]**	**8 [3]**
-Leu	**100**	**100**	**0.001**	**+**	**14 [8]**	**3 [2]**
Gln335-Lys	10	100	10	+	12.9[7.1]	27.7[17]
Trp371-Ser	100	100	10	+	46.4[6.7]	40.5[6.7]
-Gln	100	100	10	+	16.9[9.3]	9.5[7.7]
-Tyr	**100**	**100**	**100**	**+**	**197[94]**	**89.9[9]**
-Lys	**3**	**100**	**0.0001**	**(+)**	**4 [5.5]**	**2.7[2.9]**
-Leu	100	100	10	+	44.9[30]	17.3[13]
Gln400-Lys	1	10	0.1	(+)	2.9 [1.7]	3.7[1.7]
Tyr505-Lys	3	10	0.0001	(+)	1.6[1.8]	3.9[0.2]
Tyr579-Lys	**3**	**100**	**0.0001**	**(+)**	**2.5[1]**	**0 [0]**
Gln580-Lys	**3**	**100**	**0.0001**	**(+)**	**5.4[4.4]**	**0 [0]**
Leu588-Ser	100	100	100	+	17[12]	48[16]
-Gln	**0.3**	**0.1**	**0.00001**	**(+)**	**0 [0]**	**0.6 [0]**
-Tyr	100	100	100	+	110[58]	191[58]
-Lys	**<0.01**	**<0.01**	**0.00001**	**(+)**	**0 [0]**	**0 [0]**
-Leu	100	100	100	+	49.3[35]	68[2.6]

Mutations are either amber or ochre nonsense mutations, the latter being those with only one suppressed derivative listed. All values are expressed as percentages of wild type.
+ = Good growth on vitamin B_{12}, - = no growth on vitamin B_{12}, (+) = Poor growth on vitamin B_{12}.
[] = Values for standard deviations.

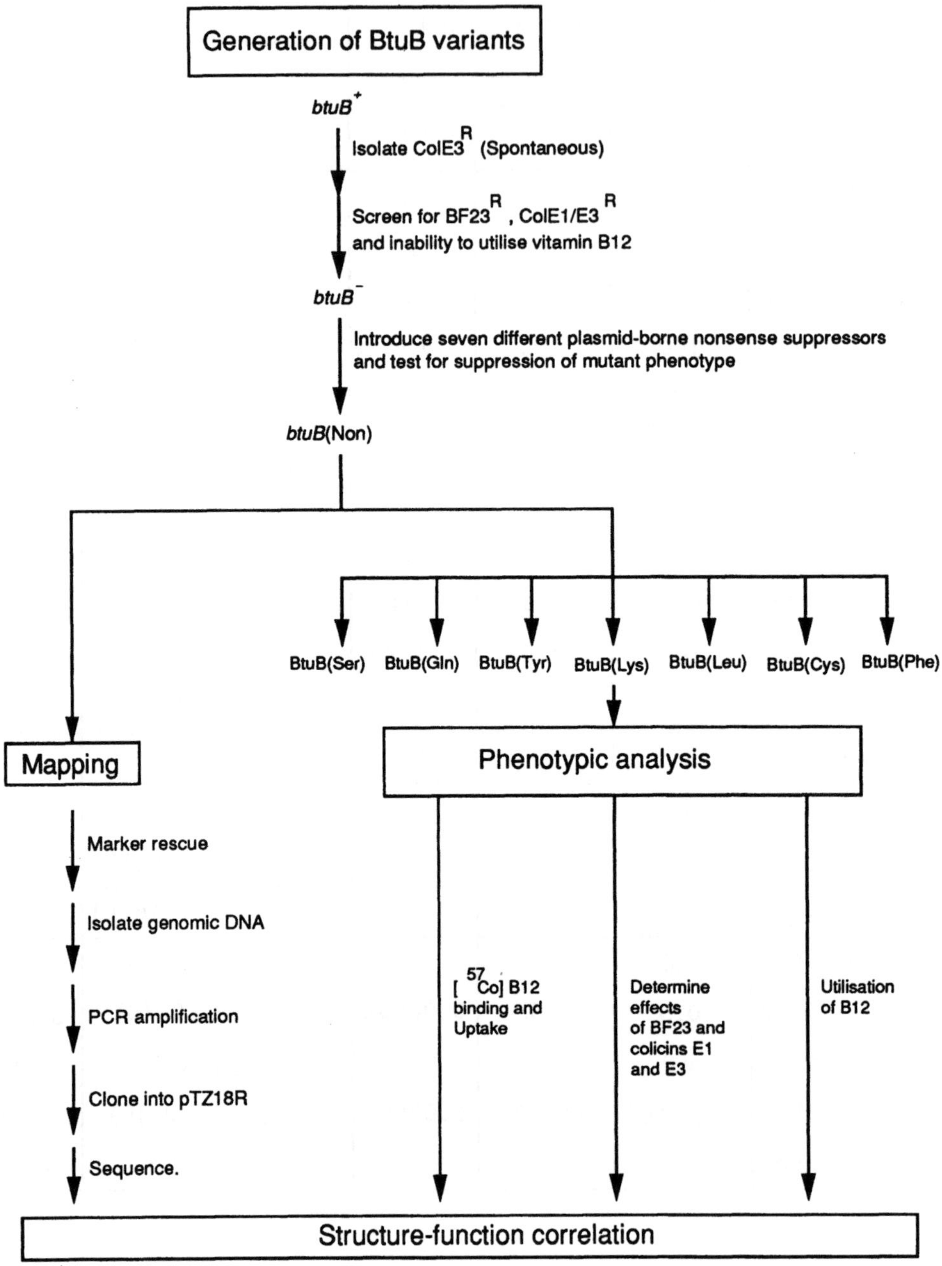

Figure 6. Isolation and analysis of BtuB variants.

DISCUSSION.

Phenotypic Analysis.

The results of the phenotypic analyses carried out on these BtuB variants (Table 3) highlight a number of points about the experimental approach and the techniques used for the analyses themselves. The only drawback of our genetic approach concerns the question of suppressor efficiency which has already been mentioned. The different suppressors insert their particular amino acids with different efficiencies as is shown in Table 2. Thus, the Su3 suppressor (an amber suppressor inserting tyrosine) is the most efficient while the least is Su5 (an ochre suppressor inserting lysine). It should be stated that the amber suppressors can only suppress amber mutations while the ochre is able to suppress both amber and ochre codons because of wobble at the third nucleotide of the anticodon. To investigate the effect of suppressor efficiency, work is at present under way to immunoblot total protein and outer membrane (Triton insoluble) preparations for the various suppressed derivatives, non-suppressed mutants, wild-type and deletion strains. This work is still to be completed but will be referred to where relevant during the discussion. In addition to this, clues about the effects of suppressor efficiency can be obtained from those substitutions which recreate the wild-type amino acid sequence. This occurs in the case of four suppressed derivatives, in three of which the Su2 suppressor inserts glutamine and in the fourth the Su6 suppressor inserts leucine. The Su2 derivatives all show a similar phenotype, wild-type colicin sensitivity and growth on vitamin B_{12} but are but are up to 90% disabled for BF23 sensitivity and vitamin B_{12} uptake and binding. This reflects the known low efficiency of the suppressor (see Table 2) and also the comparative insensitivity of the colicin titres which will be discussed later. These effects can be contrasted with the Leu588-Leu substitution (wild-type amino acid sequence recreated by the Su6 suppressor) which exhibits a near wild-type phenotype as would be expected from the greater efficiency of the suppressor.

The various phenotypic analyses carried out also have their particular advantages and disadvantages. Growth on vitamin B_{12} is very sensitive in that even the most disabled of the protein variants will grow, albeit poorly and over a long period of time. The deletion strain is unable to utilise the test concentration of vitamin B_{12}, that is, 10^{-10}M. The results of these experiments are recorded as good, poor or absent growth after three days at 30°C in Table 3.

Colicin titering appears to be a rather insensitive means of assessing the mutant phenotype. The colicin sensitivity of the various BtuB variants remained high except in the cases of some of the very disabled strains. There is a quite obvious correlation between poor growth on vitamin B_{12} and reduced colicin sensitivity. This may be due to the conditions used to isolate these mutants in that initial detection was on the basis of colicin resistance, followed by screening for suppressor-dependent growth on vitamin

B_{12} plates. It was observed that colicin E1 was able to have a slight effect on the *btuB* deletion strain, but only when used as an undiluted preparation. Since this colicin is an ionophore and acts on the inner membrane, it is possible that slight envelope damage may allow its insertion in the absence of BtuB. This effect, once noted, was taken into account during the interpretation of the overall results.

The BF23 assays, by contrast, show a consistently high degree of difference from one derivative to another and appear to be a much more sensitive indication of the effect of an amino acid substitution. In addition, from a practical point of view, these assays are much easier to score as they give clearer and more defined areas of killing.

The vitamin B_{12} uptake and binding experiments provided surprisingly little useful information in our hands due to the day-to-day variations in these assays. It is difficult, even after the calculation of standard deviation values, to assess what is a significant reduction, or in some cases increase, in binding or uptake. The consequence of this is that we cannot place as much emphasis as we would like upon these results.

The phenotypic analysis shows that these derivatives fall into two broad categories: 1. Those which are either disabled in all BtuB functions (pleiotropic mutants) or which exhibit a largely wild-type phenotype. 2. Those which are disabled in one or more functions but are of wild-type or improved function for the others. The later category are of the greatest interest as these sites are possibly involved in specific ligand binding and/or uptake and as such, in the case of binding, are likely to be external in orientation. Thus they provide valuable information to support or refute the proposed folding of the polypeptide chain. The remaining substitutions are at sites which would appear to have little effect upon the phenotypic functions observed.

The particular mutants that will now be discussed are **emboldened** in Table 3.

Gln62-Lys.

This amino acid substitution has wild-type levels of colicin E1 and E3 sensitivity, growth on vitamin B_{12} and, allowing for the low efficiency of this suppressor, reasonable levels of vitamin B_{12} binding and uptake. However, the BF23 sensitivity is very low, being **reduced** by a factor of 10^{-7}. This would suggest that the site may be specifically involved in the binding or uptake of BF23. The Gln62-Lys variant can be visualised by western blotting in both total protein and outer membrane preparations. It exhibits reduced amounts of protein compared to the wild-type but this does not appear to be sufficient to seriously impair BtuB function in this case. The reduced BF23 sensitivity of the Gln62-Lys substitution is clearly in excess of any reduction in sensitivity which could be explained by suppressor efficiency effects. The reduction is by a factor of 10^{-7} compared to the suppressor efficiency which is approximately 1% of that of a wild-type tRNA.

Gln150-Lys

This substitution exhibits a wild-type phenotype with respect to colicin sensitivity and growth on vitamin B_{12}, but reduced sensitivity to BF23 and reduced vitamin B_{12} uptake. The vitamin B_{12} binding is not as seriously affected as the uptake. Thus the site may be involved in the binding and/or uptake of BF23 and possibly of vitamin B_{12}. Western blotting shows that the polypeptide is stable and inserted into the outer membrane.

Trp214-Tyr

This is a substitution whose phenotype is almost wild-type with the exception of a marked **increase** in vitamin B_{12} binding and uptake activity. That the substitution has little effect is not surprising due to the comparable characteristics (large and aromatic side chains) of the amino acids involved. The western blotting analysis shows that the variant is produced and inserted into the outer membrane at a physiological level.

Gln299-Lys

This substitution is largely unaffected in BtuB functions with the exception of BF23 sensitivity, which is completely **absent**. The same phenotype is repeated for the leucine substitution at this site though the effect is not as marked. These variants are both seen in western blots and appear to be exported into the outer membrane in physiological quantities. The BF23 effect would appear to indicate the possible involvement of this site in BF23 binding and/or uptake.

Trp371-Tyr

This substitution gives not dissimilar results to the tyrosine variant at position 214. A difference in the behaviour of substitutions at this site is that the Trp317-Lys mutant has a wild-type response to colicin E3.

Tyr579-Lys and Gln580-Lys

These adjacent substitutions have similar phenotypes in that they show **reduced** sensitivity to colicin E1 and BF23, but produce a wild-type response upon exposure to colicin E3. Both exhibit low vitamin B_{12} binding and do not transport measurable levels of vitamin B_{12}. The western blotting analysis indicates that both BtuB variants are present in the cell but that only Tyr579-Lys is exported into the outer membrane. This is under investigation as it appears inconsistent with the wild-type colicin E3 sensitivity. Overall the results would appear to indicate that the two sites may possibly be involved in both BF23 and colicin E1 uptake and/or binding.

Leu588

The five substitutions made at this site can be split into two groups according to their phenotypes. Serine-, tyrosine- and leucine-variants are almost unaffected while the glutamine and lysine derivatives are both completely **disabled**. It is known from the behaviour of the non-suppressed amber mutant at this site that the last 7 residues are essential for the function of BtuB so such a dramatic effect is perhaps not unexpected. Western blotting shows that the unaffected variants are present at approaching wild-type levels while the glutamine and lysine substitutions are not visible in either total protein or outer membrane preparations and therefore appear to be unstable.

A Folding Model for BtuB

Our model for the transmembranous folding of the BtuB protein was derived using a modification of the method of Vogel and Jahnig (1985), which predicts potential membrane spanning amphipathic β-strands (G.C. Rowland, unpublished data). In a β-strand, the side chains of sequential residues lie on alternate sides of the plane of the interstrand hydrogen bonding and consequently every other residue is expected to be exposed to the hydrophobic interior of the membrane. The prediction uses a nine-residue window (9 to 11 residues in extended conformation are required to span the hydrophobic core of the outer membrane), centred on residue n and plots the average hydrophobicity of residues n, n+/-2, n+/-4. In addition primary structure homologies between BtuB and other outer membrane receptors with a similar function (TonB-dependent specific transport) were taken into account. The model predicted by this method is in good agreement with structural and topological information for other outer membrane proteins (Charbit *et al.*, 1986; Van de Ley *et al.*, 1986). The model is shown in Figure 7, with sites of the mutations under current investigation marked. The numbers refer to amino acid positions in the proposed mature polypeptide.

One assumption is that variants which are non-pleiotropic, that is, particularly disabled in one or more functions, are likely to be located at the binding site of the ligand whose function is affected. Hence it follows that these sites are probably located externally to the membrane, perhaps in loops exposed on the cell surface. This does not, of course, take into account the possibility that the sites may actually be involved in transporting the ligand across the outer membrane and could therefore be located in membrane-spanning domains or in loops extending into the periplasm. The majority of the non-pleiotropic substitutions are, however disabled with respect to BF23 sensitivity, a function which may not require the BtuB protein to act as anything other than an outer membrane receptor as the phage appears to be equiped with a mechanism for injecting DNA *per se*. Therefore, in the case of this ligand, the assumption may still have some validity.

The model predicts that seven of the sites at which substitutions occur will be external to the membrane. Of these, four are non-pleiotropic with respect to BF23 (Lys62, Gln150, Tyr579 and Gln580) and as such could be looked upon as supporting the model. The orientation of the external loop containing Lys62 is also supported by the work of Bradbeer and Gudmunsdottir (1990) who proposed that a cyanocobalamin-dependent calcium binding site may exist in this region.

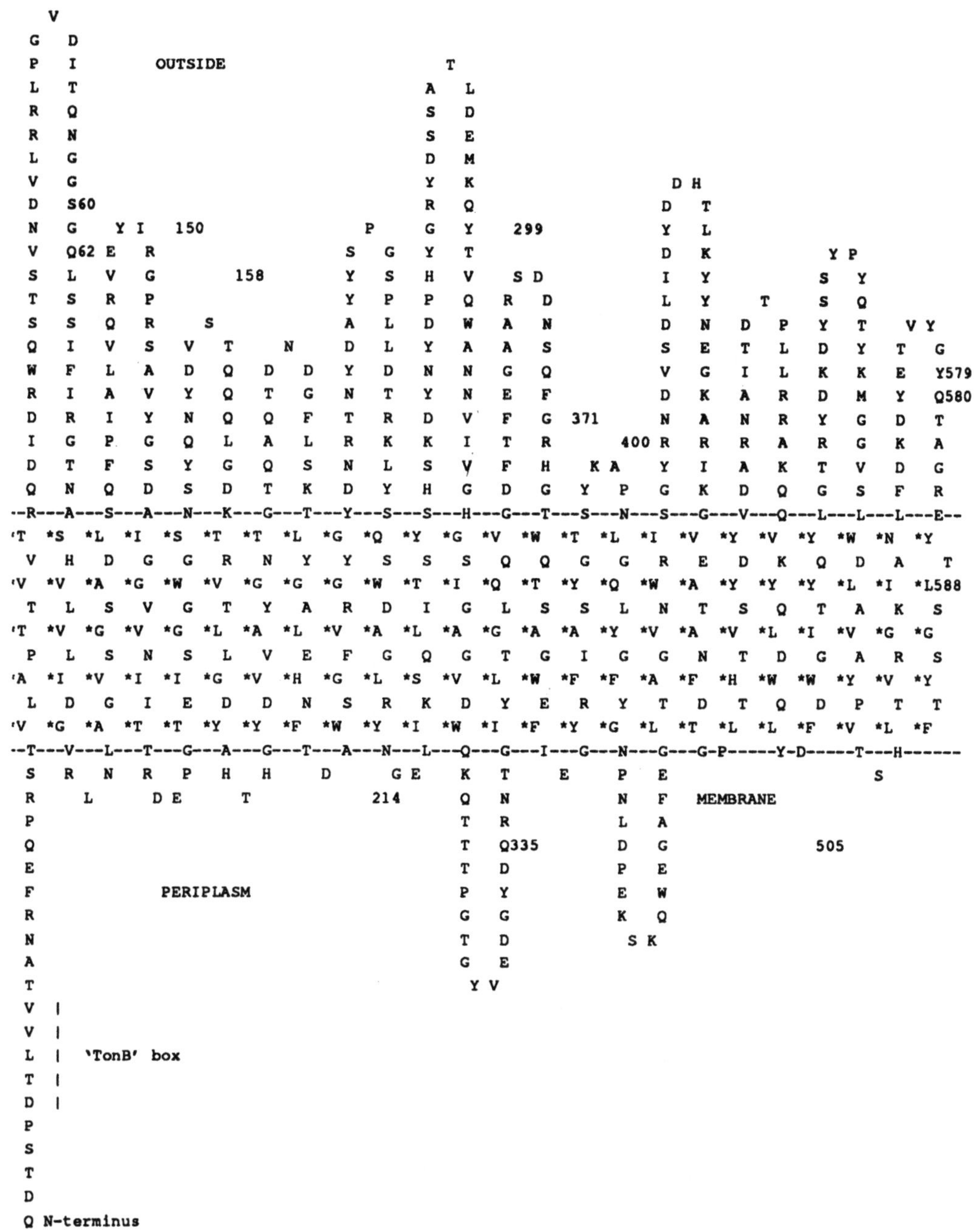

Figure 7. The folding model of BtuB. * residues forming the hydrophobic surface of the membrane-spanning domain. The figures refer to the number of residues from the N-terminus of the mature protein.

The final set of substitutions at Leu588, while displaying no differentiation between the individual ligands, do show two completely different effects for the five amino acids inserted. Three give almost wild-type function whereas the remaining two destroy BtuB function completely, even preventing insertion into the outer membrane. Thus, this site would appear to be important for the correct folding of the polypeptide and/or its export into the outer membrane. The model positions amino acid 588 on the hydrophobic surface of the final, carboxy-terminal, membrane spanning domain. The amino acids which abolish BtuB function at this position are glutamine and lysine, both of which are normally considered hydrophilic, so their effect is not surprising. The amino acids which are tolerated (serine, tyrosine and leucine), that is, allow near-normal BtuB function, are either hydrophobic or of mixed character and as such appear suitable for the position. Overall the results of the substitutions at Leu588 can be interpreted to support the model.

CONCLUSIONS.

The exploitation of nonsense suppression, combined with the generation of a collection of random, chromosomal nonsense mutants has permitted the analysis of a large number of BtuB variants, all altered by one known amino acid. The use of chromosomal mutants has allowed the variants to be analysed under conditions as close to physiological as possible. This approach has produced many protein variants which differ only slightly in sequence but which have in some cases significant phenotypes yielding structure-function information. This can be contrasted with other work carried out on the BtuB protein involving the generation of insertions and deletions within the structural gene which have yielded little information regarding **function**, but more so regarding the ***insertion*** of BtuB into the outer membrane (Koster *et al.*, 1991).

Phenotypic analysis has shown that the majority of the amino acid variants in BtuB have a pleiotropic phenotype. However, a small number appear to be disabled in only one or two of the receptor functions and it is this important information which is being used to support and update our BtuB folding model. The application of the results to the folding model has yielded mixed results. In some cases amino acids, which from the phenotypes of their substitutions could be interpreted to be on the external surface, if they are assumed to be involved in binding rather than uptake, are indeed predicted to be in this orientation: Others however are more inconsistent. The phenotypic analysis of the variants has shown that the vitamin B_{12} binding and uptake measurements are, in our hands, perhaps the least informative of the techniques employed. Growth on vitamin B_{12} appears to be a much more sensitive measure of the ability to utilise this nutrient, reflecting the small number of molecules required as cofactor in methionine production.

At present, the effect of suppressor efficiency and the export of the variants into the outer membrane are being examined through immunoblotting total cell protein and outer membrane preparations. In addition, the analysis is being extended by the use of other, plasmid-borne, suppressors. In the longer term, the main priority is the isolation and characterisation of new mutants effected in single functions based upon the information provided by this and other work.

ACKNOWLEDGEMENTS

The work reported was supported by grants from the SERC and Wellcome Trust. We would like to thank Tony Pugsley and Richard James for colicin-producing strains. We also wish to acknowledge the contribution of Geoff Rowland for allowing us to use his unpublished BtuB folding model and for many valuable discussions.

REFERENCES.

Armstrong J (1988) The multivalent vitamin B_{12} receptor of *E.coli*. A structure function analysis. Ph.D Thesis. University of Nottingham

Bassford PJ, Kadner RJ, Schnaitman CA (1977) Biosynthesis of the outer membrane receptor for vitamin B_{12}, the E colicins and bacteriophage BF23 by *E.coli*. Kinetics of phenotypic expression after the introduction of bfe^{+} and bfe alleles. J Bact 129:265-275

Bassford PJ, Kadner RJ (1977) Genetic analysis of the components involved in vitamin B_{12} uptake in *E.coli*. J Bact 132:796-805

Baty D, Frenette M, Lloubès R, Géli V, Howard SP, Pattus F, Ladzunski C (1988) Functional domains of colicin A. Mol Micro 2:807-811

Beck WS (1982) Biological and medical aspects of vitamin B_{12}. In: Dolphin D (ed) Vitamin B_{12} volume 2. Wiley, New York, p1-30

Bell PE, Nau CD, Brown JT, Konisky J, Kadner RJ (1990) Genetic suppression demonstrates interaction of TonB protein with outer membrane transport proteins in *E.coli*. J Bact 171:6526-6533

Bradbeer C, Kenley JS, DiMasi DR, Leighton JBC (1978) Transport of vitamin B_{12} in *E.coli*; Corrinoid specificity of the B_{12} binding protein and of energy-dependent B_{12} binding. J Biol Chem 253:1347-1352

Bradbeer C (1982) Cobalamin transport in microorganisms. In: Dolphin D (ed) Vitamin B12 volume 2. Wiley, New York, p31-55

Bradbeer C, Reynolds PR, Bauler GM, Fernandez MT (1986) A requirement for calcium in the transport of cobalamin across the outer membrane of *E.coli*. J Biol Chem 261:2520-2523

Bradbeer C, Gudmundsdottir A (1990) Interdependence of calcium and cobalamin binding by wild-type and mutant BtuB protein in the outer membrane of *E.coli*. J Bact 172:4919-4926

Bradbeer C (1991) Cobalamin transport in *E.coli*. Biofactors 3:11-19

Braun V (1989) The structurally related *exbB* and *tolQ* genes are interchangable in confering *tonB*-dependent colicin, bacteriophage and albomycin sensitivity. J Bact 171:6387-6390

Bonnett R, Cannon JR, Clark VM, Johnson AW, Parker LFJ, Smith EL, Todd AR (1957) Chemistry of the vitamin B_{12} group, part V. The structure of the chromophoric grouping. J Chem Soc 1:1158-1168

Brey RN (1982) Fragmentation of colicins A and E1 by cell surface proteases. J Bact 149:306-315

Charbit A, Boulain JC, Ryter A, Hofnung M (1986) Probing the topology of a bacterial outer membrane protein by genetic insertion of a foreign epitope: expression at the cell surface. EMBO J 5:3029-3037

Davis JK, Reeves P (1975) Genetics of resistance to colicins in *E.coli* K-12; Cross resistance of colicins amongst colicins of group B. J Bact 123:96-101

DeVeaux LC, Clevenson DS, Bradbeer C, Kadner RJ (1986) Identification of the BtuCED polypeptides and evidence for their role in vitamin B_{12} transport in *E.coli.* J Bact 167:920-927

DiGirolamo PM, Bradbeer C (1971) Transport of vitamin B_{12} in *E.coli.* J Bact 106:745-750

Eick-Helmerich K, Braun K (1989) Import of biopolymers into *E.coli*: Nucleotide sequences of the *exbB* and *exbD* genes are homologous to the *tolQ* and *tolR* genes respectively. J Bact 171:5117-5126

Fischer E, Günter K, Braun V (1989) Involvement of ExbB and TonB in transport across the outer membrane of *E.coli*: Phenotypic complementation of *exbB* mutants by overexpressed TonB and physical stabilisation of TonB by ExbB. J Bact 171:5127-5134

Ford JE, Holdsworth ES, Kon SK (1955) The biosynthesis of vitamin B_{12}-like compounds. Biochem J 59:86-93

Folkers K, Wolfe DE (1954) Chemistry of vitamin B_{12}: Vitamins and Hormones 12:1-51

Friedmann HC, Cagen LM (1970) Microbial biosynthesis of vitamin B_{12}-like compounds. Ann Rev Microbiol 24:159-208

Friedrich MJ, DeVeaux LC, Kadner RJ (1986) Nucleotide sequence of the *btuCED* genes involved in B_{12} transport in *E.coli* and homology with components of periplasmic binding protein-dependent transport systems. J Bact 167:928-934

Géli V, Baty D, Pattus F, Lazdunski C (1989) Topology and function of the integral membrane protein confering immunity to colicin A. Mol Microbiol 3:679-687

Glass RE, Hunter MG, Nene V (1982) Informational suppression as a tool for the investigation of gene structure and function. Biochem J 203:1-13

Glass RE, (1982) Gene function: *E.coli* and its heritable elements. Croom Helm Ltd, London

Hankte K, Zimmerman L (1981) The importance of the *exbB* gene for vitamin B_{12} and ferric iron transport. FEMS Microbiol Letts 12:31-35

Heller K (1984) Identification of the gene for host receptor specificity by analysing hybrid phages of T5 and BF23. Virol 139:11-21

Heller K, Mann BJ, Kadner RJ (1985) Cloning and expression of the gene for the vitamin B_{12} receptor protein in the outer membrane of *E.coli.* J Bact 161:896-903

Heller K, Kadner RJ (1985) Nucleotide sequence of the gene for the vitamin B_{12} receptor protein in the outer membrane of *E.coli.* J Bact 161:904-908

Higgins CF, Hiles ID, Whalley K, Jamieson DJ (1985) Nucleotide binding by membrane components of bacterial periplasmic binding protein-dependent transport systems.EMBO J 4:1033-1040

Hodgkin DC, Kamper J, Mackay M, Pickworth J Trueblood KN, White JG (1956) Structure of vitamin B_{12}. Nature 178:64-66

Hufton SE (1991) A structure-function analysis of the *E.coli* vitamin B_{12} receptor. Ph.D Thesis. University of Nottingham

Jeter RM, Olivera BM, Roth JR (1984) *Salmonella typhimurium* synthesises cobalamin *de novo* under anaerobic growth conditions. J Bact 159:206-213

Jeter RM, Roth JR (1987) Cobalamin biosynthetic genes of *E.coli.* J Bact 169:3189-3198

Kadner RJ (1978) Repression of synthesis of the vitamin B_{12} receptor in *E.coli.* J Bact 136:1050-1057

Kellenberger E (1990) The Bayer bridges confronted with results from improved electron microscopy methods. Mol Micro 4:697-705

Kenley JS, Leighton M, Bradbeer C (1978) Transport of vitamin B_{12} in *E.coli*: Corrinoid specificity of the outer membrane receptor. J Biol Chem 253:1341-1346

Koster W, Gudmundsdottir A, Lundrigan MD, Seiffert A, Kadner RJ (1991) Deletions or duplications in the BtuB protein affect its level in the outer membranes of *E.coli.* J Bact 173:5639-5647

Lundrigan MD, DeVeaux LC, Mann BJ, Kadner RJ (1987) Separate regulatory systems for the repression of *metE* and *btuB* in *E.coli*. Mol Gen Genet 206:401-407

Lundrigan MD, Kadner RJ (1988) Regulation of expression of the *E.coli* vitamin B_{12} outer membrane receptor BtuB, involves adenosylcobalamin. American Society for Microbiology Proceedings. p173

Lundrigan MD, Kadner RJ (1989) Altered cobalamin metabolism in *E.coli btuR* mutants affects *btuB* gene regulation. J Bact 171:154-161

Lundrigan MD, Koster W, Kadner RJ (1991) Transcribed sequences of the *E.coli btuB* gene control its expression and regulation by vitamin B_{12}. Proc Natl Acad Sci USA 88:1497-1483

Luria SE, Suit JL (1987) Colicins and Col plasmids. In Neidhardt FC, Ingraham JL, Brooks Low K, Magasnik B,Schaechter M, Umbarger HE (eds) *E.coli* and *Salmonella typhimurium*. Cellular and molecular biology. Volume 2. American Society for Microbiology , Washington DC

McCorquodale DT (1975) The T-odd bacteriophages. Crit Rev Microbiol 4:101-159

Michaelis S, Beckwith J (1982) Mechanism of incorporation of cell envelope proteins in *E.coli*. Ann Rev Microbiol 36:435-465

Moir PD , Hunter MG, Armstrong JT, Glass RE (1987) Studies on the gene for the multivalent vitamin B_{12} receptor of *E.coli*. FEMS Microbiol Letts 41:103-108

Nau CD, Konisky J (1989) Evolutionary relationship between the TonB-dependent outer membrane transport proteins: nucleotide and amino acid sequences of the *E.coli* colicin I receptor gene. J Bact 171:1041-1047

Neville DM, Hudson TH (1986) Transmembrane transport of diptheria toxin, related toxins and colicins. Ann Rev Biochem 55:195-224

Nomura M (1964) Mechanism of action of colicins. Proc Natl Acad Sci USA 52:1514-1521

Postle K, Good RF (1983) DNA sequence of the *E.coli tonB* gene. Proc Natl Acad Sci USA 80:5235-5239

Postle K, Skare JT (1988) *E.coli* TonB protein is exported from the cytoplasm without the proteolytic cleavage of its amino terminus. J Biol Chem 263:11000-11007

Postle K (1990) *E.coli* and the Gram-negative dilemma. Mol Micro 4:2019-2025

Pugsley AP (1985) *E.coli* K-12 strains for use in the identification and characterisation of colicins. J Gen Micro 131:369-376

Pugsley AP, Oudega B (1987) Methods for studying colicins and their plasmids. In Hardy KG (ed) Plasmids, a practical approach. IRL Press, Oxford. p105-162

Reynolds PR, Mottur GP, Bradbeer C (1980) Transport of vitamin B_{12} in *E.coli*: Some observations on the roles of the gene products of *btuC* and *tonB*. J Biol Chem 255:4313-4319

Rioux CR, Kadner RJ (1989) Vitamin B_{12} transport in *E.coli* does not require the *btuE* gene of the *btuCED* operon. Mol Gen Genet 217:301-308

Roof DM, Roth JR (1989) Functions required for vitamin B_{12}-dependent ethanolamine utilisation in *S.typhimurium*. J Bact 171:3316-3323

Sambrook J, Fritsch EF, Maniatis T (1989) Molecular Cloning. A Laboratory Manual, Second edition, Cold Spring Harbour Laboratory Press

Schneider Z, Stroinski A (1987) Comprehensive B_{12}, Second edition, Verlag Walter deGruyter and Co, Berlin New York

Schwartz M, (1976) The adsorption of coliphage l to its host: effect of variations in surface density of receptor and in phage receptor affinity. J Mol Biol 103:521-536

Sennett C, Rosenberg LE, Mellman IR (1981) Transmembrane transport of cobalamin in prokaryotic and eukaryotic cells. Ann Rev Biochem 50:1053-1086

Skare JT, Postle K (1991) Evidence for a TonB-dependent energy transduction complex in *Escherichia coli*.. Molec Microb 5:2883-2890

Staudenmaier H, Van Howe B, Yaraghi Z, Braun V (1989) Nucleotide sequences of the *fecBCDE* genes and locations of proteins suggest a periplasmic binding protein dependent transport mechanism for iron dicitrate in *E.coli*. J Bact 171:2626-2633

Van de Ley P, Struyve M , Tommassen J (1986) Topology of the outer membrane pore protein PhoE of

E.coli. J Biol Chem 216:12222-12225

Vogel H, Jahnig F (1986) Models for the structure of outer membrane proteins of *E.coli* derived from Raman spectroscopy and prediction methods. J Biol Chem 190:191-199

Volcani BE, Toohey JI, Barker HA (1961) Detection of cobamidecoenzymes in microorganisms by the ionophoretic-bioautographic method. Arch Biochem Biophys 92:381-391

White JC, DiGirolamo PM, May Lay Fu, Preston YA, Bradbeer C (1973) Transport of vitamin B_{12} in *E.coli*. J Biol Chem 248:3978-3986

Williams PH (1979) Novel iron uptake system specified by ColV plasmids: an important component in the virulence of invasive strains in *E.coli*. Infect Immun 26:925-932

GENERAL INTRODUCTION TO THE SECRETION OF BACTERIOCINS

D. Cavard and B. Oudega*
CBM—CNRS
31 chemin Joseph Aiguier
13402 Marseille Cedex 9
France

INTRODUCTION

Bacteriocins have been considered for many years as the only type of proteins secreted by bacteria. The first ones described have been isolated, identified and characterized from the extracellular milieu of both gram-negative and gram-positive bacteria. Later, it was shown that some bacteriocins are not exported. For instance, colicins have been classified into two taxonomic groups that differ, among other properties, by the fact that they are secreted (group A) or not (group B) by the producing cells of *Escherichia coli*. On the other hand, numerous proteins besides bacteriocins have been shown to be secreted in the medium. These secreted proteins are either toxins, enzymes like proteases, lipases, nucleases, or structural proteins like fimbriae. However, the bacteriocins still constitute a peculiar group of exported proteins that cannot be compared to any other group.

The main characteristics of the secreted bacteriocins are the following :

1) They do not contain an amino-terminal signal sequence or leader peptide. They share this property with other exported proteins as, for instance, the hemolysins.

2) They do not contain an internal domain that can act as an export signal. Mutations in the COOH-terminal domain of colicin A (Baty *et al.*, 1987a), colicin El (Yamada & Nagazawa, 1984) and colicin E3 (Mock & Schwartz, 1978) and in cloacin DF13 (Andreoli & Nijkamp, 1976; Van Putten *et al.*, 1988) block their release, suggesting the presence of an export signal. In fact, these mutations have provoked a change of conformation that renders the protein unsuitable for both release and uptake as these mutants are inactive. Studies on the domains of colicin A (Baty *et al.*, 1987b) did not reveal a specific domain required for release.

3) Their export relies on the expression of only one gene. It encodes a bacteriocin release protein (BRP), also called either lysis protein or Kil protein. In contrast, export of other secreted proteins relies on the activity of at least three genes (Wandersman & Delepelaire, 1990), except the hemolysin of *Serratia marcescens* (Poole *et al.*, 1988).

* Biologish Laboratorium, Vrije Universiteit van Amsterdam, De Boelelaan 1087, 1081 HV Amsterdam, The Netherlands.

NATO ASI Series, Vol. H 65
Bacteriocins, Microcins and Lantibiotics
Edited by R. James, C. Lazdunski and F. Pattus

4) They are exported late after synthesis. They accumulate in the cytoplasm in a soluble form before they are released.

5) Their release is not completely specific and occurs with that of many cytoplasmic and periplasmic proteins. However, not all soluble cytoplasmic and periplasmic proteins are released at this stage, indicating a certain specificity of the system. Possibly, a subset of soluble proteins possesses the proper structure to be released by the export system.

PROPERTIES OF THE BACTERIOCIN RELEASE PROTEINS/LYSIS PROTEINS

The bacteriocin release proteins or lysis proteins are thus responsible for the secretion of bacteriocins. The colicin lysis proteins have been the most extensively studied, together with the bacteriocin release protein encoded by the CloDF13 plasmid (hereafter also called lysis protein). They share many properties.

1) Genetic organization

The gene encoding the lysis protein is part of the colicin operon in the colicinogenic plasmid. It is always located downstream of the colicin and the immunity structural genes. Two different arrangements of colicin operons may be drawn according to the function of the encoded colicin (Figure 1). For the pore-forming colicins such as colicin A (Lloubès *at al.*, 1986), El (Sabik *et al.*, 1983) and N (Pugsley, 1989), the lysis gene is separated from the colicin structural gene by a large intercistronic area of about 550 base pairs that contains the gene encoding the immunity protein on the opposite strand. For the nuclease colicins as colicin E2 to E9 (Cole *et al.*, 1985; Cooper & James, 1984; Lawrence & James, 1984; Chak & James, 1984; Chak *et al.*, 1991) and cloacin DF13 (Hakaart *et al.*, 1981), the lysis gene is located only a few base pairs from the immunity gene. Some exceptions are known, as for example the colicin E3 operon that contains two immunity genes (Chak & James, 1984) and the colicin E9 operon which includes two immunity genes and two lysis genes one of them being truncated (James *et al.*, 1987). To explain the structure of the colicin E9 operon, an exchange between two incompatible plasmids has been suggested (Curtis *et al.*, 1989), as well as the presence of an insertion sequence in the middle of the lysis gene (Lau & Condie, 1989). The colicin D operon shares the structure of the nuclease-colicin operon (Roos *et al.*, 1989): colicin D is the only colicin of group B to possess a lysis protein (Pshennikova *et al.*, 1988).

2) Transcription

The transcription of the lysis genes is under the regulation of the colicin promoter that is subject to SOS regulation. Its expression depends on transcript readthrough from the promoter of the colicin structural gene. No lysis protein is synthesized without a colicin. But the rates of expression of the colicin

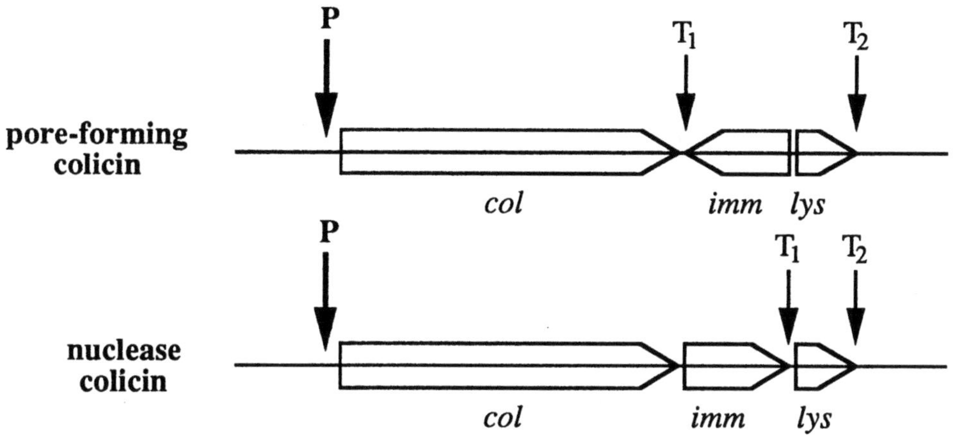

Figure 1. Localization of the lysis gene in colicin operons. *col:* colicin structural gene; *imm:* colicin immunity gene; *lys:* colicin lysis protein gene; P : promoter; T : transcription terminator.

and the lysis genes are different due to the presence of a transcription terminator located immediately after the colicin structural gene in the pore-forming colicin operon, and just after the immunity gene in the nuclease-colicin one (Figure 1) (Van Tiel-Menkveld *et al.*, 1981; Ebina *et al.*, 1983; Lloubès *et al.*, 1986)

3) Similarity of the colicin lysis genes and of the lysis proteins

Every colicin is synthesized with its own lysis protein. There are as many lysis proteins as colicins. However, both the nucleotide sequence of the lysis genes and the amino-acid sequence of the gene products (Figure 2) (De Graaf & Oudega, 1986; Pugsley, 1988) are highly similar in structure. They are not specific for one colicin and they have been shown to be interchangeable (Pugsley & Schwartz, 1983).

4) Synthesis

The lysis proteins are synthesized as precursors with a signal sequence. They contain a lipoprotein consensus box : Leu-X-Y-Cys, in which X is either Ala, Val or Ser and Y is Ala or Gly. They are lipidated and processed, as are all bacterial lipoproteins, in many steps (Wu & Tokunaga, 1986). The precursor form is acylated on the cysteine residue before it is processed by the signal peptidase II that is specific for lipoproteins. After removal of the signal sequence, the amino group of the acylated amino-terminal cysteine residue is modified by a fatty acid residue. The lipoprotein nature of the lysis proteins of cloacin DF13, colicin A, El, E2 and N has been demonstrated (Oudega *et al.*, 1984; Cavard *et al.*, 1987; Cavard, 1991; Pugsley & Cole, 1987; Pugsley, 1988).

Colicin lysis proteins

ColA	-18	MKK	II ICVILLAIMLLAA	1 CQVNNVRDT	GGGSVSPSSI	VTGVSMGSDGVGNP$_{33}$
ColE1	-17	*R*K	FF *GIF**N**VG	**A*YI**V	Q**TIA***SSKL**IAVV$_{28}$	
ColE2	-19	***	*TG*ILL***VII*S*	**A*YI**V	Q**T*****TAE***LATQ$_{28}$	
ColE3	-19	***	*TG*ILL***VII*S*	**A*YI**V	Q**T*****TAE***LATQ$_{28}$	
ColE5	-19	***	*TW*ILL***AII***	**A*YIH*V	Q**T*****SAEL**LATQ$_{28}$	
ColE6	-19	***	*TG*ILL***VII***	**A*YI**V	Q**T*****TAEL***ETQ$_{28}$	
ColE7	-19	***	*TG*ILL***AII***	**A*YI**V	Q**T*****TAEL***ETQ$_{28}$	
ColE8	-19	***	*TG*ILL***VII***	**A*YI**V	Q**T*****TAE***LATQ$_{28}$	
ColE9	1	***	*TG*ILL***VII*S*WGSKPKT$_{26}$			
CloDF13	-21	***AKA*FLFILIVSGF**V*		**A*YI**V	Q**T*A***SSEL**IAVQ$_{28}$	
ColN	-17	*CGK	*LL**FF**T*S*	****HA**V	K**T*A***SSRL**LKLSKRSKDPL$_{35}$	
ColD	-n		VV*VLC*T*	**A*YM*FREEQ****LP***SSKLI* *AIQ$_{31}$		

Figure 2. Aminoacid sequences of the colicin lysis proteins. They are aligned to that of the colicin A lysis protein: aminoacids are indicated when they are different from that of ColA; when not, they are replaced by stars.

The maturation and processing of these various lysis proteins are unique among lipoproteins. The process occurs slowly in such a way that every intermediate form can be observed at any time. Futhermore, both the mature form and the signal sequence accumulate in the cells after processing. The colicin El lysis protein constitutes an exception, as its precursor form is processed as fast as the murein-lipoprotein precursor and its signal peptide seems to be hydrolyzed immediately after processing (Cavard, 1991).

5) Assembly

The lysis proteins are cell envelope proteins, as are all lipoproteins. Their mature form has shown to be localized in both the inner and outer membranes of cloacin DF13 (Oudega *et al.*, 1894), colicin A (Howard *et al.*, 1991), colicin E2 (Cole *et al.*, 1985) and colicin E3 (Jakes & Zinder, 1984) producing cells. The colicin N lysis protein is localized only in the outer membane (Pugsley, 1988). A kinetics analysis carried out on the colicin A lysis protein has demonstrated that, just after synthesis, the precursor form, mature form and signal sequence are located in the inner membrane. Later on, the mature form is observed in the outer membrane (Howard *et al.*, 1991) indicating a slow assembly and targetting of the lysis protein.

6) Functions

The lysis proteins fulfill various functions. They provoke i) the non-specific release of colicins, that is of a subset of proteins; ii) the quasi-lysis of the culture (quasi-lysis is not a real lysis as no degradation of peptidoglycan is observed, but more a permeabilisation of the cell envelope resulting in a decrease of culture turbidity); iii) the killing of the producing cells; iv) the activation of the phospholipase A in the outer membrane. These functions may not be coupled. For instance, they provoke the release of proteins without quasi-lysis in the presence of divalent cations such as Mg_{2+} or Ca_{2+} (Oudega *et al.*, 1984; Pugsley & Schwartz, 1984; Luirink *et al.*, 1985; Cavard *et al.*, 1989). Furthermore, it is also not clear whether quasi-lysis is the result of killing of cells by induction of the phospholipase A, or by some other unknown event possibly caused by the accumulation of stable signal peptides. In any case, activation of phospholipase is not really required for release of proteins, since non-lethal mutations which provoke the permeabilization of the outer membrane also result in release of proteins (colicins) in the presence of lysis proteins (Cavard *et al.*, 1989).

STRUCTURE-FUNCTION RELATIONSHIP OF LYSIS PROTEINS

l) Role of acylation and processing

To be active, the lysis protein has to be acylated and processed. Addition of globomycin blocks the processing and inhibits protein release.

Directed mutagenesis has been performed on the lipobox. When acylation does not take place, the resulting protein is non-functional (Pugsley & Cole, 1987; Cavard *et al.*, 1987; Luirink *et al.*, 1988). The colicin E9 lysis protein constitutes an exception. It does not contain a lipobox, and it is not known at the moment whether it is functional or not (James *et al.*, 1987). Some protein release has been obtained by a colicin E2 lysis protein in which the cysteine has been replaced by a glutamine (Pugsley & Cole, 1987).

Colicin lysis proteins are unique among lysis proteins in that they are lipoproteins. Phage lysis proteins are not acylated. Some phage lysis proteins share similar functions as their colicin counterparts and sequence homology between them and the colicin lysis proteins has been noticed (Lau *et al.*, 1989).

2) Length and composition of the mature form

Work done on colicin A, E2, E3 and cloacin DF13 lysis proteins indicates that a critical length of about 20 amino-acid residues is required to fulfill every function. Neither a colicin A lysis protein of 18 amino-acid residues (Howard *et al.*, 1989) nor a cloacin DF13 BRP of 16 residues (Luirink *et al.*, 1989) are able to provoke protein release, but are still able to kill the producing cells, whereas the colicin E2 and colicin E3 lysis proteins of 20 amino-acid residues are still active (Toba *et al.*, 1986).

The length of the mature form seems to alter the acylation and maturation of the colicin A lysis protein (Howard *et al.*, 1989) but not that of the cloacin DF13 (Luirink *et al.*, 1989). It seems that shortening the mature form provokes a slow-down of the modification and processing of the lysis protein. Maturation may be evidenced on truncated colicin A lysis protein after overproduction (unpublished results). Acylation is not sufficient *per se* to promote functioning.

The sequence of the conserved amino-acid residues of the amino-terminal half of the protein is exceedingly relaxed. This has been evidenced by directed mutagenesis done on the colicin A lysis protein (Howard *et al.*, 1989). The homology of the various colicin lysis proteins could play another role rather than a functional one. A role in plasmid recombination and in the evolution of colicins has been suggested (Lau *et al.*, 1989).

3) Changes of the signal sequence

Replacement of the signal peptide of the cloacin DF13 lysis protein by the signal sequence of the murein-lipoprotein has been carried out by genetic engineering. The hybrid protein is unable to release cloacin DF13 but is still able to provoke lysis and to cause lethality (Luirink *et al.*, 1991). The stability of the sequence signal thus plays a significant role in the killing and in the release of proteins as suggested by Luirink *et al.* 1989 and by Howard *et al.* 1989).

MODE OF ACTION OF LYSIS PROTEINS

A critical concentration of the lysis protein seems to be needed for functioning, explaining the delay between the synthesis of the lysis protein and the export of colicin (Cavard *et al.*, 1989).

It has been proposed that colicin lysis proteins function by activating the phospholipase A of the outer membrane (Pugsley & Schwartz, 1984). This activation would alter the permeability of the outer membrane and would provoke release of proteins and cause quasi-lysis of the culture. This model is based on the observation that *pldA* mutants do not release protein or show quasi-lysis. In these mutants, the colicin A lysis protein is significantly degraded by the DegP protease (Cavard *et al.*, 1989). The level of the lysis protein would, therefore, not reach the sufficient concentration necessary for acting. On the other hand, it seems that lysis proteins provoke significant perturbation of the cell envelope and that the activation of the phospholipase A would be one of the last steps of their action (Howard *et al.*, 1991).

Another model has been proposed by De Graaf and Oudega (1986). The bacteriocin release proteins would form pores through both the inner and the outer membranes. The bacteriocin and other proteins would cross both membranes through these pores. Formation of these pores would disorganize the cell envelope and provoke the activation of the phospholipase A.

A third model may be suggested. Lysis proteins would form vesicles that provoke fusion of the two membranes, allowing them to cross the cell envelope. These vesicles would carry soluble proteins and would be released in the medium, explaining the presence of lysis proteins in the extracellular medium altogether with the colicin (Cavard *et al.*, 1985). The exported lysis protein behaves like the lysis proteins present in the cells, that is like an outer membrane protein; it was observed on gels only in heated samples and not in unheated samples (Cavard, 1991), suggesting that they have a similar polymeric organization

A lot of work has still to be done to understand the mechanism of action of these tiny lipoproteins.

REFERENCES

Akutsu A, Mazaki H, Ohta T (1989) Molecular stucture and immunity specificity of colicin E6, an evolutionary intermediate between E-group colicins and cloacin DF13. J Bacteriol 171:6430-6436

Andreoli PM, Nijkamp HJJ (1976) Mutants of the CloDF13 plasmid in *Escherichia coli* with decreased bacteriocinogenic activity. Mol Gen Genet 144:159-170

Baty D, Knibiehler M, Verheij H, Pattus F, Shire D, Bernadac A, Lazdunski C (1987a) Site-directed mutagenesis of the COOH-terminal region of colicin A: effect on secretion and voltage-dependent channel activity. Proc Natl Acad Sci USA 84:1152-1156

Baty D, Lloubès R, Géli V, Lazdunski C, Howard SP (1987b) Extracellular release of colicin A is non-specific. EMBO J 6:2463-2468

Cavard D, Lloubès R, Morlon J, Chartier M, Lazdunski C (1985) Lysis protein encoded by plasmid ColA-CA31. Gene sequence and export. Mol Gen Genet 199:95-100

Cavard D, Baty D, Howard SP, Verheij HM, Lazdunski C (1987) Lipoprotein nature of the colicin A lysis protein: effect of amino acid substitutions at the site of modification and processing. J Bacteriol 169:2187-2194

Cavard D, Lazdunski C, Howard SP (1989) The acylated precursor form of the colicin A lysis protein is a natural substrate of the DegP protease. J Bacteriol 171:6316-6322

Cavard D (1991) Synthesis and functioning of the colicin El lysis protein: comparison with the colicin A lysis protein. J Bacteriol 173:191-196

Chak KF, James R (1986) Characterization of the ColE9-J plasmid and analysis of its genetic organization. J Gen Microbiol 132:61-71

Chak KF, Kuo WS, Lu FM, James R (1991) Cloning and characterization of the ColE7 plasmid. J Gen Microbiol 137:91-100

Curtis MD, James R, Coddington A (1989) An evolutionary relationship between the ColE5-099 and the ColE9-J plasmids revealed by nucleotide sequencing. J Gen Microbiol 135:2783-2788

De Graaf FK, Oudega B (1986). Production and release of cloacin DF13 and related colicins. Curr Top Microbiol Immunol 125:183-205

Hakkaart MJJ, Veltkamp E, Nijkamp HJJ (1981) Protein H encoded by plasmid CloDF13 involved in lysis of the bacterial host: 1. Localization of the gene and identification and subcellular localization of the gene H product. Mol Gen Genet 183:318-325

Howard SP, Cavard D, Lazdunski C (1989) Amino acid sequence and length requirements for assembly and function of the colicin A lysis protein. J Bacteriol 171:410-418

Howard SP, Cavard D, Lazdunski C (1991) Phospholipase A-independent damage caused by the colicin A lysis protein during its assembly into the inner and outer membranes of *Escherichia coli*. J Gen Microbiol 137:81-89

Jakes K, Zinder ND (1984) Plasmid ColE3 specifies a lysis protein. J Bacteriol 157:582-590

James R, Jarvis M, Barker DF (1987) Nucleotide sequence of the immunity and lysis region of the ColE9-J plasmid. J Gen Microbiol 133:1553-1562

Lau PCK, Hefford MA, Klein P (1987) Structural relatedness of lysis proteins from colicinogenic plasmids and icosahedral coliphages. Mol Biol Evol 4:544-55

Lau PCK, Condie JA (1989) Nucleotide sequences from the colicin E5, E6 and E9 operons: presence of a degenerate transposon-like structure in the ColE9-J plasmid. Mol Gen Genet 217:269-277

Luirink J, Mayashi S, Wu HC, Kater MM, De Graaf FK, Oudega B (1988) Effect of a mutation preventing lipid modification on localization of the pCloDF13-encoded bacteriocin release protein and on release of cloacin DF13. J Bacteriol 170:4153-4160

Luirink J, Clark DM, Ras J, Verschoor EJ, Stegehuis F, de Graaf FK, Oudega B (1989) pCloDF13-encoded bacteriocin release proteins with shortened carboxyl-terminal segments are lipid modified and processed and function in release of cloacin DF13 and apparent host cell lysis. J Bacteriol 171:2673-2679

Luirink J, Duim B, de Gier JWL, Oudega B (1991) Functioning of the stable signal peptide of the pCloDF13-encoded bacteriocin release protein. Mol Microbiol 5:393-399

Mock M, Schwartz M (1978) Mechanism of colicin E3 production in strains harboring wild-type or mutant plasmids. J Bacteriol 136:700-707

Oudega B, Ykema A, Stegehuis F, de Graaf FK (1984) Detection and subcellular localization of mature protein H involved in excretion of cloacin DF13. FEMS Microbiol Lett 22:101-108

Poole K, Schiebel E, Braun V (1988) Molecular characterization of the hemolysin determinant of *Serratia marcescens*. J Bacteriol 170:3177-3188

Pshennikova ES, Kolot MN, Khmel IA (1987) Structural and functional organization of the colicin operon of the ColD-CA23 plasmid. Mol Biol 22:1273-1278

Pugsley AP (1988) The immunity and lysis genes of ColN plasmid pCHAP4. Mol Gen Genet 211:335-341

Pugsley AP, Cole ST (1987) An unmodified form of the ColE2 lysis protein, an envelope lipoprotein, retains reduced ability to promote colicin E2 release and lysis of producing cells. J Gen Microbiol 133:2411-2420

Pugsley AP, Schwartz M (1983) A genetic approach to the study of mitomycin-induced lysis of *Escherichia coli* K12 strains which produce colicin E2. Mol Gen Genet 190:366-372

Pugsley AP, Schwartz M (1984) Colicin E2 release: lysis, leakage or secretion? Possible role of a phospholipase. EMBO J 3:2393-2397

Roos U, Harkness RE, Braun V (1989) Assembly of colicin genes from a few DNA fragments. Nucleotide sequence of colicin D. Mol Microbiol 3:891-902

Sabik JF, Suit JL, Luria SE (1983) *cea-kil* operon of the ColEl plasmid. J Bacteriol 153:1479-1485

Toba M, Masaki H, Ohta T (1986) Primary structures of the ColE2-P9 and ColE3-CA38 lysis genes. J Biochem 99:591-596

Van Putten AJ, Stegehuis F, Van Bergen en Menengouwen PMP, De Graaf FK, Oudega B (1988) Alterations in the carboxy-terminal half of cloacin destabilize the protein and prevent its export by *Escherichia coli*. Mol Microbiol 2:553-562

Wandersman C, Delepelaire P (1990) TolC, an *Escherichia coli* outer membrane protein required for hemolysin secretion. Proc Natl Acad Sci USA 87:4776-4780

Yamada M, Nakazawa A (1984) Factors necessary for the export process of colicin El across cytoplasmic membrane of *Escherichia coli*. Eur J Biochem 140:249-255

FUNCTIONING OF THE pCLODF13 ENCODED BRP

J. Luirink, O. Mol and B. Oudega
Department of Molecular Microbiology
Vrije Universiteit Amsterdam
De Boelelaan 1087
1081 HV Amsterdam
The Netherlands

INTRODUCTION

Bacteriocin release proteins (BRP's) or "lysis" proteins are small membrane associated proteins required for the translocation of bacteriocins across the cell envelope of *E. coli* (De Graaf & Oudega, 1986). The gene encoding the "lysis" protein is transcribed from the same promoter as the gene encoding its corresponding bacteriocin (De Graaf & Oudega, 1986). Subcloning of several "lysis" protein genes in expression vectors under separate promoter control has facilitated the study of their structure and function (Pugsley & Schwartz, 1983; Altieri *et al.*, 1986; Luirink *et al.*, 1987a; Cavard *et al.*, 1989).

The main effects of "lysis" protein expression are:

(i) Release of its corresponding bacteriocin into the culture medium.

(ii) "Lysis", defined as a decrease in optical density in liquid cultures.

(iii) Lethality, defined as a decrease in the colony forming ability (measured on solid media).

(iv) Leakage of periplasmic proteins into the culture medium.

(v) Activation of phospholipase A located in the outer membrane of *E. coli.*

The relationships between these effects are complex and depend on the expression level of the "lysis" protein. Specific culture conditions can be chosen at which the translocation of the bacteriocin is quite specific at the level of the cytoplasmic membrane; only a small subset of cytoplasmic proteins is then co-released with the bacteriocin in a BRP-dependent way. It remains to be determined whether these proteins that "leak out" share structural features with the bacteriocins. In contrast, the translocation of bacteriocins across the outer membrane is much less specific; many periplasmic proteins are efficiently released into the culture medium, indicating that the integrity of the outer membrane is affected by the BRP, even at conditions of maintained viability of the cultured cells.

This latter observation has led to the use of the pCloDF13 encoded BRP for the secretion of human growth hormone (hGH) (Hsiung *et al.*, 1989). The hGH was genetically fused to a prokaryotic signal peptide to trigger its translocation across the cytoplasmic membrane. Subsequently, the BRP was

NATO ASI Series, Vol. H 65
Bacteriocins, Microcins and Lantibiotics
Edited by R. James, C. Lazdunski and F. Pattus

expressed in the same cells to permeabilize the outer membrane, thereby allowing the release of mature hGH from the periplasm into the culture medium. The general applicability of this approach is now being investigated using heterologous proteins of different source, size and complexity.

The efficient release of bacteriocins and "lysis" of host cells are dependent on the functioning of the detergent-resistant phospolipase A (Pugsley & Schwartz, 1984; Luirink *et al.*, 1986; Cavard *et al.*, 1987). However, it has been shown that the activation of this phospholipase A is an indirect effect of BRP-induced membrane damage (Howard *et al.*, 1991).

The "lysis" proteins of the various bacteriocin operons studied share some unusual structural properties:

i) Small size; ca. 50 amino acid residues including a typical signal peptide (De Graaf & Oudega, 1986).

ii) Lipid modification; the "lysis" proteins are lipoproteins and are processed by signal peptidase II, which specifically cleaves prolipoproteins. Lipid modification was shown to be essential for processing and functioning of "lysis" proteins in bacteriocin release and "lysis" (Cavard *et al.*, 1987; Pugsley & Cole, 1987; Luirink *et al.*, 1987b).

iii) Slow processing and a stable signal peptide which accumulates in the cytoplasmic membrane (Cavard *et al.*, 1987; Luirink *et al.*, 1989a). Normally, signal peptides of translocated proteins are very rapidly degraded after cleavage from their respective precursors by signal peptide peptidases located in the cytoplasmic membrane and in the cytoplasm (Novak & Dew, 1988).

The study described here deals with the localization of the pCloDF13 encoded BRP in whole cells, the functioning of its stable signal peptide and the mechanism of its targeting.

Localization of the pCloDF13 encoded BRP

The mature BRP has been localized previously in both inner and outer membranes by fractionation of *E. coli* minicells (Oudega *et al.*, 1984). In accordance with this result, the same distribution was found when a hybrid protein consisting of the BRP signal peptide, 25 amino acid residues of the mature BRP and the mature portion of β-lactamase (BLA) was localized by fractionation of whole cells (Luirink *et al.*, 1989b). Moreover, this localization of the hybrid BRP-BLA could be confirmed by immunoelectron microscopy (immuno-EM) using ultrathin cryosections (Luirink *et al.*, 1989b). The techniques used, fractionation, expression in minicells, gene fusion, etc. can give rise to artefacts in localization studies. Therefore, immuno-EM was used to study the localization of the native BRP in whole cells using a polyclonal mouse antiserum directed against a synthetic peptide encompassing the complete mature BRP.

The preliminary results, shown in Figure 1, indicated that cells expressing the native BRP are weakly

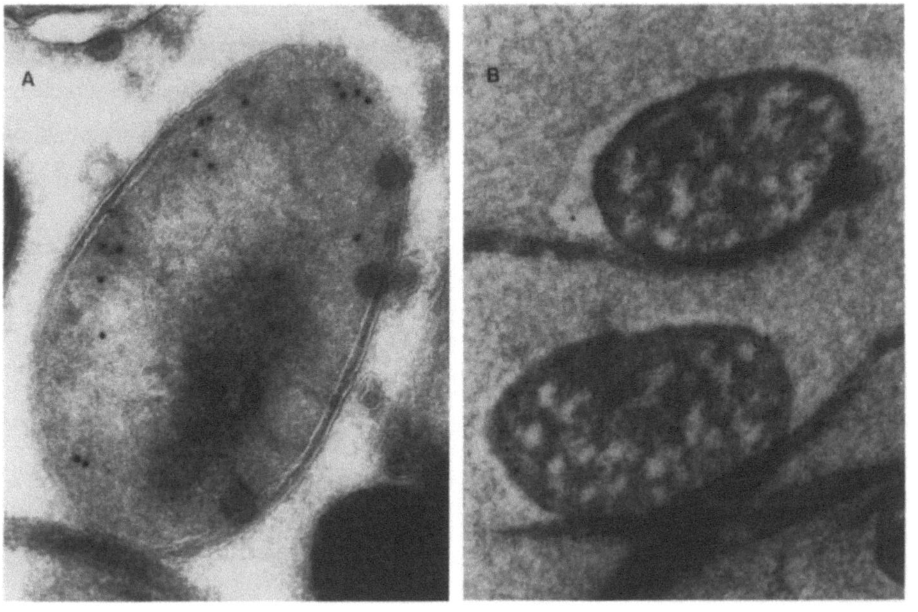

Figure 1. Immunoelectron microscopy of *E. coli* cells expressing the native mature and lipid modified BRP (A) and a mutant BRP which is not lipid-modified (B).

labelled. Most of the label was found in the cytoplasm and in the inner membranes (Figure 1A). Repeated attempts to improve the labeling, using different antibody concentrations, various labeling conditions and BRP-expression levels, failed to improve the labeling efficiency, nor did it change the apparent localization of the BRP. In contrast, under the same conditions a nonlipid modified BRP variant (mBRP) which is not processed, was efficiently labeled (Figure 1B). In this case, most of the label was found to be associated with the inner membranes, which is in agreement with the localization of a mutant, non-lipid-modified BRP-BLA hybrid protein by fractionation and immuno-EM (Luirink *et al.*, 1989b). This mBRP is expressed at the same level as its wild type counterpart, as judged by immunoblotting (not shown) using the BRP peptide antibodies to visualize the BRP. Probably, the wild type BRP is not accessible for BRP-peptide antibodies in immuno-EM. The mature, lipid-modified BRP might be an integral membrane protein which is therefore not accessible to antibodies. This could also account for the apparent resistance of the BRP (not shown) and other "lysis" proteins (Cavard, 1991) to extraction from membrane fractions using various detergents.

To test this hypothesis, the accessibility of the wild-type and mBRP to proteases was studied. Crude membranes derived from cells expressing wild-type and mBRP were treated with two different, non-specific proteases. Subsequently, the treated membrane samples were analysed by tricine SDS-PAGE and immunoblotting. The result showed that most of the mBRP precursor is degraded by both proteases (Figure

2). Essentially the same result is obtained with the wild-type (lipid-modified) BRP-precursor. Possibly, both precursors are anchored in the cytoplasmic membrane by their uncleaved signal peptides and the mature parts protrude into the cytoplasm or periplasm and are accessible to antibodies/proteases. In contrast, the wild-type mature BRP is completely resistant to added protease, which again indicates an integral membrane localization.

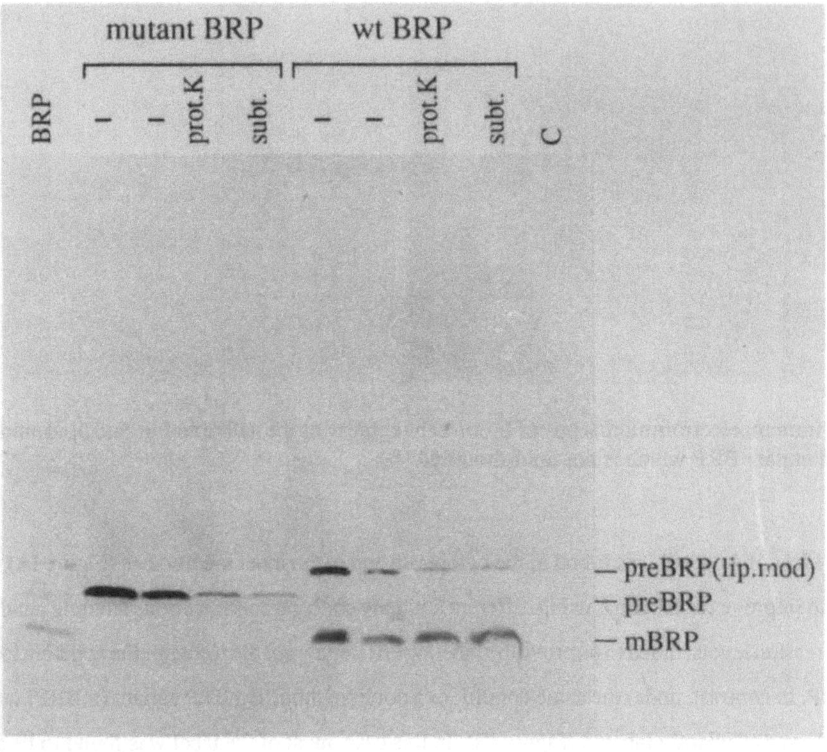

Figure 2. Protease accessibility of the BRP in *E. coli* membranes. Crude membranes were isolated from cells expressing wild-type and mutant ($Cys^{+1} \rightarrow Gly^{+1}$) BRP (non-lipid-modified) by French press treatment of cells followed by ultracentrifugation using standard procedures. Subsequently, the membranes were treated with proteinase K (final conc. 300 µg/ml; 60', 0°C) or subtilisin (final conc. 200 µg/ml; 30', 37°C). Mock treated membranes were applied in lanes 2 and 6 (no protease; 60', 0°C) and lanes 3 and 7 (no protease; 30', 37°C), respectively. A whole cell sample of cells expressing the wild-type BRP was run in lane 1. Control cells (devoid of BRP) were run in lane 10. Abbreviations: prot. K, proteinase K; subt., subtilisin; pre BRP (lip. mod.), lipid-modified precursor BRP; pre BRP, precursor BRP (non-lipid-modified); mBRP, mature BRP.

Considering the results of the different approaches to localize the BRP, it can be concluded that the BRP is located in both the inner and the outer membrane (possibly in part integral). This is consistent with the distribution of the colicin E2 and colicin A "lysis" proteins (Cole *et al.*, 1985; Cavard *et al.*, 1991). In contrast, the colicin N lysis protein was found exclusively in the outer membrane, when phospholipase-negative cells harbouring the colicin N plasmid were examined by fractionation (Pugsley, 1988).

Functioning of the stable signal peptide of the pCloDF13 encoded BRP

The pCloDF13 encoded BRP is synthesized as a precursor with an amino-terminal signal peptide which is stable after cleavage by SPaseII (Luirink *et al.*, 1989b). A possible role for this unusual signal peptide in the functioning of the BRP has been investigated by exchanging the BRP signal peptide for the unstable signal peptide of the murein lipoprotein, using site-directed mutagenesis to create unique restriction sites at the precise borders of signal sequence and mature sequence (Luirink *et al.*, 1991). The resulting hybrid protein (LPP-BRP) was normally acylated and processed by SPaseII and no stable signal peptide accumulated, suggesting that stability is an intrinsic property of the BRP signal peptide. The hybrid LPP-BRP was fully functional in causing "lysis", lethality and the release of the periplasmic enzyme β-lactamase, implying that the stable signal sequence is not essential for these effects (Figure 3). The release of cloacin DF13 however, was strongly inhibited which suggests that the BRP signal peptide is necessary for the efficient release of cloacin DF13 (Figure 3).

Figure 3. Role of the stable BRP signal peptide in BRP functioning. Mutant BRP constructs and control constructs were tested for functioning in export of cloacin DF13, "lysis", lethality and release of the periplasmic marker enzyme β-lactamase using published procedures (Luirink *et al.*, 1991). Abbreviations: LPP, murein lipoprotein; ss, signal sequence; bla, β-lactamase.

To investigate the role of the BRP signal peptide in more detail, new BRP variants were constructed. Firstly, the reverse form of the above mentioned LPP-BRP was constructed, using the same approach, to hook the stable BRP signal peptide to the mature murein lipoprotein (BRP-LPP). Secondly, the BRP signal peptide was expressed on its own by changing the triplet encoding the first amino acid of the mature BRP into a stopcodon. And finally, the unstable signal peptide of the murein lipoprotein was separated from its mature portion in the same way to serve as a control.

The expression of the encoded proteins was studied by labeling of induced cells harbouring these constructs in the presence and absence of globomycin, which inhibits the processing of prolipoproteins. The results, shown in Figure 4, indicated that the hybrid BRP-LPP is expressed and processed by SPaseII leaving a stable signal peptide. Also, the BRP signal peptide, expressed on its own, accumulated in contrast to the control signal peptide. These results again strongly suggested that stability is an intrinsic property of the BRP signal peptide itself.

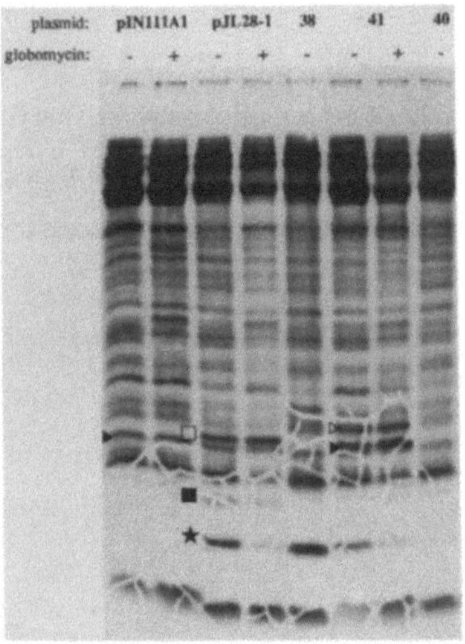

Figure 4. Expression, processing and accumulation of the stable BRP signal peptide. Cells containing different constructs were labeled with [^{3}H] amino acids, after induction with IPTG, and the resulting proteins were analysed by Tricine-SDS-PAGE. The various constructs used are as indicated. pINIIIAl, control plasmid without insert. pJL28-1, pINIIIAl with wild-type BRP gene. 38 (=pJL38), pINIIIAl derivative expressing only the stable BRP signal peptide (stop codon at +1). 41 (= pJL41), pINIIIAl derivative containing and expressing hybrid BRPLPP construct (stable signal peptide of BRP coupled to mature LPP. 40 (= pJL40), pINIIIAl derivative expressing only the unstable LPP signal peptide. Presence of globomycin is indicated by - or +. Other symbols: □, wild-type BRP precursor; ■, mature lipidated BRP; *, stable BRP signal peptide; ▷, precursor of hybrid BRP-LPP protein; ▶, mature LPP.

The constructs were also tested for functioning in the release of cloacin DF13 and β-lactamase (as a periplasmic marker enzyme) and for their ability to cause lethality and "lysis" upon overproduction. For clarity, the results (some of which are preliminary) are summarised in Figure 3. Based on these results the following conclusions can be drawn:

i) The BRP signal peptide is involved in the release of cloacin DF13. In addition, it is sufficient but not essential for causing "lysis" and lethality.

ii) The mature BRP is involved in the release of cloacin DF13 and it is sufficient for the leakage of periplasmic proteins as monitored by the release of β-lactamase. Furthermore, it is sufficient although not essential for causing "lysis" and lethality.

All the results obtained indicate that the stable BRP signal peptide plays a dual role. Firstly, it acts as a "normal" lipoprotein signal peptide in targeting, translocation, lipid-modification and processing of the attached mature BRP. Secondly, it cooperates with the mature BRP in releasing cloacin DF13 and in provoking lethality and "lysis" when highly expressed.

Further studies are necessary to find a molecular basis for these unique features of the BRP signal peptide. Several questions arise. What are the structural features that determine the stability of the BRP signal peptide? How does it interact with the inner membrane? Does it form oligomers or complexes with the mature BRP or other membrane components? Does it perhaps induce non-bilayer conformations in the inner membrane as was shown for the PhoE signal peptide (reviewed by Vrije *et al.*, 1990). Is the signal peptide involved in the formation of protein pores which allow the more or less specific passage of cloacin DF13 across the inner membrane? If so, is it directly involved in the formation of these pores or does it keep the mature BRP in a conformation which is suitable for pore formation?

To answer some of these questions, the effects of purified signal peptide and mature BRP on inner membrane vesicles and artificial membranes have to be studied. Furthermore, mutant signal peptides have to be constructed to unravel the structural features underlying stability and functioning of the BRP signal peptide.

The involvement of Sec-proteins in the translocation of the pCloDF13 encoded BRP

Since the BRP appears to be an unusual membrane protein with an unusual signal peptide and a slow rate of processing, it is interesting to know whether the BRP follows the general, "Sec-dependent" pathway for translocation targeting across the cytoplasmic membrane. Therefore, the processing of the wild-type BRP in conditional lethal *secA* and *secY* strains (temperature sensitive) was investigated. When shifted to the non-permissive temperature, these mutant strains accumulate protein precursors of most envelope proteins (for a review see Schatz *et al.*, 1990; Wickner *et al.*, 1991). SecA and SecY are the key components of the general export pathway. SecA is an inner membrane associated ATPase involved in the targeting of precursor molecules to the inner membrane and in the initial stages of their insertion into the inner membrane (Schiebel *et al.*, 1991). SecY is an integral inner membrane protein which is considered to play a role in the actual translocation of precursor proteins across the cytoplasmic membrane (Brundage *et al.*, 1990).

Processing of the BRP was severely inhibited in a *secA* strain grown at the non-permissive temperature (42°C) as compared with growth at the permissive temperature (30°C), and as compared with the isogenic wild-type strain grown at 42°C (Figure 5). The lipid-modified precursor seems to accumulate to some extent under SecA depleted conditions (Figure 5, lanes 3 and 4). Probably, a fraction of the BRP which inserts spontaneously into the inner membrane, becomes lipid-modified but is not processed for unknown reasons. Basically, the same results were obtained when a *secY* strain was tested for processing of the BRP after expression at the permissive and non-permissive temperature (not shown).

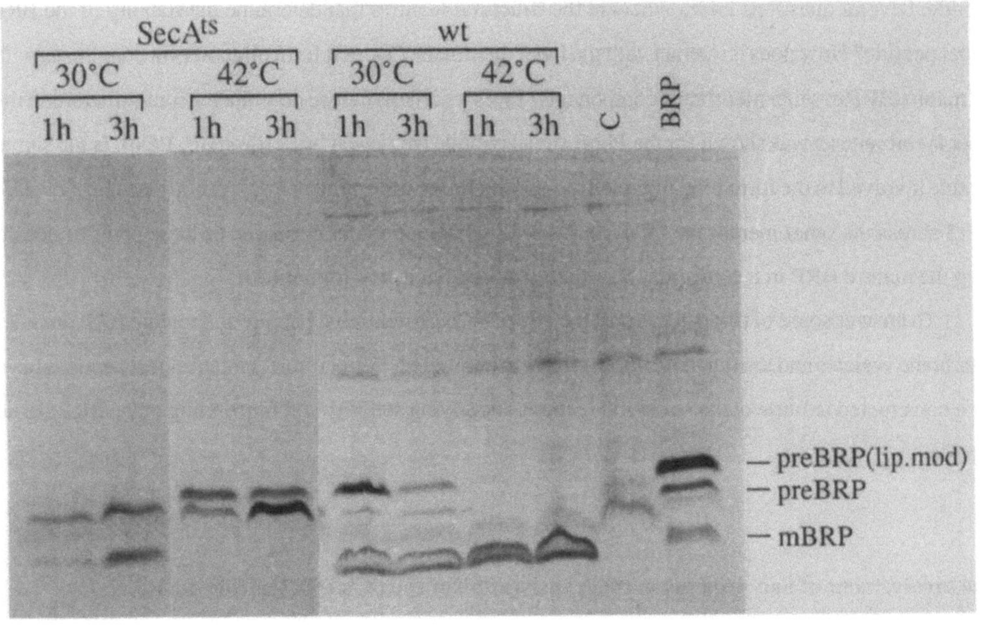

Figure 5. Expression and processing of the pCloDF13 encoded BRP in a temperature sensitive *secA* strain and in an isogenic wild-type parent strain. Conditions of growth of the cultures are indicated at the top of the figure. Samples were analysed by immunoblotting using a specific monoclonal antibody directed against the mature portion of the BRP. Symbols: preBRP (lip. mod), lipid-modified precursor of BRP; preBRP, precursor of BRP, not lipid-modified; mBRP, mature and lipidmodified BRP; C, control without BRP.

Taken together, the results indicate that the BRP follows the general route for exported proteins in *E. coli*. Given its small size, the association of the BRP with the Sec proteins is likely to occur entirely post-translational.

Preliminary evidence also showed that the release of cloacin DF13 into the culture medium is decreased in temperature sensitive *secA* and *secY* strains grown at the non-permissive temperature (not shown). Since cloacin DF13 does not contain a cleavable signal peptide, this is likely to be an indirect effect of aberrant processing of the BRP precursor and targeting under these conditions.

REFERENCES

Altieri M, Smit JL, Fan MLJ, Luria SE (1986) Expression of the cloned ColEl *kil* gene in normal and Kil^R *Escherichia coli*. J Bacteriol 168:648-654

Brundage L, Hendrick JP, Schiebel E, Driessen AJM, Wickner W (1990) The purified *E. coli* integral membrane protein SecY/E is sufficient for reconstitution of SecA-dependent precursor protein translocation. Cell 62:649-657

Cavard D (1991) Synthesis and functioning of the Colicin El lysis protein: comparison with the colicin A lysis protein. J Bacteriol 173:191-196

Cavard D, Baty D, Howard SP, Verheij HM, Lazdunski C (1987) Lipoprotein nature of the colicin A lysis protein: effect of amino-acid substitutions at the site of modification and processing. J Bacteriol 169:2187-2194

Cavard D, Howard SP, Lloubes R, Lazdunski C (1989) High level expression of the colicin A lysis protein. Mol Gen Genet 217:511-519

Cole ST, Saint-Joanis B, Pugsley AP (1985) Molecular characterization of the colicin E2 operon and identification of its products. Mol Gen Genet 198:465-472.

De Graaf FK, Oudega B (1986) Production and release of cloacin DF13 and related colicins. Curr Top Microbiol Immunol 125:183-205

De Vrije GJ, Batenburg AM, Killian JA and De Kruijff B (1990) Lipid involvement in protein translocation in *Escherichia coli*. Mol Microbiol 4:143-150

Howard P, Cavard D, Lazdunski C (1991) Phospholipase A-independent damage caused by the colicin A lysis protein during its assembly into the inner and outer membranes of *Escherichia coli*. J Gen Microbiol 137:81-89

Hsiung HM, Cantrell A, Luirink J, Oudega B, Veros AJ, Becker GW (1989) Use of bacteriocin release protein in *E. coli* for excretion of human growth hormone into the culture medium. Biotechnol 7:267-271

Luirink J, Clark DM, Ras J, Verschoor EJ, Stegehuis F, De Graaf FK, Oudega B (1989a) pCloDF13-encoded bacteriocin release proteins with shortened carboxyl-terminal segments are lipid modified and processed and function in release of cloacin DF13 and apparent host cell lysis. J Bacteriol 171:2673-2679

Luirink J, De Graaf FK, Oudega B (1987a) Uncoupling of synthesis and release of cloacin DF13 and its immunity protein by *Escherichia coli*. Mol Gen Genet 206:126-132

Luirink J, de Jong J, van Putten AJ, De Graaf FK, Oudega B (1989b) Overproduction of hybrid BRP-β-lactamase proteins: effects on functioning and localization. FEMS Microbiol Lett 58:25-32

Luirink J, Van der Sande C, Tommassen J, Veltkamp E, De Graaf FK, Oudega B (1986) Mode of action of protein H encoded by plasmid CloDF13: effects of culture conditions and of mutations affecting phospholipase A activity on excretion of cloacin DF13 and on growth and lysis of host cells. J Gen Microbiol 132:825-834

Luirink J, Watanabe T, Wu HC, Stegehuis F, De Graaf FK, Oudega B (1987b) Modification, processing and subcellular localization in *Escherichia coli* of the pCloDF13-encoded bacteriocin release protein fused to the mature portion of β-lactamase. J Bacteriol 169:2245-2250

Novak P, Dev IK (1988) Degradation of a signal peptide by protease IV and oligopeptidase A. J Bacteriol 170:5067-5075

Oudega B, Ykema A, Stegehuis F, De Graaf FK (1984) Detection and subcellular localization of mature protein H involved in excretion of cloacin DF13. FEMS Microbiol Lett 22:101-108

Pugsley AP (1988) The immunity and lysis genes of ColN plasmid pCHAP4. Mol Gen Genet 211:335-341

Pugsley AP, Cole ST (1987) An unmodified form of the ColE2 lysis protein, an envelope lipoprotein, retains reduced ability to promote colicin E2 release and lysis of producing cells. J Gen Microbiol 133: 2411-2420

Pugsley AP, Schwartz M (1983) Expression of a gene in a 400-base pair fragment of colicin plasmid ColE2-P9 is sufficient to cause host cell lysis. J Bacteriol 156:109-114

Pugsley AP, Schwartz M (1984) Colicin E2 release: lysis, leakage or secretion? Possible role of a phospholipase. EMBO J 3:2393-2397

Schatz PJ, Beckwith J (1990) Genetic analysis of protein export in *Escherichia coli.* Ann Rev Genet 24: 215-248

Schiebel E, Driessen AJM, Hartl F-U, Wickner W (1991) $\Delta\mu H+$ and ATP function at different steps of the catalytic cycle of preprotein translocase. Cell 64:927-939

Wickner W, Driessen AJM, Hartl F-U (1991) The enzymology of protein translocation across the *Escherichia coli* plasma membrane. Annu Rev Biochem 60:101-124

STRUCTURE/FUNCTION RELATIONSHIPS IN THE SIGNAL SEQUENCE OF THE COLICIN A LYSIS PROTEIN

S. P. Howard and L. Lindsay
Faculty of Medicine
Memorial University of Newfoundland
St. John's, Newfoundland AlB 3V6
Canada

INTRODUCTION

Colicins are antibacterial protein toxins which are synthesized and exported during the SOS response of *Escherichia coli* cells which bear a colicin plasmid. The export of colicins requires the action of small lipoproteins (lysis or release proteins) which are also encoded by the plasmid-borne colicin operons. The lysis protein causes a partial decrease in the optical density of the producing culture called quasi-lysis which accompanies the synthesis of the colicin and its slow release to the environment (de Graaf & Oudega, 1986).

The colicin A lysis protein (Cal), like all bacterial lipoproteins (Hayashi & Wu, 1990), is synthesized in precursor form (pCal), which is converted to a modified precursor form ($pCal^m$) by fixation of a diglyceride on the +l cysteine residue (Cavard *et al.*, 1987; Cavard *et al.*, 1989b). $pCal^m$ is then processed by signal peptidase II to yield the 18 amino acid signal peptide and the 33 amino acid mature Cal. The metabolism of Cal is very unusual however in that the lipid modification and processing steps are inordinately slow. It takes approximately 30 min for the Cal synthesized during a 1 min pulse with radioactive methionine to be completely processed. Perhaps even more unusual is the fact that the signal peptide which is cleaved from $pCal^m$ appears to be completely stable (Cavard *et al.*, 1989a; Howard *et al.*, 1989). It has also been demonstrated that the recently discovered DegP protease plays a significant role in Cal metabolism (Strauch & Beckwith, 1988; Cavard *et al.*, 1989b). When cells are treated with the antibiotic globomycin, which results in the accumulation of $pCal^m$, much of it is degraded to a smaller form of approximately the same molecular weight as mature Cal. This protease can also be expected to degrade $pCal^m$ in cells which have not been treated with globomycin, since the slow processing results in substantial quantities of $pCal^m$ accumulation even in untreated cells.

It has been established that the induction of the lysis protein causes activation of the outer membrane phospholipase A and that this activation is required for quasi-lysis and the export of the colicin (Pugsley & Schwartz, 1984; Luirink *et al.*, 1986; Cavard *et al.*, 1987). This phospholipase is an outer membrane protein however, and its activation is required only for the breach of the permeability barrier of this

NATO ASI Series, Vol. H 65
Bacteriocins, Microcins and Lantibiotics
Edited by R. James, C. Lazdunski and F. Pattus

membrane. This was shown by experiments in which colicin release was examined in *tolQ* cells, which are periplasmic-leaky (Howard *et al.*, 1991). In these cells, colicin A export did not depend on activation of phospholipase A, but still depended on the synthesis of a functioning Cal protein. This indicated that Cal caused other modifications (damage) of the envelope to effect colicin A release. Density gradient analysis of membranes isolated from cells in which Cal had been induced provided evidence for such damage. The accumulation of Cal resulted in an inability to separate the inner and outer membranes by this technique, suggesting that the lysis protein was causing fusion of the inner and outer membranes.

We have isolated a monoclonal antibody directed against Cal *via* construction of a fusion protein so that we are able to more precisely follow the processing, lipid modification, and assembly reactions which it undergoes (manuscript submitted). The availability of this antibody will allow us to determine the elements of the primary structure of precursor Cal which are responsible for the stable signal peptide and the extremely slow rate at which the post-translational lipid modification and processing reactions take place. In addition the antibody permits us to examine the localization of Cal and changes in its metabolism and localization which result from synthesis in different genetic backgrounds and in the presence of antibiotics such as globomycin.

MATERIALS AND METHODS

Bacterial strains and plasmids. Escherichia coli K12 strains W3110 and CBM were used in all experiments. CBM is strain W3110 with a *pldA* mutation. Plasmids pKA (Howard *et al.*, 1989) and pKAd were used as the source of the Cal gene. pKAd contains a deletion in the Cal operon similar to that in pATl which results in increased synthesis of Cal (Cavard *et al.*, 1989a).

Growth conditions. Strains were grown at 37°C with good aeration in either minimal media or Luria broth, supplemented with 50 μg/ml ampicillin. Cultures having an optical density of ~1.0 at 600 nm were induced with 300 ng/ml mitomycin C (Sigma).

Method of Cell Breakage. After two hours of mitomycin C induction, 10 ml cell cultures were harvested at 6,000 rpm for 5 min. The pellet was resuspended on ice in 1 ml 25% sucrose (w/w) in 10 mM Hepes pH 7.8, 5mM EDTA. Lysozyme (Sigma) was added to a final concentration of 200 μg/ml. After 20 min on ice, 9 ml of ice-cold dH_20 was added to cause osmotic cell lysis. The suspension was briefly sonicated followed by centrifugation at 2000 rpm for 10 min just prior to loading onto the sucrose gradient.

Isopycnic sucrose density gradient centrifugation. SGO, SGl, SG2 were performed as previously described (Ishidate *et al.*, 1986).

Immunoblotting. Gradient fractions were diluted 1:1 in 2X SDS-PAGE sample buffer and heated at 95°C for 5 min before being loaded onto a SDS-PAGE gel (Schagger & von Jagow, 1987), from which the proteins were transferred to nitrocellulose after electrophoresis (Towbin *et al.*, 1979). All protein transfers were done using a semi-dry transfer apparatus (Gelman). The color reaction was developed using .05 mg/ml BCIP and 0.1 mg/ml NBT diluted in 50 mM $MgCl_2$, 100 mM NaCl, 100 mM Tris, pH 9.5.

SDS-PAGE of Membrane Fractions. Sucrose density gradient sample fractions were first concentrated 10X via TCA precipitation. TCA was added to a final concentration of 10%, and the samples left for 30 min on ice. They were then centrifuged at 15,000xg at 4°C for 30 min. The supernatant was aspirated and the pellets were washed in ice-cold 90% acetone and recentrifuged for 30 min at 4°C. The supernatant was again aspirated and any excess acetone evaporated. The pellets were resuspended in sample buffer and electrophoresed on SDS-PAGE gels which were stained with Coomassie brilliant blue.

RESULTS

SYNTHESIS, LOCALIZATION AND PURIFICATION OF THE FUSION PROTEIN

The structure of the fusion protein NS is shown in Figure 1.

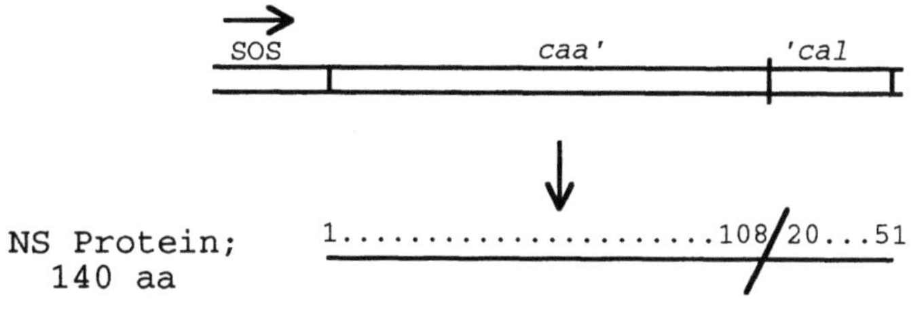

Figure 1. Structure of the NS fusion gene and protein. The fusion was constructed by restriction and ligation of pKA as previously described (submitted). The numbers refer to the amino acid residues of the original proteins.

When the NS protein was induced, a protein of the expected molecular weight (14 kDa) accumulated in the cells, and crude fractionation studies showed that it was localized entirely in the cytoplasm as a soluble protein (Figure 2).

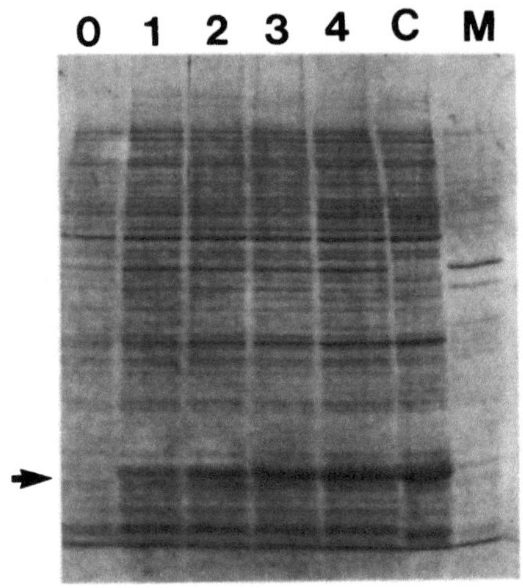

Figure 2. Synthesis and localization of the NS fusion protein. Cells containing pNS were induced and sampled after the time in hours indicated. The cells were then disrupted and centrifuged to obtain a cytoplasmic (C) and membrane (M) fraction. The arrow shows the position of NS.

A preparation of the cytoplasmic fraction was prepared and the NS protein purified using the anti-colicin A monoclonal antibody lCll, as shown in Figure 3. This was then used to raise a monoclonal antibody (CAl) against mature Cal.

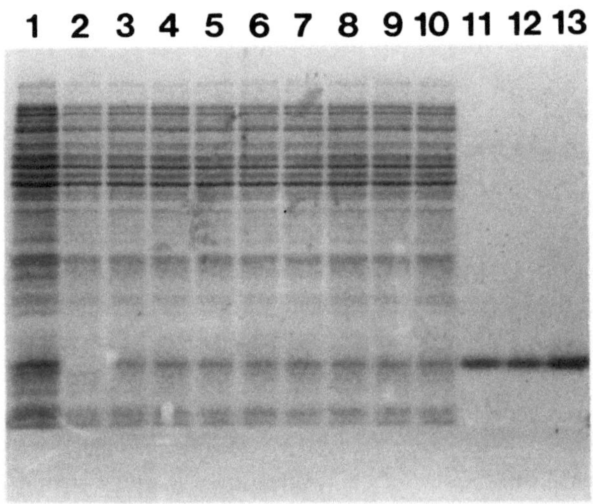

Figure 3. Purification of the NS fusion protein. An affinity column made with the monoclonal antibody lCll was used to purify NS from cell free extracts of induced W3110(pNS). The sample is in Lane 1; the pass-through fractions in Lanes 2-10, and the material eluted with 0.lM glycine, pH 2.5 in Lanes 11-13.

SYNTHESIS AND LOCALIZATION OF CAL IN WILD TYPE AND *pldA* CELLS

Isopycnic sucrose density gradient centrifugation was performed to analyze the inner and outer membranes of cells which had been induced or left uninduced for the synthesis of Cal, in both minimal media and Luria broth. Western blots using the α-Cal monoclonal antibody CAl were done to detect Cal in individual fractions of the gradients, in order to verify its identity, to determine its location and to estimate its quantity.

The gradient profile of *E. coli* cells grown in minimal media consistently contained an inner membrane peak at a density of 1.14 g/cm^3 and an outer membrane peak of 1.24 g/cm^3 These densities remained consistent among both *E. coli* strains W3110 and CBM, regardless of whether the strain carried the plasmids pKA or pKAd, or no plasmid at all. However, when comparing the gradient profiles of W3110 and CBM, there were some slight differences. The inner membrane and outer membrane peaks were more sharply defined in W3110 than in CBM, in both non-induced and induced cells. In both strains, upon induction, the height of the inner membrane and outer membrane peaks decreased and a middle peak at a density of ~1..19 g/cm^3 appeared, as shown in Figure 4.

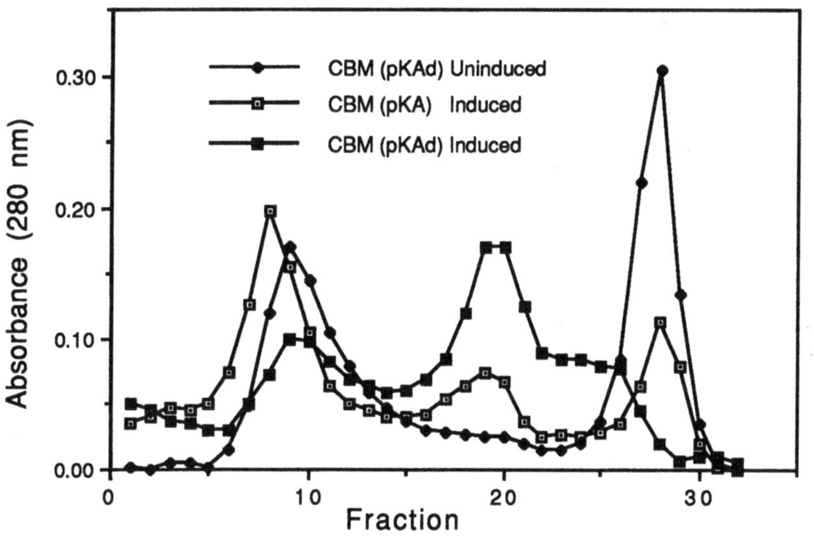

Figure 4. Effect of Cal induction on the sucrose density gradient separation of inner and outer membranes. CBM cells containing the plasmids pKA and pKAd were induced with mitomycin C. After a further 2 hr incubation, the cells were harvested, lysed and their membranes separated on an SGl gradient.

The extent of the decrease in the inner and outer membrane, as well as the increase in the height of the middle peak, was found to be slightly more pronounced in W3110 than in CBM. In addition, cells carrying the pKAd plasmid showed a much more dramatic change in the shape and location of the peaks compared to cells containing pKA (Figure 4). The effects of Cal induction on the sucrose gradient profiles in decreasing order of alteration can be summarized as:

W3110(pKAd) > CBM(pKAd) > W3110(pKA) > CBM(pKA)

The membrane profile of cells grown in LB showed a much more marked effect after Cal induction compared to cells grown in minimal media. As in minimal media, the same relationships were observed in LB with respect to the relative degree of damage done to the membrane in W3110 versus CBM cells containing pKAd versus pKA. A sample gradient profile is shown in Figure 5.

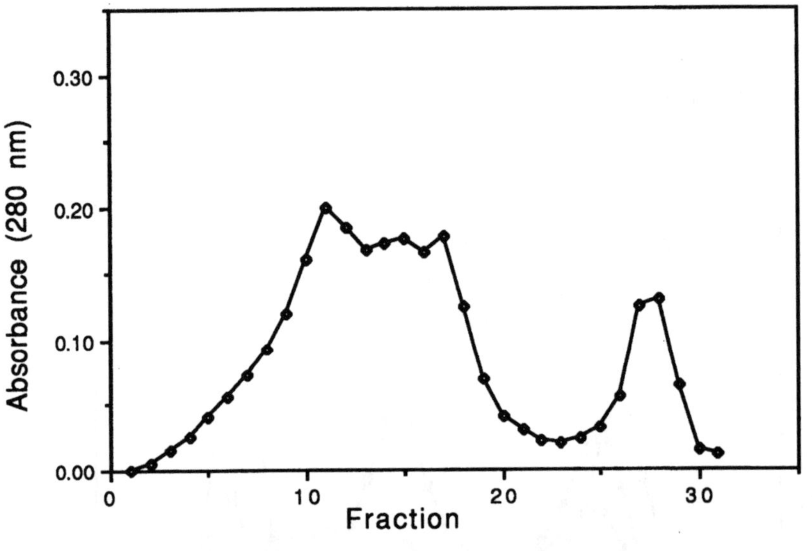

Figure 5. Membrane profile of W3110(pKA) cells grown in LB. The cells were induced and their membranes fractionated as described in the legend to Figure 4.

As a control for these experiments, cells lacking a plasmid were also induced and fractionated. The results showed no differences between the gradient profiles of non-induced and induced cells, verifying that Cal was causing the membrane damage observed. In addition, SG2 floatation gradients were done to verify that the middle peak was real and not an artifact of aggregation or co-sedimentation. Gradient

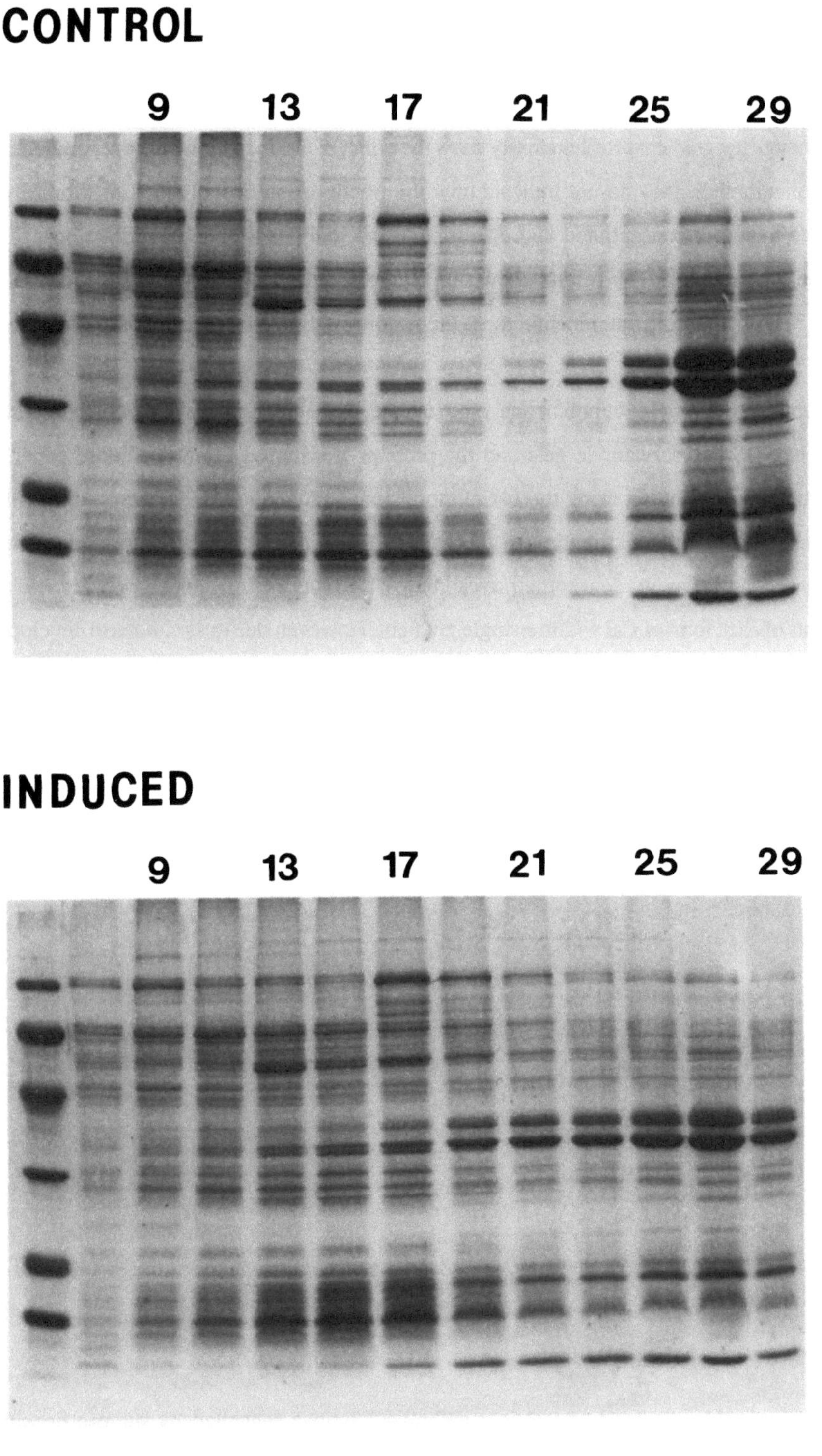

Figure 6. SDS-PAGE of membrane fractions of CBM(pKAd) cells. Membrane fractions were TCA precipitated and electrophoresed on a 10% SDS-PAGE gel.

fractions containing each of the peaks were pooled, adjusted to a density of 1.25 g/cm^3 with powdered sucrose, and loaded onto SG2 floatation gradients (Ishidate *et al.*, 1986). In all instances, the peaks reappeared at the same densities (results not shown).

Although the gradient profiles clearly show the effect of Cal induction on the separation of the inner and outer membranes, they do not indicate how the fractionation of individual membrane proteins is altered. We therefore concentrated and electrophoresed individual membrane fractions, followed by Coomassie blue staining. The results shown in Figure 6 indicated that Cal induction resulted in the spreading of both inner and outer membrane proteins toward the center of the gradient from their normal locations.

In order to perform immunoblot analysis, sucrose gradient fractions were electrophoresed on 1-70 kDa Tricine SDS polyacrylamide gels and the proteins transferred to nitrocellulose paper. This was followed by immunoblotting using the α-Cal monoclonal antibody CAl. This procedure enabled both qualitative analysis of the distribution of the various forms of Cal present across the membrane and quantitative comparisons, via densitometric scanning of the blots. In this manner, we were able to compare the amounts of each form of Cal within a single gradient. However, due to variations in development from one Western blot to another, quantitative comparisons could not be accomplished between two separate gradients. A blot of a CBM(pKA) gradient is shown in Figure 7.

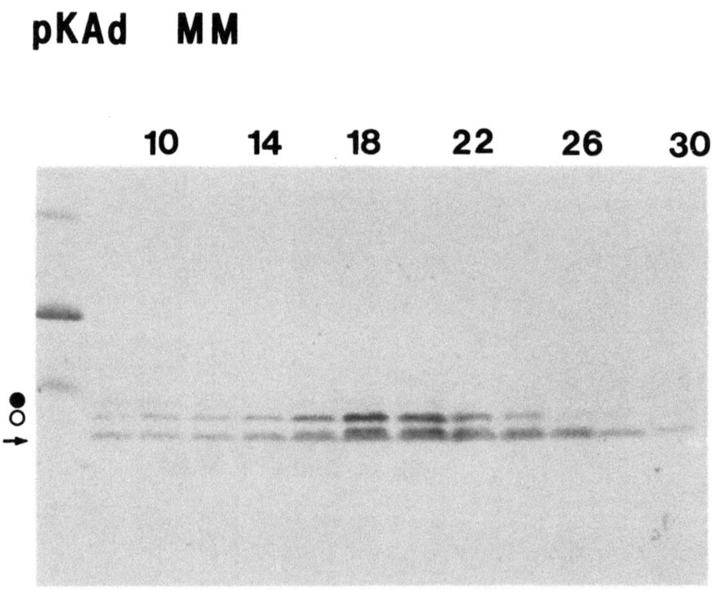

Figure 7. Immunoblot analysis of Cal in membrane fractions separated by sucrose density gradient centrifugation. The gradient fractions of induced CBM(pKA) cells were electrophoresed and immunoblotted as described in Materials and Methods. (●), pCalm; (o), pCal; (→), mature Cal.

Mature Cal accumulated at a density of 1.14 g/cm^3, which corresponds to the inner membrane peak, and as well at 1.24 g/cm^3, corresponding to the outer membrane. In cells grown in minimal media, pKA-containing strains (both W3110 and CBM) showed pCal and pCalm accumulation only in the inner membrane at a density of 1.14 g/cm^3.

When pKAd was induced in cells grown in minimal media, the mature as well as precursor forms of Cal were observed across the entire gradient, accumulating in a region of density ~1.2 g/ cm^3 . A blot of CBM(pKAd) membranes after induction in minimal medium is shown in Figure 8. Extensive degradation of Cal was also observed, resulting in the appearance of a new band just above the position of mature Cal. This degradation presumably resulted from DegP hydrolysis of pCal^m, and was always more extensive in the membranes of W3110 cells, which also consistently appeared to have more material in the center of the gradient and more total Cal present. Growth and induction of cells grown in LB also resulted in pCal and pCal^m accumulation in the center of the gradient, regardless of the plasmid present. Western blotting analysis was also done on the SG2 floatation gradient fractions which confirmed the localization of the various forms of Cal within the distinct membrane peaks (results not shown).

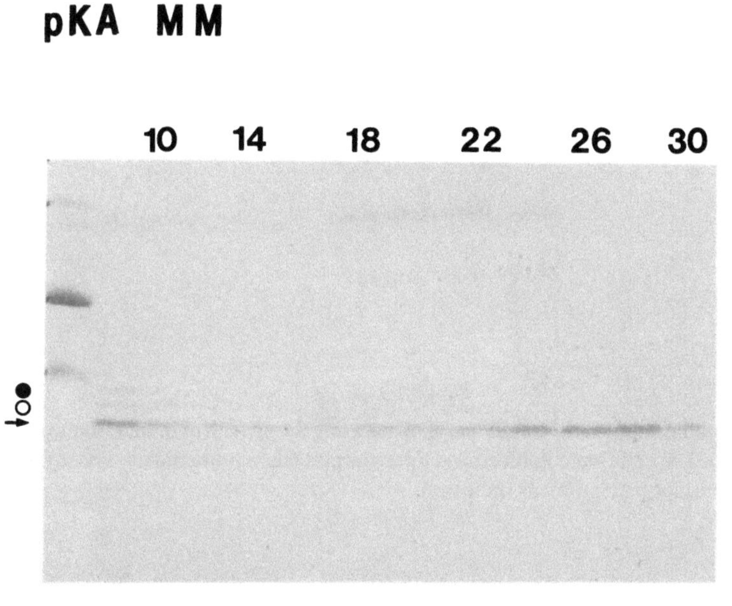

Figure 8. Immunoblot analysis of overproduced Cal in membrane fractions. The gradient fractions of induced CBM(pKAd) cells were electrophoresed and immunoblotted with CAl. (o), pCalm; (●), pCal; (→), mature Cal.

ANALYSIS OF CAL ACCUMULATION IN W3110 AND CBM CELLS

It was uncertain whether the smaller amount of membrane damage observed in cells containing the *pld*A mutation (CBM), compared to the degree of damage caused in wild type cells, was a consequence of decreased Cal production or phospholipase A inactivation. The α-Cal monoclonal antibody was therefore used to examine the accumulation of Cal in whole cell extracts of growth curve samples after induction of each strain. The intensity of the Western blot bands provided an indication of the amount of Cal accumulating after induction, as shown in Figure 9. This analysis revealed that in fact CBM cells produce lower amounts of Cal than do the W3110 cells. This thus suggests that the damage we observe on these gradients is both qualitatively and quantitatively similar between the two strains and is therefore completely independent of phospholipase activation.

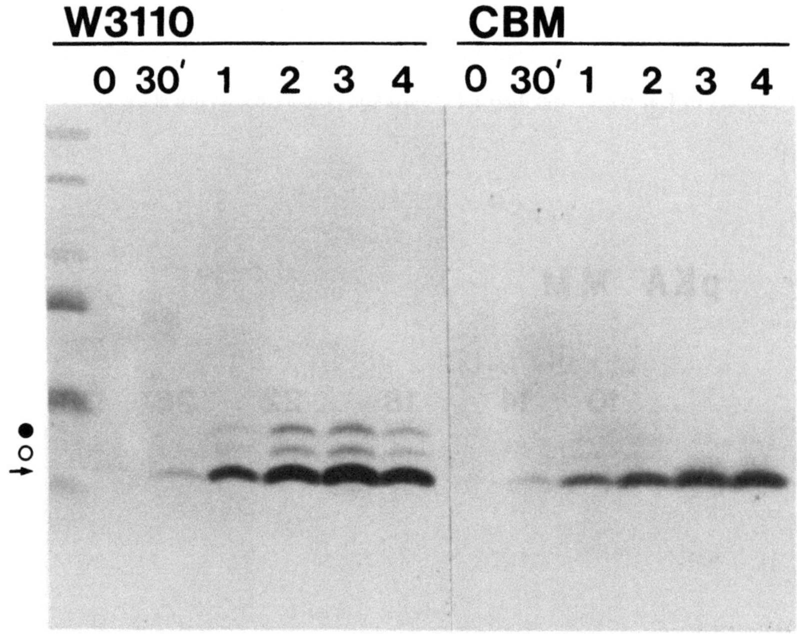

Figure 9. Immunoblot analysis of Cal accumulation during the growth of induced cultures of W3110(pKA) and CBM(pKA). The cells were induced and then samples taken immediately and after 30 min and 1,2,3, and 4 hours of further incubation as indicated.

DISCUSSION

The α-Cal monoclonal antibody we have isolated permits detailed investigation of the assembly and processing of Cal. In previous studies of Cal, localization and assembly within the cell envelope could only be studied by pulse-chase labelling followed by fractionation and was imprecise, since the mature form of the *E. coli* major lipoprotein migrated to the same position on the SDS-PAGE gels as pCal, making it impossible to verify the presence or absence of precursor Cal in the membrane fractions (Howard *et al.*, 1991). In this report we have localized the Cal intermediates without the interference of the major lipoprotein. The specific 2 hour induction time was chosen on the basis that cells which are induced for colicin A synthesis accumulate the colicin for approximately 2 hours following induction, at which time they begin to release it and other soluble proteins to the medium (Cavard *et al.*, 1989a). It was of interest to determine the precise location of the Cal intermediates at the time in which both quasi-lysis and colicin release occur.

The results of this study show that both localization and membrane damage are dependent on the amount of Cal produced. Under conditions of slow cell growth in minimal media, corresponding to lower levels of Cal production after 2 hours of induction, pCal and $pCal^{m}$ were localized only in the inner membrane. However, when grown in rich media such as LB, or even when grown in minimal media but with pKAd, which caused increased levels of Cal production, each of the forms of Cal accumulated throughout the gradients. Under conditions of decreased Cal production, which allowed almost normal fractionation of the membranes, mature Cal still accumulated in both the inner and outer membranes. We conclude that Cal is assembled into both membranes, while its processing intermediates remain in the inner membrane until such time as the accumulation of Cal causes a complete disruption of the normal structure of the envelope, perhaps involving fusion of the two membranes.

Ishidate *et al.* (1986) describe an intermediate density peak B (1.19 g/cm^3) as a membrane fraction which contains inner membrane proteins, phospholipids, and some LPS but no murein. The 1.22 g/cm^3 peak has been previously designated as OML. OML is said to be an adhesion zone between the inner and outer membranes, utilized to translocate newly synthesized LPS from the inner membrane to the outer membrane. This OML region contains both inner membrane proteins and outer membrane LPS and protein components (Ishidate *et al.*, 1986). Phospholipase A has been localized to such an intermediate membrane fraction (Bayer *et al.*, 1982) and for this reason, it is tempting to speculate that the observed disorganization of the envelope by Cal is directly responsible for phospholipase A activation.

Immunoblotting has provided information specifically about the quantity and location of Cal itself. Cal appears to disrupt the membrane in a quantitative manner such that increased Cal production leads to increased membrane disruption. Among the strains tested, there was a definite hierarchy with regard to the extent of membrane damage: W(pKAd) > C(pKAd) > W(pKA) > C(pKA). The plasmid pKAd produces more Cal than pKA due to a deletion of transcription terminator Tl (Lloubès *et al.*, 1988) in the Cal operon and our results showed significantly greater disruption of the cell envelope in cells containing pKAd. The growth curve analysis revealed that CBM cells produced lower levels of Cal compared to the wild type,

W3110, which may have accounted for the lesser amount of membrane damage observed with these cells. A number of studies have found that derivatives of the major lipoprotein, which contain signal sequence mutations inhibiting modification and processing reactions or the accumulation of modified prolipoprotein, can be harmful to the cell (Gennity *et al.*, 1990). These studies have shown that both modified and unmodified lipoprotein precursors can cause poor growth and even cell death. It is thought that this results from destructive effects of precursor accumulation in the inner membrane. These results could have important implications for the function of Cal, since its slow processing should permit both precursor and modified precursor to accumulate in the inner membrane for an extended period of time during induction of the colicin A operon. This is demonstrated conclusively here by the western blot analysis of induced cells. Damage caused by the presence of these intermediates would be expected to be a non-specific event. This would thus be in accord with previous studies of colicin A synthesis and release in which it was concluded that while colicin A release does not involve complete destruction of the cell, it does involve a permeabilization of the envelope which allows many soluble cytoplasmic and periplasmic proteins to non-specifically leak out of the cell in a diffusion-controlled manner (Baty *et al.*, 1987). It should be noted however that preliminary results from current studies on the slow processing of Cal do not support the idea that either processing intermediates or the stable signal sequence play a significant role in the release of colicin A. Mutations which result in accelerated processing and/or decreased stability of the signal sequence appear to have no effect on the efficiency with which colicin A is released from the cells (manuscript submitted). Although this does not rule out a role for these intermediates in colicin export, it does suggest that mature Cal in and of itself is sufficient for the process. Current studies are also aimed at using the monoclonal antibody to purify Cal to allow *in vitro* studies of its effect on membranes.

ACKNOWLEDGEMENTS

The authors are grateful to D. Cavard and C. Lazdunski for many helpful discussions. This research was supported by the Medical Research Council of Canada.

REFERENCES

Baty D, Lloubès R, Géli V, Lazdunski C, Howard SP (1987) Extracellular release of colicin A is non-specific. EMBO J 6:2463-2468

Bayer MH, Costello GP, Bayer ME (1982) Isolation and partial characterization of membrane vesicles carrying markers of the membrane adhesion sites. J Bacteriol 149: 758-767

Cavard D, Baty D, Howard SP, Verheij, HM, Lazdunski, C (1987) Lipoprotein nature of the colicin A lysis protein: effect of amino acid substitutions at the site of modification and processing. J Bacteriol 169 2187-2194

Cavard D, Howard SP, Lloubès R, Lazdunski C (1989a) High-level expression of the colicin A lysis protein. Mol Gen Genet 217:511-519

Cavard D, Lazdunski C, Howard SP (1989b) The acylated precursor form of the colicin A lysis protein is a natural substrate of the DegP protease. J Bacteriol 171:6316-6322

de Graaf FK, Oudega B (1986) Production and release of cloacin DF13 and related colicins. Curr Top Microbiol Immunol 125:183-205

Gennity J, Goldstein J, Inouye M (1990) Signal peptide mutants of *Escherichia coli*. J Bioenerg Biomembr *22*:233-269

Hayashi S, Wu HC (1990) Lipoproteins in bacteria. J Bioenerg Biomembr *22*:451-471

Howard SP, Cavard D, Lazdunski C (1989) Amino acid sequence and length requirements for the assembly and function of the colicin A lysis protein. J Bacteriol 171:410-418

Howard SP, Cavard D, Lazdunski C (1991) Phospholipase-A-independent damage caused by the colicin A lysis protein during its assembly into the inner and outer membranes of *Escherichia coli*. J Gen Microbiol 137:81-89

Ishidate K, Creeger ES, Zrike J, Deb S, Glauner B, McAlister TJ, Rothfield L (1986) Isolation of differentiated membrane domains from *Escherichia coli* and *Salmonella typhimurium*, including a fraction containing attachment sites between the inner and outer membranes and the murein skeleton of the cell envelop. J Biol Chem 261:428-443

Lloubès R, Baty D, Lazdunski C (1988) Transcriptional terminators in the *caa-cal* operon and *cai* gene. Nucleic Acids Res 16:3739-3749

Luirink J, van der Sande C, Tommassen J, Veltkamp E, de Graaf FK, Oudega B (1986) Effects of divalent cations and of phospholipase A activity on excretion of cloacin DF13 and lysis of host cells. J Gen Microbiol 132:825-834

Pugsley AP, Schwartz M (1984) Colicin E2 release: lysis, leakage or secretion? Possible role of a phospholipase. EMBO J 3:2393-2397

Schagger H, von Jagow F (1987) A tricine-sodium dodecyl sulfate polyacrylamide gel electrophoresis for the separation of proteins in the range from 1 to 100 kDa. Anal Biochem 166:368-379

Strauch KL, Beckwith J (1988) An *Escherichia coli* mutation preventing degradation of abnormal periplasmic proteins. Proc Natl Acad Sci USA 85:1576-1580

Towbin H, Staehelin T, Gordon J (1979) Electrophoretic transfer of proteins from polyacrylamide gels to nictrocellulose sheets: procedure and some applications. Proc Natl Acad Sci USA 76:4350-4354

THE SECRETION OF COLICIN V

Michael J. Fath, Rachel Skvirsky, Lynne Gilson, Hare Krishna Mahanty and Roberto Kolter
Department of Microbiology and Molecular Genetics
Harvard Medical School
200 Longwood Avenue
Boston, MA 02115
U.S.A

Colicin V was the first colicin to be described in the literature. In 1925, Gratia described a factor which he called factor V which was biologically active against "coli" (Gratia, 1925). This factor was studied further and given the name "colicin V" (Fredericq *et al.*, 1949), but complete characterization was hindered by the apparent instability of the protein. As more was learned about the properties of other colicins such as A, E1, and I, it became evident that colicin V (ColV) did not share most of the properties which had come to be associated with colicins. ColV is not SOS inducible, it is much smaller (6 kd) than other colicins, and it does not utilize a lysis protein for its release (Gilson *et al.*, 1990). Instead, it requires a set of dedicated export proteins. By these criteria, ColV more appropriately belongs to the related family of antimicrobial agents called microcins. For these reasons, we classify ColV as a microcin. But, for historical reasons, we feel it appropriate that its name remain unchanged.

Dedicated export systems in gram-negative bacteria

Proteins which are localized to the periplasm or the outer membrane of *E. coli* typically are synthesized as a precursor containing an N-terminal "signal sequence" and are translocated by way of the "Sec" apparatus, which is a complex of at least five proteins (Wickner *et al.*, 1991). Cleavage of the signal sequence occurs concomitant with translocation of the protein across the membrane. For a protein to be secreted into the extracellular medium it must pass through the additional barrier of the outer membrane. The number of characterized proteins and other biological molecules that are secreted out into the extracellular medium is small. However, in almost every case, extracellular secretion has been shown to require the presence of a dedicated export system. Perhaps the best characterized dedicated export system is that of α-hemolysin (HlyA) from *E. coli*, which has been shown to include three gene products, HlyB, HlyD and TolC (Holland *et al.*, 1990). The most complicated secretion system known is that for the secreted protein pullulanase, which requires at least sixteen proteins for its processing and secretion (Pugsley *et al.*, 1990).

We have characterized the export system for ColV from plasmid pColV-K30 and found that at least

NATO ASI Series, Vol. H 65
Bacteriocins, Microcins and Lantibiotics
Edited by R. James, C. Lazdunski and F. Pattus

three proteins are required for ColV secretion (Gilson *et al.*, 1987; Gilson *et al.*, 1990). These proteins are CvaA, CvaB and TolC. The complete ColV system also includes the ColV toxin itself (CvaC), the ColV outer membrane receptor (Cir) and the ColV immunity protein (Cvi). Our working model shown in Figure 1, presents a schematic of the ColV system and shows the proposed location of the six known ColV related proteins. In this presentation, we discuss our current understanding of the processes of ColV production, export, immunity, and activity.

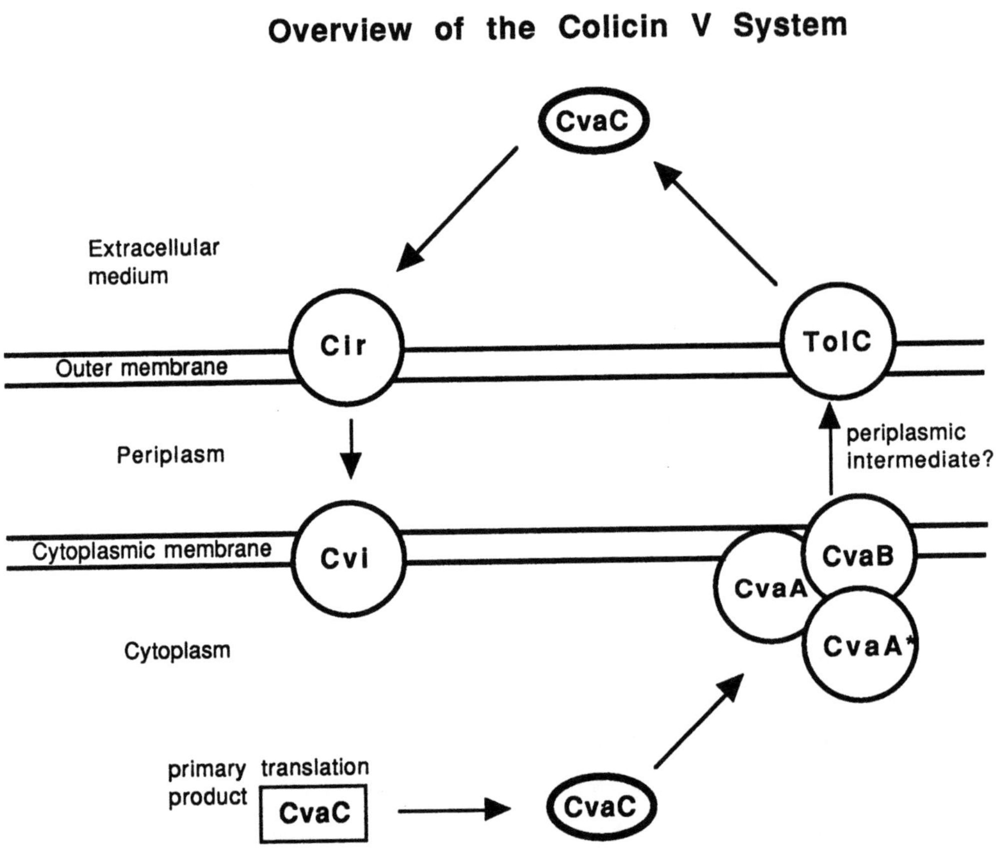

Figure 1. The current model for colicin V production, export, immunity and activity.

ColV activity

The ColV toxin itself is active against a number of *Enterobacteriaceae*, including *E. coli*, *Shigella* and *Salmonella*. The range of ColV activity is thought to be determined by the presence of the outer membrane receptor (Cir or Cir-like proteins) which is required for ColV uptake (Davies & Reeves, 1975) and therefore necessary for ColV to reach its presumed target - - the inner membrane of the sensitive cell. Since ColV is secreted into the extracellular medium, it is able to diffuse into the surrounding medium and prevent growth of nearby sensitive cells. This provides the convenient biological assay which we use to determine the level of functional extracellular toxin.

ColV activity has been found to be greatest during the late log phase when the producing cells are grown in minimal media such as M63 glucose. Activity is greatly reduced when cells are grown in rich medium. ColV exhibits its biological activity by disrupting the membrane potential of the target cell although it is not believed to form channels in the membrane (Yang & Konisky, 1984)

Cloning and Mutagenesis of ColV Determinant from pColV-K30

A 900 bp region containing the ColV genetic determinant was originally cloned from pColV-B188, but the clone did not produce extracellular ColV (Frick *et al.*, 1981). In subsequent experiments, we cloned a 4.4 kb region from the related ColV plasmid pColV-K30 and demonstrated that this clone contained the genes required for Colicin V production, export, and immunity (Gilson *et al.*, 1987). In order to map the ColV genes, the 4.4 kb region was cloned into both pBR322 and pACYC184 vectors and these plasmids (pHK11 and pHK22, respectively) were subjected to Tn*5* mutagenesis.

Characterization of the mutant ColV clones identified four complementation groups which are called *cvaA*, *cvaB*, *cvaC*, and *cvi*. Figure 2 shows a restriction map of the cloned ColV region with the location of the Tn*5* insertions. The Tn*5* insertion mutations in one gene were not polar on the activity of the other genes, since *trans*-complementation of single mutants resulted in full levels of complementation. Tn*5* insertions were not obtained in *cvi*, but the *cvi* region was identified by subcloning a 700 bp region adjacent to *cvaC* and demonstrating that it was sufficient in conferring immunity to cells which express it. It is not surprising that no insertions were obtained in *cvi* since they should be lethal to the producing cell.

Two linked genes are involved in ColV export

The four complementation groups were found to be phenotypically distinct. Mutations in *cvi* could not be obtained. Mutations in *cvaA* or *cvaB* prevented detectable ColV activity in the supernatant but led to an accumulation of active ColV in the cytoplasm of the producing cell. The CvaC protein is the ColV toxin itself, because mutations in *cvaC* abolished all ColV activity, external and internal. The fact that mutations in *cvaA* and *cvaB* do not prevent the production of active, internal Colicin V from *cvaC* demonstrates that *cvaA* and *cvaB* are not involved in activating a CvaC gene product into an active toxin. While processing of CvaC may occur during export, it is not required for making active ColV toxin.

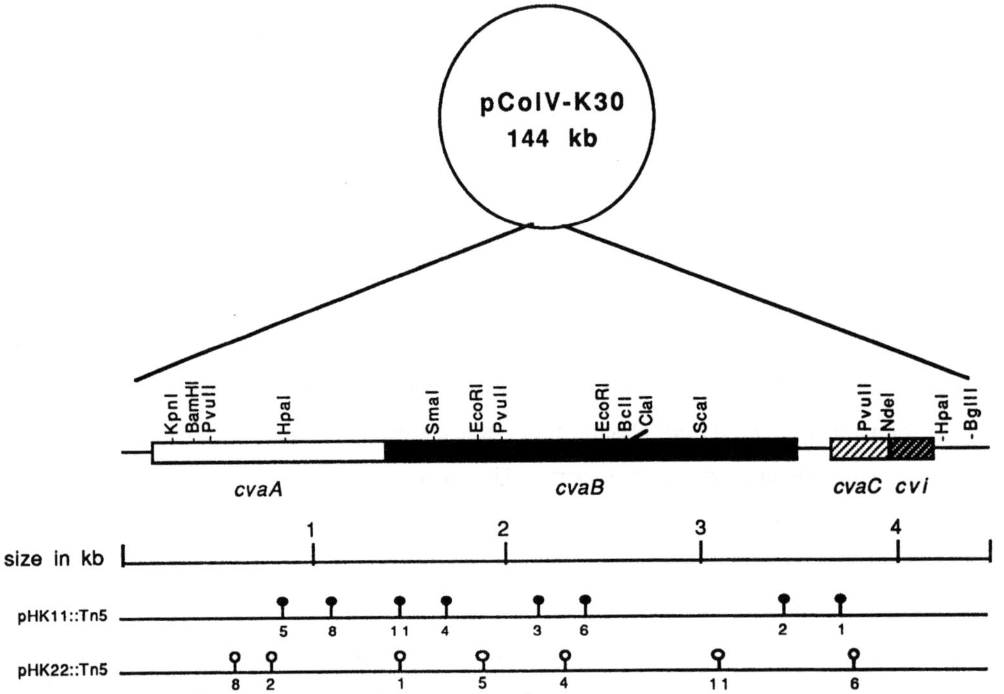

Figure 2. Restriction map of the colicin V region from pColV-K30 and location of Tn*5* insertions into the colicin V plasmids pHK11 and pHK22.

We have found that external ColV is usually quite stable - - supernatants from ColV-producing strains retain their biological activity after 65°C heat treatment, or after storage at 4°C for over one month. ColV is also active in the presence of 6M urea. In contrast, internal ColV is extremely labile. ColV activity can be found in cell lysates from *cvaA* or *cvaB* mutants when grown to late log but, after overnight growth, cells lysates show no ColV activity. This suggests that internal ColV is rapidly turned over by the cell if not exported.

We were somewhat surprised to observe that the ColV protein can be expressed by itself in the absence of *cvaAB* and *cvi*, and that the producing cells are viable. Cells which express *cvaC* have internal

ColV which can be isolated from cell lysates and shown to be active. But when these *cvi*-defective cells are transformed with the export genes, *cvaAB*, no viable transformants are obtained. This provides evidence that ColV acts directionally on the inner membrane and is only able to kill when presented to the periplasmic face of the inner membrane. It also suggests that no non-specific leakage of ColV occurs, since any such leakage would result in active external ColV which would be lethal to the producing cell.

Other preliminary experiments looking at ColV production in the absence of immunity suggest that cells with a mutation in the outer membrane receptor, Cir, can grow and produce externally active ColV when they express *cvaC* in the presence of the export genes *cvaA* and *cvaB*. If active ColV was released into the periplasm by the CvaAB exporter, it would be able to insert into the inner membrane and exert its effect in the absence of the Cir receptor. This does not appear to be the case. Therefore, this result provides genetic evidence that ColV is secreted directly into the extracellular medium without a periplasmic intermediate (Fath & Kolter, unpublished results). Such a result is consistent with the current model for α-hemolysin secretion from *E. coli*, which is also thought to occur without a periplasmic intermediate (Gray *et al.*, 1989).

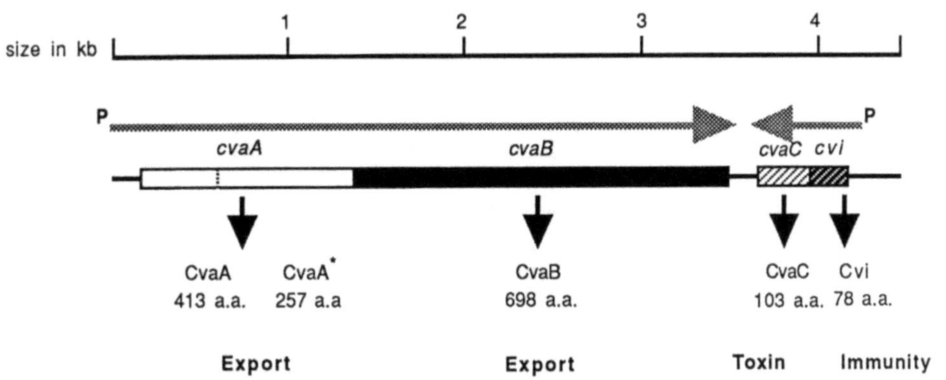

Figure 3. The colicin V region from pColV-K30 and the size and function of the four colicin V gene products.

Sequence Analysis of the ColV Operon

The DNA sequence of the entire colicin V determinant has been determined (Gilson *et al.*, 1990). Four open reading frames were identified which correspond to the four complementation groups originally identified. The genes are encoded in two converging operons, one containing *cvaA* and *cvaB* and the other containing *cvi* and *cvaC*. The sequences are identified in GenBank by the accession numbers X57524 for *cvaAB* and X57525 for *cvi/cvaC*. The direction of transcription and reading frame for each of the ColV genes was confirmed using gene fusions (Gilson *et al.*, 1990). Each gene and gene product is described in further detail below. A summary of the sequence information is provided in Figure 3.

Several lines of experimentation have shown that Colicin V production, while not inducible by SOS, is iron-regulated. It was shown that ColV expression increased in iron starved media, or in *fur* mutants (Chehade & Braun, 1988), and that β-galactosidase expression from fusions *cvi78-lacZ* (Mud51) and *cvaB55-lacZ* (Mud56) increase 12-29 fold upon addition of the iron-chelator dipyridyl (Gilson, 1990). Sequence analysis of the ColV operons showed that the region upstream of both *cvi* and *cvaC* contained a dyad symmetry element similar to the consensus sequence for other known iron regulated genes (Gilson, 1990).

Another feature of the ColV region which may play a role in ColV regulation is the 169 bp intergenic sequence located between *cvaB* and *cvaC* stop codons. Computer analysis of this sequence shows that the RNA can be folded into a highly stable, energetically-favored structure shown in Figure 4. The stem-loop structure is not preceded on either side by a poly-A sequence, which argues against it being a rho-independent terminator. The function of this region remains to be elucidated.

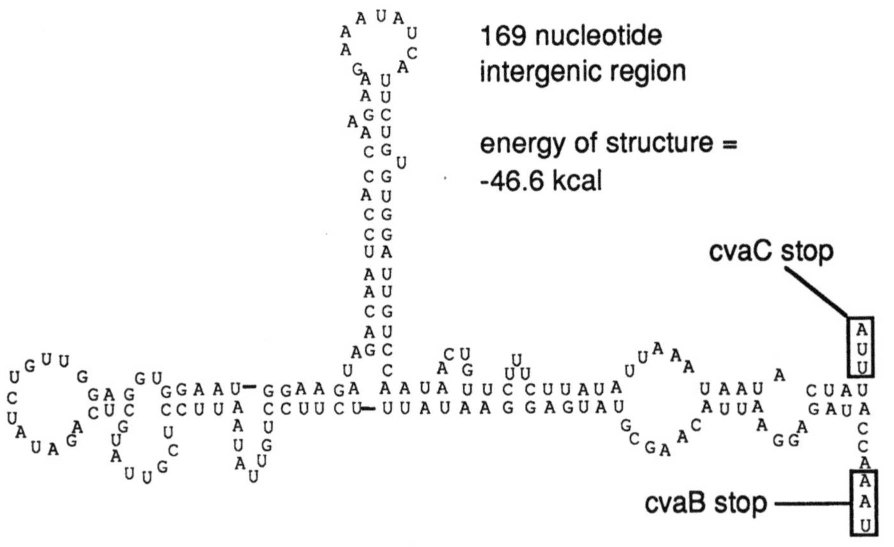

Figure 4. Putative 169 nucleotide RNA structure from the colicin V intergenic region.

Cvi, the immunity protein

Immunity to the action of ColV is provided by the small Cvi protein. Cvi is necessary and sufficient for conferring ColV immunity to *E. coli* which express it. Subclones containing only the *cvi* gene are fully immune to ColV, and the expression of the *cvaA* and *cvaB*, in the absence of *cvi*, does not result in any detectable immunity (Fath & Kolter, unpublished results). This is in contrast to the microcin B17 system, where it was found that the exporters did provide partial immunity to the cells (Garrido *et al.*, 1988).

The DNA sequence for the *cvi* gene predicts a 78 amino acid protein product with a molecular weight of 9.1 kDa (Gilson *et al.*, 1990). In mini-cells, the protein migrates at around 7 kDa (Gilson *et al.*, 1987). Sequence analysis shows that the Cvi protein has two hydrophobic stretches in the N-terminal region of the protein that could constitute membrane spanning domains. Initial purification of the *cvi78-1acZ* fusion, Mu*d*51, also provided evidence that *cvi* is localized to the inner membrane. β-galactosidase activity could only be recovered from cell lysates after the addition of Triton X-100, suggesting Mu*d*51 is membrane associated (Fath & Kolter, unpublished results).

The Mu*d*51 fusion also provides information about Cvi membrane topology. The fusion, which replaces the last codon of *cvi*, produces a bifunctional protein which is both *lacZ*$^+$ and *cvi*$^+$. This result localizes the C-terminus of Cvi to the cytoplasmic side of the membrane. Cvi is not predicted to have an N-terminal signal sequence, so it is likely that the N-terminal is also cytoplasmically localized. Based on the cytoplasmic localization of the N- and C-terminal regions and on the presence of two putative membrane spanning domains, we propose the following model for Cvi protein structure, shown in Figure 5.

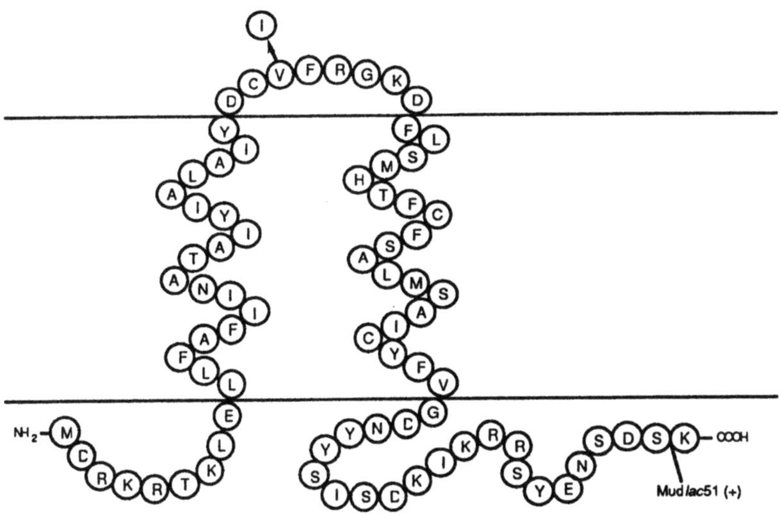

Figure 5. Model for membrane topology of Cvi.

An additional piece of data on *cvi* comes from analysis of a *cvaC* point mutation (G14D) which is defective in ColV export. This mutant was also found to have a slight defect in immunity. A spontaneous suppressor of this immunity defect was identified as a Val 31 to Ile alteration in Cvi. This altered residue is in the putative periplasmic face and would be likely to interact with ColV as it attempts to associate with the inner membrane. This result is consistent with the model of *cvi* topology.

The cvaC gene encodes the ColV toxin

The gene which encodes the ColV toxin has been designated *cvaC*. We use the terms CvaC and ColV interchangeably but tend to use CvaC when talking about the genetics of the toxin and ColV when describing the biological activity. The *cvaC* gene is located directly downstream of *cvi*, and twenty nucleotides at the end of *cvi* overlap with the start of *cvaC*.

Sequence analysis indicates that the *cvaC* gene should encode a 103 amino acid protein with a molecular weight of 10,304 Daltons (Gilson *et al.*, 1990). Mini-cell analysis identifies the ColV protein as a 6 kDa polypeptide (Gilson *et al.*, 1987). The amino acid composition is somewhat unusual. The protein is composed of 17% glycine, 14% alanine and 10% serine residues. Additionally, CvaC does not appear to contain a typical signal sequence. Hydropathy analysis (included in Figure 9) shows that the N-terminal region contains a large hydrophobic sequence which is predicted to be membrane associated. We speculate that this region of the protein is involved in the membrane insertion and killing activity of ColV.

We have shown that ColV is internally active in the absence of *cvaAB*, but we do have evidence that there may be some CvaAB-dependent processing of the ColV protein. Specifically, CvaC-PhoA fusion proteins migrate faster on SDS-PAGE gels in the presence of *cvaAB* then in the absence of *cvaAB* (Gilson *et al.*, 1990). This mobility change may be due to CvaAB-dependent processing, but the nature of this processing is not known.

Three proteins comprise a dedicated export system for ColV

The linked genes, *cvaA* and *cvaB*, are located directly adjacent to the *cvi/cvaC* operon and are transcribed by a single promoter. Similar to *cvi* and *cvaC*, the open reading frames of *cvaA* and *cvaB* overlap; in this case by five nucleotides, also suggesting translational coupling in this operon. TolC, the third gene implicated in ColV export, is unlinked to other ColV genes and maps to 66.4 minutes on the *E. coli* chromosome (Bachmann, 1990).

CvaA.

The upstream gene in the operon is *cvaA*, which encodes a protein with a predicted molecular weight of 47 kDa. Mini-cell analysis shows that two proteins are made from the *cvaA* gene; one of 43 kDa which we believe to be the complete protein, and one of 27 kDa which we call CvaA*. Sequence analysis of *cvaA* shows that there are two in-frame methionine codons at positions 156 and 160, which could act as internal start sites and give rise to a protein with a molecular weight of approximately 27 kDa. We confirmed that the 27 kDa protein was a translational restart and not a degradation product by filling in a *Bam*HI site between the *cvaA* and the *cvaA** start sites. The *Bam*HI fill-in was able to produce the 27 kDa protein, but not the 43 kDa protein seen produced from the wild type gene. The *cvaA Bam*HI fill-in construct did not produce any external ColV, demonstrating that the 43 kDa CvaA protein is required for ColV export. At this time, we do not know whether CvaA* is also required for ColV export.

Hydropathy plots of CvaA indicate that the protein is largely hydrophilic with only a small hydrophobic domain in the N-terminal 39 amino acids. The N-terminal domain does exhibit some properties of a signal sequence - a hydrophobic core and N-terminal positive charges, but it does not have a typical leader peptidase cleavage site. The possibility exists that CvaA is translocated by the "Sec" pathway and is localized to the periplasm or outer membrane. Alternately, it is possible that the N-terminal region of CvaA is a membrane spanning domain which causes CvaA to associate with the inner membrane. In contrast, CvaA* lacks the possible lipid interacting N-terminal domain and would appear to be a cytoplasmic version of CvaA. At this point however, there is no direct evidence for the localization of either *cvaA* gene product.

CvaB

The *cvaB* gene is predicted to encode a 698 amino acid protein with a molecular weight of 78 kDa (Gilson *et al.*, 1990). Amino acid and hydropathy analysis of the CvaB protein indicate that CvaB lacks a typical signal sequence but could have as many as six transmembrane domains localized in the central region of the protein between residues 179 and 438. In addition, CvaB contains sequences in the C-terminal domain which are similar to the ATP-binding domains of many other energy-transducing proteins (Blight & Holland, 1990). These characteristics place CvaB into a family of ATP-binding proteins implicated in protein trafficking which will be discussed further below.

Our working model for CvaB protein topology is shown in Figure 6. This model illustrates several interesting features of the CvaB protein. CvaB contains four large cytoplasmic loops which consist of, from N- to C-terminal, 179, 52, 63 and 261 residues. The largest cytoplasmic domain is at the C-terminal of CvaB and includes the putative ATP-binding domain. This C-terminal region shares the highest degree of sequence similarity with the other related transport proteins. The other cytoplasmic domains are likely to interact directly with the internal ColV and/or CvaA and facilitate the extracellular secretion of ColV. The periplasmic domains of CvaB are extremely small and probably serve no other function than to anchor the transmembrane domain in the membrane. The putative ATP-binding domain may serve as the energy source to facilitate the secretion of ColV out of the cell, but there is no direct evidence yet that CvaB binds to or hydrolyzes ATP.

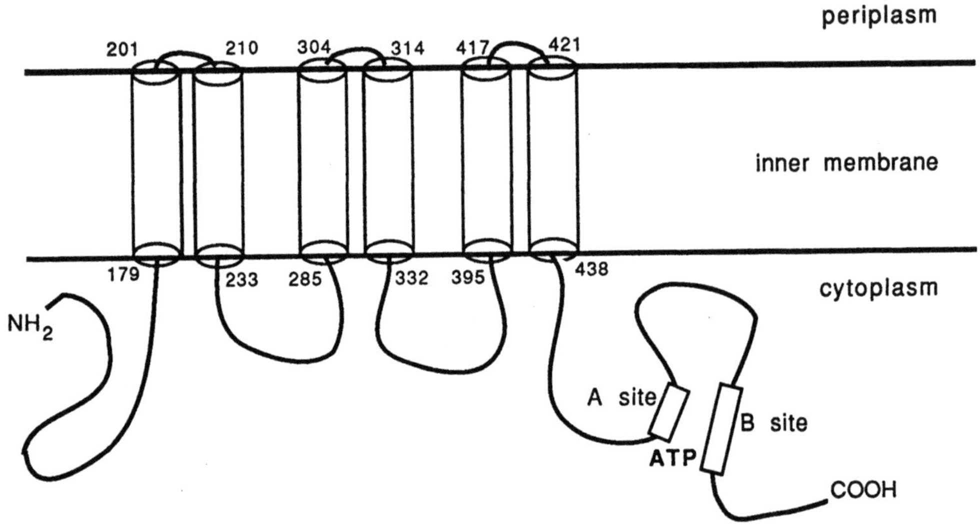

Figure 6. Model for membrane topology of CvaB.

The CvaB protein has been extremely difficult to characterize biochemically and has not yet been seen on standard SDS-PAGE gels. Several attempts have been made to overexpress *cvaB* under the control of heterologous promoters. *The cvaB* gene has been subcloned downstream of the p*tac* and the T7 promoters. Levels of CvaB, as measured by degree of ColV export, do not increase appreciably under inducing conditions, and no protein with the predicted molecular weight can be identified (Mahanty & Kolter, unpublished results). Similar problems have been encountered with the characterization of the related export protein HlyB from the *E. coli* α-hemolysin system (Mackman *et al.*, 1985). These problems have been attributed to low levels of protein expression and the inherent difficulty found in characterizing complex integral membrane proteins.

TolC

Secretion of ColV requires transport through both the inner and outer membranes of *E. coli*. Since both CvaA and CvaB are good candidates for inner membrane proteins, we suspected that a separate host protein may be involved in translocation across the outer membrane. It was recently shown that the minor outer membrane protein TolC is required for the secretion of α-hemolysin in *E. coli* (Wandersman & Delepelaire, 1990). We therefore tested to see if ColV secretion is also dependent on TolC. Cells carrying either the high copy pHK11 or native pColV-K30 plasmids were unable to export full levels of ColV in the absence of TolC (Gilson *et al.*, 1990). While TolC is important for ColV secretion, *tolC* mutations do not completely block all detectable ColV secretion. It is possible that the TolC function may be performed with greatly reduced efficiency by other proteins.

TolC has been shown to have a direct role in α-hemolysin secretion, but at this time, there is no direct evidence that TolC forms part of an export complex with either HlyBD or CvaAB. It does seem plausible though that CvaA, CvaB and TolC could form a protein complex which completely spans the inner and outer membrane and secretes ColV directly from the cytoplasm to the extracellular medium. We currently favor this model for ColV secretion, but much research will be necessary before the model should be accepted or rejected.

CvaB is a member of the MDR-like family of export proteins

When the CvaB protein sequence was compared with protein databanks, a region of approximately 200 amino acids was found which shares striking similarity with many other proteins involved in export processes. These proteins include HlyB, PrtD, NdvA/ChvA, MDR, CFTR and STE6, and were recently reviewed (Blight & Holland, 1990). HlyB is essential for hemolysin secretion in *E. coli;* PrtD is essential for protease secretion from *Erwinia chrysanthemum,* and NdvA and its homolog, ChvA are involved in polysaccharide secretion in *Rhizobium* and *Agrobacterium*. MDR is responsible for multiple drug resistance in mammalian tumor cell lines by "pumping" the drugs out of cells, the CFTR protein appears to be a novel chloride channel, and STE6 is required for export of a-factor from *Saccharomyces cerevisiae*.

Within the region of similarity are two proposed nucleotide-binding sites which make up an ATP-binding fold first described by Walker *et al.* (1982). These binding sites are found in many proteins, including those involved in export, import and other energy-transducing functions. Included in this family are many members of the periplasmic binding protein-dependent transport systems of oligopeptides (PppD), histidine (HisP), phosphate (PstB) and maltose (MalK), which utilize hydrolysis directly to energize transport (Higgins *et al.*, 1990). The structure of several ATP-binding proteins has been determined (adenylate kinase; phosphofructokinase and others), and these known structures have been used to model the ATP-binding domains for the MDR-like transporters (Hyde *et al.*, 1990; Mimura *et al.*, 1991)

We performed sequence comparisons of this conserved 200 nucleotide region to determine if the proteins involved in export are phylogenetically distinct from those involved in import or other less related functions. A phylogenetic tree was generated by computer programs based on multiple progressive alignments (Feng & Doolittle, 1987). This tree, shown in Figure 7, includes a representative subset of putative ATP-binding proteins, and shows three primary branches. One branch contains the proteins known to be involved in extracellular secretion (which we call the MDR-like subfamily); a second main branch contains components of bacterial periplasmic permeases and several gene products (Nod1, Ntr-ORF1, and FtsE) whose molecular functions are unknown. The third branch is composed of the UvrA protein (involved in DNA repair) which is only protein in this superfamily whose function is known not to be transport related. The data presented in Figure 7 support the proposition that the MDR-like exporters comprise a subfamily which is phylogenetically distinct from the other classes of characterized ATP-binding proteins.

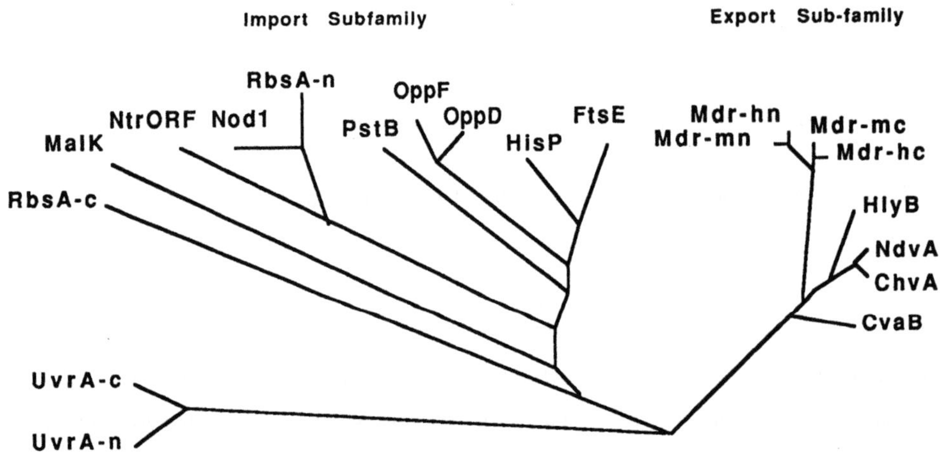

Figure 7. Phylogenetic analysis of the ATP-binding domains from the superfamily of ATPbinding proteins.

Several important differences exist between the proteins in the import and export branches. The bacterial import systems interact with a periplasmic, substrate binding protein and transport substrate into the cell. The membrane spanning and ATP-binding domains of the import systems are present as separate polypeptides rather than on a single polypeptide like the export proteins (Higgins *et al.*, 1990). Within the MDR-like export subfamily, other differences are observed between the prokaryotic and the eukaryotic members. All MDR-like export proteins contain an N-terminal hydrophobic region with six putative transmembrane domains and a conserved C-terminal nucleotide binding fold. Most eukaryotic members contain a tandem duplication of these domains within a single polypeptide. The prokaryotic members of the MDR-like family also differ from the eukaryotic members in that they require additional components to facilitate export of their substrate. For ColV secretion, *cvaA, cvaB,* and *tolC* are all required; for α-hemolysin secretion, *hlyB, hlyD* and *tolC* are required. To date, none of the eukaryotic MDR-like proteins have been found to require additional export components.

Since *cvaA* and *hlyD* play similar roles as export proteins associated with an MDR-like exporter, we tested whether they may also share some sequence similarity. Domains were found in the central regions of CvaA (a.a. 147-338) and HlyD (a.a 223-420) which are 27% identical between the proteins. Other smaller domains with higher degrees of identity were also found (Gilson *et al.*, 1990). This indicates that the second export protein in each system is also structurally conserved. Functional conservation among these bacterial export systems has also been analyzed and is described below.

ColV contains an N-terminal export signal

We have described above how the dedicated exporters CvaA, CvaB and TolC specifically interact with ColV and secrete it from the cell. Several lines of experimentation have been carried out to localize which regions of ColV interact with the export machinery and mediate secretion of ColV out of the cell. To date, three methods have been used to characterize the export signal in ColV, and the results of these experiments localize the export signal to the N-terminal 39 amino acids of the protein. A summary is shown in Figure 8.

PhoA fusions

First, a series of *cvaC-phoA* fusions was generated *in vivo* using Tn*phoA* and *in vitro* using the vector pBone1B, which contains a truncated alkaline phosphatase gene downstream of a polylinker site. These fusions were shown to be in frame with *cvaC* and shown to produce proteins with the expected mobility (Gilson *et al.*, 1990). Since the alkaline phosphatase moiety of chimeric proteins is active only when translocated across the inner membrane, the alkaline phosphatase activity of these fusions can be used as a measure of export. These fusions were assayed for alkaline phosphatase activity in the presence and the absence of CvaAB. None of the fusions gave significant activity levels in the absence of CvaAB. Fusion CvaC39-PhoA and four other fusions to more C-terminal residues of CvaC gave significant CvaAB-dependent alkaline phosphatase activity. The sixth, which fuses PhoA to the 29th amino acid of CvaC, did not give significant CvaAB-dependent PhoA activity. These results are summarized in Figure 8 and localize a major ColV export signal to the N-terminal 39 amino acids.

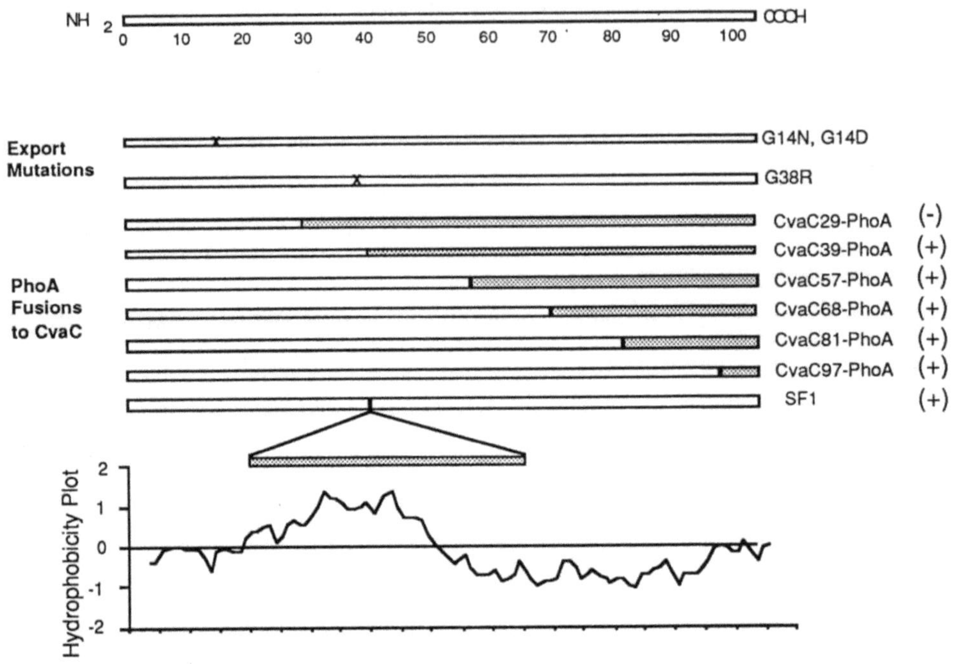

Figure 8. Summary of CvaC N-terminal export signal data

When two of the CvaC-PhoA fusions were localized, they were found to be associated with the inner membrane on the periplasmic face. This is in contrast to the extracellular localization of wild type ColV. It appears that the ColV exporter can translocate CvaC-PhoA across the inner membrane but cannot release it from the cell. One interpretation for the membrane localization of CvaC-PhoA is that additional signals, located at the C-terminal end of CvaC are required for complete extracellular secretion. In order to address this question, a sandwich fusion (SF1) was constructed which contains the entire *cvaC* gene with a truncated *phoA* gene inserted between the 39th and the 40th codon of *cvaC*. SF1 includes the N-terminal and C-terminal CvaC domains, and produces an active alkaline phosphatase fusion protein. SF1 does not appear to be secreted into the medium either, arguing that there are not specific C-terminal signals involved in complete extracellular secretion (Mahanty & Kolter, unpublished results). Instead, we believe that the membrane localization of the CvaC-PhoA fusions are due to the presence of the CvaC N-terminal hydrophobic domain in these fusions. It is possible that the bulky PhoA moiety may slow protein translocation, and the CvaC hydrophobic domain may then associate with the membrane, preventing extracellular secretion from occurring.

Export deficient point mutants

In order to identify specific amino acid residues in CvaC involved in ColV export, the *cvaC* gene was mutagenized with hydroxylamine, and mutants were isolated which were internally active but could no longer be secreted out of the cell by CvaAB. Three different mutations were identified. Two mutations were observed at Gly14 and one was found at Gly38. The Gly14Asp mutant was tight - no detectable ColV was seen. The other two mutations, Gly14Asn and Gly38Arg, were leaky and produced approximately ten percent of normal extracellular ColV activity (Gilson *et al.*, 1990). All three mutations occurred in the N-terminal region of CvaC, within the ColV export signal defined by PhoA fusion analysis.

CvaC N-terminal deletions

Proteins with typical "signal sequences" can often have their signal sequence deleted, and while they may be improperly localized, they can still retain their biological activity (Wickner *et al.*, 1991). In order to test whether deletions of the N-terminal export signal of ColV would also retain their biological activity, a series of deletions was constructed in *cvaC* which resulted in ColV proteins lacking amino acids 5 to 10, 5 to 15, 5 to 20, and 5 to 25. These deletions were inactive both externally and internally (Yeh & Kolter, unpublished results), indicating that the N-terminal domain is not exclusively involved in export. Instead, we suggest that some residues within the N-terminal domain are involved in export, while others are still required for ColV to exhibit its bactericidal activity.

Functional complementation of ColV export mutations by the α-hemolysin export system

The sequence similarities observed between CvaA/CvaB and HlyD/ HlyB led us to ask whether the α-hemolysin (Hly) export proteins could function in place of a CvaAB export system. In a series of transcomplementation experiments, we showed that the intact HlyBD could functionally substitute for CvaAB (Fath *et al.*, 1991). ColV was secreted through the HlyBD exporter at twenty percent the export level seen through CvaAB. When CvaA or CvaB was expressed along with the intact Hly export system, the level of ColV export was reduced, suggesting that the presence of a single ColV export protein interferes with heterologous export through HlyBD. The individual Hly export proteins were also able to complement mutations in their respective Cva homologue, albeit at a lower efficiency. Results of these experiments are summarized in Figure 9.

In additional experiments, export of ColV was also shown to be mediated by another MDR-like export system, the *Erwinia* protease PrtDEF proteins. We also observed that functional complementation was not reciprocal, α-hemolysin was unable to be exported by either CvaAB or PrtDEF. This could be due to the size differences between α-hemolysin, ColV and protease B, or it could be due to differences in exporter specificity.

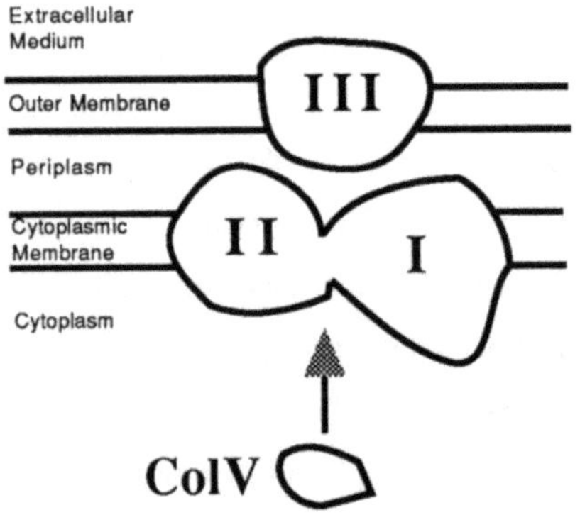

I	II	III	Export Level
MDR-like exporter	Second exporter	Outer membrane component	Halo area as a percent of wild-type
CvaB	CvaA	TolC	100
HlyB	HlyD	TolC	20
CvaB	HlyD	TolC	6
HlyB	CvaA	TolC	12
CvaB/HlyB	HlyD	TolC	12
HlyB	CvaA/HlyD	TolC	12
PrtD	PrtE	PrtF/TolC	18

Figure 9. Export of ColV through heterologous systems - transcomplementation by HlyBD and PrtDEF.

Using the CvaC-PhoA fusions and the N-terminal point mutations which were described earlier, the export signal in ColV recognized by the HlyBD exporter was characterized further. CvaC57-PhoA fusions were significantly more active in the presence of HlyBD, suggesting that the N-terminal export signal in CvaC is also being recognized by HlyBD. This is somewhat surprising, because HlyBD normally recognizes a 60 amino acid export signal at the C-terminus of α-hemolysin. There is no observable sequence similarity between the ColV export signal and the Hly export signal; yet there must be some structural similarity, since HlyBD can recognize ColV and mediate its extracellular secretion

The export-deficient point mutations in CvaC provide additional information about specific amino acids in ColV recognized by the two systems. These results are presented in Figure 10. Each of the two mutations in amino acid 14 significantly lowers ColV export through CvaAB, but these same mutations are only slightly reduced in export through HlyBD. This indicates that amino acid 14 has an important role in Cva-mediated export, but only a minor role in Hly-mediated export. In contrast, the Gly38Arg mutation reduces export through CvaAB to approximately ten percent wild-type levels and completely abolishes

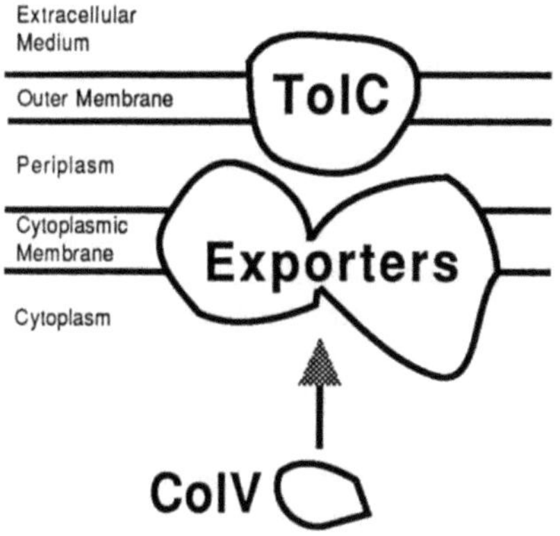

Export Proteins	Colicin V Substrate	Export Level
CvaBA	W.T.	100
HlyBD	W.T.	20
CvaBA	G14N	9
HlyBD	G14N	18
CvaBA	G14D	0
HlyBD	G14D	14
CvaBA	G38R	11
HlyBD	G38R	0

Figure 10. Export of ColV through heterologous systems - analysis of CvaC point mutations.

export through HlyBD. Therefore, this residue is an important component of the export signal for both systems. These results show that, while HlyBD and CvaAB recognize a similar N-terminal domain, specific amino acids within this domain are differentially recognized by the ColV and Hly export systems. As more export mutations are identified in the CvaC export signal, more amino acids required for ColV export by CvaAB and HlyBD will be identified. Further analysis of ColV export through the heterologous systems should provide interesting results and help further define the ColV export signals.

Acknowledgements

Support for the work presented here came from a grant from the NIH (A125944) to R.K. R.K. is the recipient of an American Cancer Society Faculty Research Award.

References

Bachmann BJ (1990) Linkage map of *Escherichia coli* K-12, Edition 8. Microbiol Rev 54:130-197

Blight MA, Holland IB (1990) Structure and function of haemolysin B, P-glycoprotein and other members of a novel family of membrane translocators. Mol Microbiol 4:873-880

Chehade H, Braun V (1988) Iron-regulated synthesis and uptake of colicin V. FEMS Microbiol Lett 52:177-182

Davies JK, Reeves P (1975) Genetics of resistance to colicins in Escherichia coli K-12: cross-resistance among colicins of group B. J Bacteriol 123:96-101

Fath MJ, Skvirsky R, Kolter R (1991) Functional complementation between bacterial MDR-like export proteins: colicin V, α-hemolysin and *Erwinia* protease. J Bacteriol 173:7549-7556

Feng D-F, Doolittle RF (1987) Progressive sequence alignment as a prerequisite to correct phylogenetic trees. J Mol Evol 25:351-360

Fredericq P, Joiris E, Betz-Barreau M, Gratia A (1949) Researche des germes proucteurs do colicins dans les selles de malades atteints de fievre paratyphoide B. C R Soc Biol 143:556-559

Frick KK, Quackenbush RL, Konisky J (1981) Cloning of immunity and structural genes for colicin V. J Bacteriol 148:498-507

Garrido MC, Herrero M, Kolter R, Moreno F (1988) The export of the DNA replication inhibitor microcin B17 provides immunity for the host cell. EMBO J 7:1853-1862

Gilson L (1990) Signal Sequence-Independent Export of Colicin V. Ph.D. Thesis, Harvard University

Gilson L, Mahanty HK, Kolter R (1987) Four plasmid genes are required for colicin V synthesis, export, and immunity. J Bacteriol 169:2466-2470

Gilson L, Mahanty HK, Kolter R (1990) Genetic analysis of an MDR-like export system: the secretion of colicin V. EMBO J 9:3875-3884

Gratia A (1925) Sur un remarquable exemple d'antagonisime entr deux souches do Colibacille. C R Soc Biol 93:1040-1041

Gray L, Baker K, Kenny B, Mackman N, Haigh R, Holland IB (1989) A novel C-terminal signal sequence targets *Escherichia coli* haemolysin directly to the medium. J Cell Sci Suppl 11:45-57

Higgins CF, Hyde SC, Mimmack MM, Gileadi U, Gill DR, Gallagher MP (1990) Binding Protein-Dependent Transport Systems. J Bioenerg Biomemb 22:571-592

Holland IB, Blight MA, Kenny B (1990) The mechanism of secretion of hemolysin and other polypeptides from Gram-negative bacteria. J Bioenerg Biomemb 22:473-491

Hyde SC, Emsley P, Hartshorn MJ, Mimmack MM, Gileadi U, Pearce SR, Gallagher MP, Gill DR, Hubbard RE, Higgins CF (1990) Structural model of ATP-binding proteins associated with cystic fibrosis, multidrug resistance and bacterial transport. Nature 346:362-365

Mackman N, Nicaud J-M, Gray L, Holland IB (1985) Identification of polypeptides required for the export of haemolysin 2001 from *E. coli*. Mol Gen Genet 210:529-536

Mimura CS, Holbrook SR, Ames GF-L (1991) Structural model of the nucleotide-binding conserved component of periplasmic permeases. Proc Nat Acad Sci USA 88:84-88

Pugsley AP, d'Enfert C, Reyss I, Kornacker MG (1990) Genetics of extracellular protein secretion by Gram-negative bacteria. Ann Rev Genet 24:67-90

Walker JE, Sarste M, Runswick MJ, Gay NJ (1982) Distantly related sequences in the α- and β- subunits of ATP synthase, myosin kinases and other ATP-requiring enzymes and a common nucleotide binding fold. EMBO J 1:945-951

Wandersman C, Delepelaire P (1990) TolC, an *Escherichia coli* outer membrane protein required for hemolysin secretion. Proc Natl Acad Sci USA 87:4776-4780

Wickner W, Driessen AJM, Hartl F-U (1991) The enzymology of protein translocation across the *Escherichia coli* plasma membrane. Ann Rev Biochem 60:101-124

Yang CC, Konisky J (1984) Colicin V-treated *Escherichia coli* does not generate membrane potential. J Bacteriol 158:757-759

INTRODUCTION TO THE SESSION ON THE EVOLUTION OF BACTERIOCINS

Richard James
School of Biological Sciences
University of East Anglia
Norwich NR4 7TJ
Norfolk
U.K

The family of E colicins provide a model system for the study of plasmid and bacteriocin evolution. The discovery of a strain of *E coli* isolated from chicken caecae by undergraduate students here at UEA in 1980 was a major breakthrough in this field. At that time 7 members of the family of E colicins had been identified and partly characterized (Watson *et al.*, 1981; Males & Stocker, 1982; Mock & Pugsley, 1982). Initial attempts by Pearl Cooper and myself to characterize the E colicin produced by *E.coli* strain J, using the classical production and immunity tests against the existing seven producing strains, failed because strain J produced two diferent E colicins. When we had independently mobilised the two different E colicin-encoding plasmids (pColE8-J and pColE9-J) to a laboratory strain, we still had great difficulty in characterizing the E colicins which they produced by production and immunity tests (Cooper & James, 1984). The problem was that, although *E.coli* carrying pColE8-J was sensitive to all other E colicins, *E.coli* carrying pColE3-CA38 was not sensitive to colicin E8. This "non-reciprocal immunity" was unusual in that there was no evidence of a second plasmid in the *E.coli* (pColE3-CA38) strain which encoded colicin E8 immunity. This situation had been described by Males and Stocker (1980) for a strain which produced colicin E7 and which also carried a small plasmid which encoded colicin E2 immunity. I remember the long discussions with Pearl Cooper as to whether the non-reciprocal immunity was due to a differential affinity of the E3 and E8 immunity proteins for colicin E8, or more likely whether there were two immunity genes on the ColE3-CA38 plasmid. A similar problem of non-reciprocal immunity arose with colicin E9, in that *E.coli* strains carrying the ColE9-J plasmid were also immune to colicin E5.

We decided that the only way to conclusively address the question of whether there were two immunity genes was to make use of the new molecular biology techniques and try to clone the immunity genes. With the help of two graduate students, Kin Chak and Mark Lawrence, we finally were able to conclusively prove that we had indeed isolated two new members of the E group colicins and that the non-reciprocal immunity we had observed was due to the presence of two immunity genes, against colicin E3 and E8 on the ColE3-CA38 plasmid (Chak & James, 1984; Lawrence & James, 1984), and against colicin E5 and E9 on the ColE9-J plasmid (Chak & James, 1986). Subsequent DNA sequencing of the relevant regions of these plasmids by a number of research groups confirmed our findings, and also revealed considerable sequence homology between the members of the DNAase group of E colicins (E2, E7, E8 and

NATO ASI Series, Vol. H 65
Bacteriocins, Microcins and Lantibiotics
Edited by R. James, C. Lazdunski and F. Pattus

E9), or between the members of the RNAase group (E3 and E6). This raises interesting questions concerning the evolution of these families of E colicins, and whether the presence of two E colicin immunity genes on the same plasmid could represent an evolutionary intermediate. It is of interest that in all cases where two immunity genes have been observed in an E colicin plasmid, one of the immunity genes is against a DNAase type E colicin, whilst the other is against an RNAase type E colicin. This suggests that some recombination event could have occurred between two plasmids, one with an RNAase killing domain, the other with a DNAase killing domain. This possibility is addressed in the contribution of Masaki *et al* in this session of the workshop.

One fundamental problem concerning the evolution of E colicins, ie. the DNAase family, is the problem of co-evolution of the colicin and the immunity gene. It has always been assumed that one role of the immunity protein is to protect the colicin-producing cell from killing by internal colicin upon its synthesis. This presents a problem in proposing a theory of evolution by the aquisition of point mutations in the colicin structural gene, in that mutations which change the phenotype of the E colicin produced will also reduce the affinity of binding of the immunity protein to the "new" colicin, and thus lead to cell death, before there is time for a parallel evolution of the immunity gene to restore immunity. The similarity in DNA sequence between colicins E2, E7, E8 and E9 however strongly support this mechanism of evolution. A possible solution to this paradox was presented earlier in the workshop by James *et al.* who suggest that DNAase type E colicins are not active inside the producing cell. An alternative explanation of how RNAase immunity genes may have evolved comes from Masaki *et al.* in their contribution to this session of the workshop in which they show that the E6 immunity gene can be mutated to encode E3 immunity whilst still retaining E6 immunity.

A mechanism of E colicin evolution which does not require co-evolution has been proposed in the case of the colicin E5 and E9 plasmids by Lau *et al.* in their contribution to this session of the workshop. In this case a transposition event is believed to have occurred between pColE5 and an E9-like plasmid, with subsequent deletion of DNA sequences giving rise to the present pColE9-J.

In their contribution ' Replicon evolution of ColE2-related plasmids', Itoh and Hiraga have compared the DNA sequence of the incompatibility regions of ten ColE2-related plasmids. The results clearly show a mosaic structure, in which the patterns of group-specific sequence homology are not correlated to either species of colicin (colicin E2 to E9), or to the type of colicin (RNAase or DNAase). They propose that this pattern must have arisen by the exchange of functional parts of the replicons by homologous recombination and/or site-specific recombination. I am particularly gratified that the results of this very comprehensive piece of work supports the findings of a preliminary incompatability study of E colicin plasmids in my laboratory (Cooper *et al.*, 1986).

REFERENCES

Chak K-F, James R (1984) Localization and characterization of a gene on the ColE3-CA38 plasmid that confers immunity to colicin E8. J Gen Microbiol 130:701-710

Chak K-F, James R (1986) Characterization of the ColE9-J plasmid and analysis of its genetic organization. J Gen Microbiol 132:61-71

Cooper PC, James R (1984) Two new E colicins, E8 and E9, produced by a strain of *Escherichia coli* J. J Gen Microbiol 130:209-215

Cooper PC, Hawkins FKL, James R (1986) Incompatability between E colicin plasmids. J Gen Microbiol 132:1859-1862

Lawrence GMP, James R (1984) Characterization of the ColE8 plasmid, a new member of the group E colicin plasmids. Gene 29:145-155

Males BM, Stocker BAD (1980) *Escherichia coli* K317, formely used to define colicin group E2, produces colicin E7, is immune to colicin E2, and carries a bacteriophage-restricting conjugative plasmid. J Bacteriol 144:524-531

Males BM, Stocker BAD (1982) Colicins E4, E5, E6, and A and properties of *btuB*$^+$ colicinogenic transconjugants. J Gen Microbiol 128:95-106

Mock M, Pugsley AP (1982) The BtuB group Col plasmids and homology between the colicins they encode. J Bacteriol 150:1069-1076

Watson R, Rowsome W, Tsao J, Visentin LP (1981) Identification and characterization of Col plasmids from classical colicin E-producing strains. J Bacteriol 147:569-577

MOLECULAR EVOLUTION OF E COLICIN PLASMIDS WITH EMPHASIS ON THE ENDONUCLEASE TYPES

Peter C.K. Lau, Michael Parsons and Tai Uchimura*
Molecular Biology Sector
Biotechnology Research Institute
National Research Council of Canada
6100 Royalmount Ave
Montreal, Quebec
Canada H4P 2R2

Introduction

The ability of a protein to counteract or neutralize the lethal action of another is an example of one of nature's fundamental biological processes. Understanding the specific nature of the protein-protein interaction that brings about the neutralization of a toxic action would be medically relevant and scientifically very rewarding.

The colicins and their immunity proteins, produced by the Col plasmids of *Escherichia coli*, are a good example of a macromolecular agonist-antagonist system. By definition, colicins are toxic proteins which kill *Escherichia coli* and closely related bacteria. The main function of the immunity proteins is to protect the colicinogenic cells from the lethal action of their 'own' toxin (for a historical account see Reeves, 1972).

Of the 23 colicin types classified thus far (Pugsley & Oudega, 1987), only two kinds have well established biochemical effects on sensitive bacterial cells. These are: i) the destruction of the cell's membrane potential as a result of the permeabilization of the cytoplasmic membrane by the pore-forming colicins (A, B, E1, Ia, Ib, K and N; for reviews see Cramer *et al.*, 1983; Lazdunski *et al.*, 1988; Pattus *et al.*, 1990), and ii) inhibition of protein synthesis by either endonucleolytic cleavage of DNA, typified by colicin E2 (Schaller & Nomura, 1976), or ribosome inactivation by specific cleavage of the 16S RNA, typified by colicin E3 (Bowman *et al.*, 1971) and cloacin DF13. The latter bacteriocin is encoded by the plasmid CloDF13, first found in *Enterobacter cloacae*, which can be stably maintained in *E. coli*. In the case of colicins E2, E3 and cloacin DF13, the immunity protein inhibits the nuclease activity by binding specifically to the carboxyl-terminus of the respective toxin. Indeed, colicin E3 was purified as a complex with its immunity protein (for reviews see Jakes, 1982; de Graaf & Oudega, 1986). In contrast to the endonucleolytic colicins, the pore-forming colicins (ionophores) are not isolated as complexes with their

*Permanent address: Department of Agricultural Chemistry, Tokyo University of Agriculture, 1-1 Sakuragaoka 1-chome, Setagayaku, Tokyo, Japan.

NATO ASI Series, Vol. H 65
Bacteriocins, Microcins and Lantibiotics
Edited by R. James, C. Lazdunski and F. Pattus

immunity proteins. The latter proteins are membrane-bound, e.g. in the ColA system the immunity protein spans the cytoplasmic membrane. Through some unknown mechanism, interaction of the immunity protein with the carboxyl-terminus of colicin A confers to the cells protection against the effects of (in this class of toxins) *exogenous* colicin A (for review see Lazdunski *et al.*, 1988).

The scope of this article includes the various evolutionary features of the nuclease-encoding ColE plasmids, limited to the DNA regions which encompass the colicin gene operons, namely *col* (colicin), *imm* (immunity) and *lys* (lysis). The *lys* gene (also known by other names, one being the bacteriocin release protein) codes for a small lipoprotein which causes colicin or bacteriocin release from the producing cell and eventual death of the host (de Graaf & Oudega, 1986). On the evolutionary aspects, we would like to accentuate the following points:

i) the modular mode of evolution of colicin nucleases with the possible involvement of a primordial microbial nuclease domain
ii) the sequence evidence of a transposition event involving an insertion sequence (IS) element
iii) coevolution of the immunity gene with its cognate colicin in a way that allows the immunity to compensate for changes in the nuclease domain of a given colicin.

The study of ColE plasmids has not been purely an academic exercise. We would like to draw examples of the biotechnological usefulness of these Col plasmids and some of the genes they encode. A possible selective advantage of the phenomenon of colicinogeny in nature is also discussed.

Col plasmids E1 to E9

The E-type colicins are defined as those which use the *btuB*-specified outer membrane protein for binding and entry into the susceptible *E. coli* cells (DiMasi *et al.*, 1973). This 614 amino-acid (a.a.) long receptor protein also carries out the active transport of vitamin B12 and the absorption of bacteriophage BF23 (Heller & Kadner, 1985). To date, nine ColE type plasmids (Table 1) have been classified on the basis of immunity tests, i.e. specific insensitivities to their own or closely related colicins (Watson *et al.*, 1981; Cooper & James, 1984; for practical details, see Pugsley & Oudega, 1987). Because a few ColE plasmids have now been found to contain double immunity genes that are non-homologous, it has been pointed out by Chak & James (1984) that the panel of indicator strains used in the classification scheme should contain only single colicin immunity genes.

All ColE plasmids that have been characterized thus far are non-conjugative and are, therefore,

Table 1: Properties of the ColE plasmids and their subtypes

PLASMID	SIZE (Kb)	INCOMPATIBILITY /COMPATIBILITY	SPECIAL CHARACTERISTICS
Col El-K30	6.646		Completely sequenced; (All Col El plasmids, but not E2 to E9, are amplifiable by chloramphenicol)
El-K47	10.5		
El-K53	5.4		
El-K321	6.5		
El-N104	6.5		
Col E2-P9	6.8	C with E3, E8	Prototype colicin E2
E2-GEI288	6.8		
E2-GEI544	6.8		Identical restriction maps but homologous to E2-P9
E2-K321	6.8		
E2-GEI602	6.8		Unique *Hinc*II site
E2-CA42	6.1		Appears to be structurally distinct
$E2_{imm}$-K317	4.3		Colicin-defective; imm $E2^{+}$; cryptic lysis
Col E3-CA38	7.2	I with E7, E8	Double immunity (E3 + E8);
Col E4-K365	6.7		Identical restriction maps
E4-CT9	6.7		
E4-284	6.7		
Col E5-099	6.7	C with E8; I with E6, E9	Unusual nuclease and imm structures
Col E6-CT14	11.2	C with E3, E7, E8; I with E5, E9	Double immunity (E6 + E8); largest Col E plasmid; double replicon
Col E7-K317	6.1	C with E2, E4, E6, E9; I with E3, E8	Parent strain carries $E2_{imm}$-K317
Col E8-J	6.8	C with E2, E4, E5, E6, E9; I with E3, E7	Has 2 near-identical imm structures in E3 and E6 plasmids
Col E9-J	7.2	C with E2, E3, E4, E7, E8 I with E5, E6	Double immunity (E9 + E5); presence of atypical lysis protein and insertion sequence (IS*E9*)

necessarily small. They range in size from 5.4-11.2 kb (kilobase pairs). Hardy *et al.* (1973) classified these plasmids as the Group I type, distinguishing them from the higher molecular weight Group II colicin plasmids which include ColB, Ib and V. In general, the copy numbers of the various ColE plasmids are estimated at about 15 per chromosome, but those of the E1-type can be substantially increased (1000-3000 copies) by the addition of chloramphenicol (Clewell, 1972). For a discussion of replication and incompatibility of ColE and related plasmids the following references would be instructive: Davidson, 1984; Chan *et al.*, 1985; Cooper *et al.*, 1986; Morlon *et al.*, 1988, Itoh and co-workers, this volume.

Among the nine representative ColE plasmids and their subtypes, the complete DNA sequence of only one, ColE1-K30, has been determined (Chan *et al.*, 1985). Extensive restriction mappings of ColE2 through ColE9 have revealed that these plasmids are highly related in structure (Watson & Visentin, 1980; Lawrence & James, 1984; Watson *et al.*, 1985; Chak & James, 1986). For example, over 80% of the restriction sites of 13 Col plasmids, representing E2 to E7 types, were found to be similarly positioned (Watson *et al.*, 1985). A classical case is that of ColE2-P9 (DNase producer) and ColE3-CA38 (RNase producer) in which the two plasmids differ only in a limited region, either by heteroduplex analysis (Inselburg, 1973) or restriction mapping (Watson & Visentin, 1980). The largest ColE plasmid is ColE6-CT14; it is some 4 kb larger than the rest of the ColE plasmids except ColE1-K47. A second replicon has been found in the extraneous 4 kb DNA fragment of ColE6 (Selvaraj, Lau and Thatte, unpublished results). Without the 4 kb fragment, the restriction map of ColE6 is otherwise virtually indistinguishable from that of ColE3-CA38 (Watson *et al.*, 1985). We will make no further attempt to align the ColE plasmid maps by common restriction sites but instead offer the following comment. What seems to be structurally distinct by restriction data in relation to the remaining E2 plasmids, e.g. ColE2-CA42 (Table 1; Watson *et al.*, 1985), may actually turn out to be rather similar in gene organization. This was the case with the ColE1-K30 and pKY-1 plasmids (Higashi *et al.*, 1986; equivalent to ColE1-K53, this lab) where it was found that although the restriction maps are completely different, the arrangements of the *col*, *imm* and *lys* genes, and the region of replication origins were the same in both cases. This is not surprising since single nucleotide changes often lead to different endonuclease recognition sites.

Representative operon gene arrangements from each of the E colicin groups are now known

With the determination of nucleotide sequences from plasmids ColE4-K365 (E4, this study and unpublished results) and ColE7-K317 (Chak *et al.*, 1991; this study and unpublished results), we complete the picture of the colicin operon gene organizations for each representative group of ColE plasmids (Figure 1). This also allows the grouping of colicin E4 (Watson *et al.*, 1981 and refs. therein) in the RNase class,

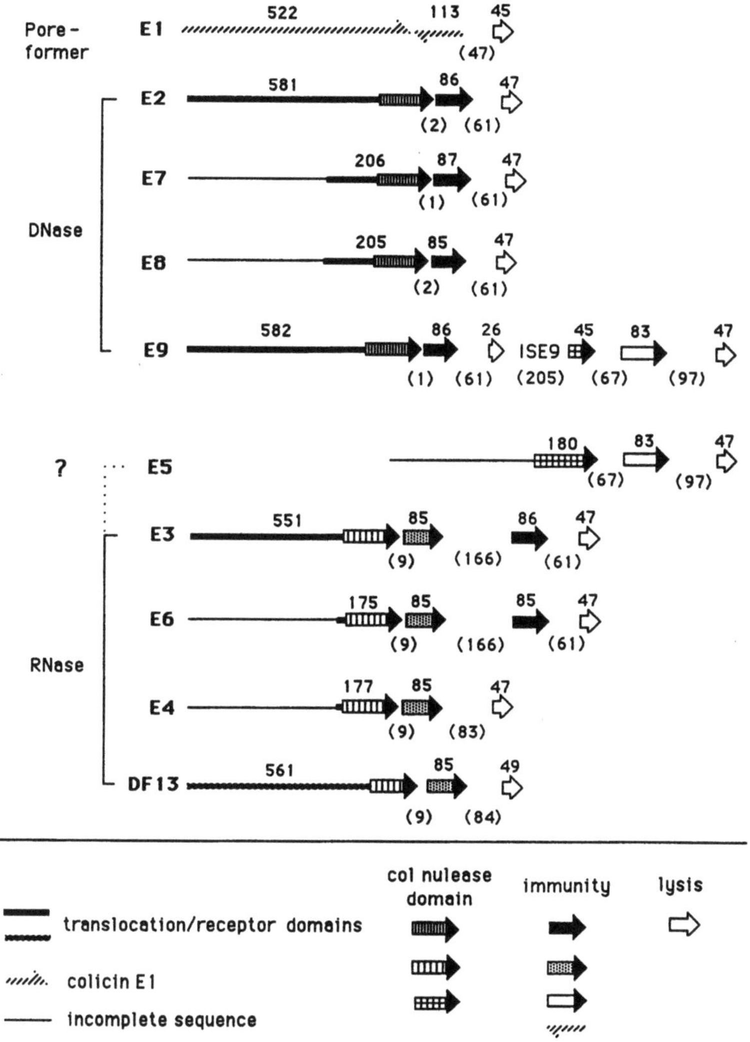

Figure 1. Gene organizations in the E-type colicin operons including the cloacin DF13 system. The plasmids are grouped according to the colicin types based on either sequence relatedness or biochemical data. The gene orders are as follows: for E1, E2, E4, E5, E7, E8 and DF13: *col-imm-lys* (see original references for other gene designations); for E3 and E6: *col-imm-E8(E3)imm-lys;* for E9: col-imm-(truncated *lys* and colE5)-E5[E9]*imm-lys*. The numbers in parentheses indicate the various intergenic spaces in base-pairs; those above the arrows indicate the sizes (in amino acids) of the Imm and Lys proteins, and either of the complete sequence or only the C-terminal nuclease domains of the colicins. No attempt was made in the diagram to demarcate the boundaries of the translocation and receptor-binding domains of the completely sequenced bacteriocins. The references are: **E1** (E1-K30; Chan *et al.*, 1985); **E2** (E2-P9; Lau *et al.*, 1984ab; Masaki *et al.*, 1985; Cole *et al.*, 1985; Toba *et al.*, 1986); **E3** (E3-CA38; Watson *et al.*, 1984; Lau *et al.*, 1984b; Masaki & Ohta, 1985; Toba *et al.*, 1986); **E4** (E4-K365; Lau *et al.*, this study); **E5** (E5-099; Lau & Condie, 1989; Curtis *et al.*, 1989); **E6** (E6-CT14; Lau & Condie, 1987); **E7** (E7-K317; Chak *et al.*, 1991; Lau *et al.*, this study); **E8** (E8-J; Uchimura & Lau, 1987; Toba *et al.*, 1988); **E9** (E9-J; James *et al.*, 1987; Lau & Condie, 1989; Eaton & James, 1989); **DF13** (CloDF13; Nijkamp *et al.*, 1986).

of which colicin E3 is the prototype (we would like to bestow cloacin DF13 an honorary colicin due to its structural and functional similarity to colicin E3 and the stable maintainance of the producing plasmid in *E. coli).* The DNase class consists of colicins E2, E7, E8 and E9. Although indirect evidence has suggested that colicin E5 acts as an RNase (Mock & Pugsley, 1982), the amino acid sequences of the immunity protein and the C-terminal portion of the colicin have been found to bear no resemblance to any of the colicin nucleases known so far (Lau & Condie, 1989; Curtis *et al.*, 1989). If the *col-imm* intergenic space could serve as an indicator of a particular class of colicin (these are consistently: 1-2 nucleotides in the E2-class; 9 nucleotides in the E3-class; converging arrangement in the E1 -type, Figure 1), then E5 is likely to represent a new group which at present has no other full members, except for the partial sequence found in ColE9-J.

In the simplest situation and exceptional to ColE1, the *col* and *lys* genes form an operon with the *imm* gene situated in between them but transcribed in the opposite direction. Co-transcription of *col* and *lys* from an SOS-regulated promoter situated upstream from the *col* gene ensures that induction of colicin synthesis will lead to its release (Sabik *et al.*, 1983). The inducers of the *col* promoter include UV irradiation and DNA damaging agents such as mitomycin C (Tyler & Sherratt, 1975). The *imm* gene is presumably transcribed constitutively from its own promoter. In the E2 class, as shown with ColE2-P9 (Lau *et al.*, 1984a; Cole *et al.*, 1985; Masaki *et al.*, 1985), the *col, imm and lys* genes form one operon. This is considered a canonical gene arrangement for the nuclease-type E colicins. Expression of these genes is controlled by an inducible promoter upstream from *col* and transcription terminators have been detected downstream from *imm* and *lys*. The expression of the lethal *lys* gene is likely to be modulated by a hairpin structure situated between the *imm* and *lys* genes (Watson *et al.* 1984; Lau *et al.*, 1984a). Presumably, the expression of the E2 *imm* gene is constitutive and independent of the SOS-response (Masaki *et al.*, 1985). Due to the short intergenic space, the possibility of a coupled translation phenomenon has been discussed (Lau *et al.*, 1984b). This phenomenon, as first described for the *E. coli* trp operon (Oppenheim & Yanofsky, 1980), is characterized by an overlap of regulatory signals in a polycistronic message so that a successful translation initiation of a distal gene depends on an efficient translation of the preceding gene. The expression of colicin genes, in general, is also subject to catabolite repression (Tyler & Sherratt, 1975; Pugsley, 1984a).

Dual immunity genes in a single plasmid

Until the discovery of colicin E8-J and colicin E9-J (Cooper & James, 1984), the occurrence of a dual immunity gene in ColE plasmids was unheard of. Chak and James (1984) reported that ColE3-CA38, in addition to its colicin-specific *imm* gene, carries an immunity gene for a heterologous colicin E8. Subsequent sequencing (Lau *et al.*, 1984b) revealed that this '*trans*-acting' E8 *imm* in E3 (referred to as E8[E3] *imm*, [XX] representing the plasmid carrier) is a homologue of ColE2-P9 *imm* gene. ColE6-CT14 also carries an E8 *imm* homologue (E8[E6] *imm*) downstream of its colicin-specific E6 *imm* (Lau& Condie, 1989). Sequence comparison of the two *trans*-acting E8 immunity proteins with the ' *cis*-acting' E8 Imm carried in ColE8-J revealed only six amino acid substitutions; E8 Imm and E8[E6] Imm differ by only 3 amino acids. Yet neither *trans*-acting immunity protein conferred complete immunity to the exogenous colicin E8 (Cooper & James, 1984). It is likely that complete complementarity between the structures of a given colicin and its immunity protein is a prerequisite for complete protection. This point will be elaborated in a later section.

The 166 bp intergenic space between E3 *imm* and E8[E3] *imm* (Figure 1) is of interest because it might provide some insight into the origin of the extraneous sequences. Sequence analysis showed that recombination or a cross-over might have taken place in the relatively AT-rich sequence in the middle of the 166 intergenic sequence (Uchimura & Lau, 1987). This DNA region coincides with the positioning of the *lys* gene seen in other plasmids (see Figure 1 for an alignment). In comparison with the CloDF13 *imm-lys* intergenic sequence, the 5'-half of the E3 166 bp sequence showed 74% base identity. This portion of the sequence includes a potential attenuator sequence for the downstream *lys* gene. Interestingly, in the E2-class plasmids, of which ColE9-J is the only example with two immunity genes (E9+E5), sequence deviation from the canonical *col* operon gene organization also appears to fall in the vicinity of *lys*. This makes the *lys* gene or its nearby sequences a hot-spot for recombination among ColE plasmids (Toba *et al.*, 1986; Lau & Condie, 1989).

A unique case of an IS element in ColE plasmids

Transposition events and DNA sequence insertions have long been considered to play a role in bacterial plasmid evolution (Cohen & Kopecko, 1976). Several classes of mobile genetic elements such as insertion sequences (IS elements) have now been found as natural constituents of the bacterial chromosome, their plasmids and phage genomes. Such elements were shown to mediate recombination processes, leading to gross DNA rearrangements such as replicon fusion and segregation, deletion and inversion. For example, IS*1* was found in R plasmids and in phage P1 DNA; IS2, IS*3* and $\gamma\delta$ sequences were present in F plasmids (For a review, see Iida *et al.*, 1983). The possibility of transposable elements being involved in colicin plasmid evolution has been entertained on several occasions (Luria & Suit, 1982; Pugsley, 1984a; Lau *et al.*, 1984a). Not until recently, has an IS-like element, designated IS*E9*, been identified in the ColE9-J plasmid (Lau & Condie, 1989).

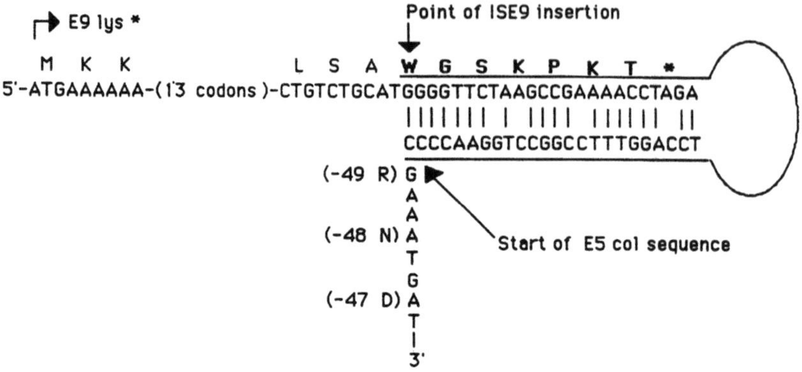

Figure 2. Gene disruption property of IS*E9* in the ColE9-J plasmid. The sequence of only the stem portion of the IS*E9* secondary structure is shown; it is potentially made up of 32 G-C, 28 A-T and 10 G-T pairings with a calculated energy of -72 kcal. The amino acid sequence in bold type is the new C-terminus for the atypical E9 lysis protein (Lys*). The negative numbers indicate amino acids from the C-terminus of colicin E5.

To the best of our knowledge, IS*E9* is a degenerate transposon of 'the third kind' since there are only two other known examples of this class, viz. IS*101* and Tn*951*. All of these elements have inverted repeat (IR) sequences at their termini which are closely related to those in the Tn*3* transposon family (Heffron, 1983). IS*101* *is* 209 base-pair long; it is derived from plasmid pSC101 and was discovered in a chimeric phage (f1') formed between f1 and pSC101 (Fischhoff *et al.*, 1980). Because of its small size it does not have the capacity to encode either a transposase or a resolvase protein needed for transposition. Instead, it only contains the target sequences necessary for both transposition and resolution by the γδ sequence, a member of the Tn*3* transposon class. Insertion of IS*101* into the f1 phage DNA generated a new carboxyl-terminus for the chimeric f1 ' phage gene IV protein (Fischhoff *et al.*, 1980). Tn*951* encodes the genes for the lactose operon; although 16.6 kb in size, it does not carry the genes for its transposition (Cornelis *et al.*, 1981). IS*E9* resembles IS*101* in several aspects including size (205 bp), sequence identity (42.4%) and an extensively base-paired secondary structure (Figure 2). Notably, insertion of IS*E9* on the 'left' end of the IR sequence generates a novel carboxyl-terminus giving rise to an atypical lysis protein of 26 amino acids. On the 'right' end of the IR sequence, IS*E9* apparently deletes out the 5'-sequence of the E5 *col* gene such that only the 49 C-terminal amino acids of colicin E5 are left in E9 (Figures 1 and 2). Beyond the 3'-end of ISE9 is the sequence encompassing the E5[E9] *imm* and *lys* genes. This DNA region of the E9 plasmid is virtually identical to that of the ColE5-099 plasmid (Figure 1; Lau & Condie, 1989; Curtis *et al.*, 1989). Clearly, at some point in time, replicon fusion between ColE5 and an E9-1ike plasmid and subsequent or concurrent deletion of some deleterious DNA sequences have occurred, thus giving rise to the presently known ColE9-J. It is noteworthy that ColE5 and ColE9-J plasmids are incompatible (Cooper *et al.*, 1986).

Common origin among endonucleolytic colicins?

A common tripartite domain structure for colicin nucleases, including cloacin DF13, has been well documented: i) an N-terminal glycine-rich sequence is involved in translocation across cell membranes, ii) a central region plays a receptor-recognition and binding role and iii) a C-terminal domain which has the killing activity and also the region where the 'homologous' immunity protein binds (Ohno-Iwashita & Imahori, 1980; Jakes, 1982; de Graaf & Oudega, 1986). Because colicins E2 and E3 recognize the same *btu*B-specified receptor protein, the two proteins share a virtually identical N-terminal sequence that spans 426 amino acids (a Glu to Ala substitution was noted at amino acid residue 240; at the nucleotide level, there are only seven third base changes in this DNA stretch). Surprisingly, the recently determined colicin E9 sequence (Eaton and James 1991) which is one a.a. residue longer than colicin E2, showed a number of substitutions in between a.a. 121-144 which is otherwise highly conserved between colicins E2, E3 and cloacin DF13. The significance, if any, of this sequence change in colicin E9 is not apparent. Nonetheless, for all intents and purposes, the N-terminal sequences of the nuclease-type colicins are virtually identical since they recognize a common receptor protein.

Whether there is an evolutionary relationship between the RNase domains of the colicin E3-class and the microbial extracellular RNases, and whether there is a common origin for the nuclease domains of colicin E2- and E3-types, are two intriguing questions. The fungal and bacterial RNases such as T1, C2, U2 and Ba (barnase) which range in size from 102-113 amino acids are known to form an evolutionarily related family with conserved catalytic and substrate binding residues (Hill *et al.*, 1983). Some of these RNases also share common tertiary structures (Heinemann & Saenger, 1982). Based on sequence alignment of the invariant residues of various microbial RNases with the 97 amino-acid long nuclease domain of colicin E3 (Figure 3), it has been proposed that these RNases possibly share a common ancestry (Lau *et al.*, 1984a; Hartley, 1989). The invariant catalytic Glu-58 residue (in T1 numbering) is replaced by Lys-333 in the E3 sequence. This substitution could be due to a single base change from GAG to AAG. Colicins E4, E6 and cloacin DF13 also contain the invariant residues at His-462, Arg-497 and H-513 in the E3 numbering (Figure 4). How catalysis takes place in colicin RNases and whether there is any resemblance to the model of RNase T1 (Heinemann & Saenger, 1982) await the elucidation of the crystal complex structure of these RNases with their substrate(s). In a recent report (Calnan *et al.*, 1991) an arginine-mediated RNA recognition mechanism (the so-called arginine fork) was described for an RNA-binding protein. Although several arginine residues are present in the basic domains of colicin RNases for involvement in bonding with the phosphates of the RNA substrate, the invariant Arg-497 (in E3 numbering) is the most likely candidate.

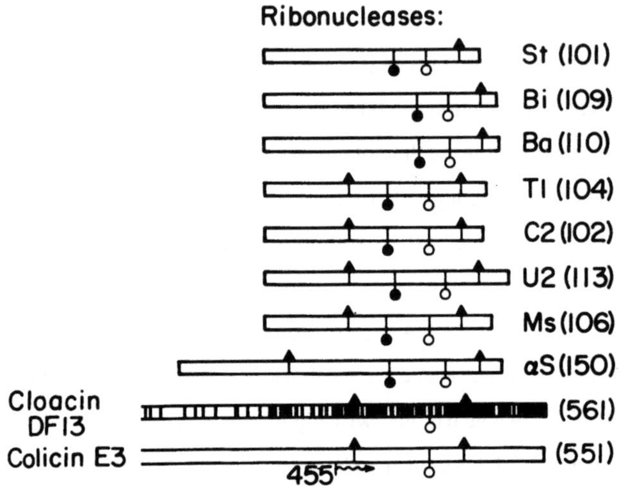

Figure 3. Schematic alignment of the catalytic and substrate binding residues of various microbial RNases with the putative candidates in the C-termini of colicin E3 and cloacin DF13. The filled triangles are the substrate binding residue His-40 and catalytic residue His-92 (in T1 numbering). The filled and open circles are catalytic residues Glu-58 and Arg-77 (Hill *et al.*, 1983). The regions of sequence identity in cloacin DF13 to that of colicin E3 are shaded. αS is α-sarcin (Wool, 1984). The size of the proteins in amino acids are in parentheses. The nuclease domain of colicin E3 is *ca* 97 amino acids.

When all the sequences of the nuclease domains from both the E2-class and the E3-class are compared (Figure 4), some 20% of the residues are found to be either common (5 scores out of 8) or invariant. Divergence from an ancestral nuclease domain would lead to the low degree of sequence identity. The acquisition of a second gene domain with the ability to enter cells would give rise to the presently known colicins and cloacin (Figure 5). Functionally, the E2-class colicins have probably evolved to recognize and cleave both single-stranded and double stranded DNA's nonspecifically. On the contrary, the E3-class colicins, demonstrated for at least colicin E3 and cloacin DF13, cleave the 16S RNA specifically. Interestingly, in the case of microbial RNases, specificity towards their substrates could either be weak (e.g. barnase which recognizes purines better than pyrimidines) or very specific (e.g. C2 which is guanine-specific). The mold nuclease, called α-sarcin, appears to be exceptional in the evolutionary of RNases. It is cytotoxic and is merely 150 a.a. long; this sequence shares 34% identity with U2 RNase

(Wool, 1984). Like colicin E3, α-sarcin inhibits protein synthesis but it cleaves the eukaryotic 28S RNA specifically.

```
a)

  462                                                  516  521
E2   GKPVGDKWLDDAGKDSGAPIPDRIADKLRDKEFKNFDDFRKKFWEEVSKDPDLSKQFKGS
E8   ----------------------------------------R----------E-----NPG
E7   ----NN---NN----L-S-V-----N--------S----------------E-----SRN
E9   ----------------------------------S------AV--------E---NLLP-
     GKPV  KWL  AGKD G P PDRIA KLRDKEFK FDDFR  WEEVSKDP LSK

           455   462                                        500
E3         KGFKDYGHDY*HPAP*******KTENIKGLGDLKPGIPKTPKQNGGGKRKRWTG
E6         --V-------*--D-*******---D-----E--E-K------G-----A--Y-
DF13       --T-------*F-D-*******---D-----E--E-K------G-----A--Y-
E4         --V-------*----*******---E-----E--KAPK-----G---R-D--I-
           KG KDYGHDY* P P*******KTE IKGLG LK   KTPKQ GGG R RW G

           K   D G D   P P         E K   D                    G

                        547                               581
E2   NKTNIQKGKAPFARKKDQVGGRERFELHHDKPISQDGGVYDMNNIRVTTPK*RHIDIHRGK
E8   --KRLSQ-L--R--N--T----RS------------------D-L-I----*--------Q
E7   -NDRMKV----KT-TQ-VS-K-TS-----E-----N------D--SVV---*---------
E9   --SSVS--YS--TP-NQ-----KVY----------G-E----D--------*---------
     N      G  P      G R  ELHH KPISQ G VYDM N    TPK*RHIDIHRG

                 513                                    551
E3   DKGRKIYE***WDSQHGELEGYRASDGQHLGSFDPKTG****NQLKGPDPK*RNIKKYL
E6   --------***----------------------E----****---------*-------
DF13 --K-----***----------------E---A------****K-V------*-------
E4   --------***----------------E-I-A------****K-I------G-------
     DK RKIYE***WDSQHGELEGYRASDG H G F PKTG**** Q KGPDPK RNIKKYL

      K R                 G R    H       G          PK R I

b)

E5 NKGKIPGLKIDQKIRGQMPERGWTEDDIKNTVSNGATGTSFDKRSPKKTPPDYLGRNDPATV

   YGSPGKYVVVNDRTGEVTQISDKTDPGWVDDSRIQWGNKNDQ
```

Figure 4 a). Sequence alignments of the C-terminal nuclease domains of E2-class and E3-class colicins. The domains are defined by the tryptic T2A fragments (Ohno-Iwashita & Imahori, 1980; Lau *et al.*, 1984b). In each sub-set, only those amino acids which are different from the E2 or E3 sequences are shown. The consensus sequences are in bold type. The asterisks indicates gaps introduced by the Corpet (1988) program for optimal alignment of the two colicin classes. Common (5 scores out of 8) or invariant residues between the two classes are underlined. Some of the numbered residues are discussed in the text. b). Unique sequence of the C-terminal 104 amino acids of colicin E5. The references to all the E-type sequences are from this lab cited in the legend of Figure 1 (also consult other references cited theirin); DF13 sequence is from Nijkamp *et al.* (1986).

To support the notion of domain shuffling of the colicin building blocks, it is worth noting that the structures of pyocins, which are chromosome-encoded bacteriocins of *Pseudomonas aeruginosa* (Sano *et al.*, 1990), are found to contain regions of high sequence identity to colicins E2 and E3 and cloacin DF13. Similar gene fusions have been proposed for the evolution of colicin D and its related toxins (Roos *et al.*, 1989). To further illustrate that the RNase domain has moved around, a very recent report described the presence of an RNase-like domain near the N-terminus of the RNA polymerase II from the fruit fly and yeast (Shirai & Go, 1991). The region of sequence similarity includes the catalytic sites Glu, Arg and His.

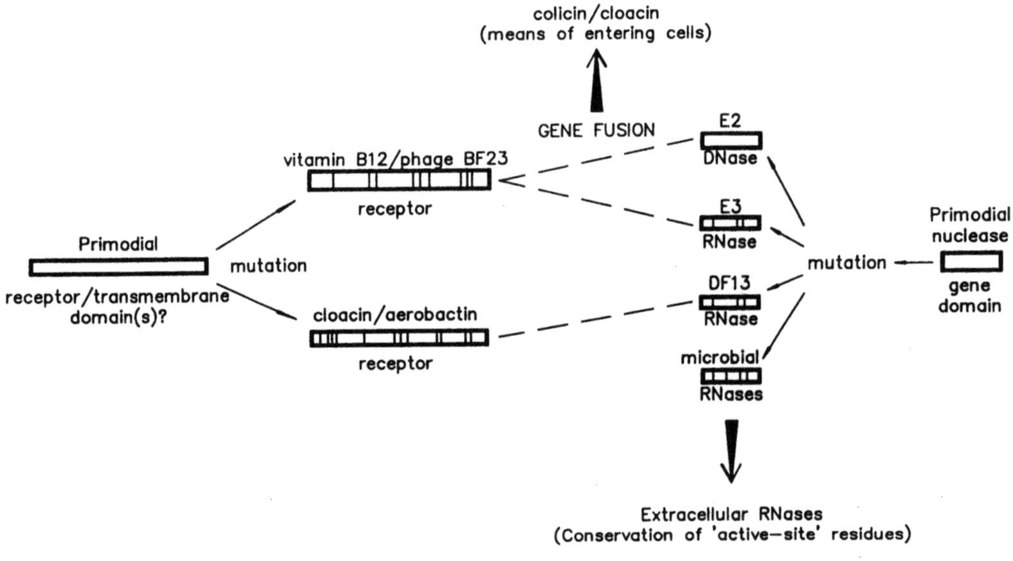

Figure 5. A gene fusion model for colicin nuclease evolution.

Coevolution of immunity proteins and colicin nuclease domains

Figure 6 shows a sequence comparison of immunity proteins from a) the E2-class and b) the E3-class. The degree of differences in each immunity class together with a comparison of the corresponding nuclease domains is summarized in Tables 2 and 3. In the E2-class, E9 Imm is most related to E2, followed by E7 and E8 and its *trans*-acting homologues. A hypervariable region appears to exist in these protein sequences, concentrated near the N-terminal portions of the proteins, approximately flanked by the two Phe residues at positions 15 and 40 in the E2 numbering. This region includes the single cysteine (except E7 which has none) and the a.a. insertions or deletions which are found in these proteins.

All the immunity proteins in the E3-class are of the same size (Figure 6b) and, generally, their primary sequences are less variable (ranging from 13-30%) than those of the E2-class (31-45%). Towards the middle of these polypeptides, a stretch of 12 residues is totally conserved. The E3 Imm is unique in this group in having a cysteine residue. If the formation of an inter-molecular disufide bond between two E3 immunity molecules occurs, this necessarily implies that the other immunity proteins within this group would function in a different way. In general, all the immunity proteins, including the unrelated E5 Imm (Figure 6c), are acidic with a calculated isoelectric point of approximately 4.5-5.9. The 83 a.a. long E5 Imm also has a cysteine.

```
a)

E2        MELKHSISDYTEAEFLEFVKKICRAEGATEEDD*NKLVREFERLTEHPDGS 50
E7        ----N-----------VQLL-E-EKENV-ATD-VLDV-LEH-VKI------T 51
E9        ----------------QL-TT--N-DTSS--EL*V---TH--EM----S-- 50
E8        ----N-------T--KKIIED-INC--DEKKQ**DDNLEH-ISV----S-- 49
E8[E6]    V---K--G----T--KKIIEN-INC--DEKKQ**DDNLEH-ISV----S-- 49
E8[E3]    V---K--G----T--KK-IED-INC--DEKKQ**DDNLEY-INV----S-- 49
           ELK SI DYTE EF       I                  F   TEHP G

E2        DLIYYPRDDREDSPEGIVKEIKEWRAANGKSGFKQG 86
E7        ------S-N-D-------------------P----- 87
E9        ------KEGDD---S---NTV-Q------------- 86
E8        ------EGNNDG---AVI------------------ 85
E8[E3]    ------EGNNDG---AVI------------------ 85
E8[E6]    ------EGNNDG----VI------------------ 85
          DLIYYP      SP       K WRAANGK GFKQG

b)

E3        MGLKLDLTWFDKSTEEDKGEEYSKDFGDDGSVMESLGVPFKDNVNNGCFD 50
E6        -----HIN----T---F--G------------I-R--M-----I---W-- 50
E4        -----N------K--DF---------------I----M-L--II---G-- 50
DF13      -----HIH----K---F--G------------I----M-L---I---W-- 50
          MGLKL   WFDK TE  KG EYSKDFGDDGSV E LG P KD  NNG FD

E3        VIAEWVPLLQPYFNHQIDISDNEYFVSFDYRDGDW 85
E6        ----------------------------------- 85
E4        -KKS---------KNK-EVDKYW-QI-------N- 85
DF13      -EKP--SI---H-KNV----KFD-----V----N- 85
          V   WV  LQP F   I      Y  SF YRDG W

c)

E5        MKLSPKAAIEVCNEAAKKGLWILGIDGGHWLNPGFRIDSSASWTYDMPEE 50
E5[E9]    -------------------------------------------------- 50

E5        YKSKIPENNRLAIENIKDDIENGYTAFIITLKM 83
E5[E9]    ----T---------------------------- 83
```

Figure 6. Aligned amino acid sequences of a) E2-class and b) E3-class immunity proteins. The symbols are as described in Figure 4. E8[E3], e.g. indicates the ColE3 plasmid carrying the E8 immunity gene. c) Unique sequence of E5 Imm in comparison with its single amino acid variant in E9. The references to the sequences are as described in the legend of Figure 4.

Table 2. % difference between the amino acid sequences of (a) the E2-class immunity proteins and (b) the C-terminal 120 residues of the corresponding DNase colicins

(a)	E2	E7	E8	E9	(b)	E2	E7	E8	E9
E2	--					--			
E7	33.3	--				27.5	--		
E8	42.0	38.0	--			16.7	28.3	--	
E9	31.4	41.4	45.3	--		20.0	32.5	23.3	--

Table 3. % difference between the amino acid sequences of (a) the E3-class immunity proteins and (b) the C-terminal 119 residues of the corresponding RNase colicins and cloacin DF13

(a)	E3	E4	E6	DF13	(b)	E3	E4	E6	DF13
E3	--					--			
E4	29.4	--				16.0	--		
E6	12.9	29.4	--			9.2	12.6	--	
DF13	29.4	24.7	23.5	--		20.2	18.5	11.8	--

Like the immunity proteins, the nuclease domains of the E2-class colicins are more variable (17-32%) than those of the E3-class (9-20%). It also appears that a hypervariable region exists in the sequences of the E2-class nucleases. This region spans from Lys-516 to Glu-547 in the E2 numbering (Figure 4a). In the case of the E3-class RNase domains, the amino acid substitutions are not clustered which is a characteristic of the respective cognate immunity proteins. There is a good degree of sequence conservation in certain regions, one being from Gly-500 to Gly-524 (except for a Lys interruption in DF13). It has been found in E3 that Gly-524 and a less conserved residue (Ser-529) are important in RNase activity (Escuyer & Mock, 1987). Removal of the C-terminal 7 to 10 amino acids of the colicin E3 and cloacin DF13 sequences resulted in either a drop in RNase activity or inability to bind to its immunity protein (Gaastra *et al.*, 1978; Ohno *et al.*, 1980). Colicin E4 is unique in the E3-class in having a Gly insertion at the eigthth position from the C-terminus. Like E6 and cloacin DF13, E4 has a potentially flexible glycyl tetrapeptide that is not found in the E3 sequence. Whether this difference in E3 has any bearing with its immunity binding function in relation to the other colicins in this group remains to be determined. Unlike the immunity proteins, none of the nuclease domains has a cysteine residue. All the nuclease domains are hydrophilic and basic in character; the acidic and basic contents sum up to 38.8-44.3%, with lysines predominating in these sequences. The calculated isoelectric points of these short domains range from 9.8-10.3.

It has been reported that the cloacin plasmid CloDF13 carried in *E. coli* K12 confers immunity to colicin E6, implying sequence identity of their immunity proteins or of their nuclease domains (Males & Stocker, 1982). Recent comparison of the primary structures of these proteins showed that neutralization of colicin E6 by DF13 Imm may possibly be a function of the rather similar N-terminal portions of the DF13 and E6 Imm proteins. Interestingly, the N-terminal halves of colicin E6 and cloacin DF13 nuclease domains are more identical to each other than are colicins E6 and E3. E3 Imm, which shows 70% sequence identity with DF13 Imm, can only interact weakly with cloacin DF13 (Kool *et al.*, 1975); the nuclease domains of these bacteriocins are 82% homologous (Masaki & Ohta, 1982; Lau *et al.*, 1984b).

In the extreme situation, an absolute structural complementarity of the interacting colicin-immunity molecules may be required for complete neutralization of the lethal effect of the respective toxins. This is illustrated by the three variants of the E8 Imm, eg. E8[E3] Imm, which differs in only six amino acids from the *cis*-acting E8 Imm in the native ColE8-J plasmid, confers to *E. coli* cells only partial protection from the exogenously added colicin E8 (Cooper & James, 1984). Plasmid E6 also has been shown to impart partial immunity to colicin E8 (Chak & James, 1986). E8[E6] Imm differs from E8 Imm by only 3 amino acids. It has been suggested that the incomplete protection may be due to a concentration effect (Cooper & James, 1984). The production of the *trans* copies of E8 Imm in the absence of the accompanying colicin E8 may be at a lower level than when the cognate *col* gene is present. The presence of the GTG initiator codons in these two genes, instead of the canonical ATG codon, may also contribute to the presumptive low expression. To distinguish among these possibilities, it will be necessary to mutate the native *cis*-acting E8 *imm* to either of the two *trans*-acting E8 *imm* sequences and examine the 'all or partial' protective effect.

Most of the nucleotide changes in the colicin E2 and E8 operons are found to be clustered in the coding regions of the nuclease domains and the immunity proteins (Uchimura & Lau, 1987). Many of these base changes are silent in nature; others result in a.a. changes which are either chemically similar or different. This feature is also apparent in other E colicin operons (not shown). Given the intimate nature of the interactions between the acidic immunity protein and the basic nuclease domain of a colicin needed for complete protection, it is possible that changes in the nuclease domain of a given colicin are compensated in a co-ordinate manner in the immunity proteins via the process of coevolution.

The lysis proteins: a class of lethal lipoproteins

Regardless of the type of Col plasmids, the lysis proteins form a conserved group of proteins (Figure 7). The small lysis proteins are classified as lipoproteins by the following criteria: i) ability to be radiolabelled by glycerol or palmitate, a modification which is prerequisite to processing by the bacterial signal peptidase II (SPase II), ii) the presence of a short signal peptide which has, at its C-terminus, a consensus sequence of Leu-X-Y-Cys for recognition by SPase II. In the Col lysis proteins, X= Ser, Ala or Val and Y= Ala or Gly. Lipid modification (diglyceride attachment) takes place at the sulfhydryl group of the invariant Cys residue which becomes the first a.a. of the mature protein after processing. A third fatty acid is apparently added to the α-amino group of the Cys following release of the signal peptide, iii) inhibition of SPase II processing by a cyclic antibiotic peptide, globomycin (Inukai *et al.*, 1978).

```
                     lipo-box                                  size
                      ^^^ ^
E2    MKKITGIILLLLAV**IILSA CQANYIRDVQGGTVSPSSTAEVTGLATQ          47
E2K   --------------**---A- ---------------------L--VE--          47
E3    --------------**----- ----------------------------          47
E4    -----------F-A**---A- ---------------------L--VE--          47
E5    -----W--------**---A- -------H--------------------          47
E6    --------------**---A- ---------------------L--VEV-          47
E7    -------------A**---A- ---------------------L--VE--          47
E8    --------------**---A- ----------------------------          47
E9    -------------A**---A- ------------------S--L------          47

E1    -R-R**FFVGIF-I**NL-VG -------------IA---SSKL--I-V-          45
E1K   -R-R**FFVGIF-I**NL-VG --------------A---SSKL--ISV-          45
DF13  ---AKA-F-FI-I-SGFL-V- --------------A---SS-L--I-V-          49
N       MCGK-L-I-FFI**MT--- --V-HI---K----A---SSRL---KLSKRSKDPL 52
A      ----I*-CVI---I**ML-A- --V-NV--TG--S-----IV**--VSMGSDGVGNP 51

E9*    --------------**------WGSKPKT                              26
```

Figure 7. Alignment of Col lysis protein sequences. The consensus sequences are in bold type. All sequences are compared to E2 (ColE2-P9). The new sequences are E2K ($E2_{imm}$-K317) and E4 (ColE4-K365) from this study; E1K is from Higashi *et al.*, 1986; E9* is the atypical lysis protein. The invariant Cys is the first a.a. of the mature lysis protein.

The lipoprotein nature of the Col group of lysis proteins was first demonstrated by Cavard *et al.* (1987) in the ColA system. Substitution of the Cys residue of the mature protein, by site-directed mutagenesis, revealed that lipid modification is essential for efficient SPase II processing, targetting and function of the lysis protein. Before the onset of lysis, cellular phospholipase activity is elevated apparently as a consequence of membrane pertubations due, presumably, to integration of the lysis protein into the cytoplasmic membrane (de Graaf & Oudega, 1986).

Besides the obligatory N-terminal cysteine, the alignment of all the lysis protein sequences reveals 10 fully conserved a.a. residues in the mature portions (Figure 7). The sequences of the lysis proteins from E2 to E9 are very homologous. The mature portions of the lysis proteins of E2, E3 and E8, and $E2_{imm}$-K317, E4 and E7 are identical. The newly-determined E4 lysis sequence (this study) is unusual in having a Phe in its signal portion, a characteristic feature in the lysis protein of CloDF13 plasmid and those from the pore-forming ColE1-type and ColN (Pugsley, 1988). Except for the rather conserved sequences at the N-terminal portions of the mature proteins, it would appear that the sequences of these lysis proteins are distinct from those of the E2-E9 types.

The ColA lysis protein (Cavard *et al.*, 1985) is most variable (Lau *et al.*, 1987). In this respect, the study of the structure-function relationship of this protein may not be as readily extrapolated to other lysis proteins (and vice versa). To illustrate this, Howard *et al.* (1989) observed that the truncated ColA lysis protein (Cal) containing the signal sequence, plus 16 or 18 amino acids of the mature protein, are neither lipid-modified nor processed by SPase II. On the other hand, Luirink *et al.* (1989) showed that a mutant tetrapeptide of the CloDF13 lysis protein is modified and processed accordingly. Our results with the ColE2-P9 lysis gene agree with the CloDF13 system (Rioux, Bergeron, Lin, Grothe, O'Connor-McCourt and Lau (1991), manuscript submitted to Gene). Thus it is likely that in the ColA system, the unusual C-terminus of Cal somehow interferes with the proper functioning of the lipid modification step.

It is worth noting that the *lys* gene from $E2_{imm}$-K317 is *cryptic* in its native plasmid. This is to be expected since $E2_{imm}$-K317 is colicin-defective (Watson *et al.*, 1981; Males & Stocker, 1984) and since the transcription of the *lys* gene, under normal conditions, is dependent on the *col* promoter sequence which we have found to be deleted in the $E2_{imm}$-K317 plasmid (unpublished results, this lab). The reason for the presence of the atypical lysis protein in ColE9-J has been described. The substitution of the obligatory Cys by Trp in the lipo-box sequence prevents lipid modification and, accordingly, it is left unprocessed, at least by SPase II. Although a glycyl-serinyl sequence is present, it is not clear whether the bacterial SPase I (Perlman & Halvorson, 1983) would act in this region. If left structurally intact, E9 Lys* resembles the bee venom, mellitin (Terwilliger & Eisenberg, 1982). Both these proteins are 26 residues long and characterized by a hydrophobic N-terminus and a hydrophilic C-terminus. Chak and James (1986) have reported that this protein is functional. Biophysicists would find this peptide of interest since its small size renders it particularly amenable to peptide synthesis and a.a. changes for membrane-insertion studies, etc.

Previously, we have found structural relatedness of Col lysis proteins with those of the icosahedral coliphages such as ϕX174-E protein and MS2-L protein (Lau *et al.*, 1987). We reckoned that it is more than a coincidence that the essential portions of ϕX174-E and MS2-L proteins should display secondary structure and hydropathic characteristics similar to those of bacteriocinogenic proteins. Besides, there are several biochemical studies which document similarities in these systems. While the notion of coevolution of bacteria with their phages has been well documented (Levin & Lenski, 1983), is it not possible that Col plasmids are in some ways 'coatless' phages? (and for that matter, "wolves in sheeps' clothings"?).

Colicinogeny in nature

Luria and Suit, (1987) recently wrote that "colicinogeny appears to be an unnecessary complication in the life of coliform bacteria..". Indeed, whether colicin production has a decisive selective advantage over non-colicinogenic bacteria has not been obvious. It has been suggested that high proteolytic activity in the intestinal tract would degrade colicins and that the intestinal anaerobic environment would reduce colicin production (Pugsley, 1984b). Recent data, however, suggests the contrary. By immunoreactivity, the concentration of trypsin, chymotrypsin and elastase in the colon were found to be low (Bohn *et al.*, 1983). Anaerobiosis actually increased the gene expression of colicins E1, E2, E3, D and K in a significant manner (Malkhosyan *et al.*, 1991).

The occurrence of colicinogeny in normal bacteria has been found to be rather high. Indeed, this could be taken as a good indicator for the possible ecological importance of colicinogeny. In the Murray collection, 10 of the 32 *E. coli* strains collected from human samples of the pre-antibiotic era (1917-1954) were found to be Col^+ (Hughes & Datta, 1983). Pugsley, (1984a) found that over 30% of the Gram-negative bacteria isolated from the Seine river are Col^+. We have analyzed the *E. coli* standard reference (ECOR) collection of 72 strains which were collected from humans and 16 other mammals (Ochman & Selander, 1984). 48.6% (24 strains from humans and 11 strains from animals) of this collection were scored Col^+. About a third (13 strains) of these were of the ColE-type; preliminary hybridization data with known ColE plasmids indicated that most of these are of the E1-type (unpublished results, this lab). Interestingly, Bradley et al. (1991) recently reported that colicin production of *any kind* may be rare among all groups of pathogenic *E. coli*. It is postulated that in a pathogenic bacterial setting, normal *E. coli* producing a colicin would have an upper edge (Bradley, personal communication).

ColE plasmids in biotechnology

Plasmids, by virtue of their autonomous replication, are invaluable tools in genetic engineering, either as cloning vectors or sources of vectors for gene expresion purposes. The importance of ColE plasmids in these applications cannot be over emphasized. Indeed, ColE1 plasmids were one of the first cloning vehicles in recombinant DNA technology (Hershield *et al.* ,1974). To this day, new vectors based on the ColE1 replicon continue to be developed (e.g. Gil & Bouche, 1991). It should also be borne in mind that the indispensable restriction endonuclease *Eco*RI and its associated methylase are encoded on a ColE1 plasmid, called pMB1 (Betlach *et al.*, 1 976).

The promoter, regulatory region and translational start site of the colicin E1 gene have been used to construct plasmids that express gene fusions to the *lac*Z gene of *E. coli* (Weisemann & Weinstock,

1985). This study took advantage of the inducibility of the *col* promoter and its sensitivity to catabolite repression. Ozaki *et al.* (1980) and Vernet *et al.* (1985) have made use of the lethal activiites of colicins E1 and E3 respectively, to develop positive or direct selection vectors for gene expression. Positive selection is based on the inactivation of the lethal colicin activities by DNA insertions in the colicin coding sequence.

Despite the fact that *E. coli is* now recognized as an industrial microorganism, it has the short-coming of not being able to secrete proteins into the culture medium. To overcome this effect, Hsiung and co-workers, (1989) have exploited the property of the lysis protein of the CloDF13 system to cause semi-permeabilization of the *E. coli* membranes. In a dual compatible plasmid system, production and secretion of human growth hormone (hGH) was demonstrated. Both plasmids contain a hybrid *E. coli* lipoprotein promoter and *lac* operator to direct the expression of the lysis protein and hGH. The expression of the genes was tightly regulated by the *lacI* repressor gene. This study is one illustration of the usefulness of the Col lysis protein or its signal peptide in the construction of secretion vehicles (Klein *et al.*, 1988).

Recently, the ColE2 lys gene has been used in the construction of a cloning system designed for customized lipopeptide or peptide synthesis in bacteria (Lau & Rioux, patent filed; Figure 8).

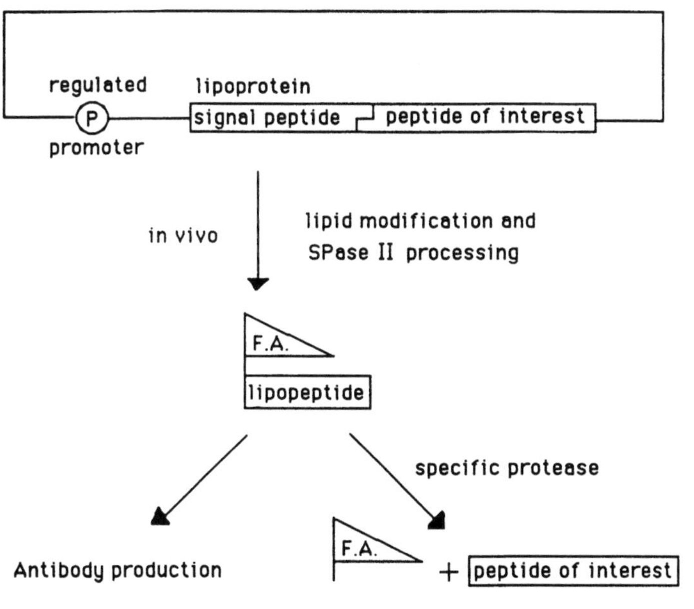

Figure 8. A schematic representation of a bacterial cloning system for customized (lipo)peptide synthesis. The peptide of interest can be released from the fatty acid (F.A.) moiety if a specific protease recognition site is incorporated in the peptide gene sequence.

This application takes advantage of the maturation process of the Col lysis lipoproteins which are presumed to be similar, if not identical, to that of the well-characterized murein lipoprotein (Wu & Tokunaga, 1986). In a generalised situation, the fusion of any DNA sequence coding for a peptide or polypeptide of interest to the gene portion coding for the signal sequence of a lipoprotein is expected to produce, after *in vivo* lipid modification and processing, a lipopeptide of interest. Since the lipoprotein signal peptides are invariably found to possess a β-turn secondary structure immediately following the SPase II recognition site (Klein *et al.*, 1988), the gene sequence coding for the first few amino acids of the mature ColE2 lysis protein was included in the synthetic DNA piece which codes for the peptide sequence. We anticipate that the production of lipopeptides via a recombinant DNA route will be useful since lipopeptides are potential vaccine candidates and they can serve as carrier-adjuvant systems (Jung, 1987; Deres *et al.*, 1989). In addition, this plasmid system is potentially useful in the production of those peptides which are otherwise difficult to synthesize chemically for reasons of a.a. composition or size. In this application, any specific protease recognition site can be incorporated into the DNA sequence that accompanies the gene specifying the peptide of interest.

The extent of possible applications of the Col plasmid genes is unlimited. When the BtuB receptor-binding domain of colicin E3 was replaced by the N-terminal portion of the phage f1 attachment protein, which recognizes the F pilus receptor, the receptor specificity of colicin E3 was switched accordingly (Jakes *et al.*,1988). This result implies that the RNase or DNase activity of the various colicins may possibly be targetted to other cells by similar genetic engineering, thus broadening the spectrum of antibiotic-like activity of colicins. In another potential application in the area of environmental concern, e.g. the SOS-regulatory system of the *col* operon can be fused to a convenient marker gene (e.g. the luminescence *lux* genes or catechol dioxygenase *xyl*E gene) for the development of a plasmid system to monitor chemical mutagens and carcinogens.

Concluding remarks

Much progress has been made in the elucidation of the primary structures of the immunity proteins and the essential nuclease-active domains of the cognate colicins from the nine representative groups of E colicins that are known to date. Unfortunately, this knowledge has not led to the basic understanding of what makes colicin E2 a nonspecific DNase and colicin E3 a specific RNase. And, what is colicin E5? Is it correct to assume that those colicins within the E2- or E3-class have specificities of their prototypes? Clearly, much biology and biochemistry is yet to be done. While it is not a priority to uncover new E colicins, effort is needed to elucidate the mechanism by which these proteins express nucleotide recognition or specificity since nucleases, in general, play an important role in biological processes.

Of paramount importance is the necessity to unravel the basis for the molecular interaction between a specific nuclease domain and its cognate immunity protein. Since crystallization and tertiary structure determination represent the most direct methods of studying structure and function of a protein, the primary structures that have been determined for the various colicin-immunity proteins are necessary groudwork towards these studies. Analyses of the comparative structures of these proteins have allowed the derivation of some testable hypotheses. Although a preliminary study of the colicin E3-immunity protein complex by small X-ray diffraction was reported sometime back (Levison *et al.*, 1983) it is gratifying to see progress made in this direction in two contributions to this Workshop. In our bias, we feel that the colicin E8 system is especially attractive for this kind of study since Mother Nature has already provided two variants of the immunity protein which differ in only a few amino acids. Yet these do not completely neutralize colicin E8. Until we understand the basis of specificity in a natural setting, premature site-mutagenesis experiments carried out on these molecules will only necessarily complicate the picture which is a complex one to begin with. If the *trans*-acting E8 immunity proteins were destined for a yet-to-be discovered colicin(s), then a molecular challenge exists for the design of a possible ' new' toxin. Undoubtedly, the colicin-immunity system of the nuclease type offers a unique opportunity to study macromolecular recognition in an interactive manner. A challenge also exists in evolving a colicin DNase to a specific DNA endonuclease.

Lipid modification of numerous eukaryotic proteins is becoming recognized as a mechanism for regulating the localization and function of these proteins (Sefton & Buss, 1987). The small lysis lipoprotein of the Col system will be invaluable towards these studies as a model system in bacteria. From a systematic study of the primary sequences of the natural variants of the lysis proteins only a small number of conserved residues are found. This information will be useful guidance for site-directed mutagenesis experiments. Certainly, the 28-35 amino acid long lipoprotein is no small challenge for crystallographers for a molecular view of this special class of proteins.

We have discussed the evolution of E colicin plasmids without the consideration of the plasmid regions necessary for replication or plasmid maintainance. Perhaps this is justifiable. Once again, to quote Luria and Suit (1987): "the three genes *(col, imm* and *lys)* could simply not be there as far as reproduction and control of the plasmids are concerned". Nevertheless, it would be instructive to learn about the subtleties of these DNA regions and of the replicons other than the well-established ColE1-type (Davidson, 1984). This area of research has been the subject of intense study by Itoh and co-workers (this volume).

Acknowledgments

Dr. LP Visentin is thanked for introducing one of us (PL) to the many wonders of colicins. MP is supported by a summer studentship program from the National Research Council of Canada (NRCC). We would like to thank JR Horton for a careful reading of the manuscript, and S Gaffney and KC Chang for the preparation of tables and some of the figures. This paper is NRCC publication number 32792.

References

Betlach M, Hershield V, Chow L, Brown W, Goodman M, Boyer HW (1976) A restriction endonuclease analysis of the bacterial plasmid controlling the *EcoRI* restriction and modification of DNA. Fed Proc 35:2037-2043

Bohe M, Borgstrom A, Genell S, Ohlsson K (1983) Determination of immunoreactive trypsin, pancreatic elastase and chymotrypsin in extracts of human feces and ileostomy drainage. Digestion 27:8-15

Bowman CM, Dahlerg JE, Ikemura T, Konisky J, Nomura M (1971) Specific inactivation of 16S ribosomal RNA induced by colicin E3 *in vitro*. Proc Natl Acad Sci USA 68:964-968

Bradley DE, Howard SP, Lior H (1991) Colicinogeny of 0157:H7 enterohemorrhagic *Escherichia coli* and the shielding of colicin and phage receptors by their 0-antigenic side chains. Can J Microbiol 37:97-104

Calnan BJ, Tidor B, Biancalana S, Hudson D, Frankel AD (1991) Arginine-mediated RNA recognition: the arginine fork. Science 252:1167-1171

Cavard D, Baty D, Howard SP, Verheij HM, Lazdunski C (1987) Lipoprotein nature of the colicin A lysis protein: effect of amino acid substitutions at the site of modification and processing. J Bacteriol 169:2187-2194

Cavard D, Lloubes J, Morlon J, Chartier M, Lazdunski C (1985) Lysis protein encoded by plasmid ColA-CA31: gene sequence and export. Mol Gen Genet 199:95-100

Chak KF, James R (1984) Localization and characterization of a gene on the ColE3CA38 plasmid that confers immunity to colicin E8. J Gen Microbiol 130:701-710

Chak KF, James (1986) Characterization of the ColE9-J plasmid and analysis of its genetic organization. J Gen Microbiol 132:61-71

Chak KF, Kuo WS, Lu FM, James R (1991) Cloning and characterization of the ColE7 plasmid. J Gen Microbiol 137:91-100

Chan PT, Ohmori H, Tomizawa J, Lebowitz J (1985) Nucleotide sequence and gene organization of ColE1 DNA. J Biol Chem 260:8925-8935

Clewell DB (1972) Nature of ColE1 plasmid replication in the presence of chloramphenicol. J Bacteriol 110:667-676

Cohen SN, Kopecko (1976) Structural evolution of bacterial plasmids: role of translocating genetic elements and DNA sequence insertions. Fed Proc 35:2031-2036

Cole ST, Saint-Joanis B, Pugsley AP (1985) Molecular characterization of the colicin E2 operon and identification of its products. Mol Gen Genet 198:465-472

Cooper PC, James R (1984) Two new colicins, E8 and E9, produced by a strain of *Escherichia coli* J. J Gen Microbiol 130:209-215

Cooper PC, Hawkins FKL, James R (1986) Incompatibility between E colicin plasmids. J Gen Microbiol 132:1859-1862

Cornelis G, Sommer H, Saedler H (1981) Transposon Tn*951* (Tn*Lac*) is defective and related to Tn*3*. Mol Gen Genet 184:241-248

Corpet F (1988) Multiple sequence alignment with hierarchical clustering. Nucleic Acids Res 16:10881-10890

Cramer WA, Dankert JR, Uratani Y (1983) The membrane channel-forming bacteriocidal protein, colicin E1. Biochim Biophys Acta 737:173-193

Curtis MD, James R, Coddington A (1989) An evolutionary relationship between the ColE5-099 and the ColE9-J plasmids revealed by nucleotide sequencing. J Gen Microbiol 135:2783-2788

Davidson J (1984) Mechanism of control of DNA replication and incompatibility in ColE1-type plasmids - a review. Gene 28:1-15

de Graaf FK, Oudega B (1986) Production and release of cloacin DF13 and related colicins. Curr Top Microbiol Immunol 125:183-205

Deres K, Schild H, Wiesmuller KH, Jung G, Rammensee HG (1989) *In vivo* priming of virus-specific cytotoxic T lymphocytes with synthetic lipopeptide vaccine. Nature 342:561-564

DiMasi RD, White J, Schnaitman CA, Bradbeer C (1973) Transport of vitamin B12 in *Escherichia coli:* common receptor sites for vitamin B12 and E colicins on the outer membrane of the cell envelop. J Bacteriol 115:506-573

Eaton T, James R (1989) Complete nucleotide sequence of the colicin E9 *(ceal)* gene. Nucleic Acids Res 17:1761-1761

Escuyer V, Mock M (1987) DNA sequence analysis of three missense mutations affecting colicin E3 bactericidal activity. Mol Microbiol 1:82-85

Fischhoff DA, Vovis GF, Zinder ND (1980) Organisation of chimeras between bacteriophage f1 and plasmid pSC101. J Mol Biol 144:247-265

Gaastra W, Oudega B, de Graaf FK (1978) The use of mutants in the study of structure-function relationships in cloacin DF13. Biochim BiophysActa 540:301-312

Gil D, Bouche P (1991) ColE1-type vectors with fully repressible replication. Gene105:17-22

Hardy KG, Meynell GG, Dowman JE, Spratt BG (1973) Two major groups of colicin factors: their evolutionary significance. Mol Gen Genet 125:217-230

Hartley RW (1989) Barnase and barstar: two small proteins to fold and fit together. TIBS 14:450-454

Heffron F (1983) Tn*3* and its relatives. In: Shapiro JA (ed) Mobile genetic elements. Academic Press Inc. New York, pp223-260

Heinemann U, Saenger W (1982) Specific protein-nucleic acid recognition in ribonuclease T1-2'-guanylic acid complex: an X-ray study. Nature 299:27-31

Heller K, Kadner RJ (1985) Nucleotide sequence of the gene for the vitamin B12 receptor protein in the outer membrane of *Escherichia coli.* J Bacteriol 161:904-908

Hershield V, Boyer HW, Yanofsky C, Lovett MA, Helinski DR (1974) Plasmid ColE1 as a molecular vehicle for cloning and amplification of DNA. Proc Natl Acad Sci USA 71:3455-3459

Higashi M, Hata M, Hase T, Yamaguchi K, Masamune (1986) The nucleotide sequence of cea* and the region of origin of plasmid pKY-1. J Gen Appl Microbiol 32:433-442

Hill C, Dodson G, Heinemann U, Saenger W, Mitsui Y, Nakamura K, Borisov S, Tischenko G, Polyakov K, Pavlovsky S (1983) The structural and sequence homology of a family of microbial ribonucleases. TIBS 8:364-369

Howard SP, Cavard D, Lazdunski C (1989) Amino acid requirements for assembly and function of the colicin A lysis protein. J Bacteriol 171:410-418

Hsiung HM, Cantrell A, Juirink J, Oudega B, Veros AJ, Becker GW (1989) Use of bacteriocin release protein in *Escherichia coli* for excretion of human growth hormone into the culture medium. Bio/tech 7:267-271

Hughes VM, Datta N (1983) Conjugative plasmids in bacteria of the "pre-antibiotic" era. Nature 302:725-726

Iida S, Meyer J, Arber W (1983) Prokaryotic IS elements. In: Shapiro JA (ed) Mobile genetic elements. Academic Press Inc. New York pp159-221

Inselburg J (1973) Colicin factor DNA: a single non-homologous region in ColE2-E3 heteroduplex molecules. Nature 241:234-237

Inukai M, Takeuchi M, Shimizu K, Arai M (1978) Mechanism of action of globomycin. J Antibiot (Tokyo) 31:1203-1205

Jakes KS (1982) The mechanism of action of colicin E2, colicin E3 and cloacin DF13. In: Cohen P, Van Heyningen S (eds) Molecular action of toxins and viruses, vol 2. Elsevier, Amsterdam, pp131-167

Jakes K, Davis NG, Zinder N (1988) A hybrid toxin from bacteriophage f1 attachment protein and colicin E3 has altered cell receptor specificity. J Bacteriol 170:4231-4238

James R, Jarvis M, Barker DF (1987) Nucleotide sequence of the immunity and lysis regions of the ColE9-J plasmid. J Gen Microbiol 133:1553-1562

Jung G (1988) Low-molecular weight lipopeptide carrier-adjuvant systems for immunization: development and results. In: Shiba T, Sakakibara S (eds) Peptide Chemistry 1987. Protein Research Foundation, Osaka pp.751-758

Klein P, Somorjai RL, Lau PCK (1988) Distinctive properties of signal sequences from bacterial lipoproteins. Protein Eng 2:15-20

Kool AJ, Pols C, Nijkamp HJJ (1975) Bacteriocinogenic CloDF13 minicells of *Escherichia coli* synthesize a protein that accounts for immunity to bacteriocin CloDF13: purification and characterization of the immunity protein. Antimicrob Agents Chemother 8:67-75

Lau PCK, Condie J (1989) Nucleotide sequences from the colicin E5, E6 and E9 operons: Presence of a degenerate transposon-like structure in the ColE9-J plasmid. Mol Gen Genet 217:269-277

Lau PCK, Hefford MA, Klein P (1987) Structural relatedness of lysis proteins from colicinogenic plasmids and icosahedral coliphages. Mol Biol Evol 4:544-556

Lau PCK, Rowsome RW, Watson RJ, Visentin LP (1984a) The immunity genes of colicins E2 and E8 are closely related. Biosci Rep 4:565-572

Lau PCK, Rowsome RW, Zuker M, Visentin LP (1984b) Comparative nucleotide sequences encoding the immunity proteins and the carboxyl-terminal peptides of colicins E2 and E3. Nucleic Acids Res 12:8733-8745

Lawrence GMP, James R (1984) Characterization of the ColE8 plasmid, a new member of the group E colicin plasmids. Gene 29:145-155

Lazdunski C, Baty D, Geli V, Cavard D, Morlon J, Lloubles R, Howard P, Knibiehler M, Chartier M, Varenne S, Frennette M, Dasseux JL, Pattus F (1988) The membrane-channel-forming colicin A: synthesis, secretion, structure, action and immunity. Biochim Biophys Acta 947:445-464

Levin BR, Lenski RE (1983) Coevolution in bacteria and their viruses and plasmids, In: Futuyma DJ, Slatkin M (eds) Coevolution. Sinauer, Sunderland, Mass. pp99-127

Levison BL, Pickover CA, Richards FM (1983) Dimerization by colicin E3* in the absence of immunity protein. J Biol Chem 258:10967-10972

Luirink J, Clark DM, Ras J, Verschoor EJ, Stegehuis F, de Graaf FK, Oudega B (1989) pCloDF13-encoded bacteriocin release protein with shortened carboxyl-terminal segments are lipid modified and processed and function in release of cloacin DF13 and apparent host lysis. J Bacteriol 171:2673-2679

Luria SE, Suit JL (1982) Transmembrane channels produced by colicin molecules. In: Martonosi AN (ed) Membranes and Transport, vol.2. Plenum Press, New York pp279-284

Luria SE, Suit JL (1987) Colicins and Col plasmids. In: Neidhardt FC (ed) *Escherichia coli* and *Salmonella typhimurium*, Cellular and molecular biology, vol. 2. American Society for Microbiology, Washinton DC, pp1615-1624

Males BM, Stocker BAD (1980) *Escherichia coli* K317, formerly used to define colicin group E2, produces colicin E7, is immune to colicin E2, and carries a bacteriophage-restricting conjugative plasmid. J Bacteriol 144:524-531

Males BM, Stocker BAD (1982) Colicins E4, E5, E6, and A and properties of btuB$^+$ colicinogenic transconjugants. J Gen Microbiol 128:95-106

Malkhosyan SR, Panchenko YA, Rekesh AN (1991) A physiological role for DNA supercoiling in the anaerobic regulation of colicin gene expression. Mol Gen Genet 225:342-345

Masaki H, Ohta T (1985) Colicin E3 and its immunity genes. J Mol Biol 182:217-227

Masaki H, Toba M, Ohta T (1985) Structure and expression of the ColE2-P9 immunity gene. Nucleic Acids Res 13:1623-1635

Mock M, Pugsley AP (1982) The BtuB group Col plasmids and homology between the colicins they encode. J Bacteriol 150:1069-1076
Morlon M, Sherratt D, Lazdunski C (1988) Identification of functional regions of the colicinogenic plasmid ColA. Mol Gen Genet 211:223-230
Nijkamp HJJ, de Lang R, Stuitje AR, van den Elzen PJM, Veltkamp E, van Putten AJ (1986) The complete nucleotide sequence of the bacteriocinogenic plasmid CloDF13. Plasmid 16:135-160
Ochman H, Selander RK (1984) Standard reference strains of *Escherichia coli* from natural populations. J Bacteriol 157:690-693
Ohno S, Saito K, Suzuki K, Imahori K (1980) The effects of carboxypeptidase digestion on the function of colicin E3. J Biochem 87:989-992
Ohno-Iwashita Y, Imahori K (1980) Assignment of the functional loci of colicin E2 and colicin E3 by the characterization of proteolytic fragments. Biochemistry 19:652-659
Oppenheim DS, Yanofsky C (1980) Translational coupling during expression of the tryptophan operon of *Escherichia coli*. Genetics 95:785-795
Ozaki LS, Maeda S, Shimada K, Takagi Y (1980) A novel ColE1::Tn*3* plasmid vector that allows direct selection of hybrid clones in *E. coli*. Gene 8:301-314
Pattus F, Massotte D, Wilmsen HU, Lakey J, Tsernoglou D, Tucker A, Parker MW (1990) Colicins: prokaryotic killer-pores. Experientia 46:180-192
Perlman D, Halvorson HO (1983) A putative signal peptidase recognition site and sequence in eukaryotic and prokaryotic signal peptides. J Mol Biol 167:391-409
Pugsley AP (1984a) The ins and outs of colicins. Part I: production and translocation across membranes. Microbiol Sci 1:168-175
Pugsley AP (1984b) The ins and outs of colicins. Part II: lethal action, immunity and ecological implications. Microbiol Sci 1:203-205
Pugsley AP (1988) The immunity and lysis genes of ColN plasmid pCHAP4. Mol Gen Genet 211 :335-341
Pugsley AP, Oudega B (1987) Methods for studying colicins and their plasmids. In: Hardy KG (ed) Plasmids- a practical approach. IRL Press Ltd, Oxford. pp105-161
Reeves P (ed) (1972) Molecular biology, biochemistry and biophysics. vol. 11. The bacteriocins. Springer, New York
Roos U, Harkness RE, Braun V (1989) Assembly of colicin genes from a few DNA fragments: nucleotide sequence of colicin D. Mol Microbiol 3:891-902
Sabik JF, Suit JL, Luria SE (1983) *cea-kil* operon of the ColE1 plasmid. J Bacteriol 153:1479-1485
Sano Y, Matsui H, Kobayashi M, Kageyama M (1990) Pyocins S1 and S2, bacteriocins of *Pseudomonas aeruginosa*. In: Silver S, Chakrabarty AM, Iglewski B, Kaplan S (eds) *Pseudomonas:* biotransformations, pathogenesis, and evolving biotechnology. American Society for Microbiology, Washington, DC. pp352-358
Schaller K, Nomura M (1976) Colicin E2 is a DNA endonuclease. Proc Natl Acad Sci USA 73:3989-3993
Sefton BM, Buss JE (1987) The covalent modification of eukaryotic proteins with lipids. J Cell Biol 104:1449-1453
Shirai T, Go M (1991) RNase-like domain in DNA-directed RNA polymerase II. Proc Natl Acad Sci USA 88:9056-9060
Terwilliger TC, Eisenberg D (1982) The structure of mellitin 1. Structure determination and partial refinement. J Biol Chem 257:6010-6015
Toba M, Masaki H, Ohta T (1986) Primary structures of the ColE2-P9 and ColE3-CA38 lysis genes. J Biochem 99:591-596
Toba M, Masaki H, Ohta T (1988) Colicin E8, a DNase which indicates an evolutionary relationship between colicins E2 and E3. J Bacteriol 170:3237-3242
Tyler J, Sherratt DJ (1975) Synthesis of E colicins in *Escherichia coli*. Mol Gen Genet 1 40 :349-353
Uchimura T, Lau PCK (1987) Nucleotide sequences from the colicin E8 operon: Homology with plasmid ColE2-P9. Mol Gen Genet 209:489-493

Vernet T, Lau PCK, Narang SA, Visentin LP (1985) A direct-selection vector derived from pColE3-CA38 and adopted for foreign gene expression. Gene 34:87-93

Watson R, Rowsome W, Tsao J, Visentin LP (1981) Identification and characterization of Col plasmids from classical colicin E-producing strains. J Bacteriol 147:569-577

Watson RJ, Lau PCK, Vernet T, Visentin LP (1984) Characterization and nucleotide sequence of a colicin-release gene in the *hic* region of plasmid ColE3-CA38. Gene 29:175-184 (Corrigendum 42:351-353 (1986)

Watson RJ, Vernet T, Visentin LP (1985) Relationships of the Col plasmids E2, E3, E4, E5, E6 and E7: Restriction mapping and colicin gene fusions. Plasmid 13:205-210

Watson R, Visentin LP (1980) Restriction endonuclease mapping of ColE2-P9 and ColE3-CA38 plasmids. Gene 10:307-318

Weisemann JM, Weinstock GM (1985) Use of transcription and translation signals from the colicin E1 gene to express DNA sequences in *Escherichia coli*. Gene Anal Techn 2:9-16

Wool IG (1984) The mechanism of action of the cytotoxic nuclease α-sarcin and its use to analyse ribosome structure. TIBS 9:14-17

Wu HC, Tokunaga M (1986) Biogenesis of lipoproteins in bacteria. Curr Top Microbiol Immunol 125:127-1 57

IMMUNITY SPECIFICITY AND EVOLUTION OF THE NUCLEASE-TYPE E COLICINS

H. Masaki, S. Yajima, A. Akutsu-Koide, T. Ohta and T. Uozumi
Department of Biotechnology
Faculty of Agriculture
The University of Tokyo
Yayoi, Bunkyo-ku, Tokyo 113
Japan

INTRODUCTION

E-Group colicins are encoded by respective ColE plasmids and kill *Escherichia coli* cells through binding to a common surface receptor, BtuB, which is also the receptor for phage BF23 and vitamin B12 uptake (Buxton, 1971; DiMasi *et al.*, 1973). Since the host *E. coli* cells are naturally sensitive to colicins, colicinogenic cells must be protected from the colicin action by means of "immunity", which is determined by the same plasmids. E-group colicins are divided into subclasses, A and El to E9, according to the immunity spectra (Table l; Cooper & James, 1984; Pugsley & Oudega, 1987). Among them, colicins A and El act as ionophores and show little relation to other E-group colicins in their modes of action and amino acid sequences, though they all have a similar domain structure arrangement, and their operons show substantial homology as to the promoter/operator regions and the distal lysis (or *kil)* genes (Chan *et al.*, 1985; Morlon *et al.*, 1988).

Other E-group colicins, in which we are interested, were shown or suggested to be nucleases; E2, E7, E8 and E9 are DNAases (Schaller & Nomura, 1976; Toba *et al.*, 1988; Chak *et al.*, 1991), and E3 and E6 (and possibly E4 and E5) act as RNAases which specifically inactivate 16S-RNA of the ribosomes (Boon, 1971; Bowman *et al.*, 1971; Ohno & Imahori, 1978; Mock& Pugsley, 1982). These activities are exclusively located on their C-terminal tryptic fragment, T2A. The E colicin plasmids protect the host cells from the action of their own colicins by synthesizing cognate inhibitor proteins (Imm). The immunity of nuclease-type colicins is thus defined by the specific binding of T2A and Imm (Jakes & Zinder, 1974; Ohno-Iwashita & Imahori, 1980; Jakes, 1982; Pugsley & Oudega, 1987).

Although colicin promoters are responsive to SOS-induction, a small amount of colicin is non-inducibly produced in the medium, enough to kill non-colicinogenic cells or other colicinogenic cells of a different immunity type. *E. coli* cells carrying a ColEl plasmid with a mutant *imm* gene (*imm*$^-$) are relatively stable; in particular, a *btuB* host strain exhibits almost normal growth. In contrast, cells carrying an *imm*$^-$ ColE2- or ColE3-derivative tend to be lethal, even with the *btuB* host mutation (Ozaki *et* al., 1982; Masaki & Ohta, 1985; our unpublished observation). Since the nuclease-type colicins, unlike the

NATO ASI Series, Vol. H 65
Bacteriocins, Microcins and Lantibiotics
Edited by R. James, C. Lazdunski and F. Pattus

ionophore-type ones, have their targets within the cytoplasm, the direct inhibition of the enzymatic activity by Imm proteins is thus particularly important for the maintenance and ecological success of the colicinogenic cells, and consequently of the Col plasmids.

Immunity specificity bears a dual meaning; the killing specificity determined by the colicin T2A domain and the protection specificity by the Imm protein. Sequence comparison of the nuclease-type E-colicins tells us that they have recently diverged through the alteration of a limited operon region encoding T2A and Imm as a set, suggesting the evolutionary importance of the coordination of T2A and Imm proteins. At the same time, T2A and Imm proteins represent a promising model for the study of protein-protein interaction, using both genetical and physicochemical approaches.

IMMUNITY SPECIFICITY AND OVERALL HOMOLOGY OF THE NUCLEASE-TYPE COLICINS

Table 1 shows the presence (+) or absence (-) of immunity of E colicin plasmids toward nuclease-type colicins in *E. coli* K12 cells. The immunities lying along the diagonal axis are self-evident from the definition of immunity. Three non-diagonal immunities, i.e., the E5 immunity of ColE9, and the E8 immunities of ColE3 and ColE6, are due to secondary immunity genes on these plasmids (Figure 1).

Table 1. Immunity of Plasmids in *E.coli* K12 to Nuclease-type E-Colicins.

	colicin								cloacin	colicin
plasmid	E2	E3	E4	E5	E6	E7	E8	E9	DF13	E3::DF13
ColE2-P9	+	-	-	-	-	-	-	-		-
ColE3-CA38	-	+	-	-	-	-	+	-		-
ColE4-CT7	-	-	+	-	-	-	-	-		-
ColE5-099	-	-	-	+	-	-	-	-		-
ColE6-CT14	-	-	-	-	+	-	+	-	?	+
ColE7-K317	-	-	-	-	-	+	-	-		-
ColE8-J	-	-	-	-	-	-	+	-		-
ColE9-J	-	-	-	+	-	-	-	+		-
CloDF13	-	-	-	-	+	-	-	-		+

+: presence, -: absence of immunity

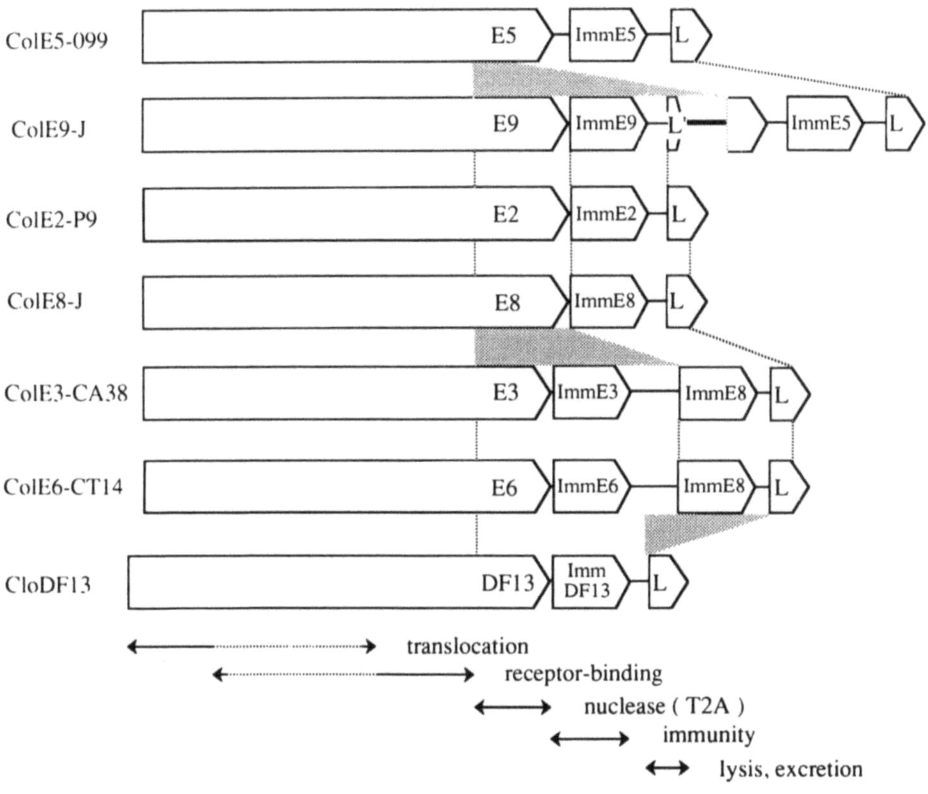

Figure 1. Operon structures and protein functions of the nuclease-type E-colicins and cloacin DF13. ColE7-K317 has a similar structure to ColE2-P9 (not shown). Homologous sites are connected with vertical or oblique lines. The C-terminal small domain is a DNAase for colicins E9, E2 and E8, and an RNAase for colicins E3, E6, cloacin DF13, and possibly colicin E5. The sequences of the remaining N-terminal parts of all the colicins are almost identical. The areas exhibiting no homology are shaded, except between colicin E6 and cloacin DF13 structural genes. The thick line in ColE9 indicates IS*E9*. L, the lysis gene (the BRP gene for CloDF13).

Comparison of the colicin operons and the deduced amino acid sequences indicates that all the nuclease-type colicins share the N-terminal four-fifth region. The remaining C-terminal nuclease region (T2A) and accompanying Imm proteins form a variable region (Figure 1). The DNAase-regions of E2, E7, E8 and E9, as well as the cognate Imm proteins, are reasonably homologous but still constitute the most varied regions of the molecules, reflecting the differences in immunity (Lau *et al.*, 1984; Masaki *et al.*, 1985; Cole *et al.*, 1985; James *et al.*, 1987a; Uchimura & Lau, 1987; Toba *et al.*, 1988; Lau & Condie, 1989; Chak *et al.*, 1991). Likewise, E3 and E6 of the RNAase-type are highly homologous, but the differences in their sequences are concentrated in the T2A region and the Imm protein (Figure 2a; Masaki & Ohta, 1985; Akutsu *et* al., 1989; Lau & Condie,1989). Although E5 is assumed to belong to the RNAase-type

(Mock & Pugsley, 1982), its T2A-region and Imm show no similarity to those of other E colicins (Lau & Condie, 1989, Curtis *et* al., 1989).

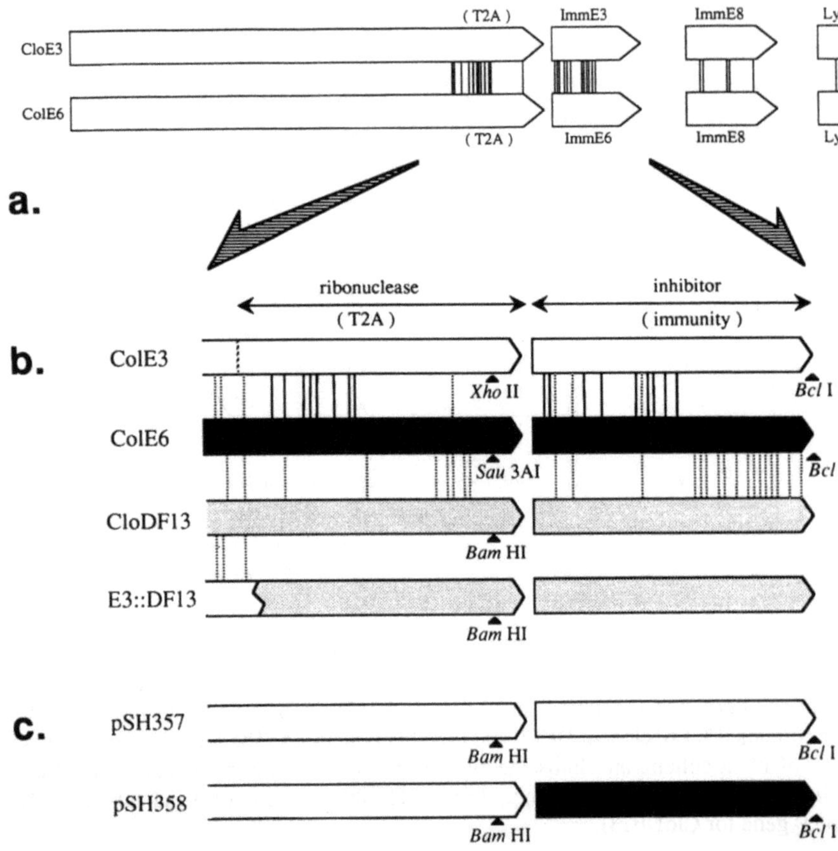

Figure 2. Sequence comparison of colicins E3, E6 and cloacin DF13, and of their immunity proteins. a) Non-identical amino-acids of ColE3 and ColE6 (connected with vertical lines) are concentrated within their T2A-Imm regions. b) T2A-Imm regions of ColE3, ColE6, CloDF13 and ColE3::DF13 plasmids. Non-identical amino-acids are connected with solid or broken vertical lines; solid lines represent those possibly involved in immunity specificity determination. c) Plasmids used in the experimental co-evolution. pSH357 is the parental *colE3-immE3* plasmid, in which a unique *Bam*HI site was introduced to punctuate the T2A and Imm regions. pSH358 is a destabilized *colE3-immE6* plasmid, from which a stabilized *imm* gene was isolated.

Cloacin DF13 encoded by plasmid CloDF13 shows nearly identical RNAase activity to colicin E3 (de Graaf *et* al., 1973), and exhibits considerable homology to E3 and E6, particularly in the T2A-corresponding region and the Imm protein (Figure 2; van den Elzen *et* al., 1983). Plasmid CloDF13 shows cross-immunity with ColE6, but not with ColE3 (Males & Stocker, 1982; James *et al.*, 1987b). *E. coli* K12 is, however, naturally resistant to cloacin DF13 (hence not "colicin" but cloacin), due to a lack of the cloacin receptor (Krone *et al.*, 1983). In order to determine the immunity property of cloacin DF13 in relation to those of E-colicins, we constructed a chimeric "colicin" E3::DF13 through homologous recombination between ColE3- and CloDF13-derived plasmids (Figure 2b; Akutsu *et al.*, 1989). The N-terminal portion of E3::DF13, including the receptor-binding domain, is derived from E3, and the C-terminal T2A domain from DF13, together with the ImmDF13 protein. An *E. coli* strain carrying ColE6 was completely immune to colicin E3::DF13, indicating that the immunity specificities of ColE6 and CloDF13 are entirely equivalent (Table 1).

SOME EVOLUTIONARY MODELS FOR THE NUCLEASE-TYPE COLICINS

Based on structural and functional data for bacteriocin operons available at present, we can tentatively propose some evolutionary models including the *colE2, colE8, colE3, colE6 and cloDF13* operons. The most striking finding of the sequence comparison is that non-homology is confined within a limited operon region involved in immunity specificity, the T2A-*imm* region. The operon region from T2A to the lysis gene seems highly susceptible to gene rearrangement (Figure 1), in particular, the T2A-*imm* region forming a hot-spot of point mutations (Figure 2). In other words, the almost identical sequences in other operon regions suggest a fairly recent divergence of these operons.

Apparently, ColE3 and ColE6 have an extra *immE8* gene and this region is absent from CloDF13 (Figure 1). But we assume that these *immE8* genes in ColE3 and ColE6 are not the result of a simple DNA transposition of *immE8* into ColE3 and ColE6.

<u>Model 1</u>

ColE2 ↔ ColE8 → ColE3 → ColE6 → CloDF13

According to model 1, the T2A region of ColE8 was replaced through a kind of gene conversion with a proto-E3 segment which encodes an RNAase and its inhibitor, most likely of a certain bacteriocin gene, giving rise to ColE3 (Figure 3a). Either when the ColE6 plasmid became completely segregated from ColE8 or when the host strain changed from *E. coli* to a certain colicin E-resistant strain, the *immE8* gene lost its evolutionary advantage (discussed below) and the 401 base pair segment carrying *immE8* was deleted from the plasmid, finally leading to CloDF13.

Model 2

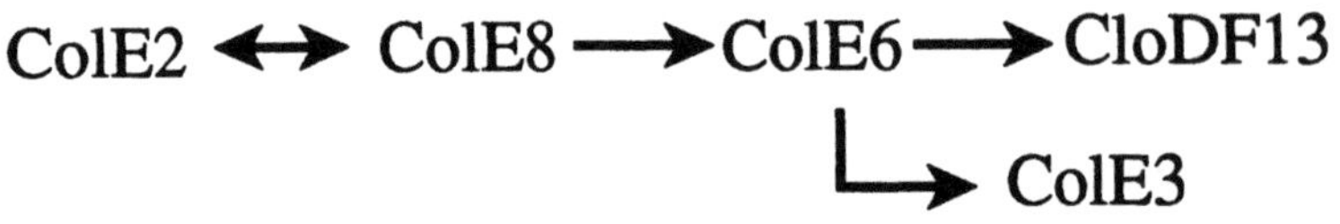

Alternatively, ColE6 might be a more proximal progeny of ColE8, and ColE3 only branched from ColE6. Among the ColE plasmids, only ColE6 has an extra DNA fragment containing a secondary replication origin which belongs to another incompatibility group (Watson *et al.*, 1985; T. Itoh, personal communication). For model 2, this extra fragment of ColE6 seems uncalled-for as an intermediate plasmid between ColE8 and ColE3. Moreover, the finding that ColE8 and ColE3 but not ColE6 belong to the same incompatibility group makes model 2 less likely (Cooper *et al.*, 1986).

Model 3

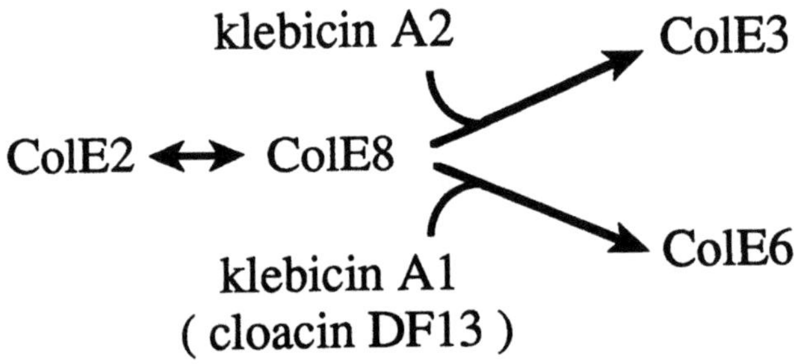

The finding of three A-group klebicins (Cooper & James, 1985; James *et al.*, 1987b) suggested that the evolutionary route of plasmids is actually more complicated. According to the results of phenotypic analysis, the klebicin Al plasmid showed immunity to colicin E6, but not to E8. Here "cloacin" DF13 could be reclassified as "klebicin" Al. Another klebicin plasmid encoding A2 showed immunity to colicin E3, but not to E8. Thus klebicin A2 appeared to be the klebicin counterpart of colicin E3, just like klebicin Al (cloacin DF13) to colicin E6. These relationships have made it difficult to connect the four bacteriocins (E3, E6, Al and A2) evolutionarily with a single line in a model. The third klebicin plasmid encoding A3 showed a more complicated phenotype.

One possible modification of the model is that ColE8 captured the T2A-*imm* regions of the klebicin

Al (cloacin DF13) and klebicin A2 plasmids, giving rise to ColE3 and ColE6, respectively (Figure 3a). This model requires two independent gene conversion events which occurred in an identical fashion, since there is no sequence gap between the present E3 and E6 operons (Akutsu *et al.*, 1989). In addition, the origin of the about 100 bp segment preceding the *immE8* gene of the present ColE3 and ColE6 then remains to be explained (Figure 1); this region has no correspondence in either ColE8 or CloDF13 (Masaki *et al.*, 1985; Toba *et al.*, 1988). Nonetheless, the *immE8* genes of ColE3 and ColE6 show almost the same degree of divergence, or evolutionary distance, from *immE8* of ColE8, making model 3 more likely than models 1 and 2.

Anyway, it is probably necessary to assume recombination between plasmids and re-shuffling of gene segments after a plasmid with a new immunity type has once been established. Thus, our evolutionary models might be first approximations. Furthermore, we must be aware that these evolutionary models should be drawn in terms of not present but ancestral bacteriocin plasmids. Further structural and functional analysis of A-group klebicins will help us to draw a more exact plasmid evolution model.

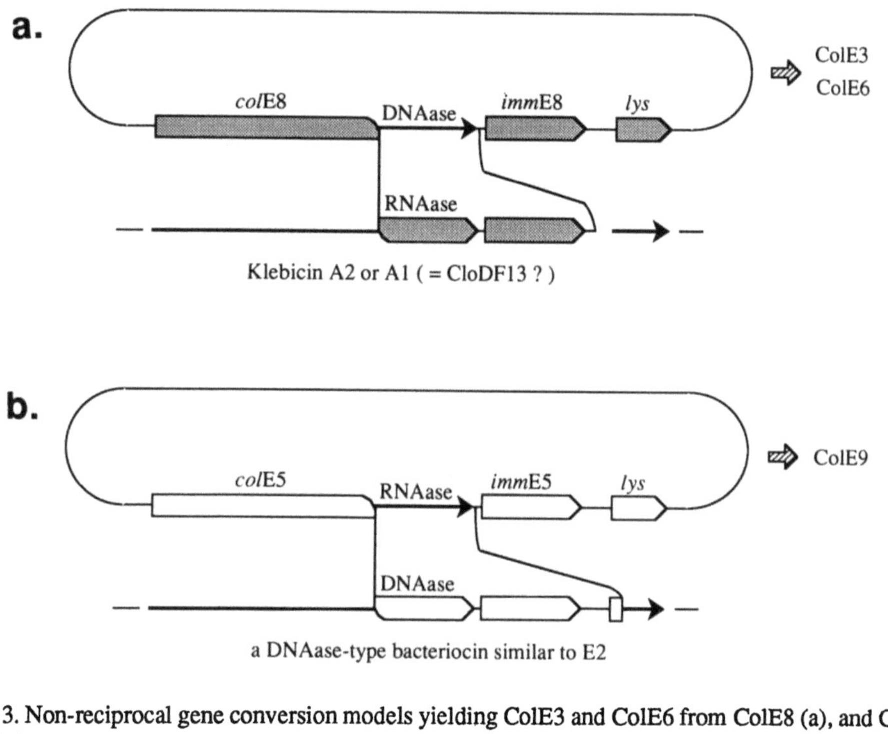

Figure 3. Non-reciprocal gene conversion models yielding ColE3 and ColE6 from ColE8 (a), and ColE9 from ColE5 (b).

SELECTIVE PRESSURES BETWEEN PLASMIDS AND WITHIN A PLASMID

Colicins are potentially lethal to *E. coli* cells, which leads to two kinds of outstanding selective pressure on plasmid evolution: one operative between Col plasmids and the other between genes within a plasmid.

Selective Pressure between Plasmids and Gene Rearrangement

Colicinogenic strains of different immunity types kill each other. During the evolution "within" the DNAase-type (E2 and E8) or RNAase-type (E3 and E6) colicins, the accumulation of point mutations could gradually shift their immunity specificities, finally leading to a new type ColE plasmid. In contrast, a one-step conversion from DNAase to RNAase, even if accompanied by a change in immunity, might be difficult, since a progeny should be swiftly killed by an excess amount of the parental colicin. In this sense, a non-reciprocal gene conversion which retains the parental *immE8* gene must have been an inevitable event; this enabled the newly formed ColE3 cell to kill the ColE8 cells without being killed by E8 (Figure 3a). The opposite situation is seen in the case of ColE5 (possibly an RNAase-type) and ColE9 (DNAase); ColE9 has a part of the E5 operon including *immE5* in the 3' region of the E9 operon (Figure 3b; Chak & James, 1986; Curtis *et al.*, 1989).

Lau and Condie (1989) found a degenerate transposon-like structure, designated as IS*E9*, at the end of the ColE5-derived segment in ColE9 (Figure 1), and the authors assumed that a conjugate plasmid molecule (a double replicon of ColE5 and proto-ColE9) had transiently appeared before the plasmid segregation producing the present ColE9. On the other hand, such a transposon-like structure is not found in the 5' region of the *immE8* gene of either ColE3 or ColE6, though some possible secondary structures are found (Lau *et al.*, 1984). A transposon-directed gene rearrangement seems feasible, but is possibly only one of the incidental ways in the evolution of colicins .

Considering that T2A and the following region, in particular, the T2A-*imm* segment, are variable within the operon (Figures 1 and 2), we cannot help supposing another recombinational hot-spot, the 5' end of the T2A-corresponding region. Figure 4 shows the c.a. 140 base pair sequence preceding the T2A regions of ColE8, ColE6 and CloDF13. The homology between ColE8 and ColE6 (or ColE3, not shown) gradually disappears along about 100 base pairs, and the homology between ColE6 and CloDF13, i.e., between the colicin and cloacin plasmids, gradually appears along nearly the same region. This suggests that this region has been subjected to a certain kind of non-site-specific recombination. We observed that this region is considerably tolerant to DNA insertion or replacement, without interference of the colicin activity (unpublished data). The corresponding protein region seems to form a hinge between functional domains, and is also highly sensitive to various proteases (Mooi & De Graaf, 1976; Ohno-Iwashita & Imahori, 1980). Although the repeatedly appearing A-tracts in this region suggested DNA bending, the delay of the electrophoretic mobility of the ColE3 DNA fragment containing this region was not so obvious compared to the mobilities of other plasmid fragments (data not shown).

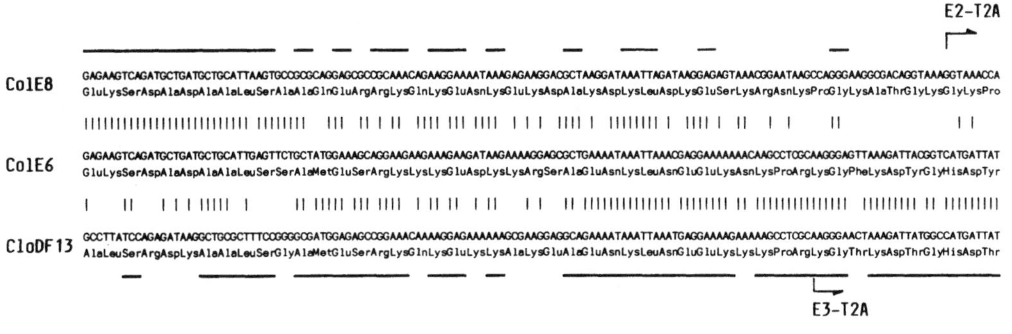

Figure 4. Nucleotide and amino acid sequences around the hinge region between the receptor binding and nuclease domains of colicins E8, E6 and cloacin DF13. Identical nucleotides are connected with vertical lines. Identical amino acids between E8 and E6 are indicated with upper lines, and those between E6 and DF13 with underlines. The putative N-terminus of the E2-T2A fragment and the N-terminus of the E3-T2A fragment are indicated by hooked arrows at the corresponding sites of the E8 and DF13 sequences, respectively.

Yasueda et al. (1988) found a putative cross-over site for plasmid resolution and stability in both ColE2 and ColE3 between the colicin operon and the replication region, whereas the corresponding sites of ColE1, ColK and CloDF13 are in the opposite regions of the plasmids (Summers & Sherratt, 1988). This resolution site is functional and most probably forms an efficient recombinational site between nuclease-type Col plasmids (T. Itoh, personal communication). Thus, we consider that the gene conversion of the downstream part of Col operons may have occurred generally at the 5' end of the nuclease-encoding region and this plasmid resolution site, with somewhere in the 3' region of *imm* being an optional preferential recombination site.

Selective Pressure between Genes

The other form of selective pressure operates between *col* and *imm* genes within a plasmid. Strict immunity is essential for a nuclease-type Col plasmid, as described above. Some mutation within either the T2A or Imm region affecting the binding specificity might have rendered the host cells lethal and consequently evoked complementary mutation(s) within the counterpart protein, giving rise to a new Col plasmid with an altered immunity specificity. We refer to this special kind of coevolution as 'hand-in-hand evolution', which maintains the mutual tight binding specificity between colicin and immunity proteins.

No differences in enzymatic activities have been noted so far in DNAase-colicins with different immunities, as well as in RNAase-colicins. Although the active site of T2A as an nuclease has not yet been identified, the Imm proteins certainly recognize amino acids of T2A other than those forming the enzymatic active site, possibly those on the surface of the T2A molecule. In this sense, the situation may be quite different from the case of a protease and its inhibitor, e.g., bacterial serine proteases and the *Streptomyces* subtilisin inhibitor (SSI), in which the inhibitor can be regarded as a non-digestible substrate-analog (Hiromi *et al.*, 1985).

Immunity specificities are determined by the combination of non-homologous amino acids in the T2A region and Imm protein. Here we paid particular attention to E3, E6 and DF13 (Figure 2). Whereas E6 and E3::DF13, as well as ImmE6 and ImmDF13, are functionally equivalent, ColE6 shows sequence diversity from CloDF13 to almost the same extent as from ColE3 in the T2A-*imm* region. This helped us to exclude some of the non-homologous amino acids from the list of candidates for immunity determinants. At most, eight of the 97 amino acids in T2A should distinguish between ImmE3 and ImmE6 (or ImmDF13), and nine of the 84 amino acids in Imm between E3 and E6 (or DF13) (Figure 2b).

IDENTIFICATION OF THE MAJOR SPECIFICITY DETERMINANT OF ImmE3 BY EXPERIMENTAL COEVOLUTION

Sequence data suggested rapid alteration of the T2A-*imm* region in the past, and sequence comparison restricted the immunity specificity determinants of E3 and E6 to a small number of amino acids.

These findings tempted us to attempt an experimental "hand-in-hand evolution" with ColE3 and ColE6 plasmids. An artificially destabilized plasmid with a non-cognate immunity gene would render most of the host cells lethal and, consequently, lead to the enrichment of stabilized variants. From among them, we might possibly isolate a new-type of colicin (when the immunity specificity shifted), or obtain some essential information as to specificity determination (when the immunity specificity resumed one of the original interactions) (Masaki *et al.*, 1991).

Experimental Coevolution with a *colE3-immE6* Hybrid Plasmid

The parental colicin E3 plasmid used was pSH357, which carries the wild-type colicin E3 operon, except that the *Sau*3AI site near the 3' end of the *colE3* gene had been converted to a *Bam*HI site without changing the encoding amino acid (Figure 2c). A *colE3-immE6* hybrid plasmid, pSH358, was constructed by replacing the *Bam*HI-*Bcl*I fragment of pSH357 *(immE3)* by the corresponding *Sau*3AI *immE6* fragment of plasmid ColE6, using an *recA* strain as a host in order to repress the colicin expression (Figure 2c). A *recA*$^+$ strain transformed with pSH358 barely grew and was highly sensitive to the addition of a low concentration of mitomycin C, an SOS-inducer.

From among this suicidal E3$^+$ImmE6$^+$ cell population, we isolated a stabilized, i.e., evolved, clone which produced colicin E3 stably and apparently exhibited immunity to both E3 and E6. The transfer of restriction fragments from this "evolved" plasmid to a set of specially designed acceptor plasmids (Masaki *et* al., 1991) restricted the responsible mutation(s) to the *Bam*HI-*Bcl*I segment containing the *imm* gene.

Sequence analysis showed that the phenotypic change of the evolved plasmid is due to a single transversion in *immE6*, TGG (Trp-48) to TGT (Cys) (Figure 5a). This result indicates that ImmE6 acquired immunity toward E3 surprisingly only through a single replacement of Trp-48 by Cys. On the contrary, the plasmid did not lose the original immunity to E6. Interestingly, Cys is the 48th residue of ImmE3, and appears only once throughout the sequences of E3, E6, ImmE3 and ImmE6 proteins.

Evaluation of the Immunity Determinants in ImmE3 and ImmE6 with Chimeric Immunity Genes

To evaluate the contributions of the amino-acids in ImmE3 and ImmE6 to the respective immunity specificities, both *immE3* and *immE6* were cloned in tandem into pUC18, and three *immE3::immE6* and three *immE6::immE3* chimeric genes were constructed through *recA*-dependent homologous recombination (Figure 5; Masaki et *al.*, 1991). The nine amino-acid pairs possibly involved in immunity determination were thus divided into four groups by three recombination sites, #1 to #3 (Figure 5a). Since the C-terminal halves of ImmE3 and ImmE6 are identical, the recombination at site #3 consequently represents a point mutation and the *immE63C* gene produces the same protein as the plasmid obtained from the above coevolution experiment.

The properties of the chimeric genes were consistent with those of the evolved plasmid. The results of phenotypic examination (Figure 5b, the cross-streak test) suggest that a single replacement of Trp-48 by Cys in ImmE6 (ImmE63C) confers the E3 immunity, the E6 immunity being retained. Coincidentally, replacement of Cys-48 by Trp in ImmE3 (Imm36C) abolishes the E3 immunity and does not acquire the E6 immunity. Thus, Cys-48 should be a unique immunity determinant for ImmE3, while Trp-48 is not the absolute immunity determinant for ImmE6.

In the spot test (Figure 5b), the apparent titers of purified E3 and E6 samples toward cells harboring various *imm* plasmids were compared quantitatively. Here again, replacement of Trp-48 by Cys in ImmE6 confers a great extent of E3 immunity (ImmE63C vs. ImmE6), and replacement of Cys-48 by Trp in ImmE3 greatly reduces the E3 immunity (ImmE36C vs. ImmE3). Moreover, ImmE63C has only Cys-48 of nine ImmE3-specific amino acids, and shows a higher E3 immunity (a lower spot test value) than ImmE36C, which has remaining eight ImmE3-specific amino acids. This means that Cys-48 alone (ImmE63C) contributes much more to the E3 immunity specificity than the eight other ImmE3-specific amino acids together (ImmE36C).

a.

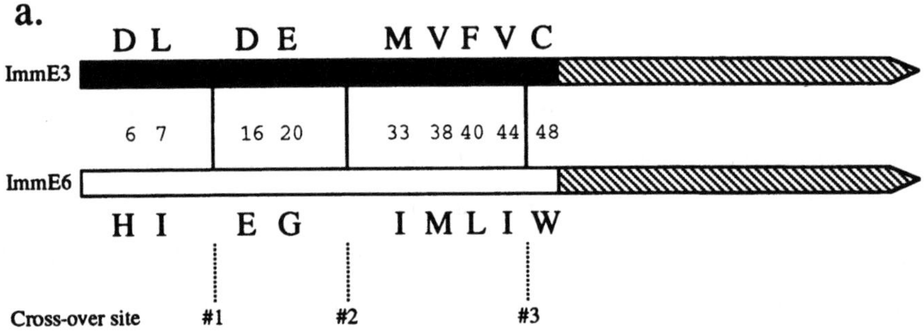

b.

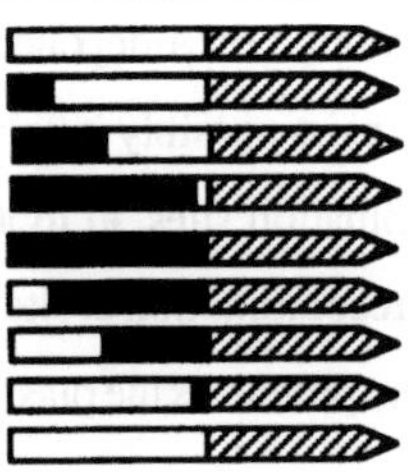

Imm protein	E3 immunity: cross streak	E3 immunity: spot test	E6 immunity: cross streak	E6 immunity: spot test
ImmE6	-	2^{15}	+	$<2^{0}$
ImmE36A	-	2^{13}	-	2^{8}
ImmE36B	-	2^{10}	-	2^{9}
ImmE36C	-	2^{10}	-	2^{12}
ImmE3	+	$<2^{0}$	-	2^{14}
ImmE63A	+	$<2^{0}$	-	2^{12}
ImmE63B	+	2^{8}	-	2^{9}
ImmE63C	+	2^{8}	+	2^{5}
ImmE6	-	2^{15}	+	$<2^{0}$
none (pUC18)	-	2^{18}	-	2^{15}

Figure 5. Construction of chimeric immunity genes, and evaluation of the specificity determinants in ImmE3 and ImmE6. The Imm proteins are represented by arrows; the solid, open and hatched areas indicate the ImmE3-specific, ImmE6-specific and C-terminal identical regions, respectively. a) Possible specificity determinants deduced in the sequence comparison (Figure 2b) are shown with residue numbers. Homologous recombinations at sites #1 to #3 gave rise to six chimeric immunity genes as illustrated in b). b) Cross-streak test; "+" denotes immunity and "-" sensitivity to each colicin. Spot test; purified E3 and E6 samples were serially diluted by a factor of two, and the apparent titers of E3 and E6 against the cells containing the respective immunity gene plasmids were measured. A value of $<2^{0}$ represents complete immunity, and 2^{18} for E3 and 2^{15} for E6 complete sensitivity in this experiment. The vector plasmid was pUC18 (Yanish-Perron et *al.*, 1985), and immunity was measured under a non-induced condition for the *lac* promoter.

The results of the spot test also suggest that other amino acids affect each immunity specificity to various extents. Alteration of Asp-16 and Glu20 of ImmE3 to those of ImmE6 seriously reduces the E3 immunity (ImmE63B), but unlike in the case of Cys-48, the presence of these two residues alone does not confer sufficient E3 immunity (ImmE36B), suggesting that Asp-16 and/or Glu-20 (probably the latter) are sub-determinants of the E3 immunity. On the other hand, the determinants of ImmE6 seem to be more dispersed over the sequence and operate well in an accumulative manner. His-6 and/or Ile-7 (probably the former) are needed but are not sufficient by themselves for the full E6 immunity (ImmE36A vs. ImmE63A). Likewise, Trp-48 is operative but not sufficient for the E6 immunity (ImmE36C vs. ImmE63C).

In this experimental coevolution, we directly identified the unique specificity determinant of ImmE3 which discriminates between colicins E3 and E6. The uniqueness of Cys-48 in ImmE3 suggests that its counterpart in the E3-T2A region is also formed from one or only a few amino acids. This will be confirmed when a colicin E6 derivative with E3 specificity is obtained through an alternative experimental coevolution starting with a *colE6-immE3* hybrid plasmid, but we have not succeeded in this type of coevolution so far. Since an "evolved" *colE6* gene would be accompanied by the native *immE3* gene, a new plasmid molecule will be, if at all, still sensitive to E6, which is abundant both within the cell and in the culture. Thus it will acquire a selective advantage only after complete segregation from its parental plasmid or clone, unlike in the case of the evolution of the immunity gene, in which a stabilizing mutation confers an immediate selective advantage.

HIGHER ORDER STRUCTURE OF THE ImmE3 PROTEIN

The binding specificities of T2A and Imm must finally be explained on the basis of their higher-order structures and in terms of physicochemical parameters. Toward this goal, we are now attempting to determine the three-dimensional structure of the ImmE3 protein by means of NMR. We have prepared and purified a large amount of the ImmE3 protein, as well as several kinds of ^{15}N- or ^{2}H-labeled derivatives of it, and obtained 2- or 3-dimensional NMR spectra (reviewed in: Oppenheimer & James, 1989a, 1989b). We have finished the assignment of the most part of the proton signals along the polypeptide chain and propose here a preliminary secondary structure model of ImmE3 (Figure 6).

The previous sequence data could not predict definite secondary structures of T2A and Imm according to the method of Nagano (Masaki & Ohta, 1985; unpublished data). The present NMR data suggest that the major part of the ImmE3 molecule is folded into an antiparallel β-sheet structure and contain only a few α-helices.

Our genetic experiments revealed that the major specificity determinant of ImmE3 is Cys-48, and that cooperative ImmE6 determinants are probably His-6 and Trp-48, as described above. The most interesting finding as to the present secondary structure is that these two amino acid positions, 6 and 48, are close and their side groups are extended to the same side of the β-sheet. Thus, the specificities are suggested to be determined by a small area on the surface of both the ImmE3 and ImmE6 proteins.

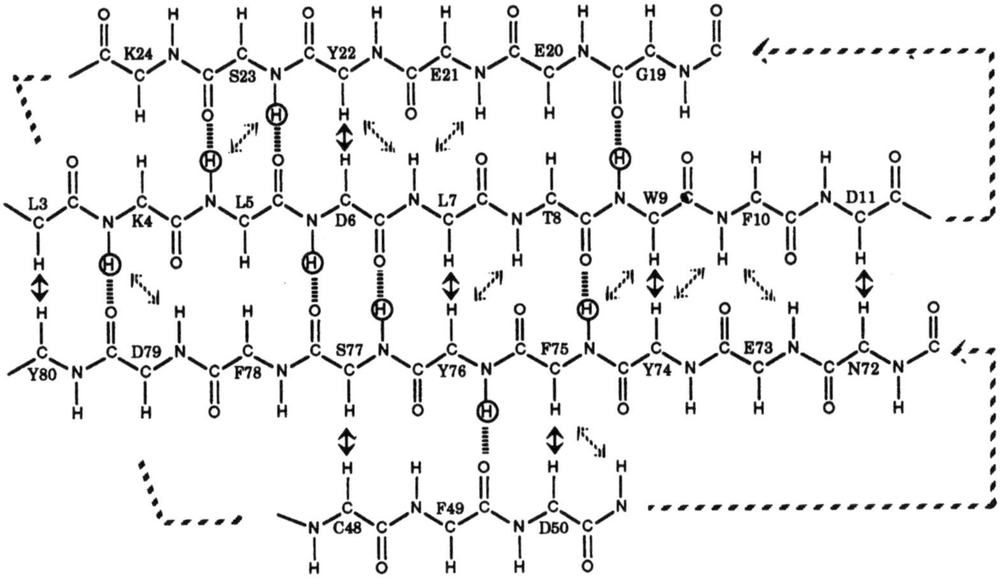

Figure 6. Central anti-parallel β-sheet structure of ImmE3 deduced from the results of NMR spectroscopy. Amino acids are presented using a one-letter code with residue numbers beside the α-carbon atoms. Numbering starts at the initiator methionine, which is actually cleaved-off. Protons forming hydrogen bonds are circled. NOEs (nuclear Overhauser effects) observed between protons of different amino acids are indicated by short double-headed arrows.

A simple sequence comparison of both the T2A and Imm regions suggested that the "colicin-" and "cloacin-specific" amino acids are rather concentrated in their C-terminal halves (Figure 2b: Masaki & Ohta, 1982). One plausible explanation is that the C-terminal regions of T2A and Imm interact with each other, as the N-terminal regions have interactions to determine the E3 and E6(DF13) specificities. This speculation seems consistent with the conclusion obtained from the experiment with a cloacin DF13 deletion mutant that the C-terminal region of cloacin DF13 interacts with ImmDF13 (Gaastra *et al.*, 1978). However, colicin E6 is perfectly blocked by, i.e., has a good affinity to, ImmDF13, as is the colicin E3::DF13 to ImmE6. Moreover, we further obtained two hybrid plasmids, containing *colE3::DF13-*

immE6 and *colE6-immDF13* operons, respectively. These two plasmids stably produced colicins like their parents (Akutsu *et al.*, 1989). It is thus more likely that the differences between colicin and cloacin sequences in both the C-terminal regions of T2A and Imm are canceled by the intramolecular folding of each protein, though this speculation does not exclude the possibility of some weak interaction between T2A and Imm at their C-termini.

CONCLUDING REMARKS

We have reviewed the conspicuous features of the immunity and evolution of the nuclease-type E-group colicin operons and briefly looked into the physical mechanism underlying the immunity specificity. Sequence comparison strongly suggests rapid divergence and evolution of the nuclease/inhibitor-encoding region within these colicin operons. But unfortunately, bacteria have no fossil as an absolute time standard in evolution and, thus, we have to solve the problem through some experimental approaches. Our forced experimental evolution may provide one clue. We obtained three kinds of T2A (those of E3, E6 and E3::DF13) and nine Imm proteins (ImmE3, ImmE6, ImmDF13, and six chimeric Imm proteins), which exhibit various extents of specificity on a similar structural basis. We believe that this set of proteins is promising for further investigations to reveal the physicochemical background of the general protein-protein interaction.

Two types of selective pressure, one operative between plasmids and the other between genes within a plasmid have been discussed. Alternatively, as the selective pressure between plasmids "within a cell", plasmid incompatibility may have been one of the most important driving forces of the evolution as well, which is discussed in detail by Dr. T. Itoh in this book. Anyway, the primary motive force making only a confined operon region a mutational hot spot remains unknown, which also must finally be determined experimentally.

REFERENCES

Akutsu A, Masaki H, Ohta T (1989) Molecular structure and immunity specificity of colicin E6, an evolutionary intermediate between E-group colicins and cloacin DF13. J Bacteriol 171:6430-6436

Boon T (1971) Inactivation of ribosomes *in vitro* by colicin E3. Proc Natl Acad Sci USA 68:2421-2425

Bowman CM, Sidikaro J, Nomura M (1971) Specific inactivation of ribosomes by colicin E3 *in vitro* and mechanism of immunity in colicinogenic cells. Nature NB 234:133-137

Buxton RS (1971) Genetic analysis of *Escherichia coli* K12 mutants resistant to bacteriophage BF23 and the E-group colicins. Mol Gen Genet 113:154-156

Chak K-F, James R (1986) Characterization of the ColE9-J plasmid and analysis of its genetic organization. J Gen Microbiol 132:61-71

Chak K-F, Kuo W-S, Lu F-M, James R (1991) Cloning and characterization of the ColE7 plasmid. J Gen Microbiol 137:91-100

Chan PT, Ohmori H, Tomizawa J, Lebowitz J (1985) Nucleotide sequence and gene organization of ColEl DNA. J Biol Chem 260:8925-8935

Cole ST, Saint-Joanis B, Pugsley AP (1985) Molecular characterization of the colicin E2 operon and identification of its products. Mol Gen Genet 198: 465-472

Cooper PC, James R (1984) Two new E colicins, E8 and E9, produced by strain of *Escherichia coli*. J Gen Microbiol 130:209-215

Cooper PC, James R (1985) Three immunity types of klebicins which use the cloacin DF13 receptor of *Klebsiella pneumoniae*. J Gen Microbiol 131:2313-2318

Cooper PC, Hawkins FKL, James R (1986) Incompatibility between E colicin plasmids. J Gen Microbiol 132:1859-1862

Curtis MD, James R, Coddington A (1989) An evolutionary relationship between the ColE5-099 and ColE9-J plasmids revealed by nucleotide sequencing. J Gen Microbiol 135:2783—2788

De Graaf FK, Niekus HGD, Klootwijk (1973) Inactivation of bacterial ribosomes *in vivo* and in *vitro* by cloacin DF13. FEBS Lett 35:161-165

DiMasi DR, White JC, Schnaitman CA, Bradbeer C (1973) Transport of vitamin B12 in *Escherichia coli:* common receptor sites for vitamin B12 and the E colicins on the outer membrane of the cell envelope. J Bacteriol 115:506-513.

Gaastra W, Oudega B, de Graaf FK (1978) The use of mutants in the study of structure-function relationship in cloacin DF13. Biochem Biophys Acta 540:301-312

Hiromi K, Akasaka K, Mitsui Y, Tonomura B, Murao S (eds) (1985) Protein protease inhibitor—The case of *Streptomyces* subtilisin inhibitor (SSI). Elsevier, Amsterdam

Jakes KS (1982) The mechanism of action of colicin E2, colicin E3 and cloacin DF13. In: Cohen P, Van Heyningen S (eds) Molecular action of toxins and viruses. Elsevier, Amsterdam, p 131

Jakes KS, Zinder ND (1974) Highly purified colicin E3 contains immunity protein. Proc Natl Acad Sci USA 71:3380-3384

James R, Jarvis M, Barker DF (1987a) Nucleotide sequence of the immunity and lysis region of the ColE9-J plasmid. J Gen Microbiol 133:1553-1562

James R, Schneider J, Cooper PC (1987b) Characterization of three group A klebicin plasmids: localization of their E colicin immunity genes. J Gen Microbiol 133:2253-2262

Krone WJA, Oudega B, Stegehuis F, de Graaf FK (1983) Cloning and expression of the cloacin DF13/ aerobactin receptor of *Escherichia* coli(ColV-K30). J Bacteriol 153:716-721

Lau PCK, Condie JA (1989) Nucleotide sequences from the colicin E5, E6 and E9 operons: presence of a degenerate transposon-like structure in the ColE9-J plasmid. Mol Gen Genet 217:269-277

Lau PCK, Rowsome RW, Watson RJ, Visentin LP (1984) The immunity genes of colicin E2 and E8 are closely related. Biosci Rep 4:565-572

Males BM, Stocker BAD (1982) Colicins E4, E5, E6 and A and properties of *btuB*$^+$ colicinogenic transconjugants. J Gen Microbiol 128:95-106

Masaki H, Akutsu A, Uozumi T, Ohta T (1991) Identification of a unique specificity determinant of the colicin E3 immunity protein. Gene 107:133-138

Masaki H, Ohta T (1982) A plasmid region encoding the active fragment and the inhibitor protein of colicin E3-CA38. FEBS Lett 149:129-132

Masaki H, Ohta T (1985) Colicin E3 and its immunity genes. J Mol Biol 182: 217-227

Masaki H, Toba M, Ohta T (1985) Structure and expression of the ColE2-P9 immunity gene. Nucl Acids Res 13:1623-1635

Mock M, Pugsley AP (1982) The BtuB group Col plasmids and homology between the colicins they encode. J Bacteriol 150:1069-1076

Mooi FR, de Graaf FK (1976) Effect of limited proteolysis on bacteriocin activity *in vivo* and *in vitro*. FEBS Lett 62:304-308

Morlon J, Chartier M, Bidaud M, Lazdunski C (1988) The complete nucleotide sequence of the colicinogenic plasmid ColA: high extent of homology with ColEl. Mol Gen Genet 211:231-243

Ohno S, Imahori K (1978) Colicin E3 is an endonuclease. J Biochem 84:1637-1640

Ohno-Iwashita Y, Imahori K (1980) Assignment of the functional loci in colicin E2 and E3 molecules by the characterization of their proteolytic fragments. Biochemistry 19:652-659

Oppenheimer NJ, James TL (eds) (1989a) Nuclear magnetic resonance, part A, Special techniques and dynamics. In: Methods in enzymology, vol 176. Academic Press, San Diego New York Berkeley Boston London Sydney Tokyo Toronto

Oppenheimer NJ, James TL (eds) (1989b) Nuclear magnetic resonance, part B, Structure and mechanism. In: Methods in enzymology, vol 177. Academic Press, San Diego New York Berkeley Boston London Sydney Tokyo Toronto

Ozaki LS, Kimura A, Shimada K, Takagi Y (1982) ColEl vectors for direct selection of cells carrying a hybrid plasmid. J Biochem 91:1155-1162

Pugsley AP, Oudega B (1987) Methods for studying colicins and their plasmids. In: Hardy KG (ed) Plasmids: A Practical Approach. IRL Press, Oxford, pl05

Summers DK, Sherratt DJ (1988) Resolution of ColEl dimers requires a DNA sequence implicated in the three-dimensional organization of the *cer* site. EMBO J 7:851-858

Schaller K, Nomura M (1976) Colicin E2 is a DNA endonuclease. Proc Natl Acad Sci USA 73:3989-3993

Toba M, Masaki H, Ohta T (1986) Primary structures of the ColE2-P9 and ColE3-CA38 lysis genes. J Biochem 99:591-596

Toba M, Masaki H, Ohta T (1988) Colicin E8, a DNase which indicates an evolutionary relationship between colicins E2 and E3. J Bacteriol 170: 3237-3242

Uchimura T, Lau PCK (1987) Nucleotide sequences from the colicin E8 operon: homology with plasmid ColE2P-9. Mol Gen Genet 209:489-493

Van den Elzen PJM, Walters HHB, Veltkamp E, Nijkamp HJJ (1983) Molecular structure and function of the bacteriocin gene and bacteriocin protein of plasmid CloDF13. Nucl Acids Res 11:2465-2477

Watson RJ, Vernet T, Visentin LP (1985) Relationship of the Col plasmids E2, E3, E4, E5, E6, and E7: restriction mapping and colicin gene fusions. Plasmid 13:205-210

Yanisch-Perron C, Vieira J, Messing J (1985) Improved M13 phage cloning vectors and host stains: nucleotide sequences of the M13mpl8 and pUCl9 vectors. Gene 33:103-119

Yasueda H, Horii T, Itoh T (1989) Structural and functional organization of ColE2 and ColE3 replicons. Mol Gen Genet 215:209-216

REPLICON EVOLUTION OF COLE2-RELATED PLASMIDS

T. Itoh and S. Hiraga
Department of Biology
Faculty of Science
Osaka University
Toyonaka, Osaka 560
Japan

The E colicins are a group of bacterial plasmid-coded toxins which use the *E. coli* cell surface receptor, the *btuB* gene product. Nine different colicins are distinguished based upon the specificity of interaction of the colicins with their immunity proteins (Males & Stocker, 1982; Cooper & James, 1984). The seventeen plasmids which carry the genes for the E colicins and their immunity proteins are small (about 7 kb, except for plasmids ColE2imm-K317 and ColE6-CT14 which are about 5 kb and 11 kb, respectively), multicopy and nonconjugative (Watson *et al.*, 1981; Mock & Pugsley, 1982; Cooper & James, 1984). Analyses of these plasmids, using various restriction enzymes, have previously shown that the plasmids ColE2 to E9 are structurally similar (Lawrence & James, 1984; Watson *et al.*, 1985a; Watson *et al.*, 1985b; Chak & James, 1986), whereas plasmid ColEl is distinct (see Chan *et al.*, 1985). More recently, biochemical and nucleotide sequencing analyses of these plasmids have revealed that colicins E2 to E9 are closely related (Toba *et al.*, 1988; Chak *et al.*, 1991; for details see papers by James *et al.*, Lau, and Masaki *et al.* of this volume). Here we refer to plasmids ColE2 to E9 as the ColE2-related plasmids.

ColEl, ColE2 and ColE3 require host DNA polymerase I for their replication (Kingsbury & Helinski, 1970; Tacon & Sherratt, 1976). ColEl continues to replicate in chloramphenicol-treated cells, resulting in accumulation of plasmid DNA (Clewell & Helinski, 1972), whereas ColE2, ColE3, ColE4 and ColE7 do not (Watson *et al.*, 1981). Mechanisms of initiation of replication and its regulation of ColEl have been studied extensively (see Tomizawa, 1987) and are quite different from those of ColE2 and ColE3 (Itoh & Horii, 1989; see below). A phenomenon in which two closely-related plasmids can not be propagated stably in the same cell line is called plasmid incompatibility. Incompatibility is a manifestation of relatedness, that is, the two plasmids belonging to the same group share at least one common element involved in plasmid replication control or stable inheritance, and therefore it has been used for classification of plasmids. Incompatibility relationships of the ColE plasmids have been little studied (Inselburg, 1974; Cooper *et al.*, 1986; Tajima *et al.*, 1988).

Here we first describe the mechanisms of initiation of replication and its control in ColE2-P9 and ColE3-CA38. We then show that all of the 17 ColE2-related plasmids so far we examined are closely related and share the mechanisms of initiation of replication and its control common to those of ColE2 and ColE3.

NATO ASI Series, Vol. H 65
Bacteriocins, Microcins and Lantibiotics
Edited by R. James, C. Lazdunski and F. Pattus

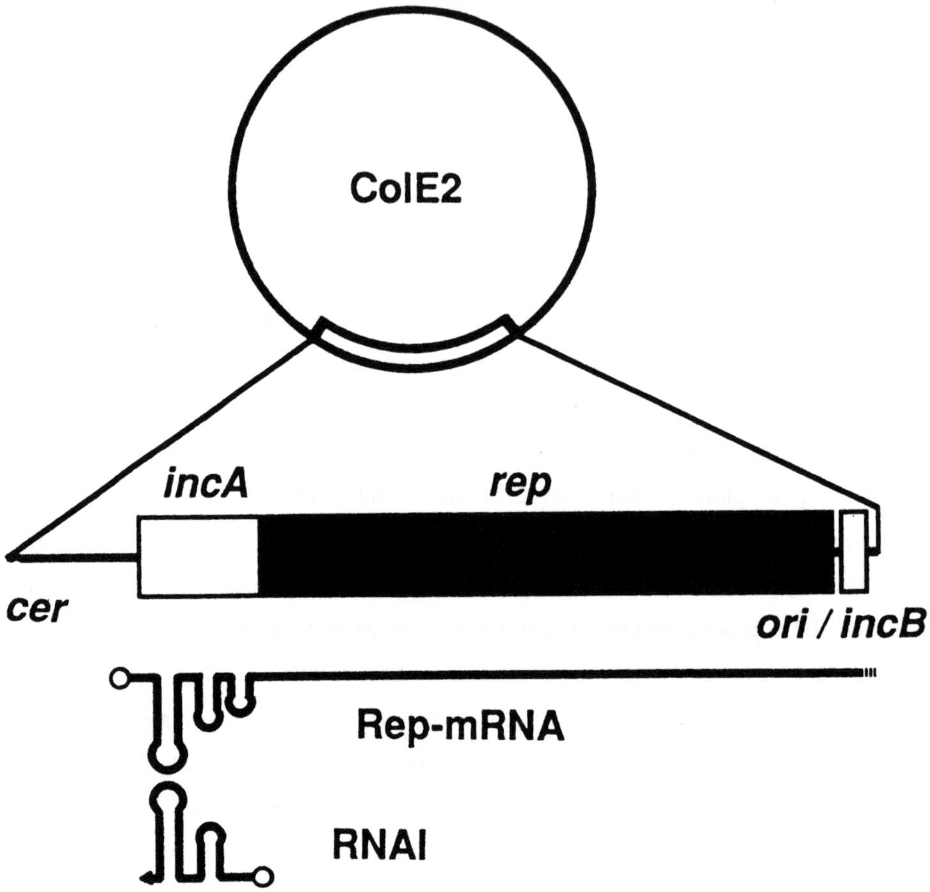

Figure 1. Schematic diagram of the ColE2-P9 replicon. *rep*, a gene encoding a plasmid-specific initiator protein of replication, Rep; *ori/incB*, an origin of replication/incompatibility by the cloned origin; *incA*, incompatibility by an antisense RNA, RNA I; *cer*, a site for site-specific recombination. Originally *cer* stands for ColEl resolution (Summers & Sherratt, 1984). We tentatively extend its usage to the corresponding sites of the ColE2-related plasmids. mRNA and RNA I are indicated.

The basic mechanisms of initiation of replication and its control in ColE2-P9 and ColE3-CA38

The basic replicon of plasmid ColE2-P9 is about 1.3 kb in length and consists of the *rep* gene, the cis-acting site (origin) and the *incA* gene (Figure l; Horii & Itoh, 1988; Tajima *et al.*, 1988; Yasueda *et al.*, 1989). The structural and functional organization of the ColE3-CA38 replicon is essentially identical to that of ColE2-P9. The plasmid-coded Rep protein (35 Kd) is essential for plasmid DNA replication (Itoh and Horii, 1989; Kido *et al.*, 1991). The minimal origin region is about 50 bp long. The cloned origin region excludes coexisting autonomously replicating plasmids with the homologous origin (Tajima *et al.*, 1988). We call this the IncB function. Replication of ColE2 is negatively regulated by a small antisense RNA, RNA I, at the level of the Rep expression (Takechi S, Yasueda H & Itoh T, submitted). RNA I is the product of the *incA* gene, transcribed in a direction opposite to that of the Rep mRNA from the region upstream of the Rep coding region. The cloned *incA* gene excludes coexisting autonomously replicating ColE2 or ColE3 (Yasueda H, Takechi S & Itoh T, submitted). We call this the IncA function. Just to the left of the *incA* gene, there is a region called *cer* required for site-specific recombination and stable maintenance of ColE2 by host bacteria (Kantake N & Itoh T, unpublished).

By using an *in vitro* replication system of ColE2 with crude extracts of *E. coli* cells we showed that replication starts from a fixed site and proceeds unidirectionally. We localized the initiation site of DNA synthesis on the nucleotide sequence of the origin region (Figure 2; Takechi S & Itoh T, in preparation). The leading strand synthesis starts from the T residue and proceeds to the right. The lagging strand synthesis stops at the C residue. The inhibition of the lagging strand synthesis beyond this residue might be the basic mechanism of unidirectional replication of this plasmid. At the 5' end of the newly synthesized leading strand DNA fragment, there remains a primer RNA, 5'ppApGpA.

Figure 2. Origin of replication of ColE2-P9. The line between the nucleotide sequences indicates the minimal origin region. Large arrowheads indicate positions which determine the specificity of interaction of the Rep proteins with the origins between ColE2 and ColE3. Primer RNA (broken line) and direction of DNA synthesis (thin arrow) are indicated. The overall replication proceeds to the right.

We have purified the Rep protein to near homogeneity (Kido *et al* 1991; Matsui H & Itoh T, in preparation). The purified Rep protein specifically binds to the origin region. The IncB incompatibility function of the origin region is explained as titration of the Rep proteins by the cloned origin regions.

We reconstituted the leading-strand DNA synthesis at the origin with purified proteins (Takechi S, Matsui H & Itoh T, in preparation). The purified Rep protein and DNA polymerase I can start the leading-strand DNA synthesis exactly at the same T residue and the primer RNA is ppApGpA. ADP is essential for the reaction and is directly incorporated into the primer RNA as the first residue. The Rep protein is a sequence-specific DNA binding protein which specifically recognizes the origin sequence and is also a primase which synthesizes a primer RNA at the specific site. The primer RNA is specifically used by DNA polymerase I to start DNA synthesis.

The expression of the Rep protein is regulated negatively by an antisense RNA, RNAI, at a post-transcriptional stage, probably at the translation step (Takechi S, Yasueda H & Itoh T, submitted). Mutations which result in a phenotype of increased copy number of plasmids in host bacteria were mapped in the coding region of RNA I, mostly at the loop portion of the stem-loop structure (Yasueda H, Takechi S & Itoh T, submitted). Each of these single-base pair substitutions greatly decreases the inhibitory activity of RNA I and the sensitivity of the Rep mRNA to RNA I action at the same time. These results indicated that RNA I interacts with the complementary region of the Rep mRNA to regulate expression of the Rep protein and that interaction at the loop portions of these RNA molecules is important for regulation.

In fact RNA I rapidly forms a stable complex with the Rep mRNA *in vitro* at near physiological conditions (Sugiyama T & Itoh T, submitted). The rate of complex formation is greatly decreased by a single-base pair substitution. Binding of RNA I causes significant changes in the secondary structure of the Rep mRNA (Sugiyama T, & Itoh T, unpublished). In the absence of RNA I the region of the Rep mRNA around the initiation codon of the Rep protein might form a certain secondary structure, which is accessible for ribosomes. Binding of RNA I might change the secondary structure of the Rep mRNA, which is inaccessible for ribosomes.

The nucleotide sequences of the replicon regions of ColE2 and ColE3 are quite homologous (Yasueda *et al*., 1989). The *incA* regions of these plasmids are completely homologous, and so they share the identical negative regulatory mechanism of initiation of replication. The Rep protein acts on the origin in a plasmid-specific manner, that is, there is no cross-reaction between the Rep proteins and origins of ColE2 and ColE3. The specificity between ColE2 and ColE3 is determined by the absence (ColE2) and the presence (ColE3) of an A/T base pair in the middle of the origin regions (Figure 2). To study specificity determinants of the Rep proteins a series of chimeric *rep* genes between ColE2 and ColE3 have been constructed by a simple and efficient method developed in our laboratory (Itoh T, unpublished). Genetic analyses using resultant chimeric *rep* genes revealed that the specificity of ColE2 is determined by a combination of one segment with 13 amino acid residues and the other with 11 amino acid residues in the carboxy terminal half of the Rep proteins. On the other hand the specificity of ColE3 is determined by a segment with 22 amino acid residues, which corresponds to the former of the two segments in ColE2.

Pairs of differently marked derivatives of ColE2, or of ColE3, which share the identical specificity types of both IncA and IncB functions, exclude each other very fast (Tajima *et al.*, 1988). However, in derivatives of ColE2 and ColE3 which share only the specificity of the IncA function mutual exclusion is very slow (Inselburg, 1974; Cooper *et al.*, 1986; Tajima *et al.*, 1988). So ColE2 and ColE3 have occasionally been judged to be compatible with each other. So far we do not know why mutual exclusion between them is so slow.

The region between the basic replicon and the colicin operon of ColE2 is involved in stable maintenance of plasmids by host bacteria (Kantake N & Itoh T, unpublished). The 200-bp region contains a site for site-specific recombination, which resolves plasmid dimer (polymer) molecules to monomer molecules. As plasmid copy number is controlled by maintaining the total number of the origins at a constant level and plasmid molecules are randomly distributed into daughter cells at cell division, an increase in proportion of monomer molecules is expected to increase the probability of inheritance of plasmids by daughter cells (Summers & Sherratt, 1984). The region of ColE2 required for the site-specific recombination is homologous to those of ColEl and related plasmids (Hakkaart *et al.*,1984; Summers *et al.*, 1985). Homology is higher in the regions containing the actual cut-and-rejoin (crossover) sites between ColE2 and ColE3 (Kantake N and Itoh T, unpublished) and between ColEl and ColK (Summers *et al.*, 1985). The site-specific recombination in ColE2 requires the active *xerA* gene which is one of the host genes required for the site-specific recombination in ColEl (Stirling *et al.*, 1988). The homology in nucleotide sequences together with the requirement for the *xerA* gene product indicate that the identical *E. coli* recombination system performs the site-specific recombination in ColEl and ColE2.

Comparison of the replicon regions of ColE2-related plasmids

The restriction maps of seventeen ColE2-related plasmids are quite similar (Lawrence & James, 1984; Watson *et al.*, 1985a; Watson *et al.*, 1985b; Chak & James, 1986; Itoh T, unpublished). ColE6 is larger than the others due to insertion of a 4-kb segment and ColE2imm-K317 is somewhat smaller. The replicon regions of ColE2-P9 and ColE3-CA38 are localized in the 1.3-kb regions of the plasmids. The restriction maps of the corresponding regions of other ColE2-related plasmids show some similarity, suggesting a possibility that the replicon regions of these plasmids are also localized in these regions.

Incompatibility between some of the ColE2-related plasmids have been examined (Cooper *et al.*, 1986). ColE3, ColE7 and ColE8 are mutually incompatible, indicating that they share a common regulatory mechanism for replication and/or maintenance. They might share a common IncB function, that is, a common set of the *rep* gene and origin. ColE5 and ColE9 are also mutually incompatible and ColE6 asymmetrically excludes ColE5 and ColE9. They might also share a common IncB function.

As we had already cloned the *incA* and *incB* genes of ColE2 -P 9 and ColE3-CA38 (Tajima et al., 1988), we examined their effect on replication of other ColE2-related plasmids (Table 1). The *incA* gene of ColE2-P9 excluded ColE2-P9 and ColE3, as expected. In addition it excluded some of the ColE2 plasmids and ColE8. So they share a common IncA function. That means they share a common negative regulatory mechanism of replication. Experiments with the cloned *incB* gene of ColE2 -P 9 showed that some of the ColE2 plasmids and all the ColE4 plasmids share a common specificity type of the Rep protein-origin interaction. The results also showed that ColE3, ColE7 and ColE8 share another specificity type of the Rep protein-origin interaction.

Table 1. Properties of various ColE plasmids

Plasmid	Type of colicin	Type of immunity against colicin	Type of incompatibility		
			IncA (RNAI)	IncB (Rep/Ori)	
ColE2-CA42	E2	E2	E2-CA42	E2-CA42	*
ColE2-K321	E2	E2	E2-P9	E2-P9	
ColE2-GEI288	E2	E2	E2-P9	E2-P9	
ColE2-GEI544	E2	E2	E2-P9	E2-P9	
ColE2-GEI602	E2	E2	E5-099	E2-CA42	*
ColE2-P9	E2	E2	E2-P9	E2-P9	*
ColE2imm-K317	—	E2	E2imm-K317	E2-CA42	
ColE3-CA38	E3	E3, E8	E2-P9	E3-CA38	*
ColE4-288	E4	E4	E2-CA42	E2-P9	
ColE4-CT9	E4	E4	E2-CA42	E2-P9	*
ColE4-K365	E4	E4	E2-CA42	E2-P9	
ColE5-099	E5	E5	E5-099	E5-099	*
ColE6-CT14	E6	E6, E8	E5-099	E5-099	*
ColE6-Ind8	E6	E6, E8	E5-099	E5-099	
ColE7-K317	E7	E7	E2-CA42	E3-CA38	*
ColE8-J	E8	E8	E2-P9	E3-CA38	*
ColE9-J	E9	E9, E5	E9-J	E5-099	*

*, nucleotide sequences of the replicon regions were compared in this paper.

We then cloned the segments of these plasmids corresponding to the replicon regions of ColE2-P9 and ColE3 (Figure 3). We easily obtained autonomously replicating plasmids from all of them. It turned out that ColE6 is a composite plasmid with the second replicon in the 4-kb segment (Itoh, unpublished). The second replicon seems to be unrelated to ColE-related replicons. Asymmetry of incompatibility exhibited by ColE6 is readily explained by the fact that ColE6 is a composite plasmid.

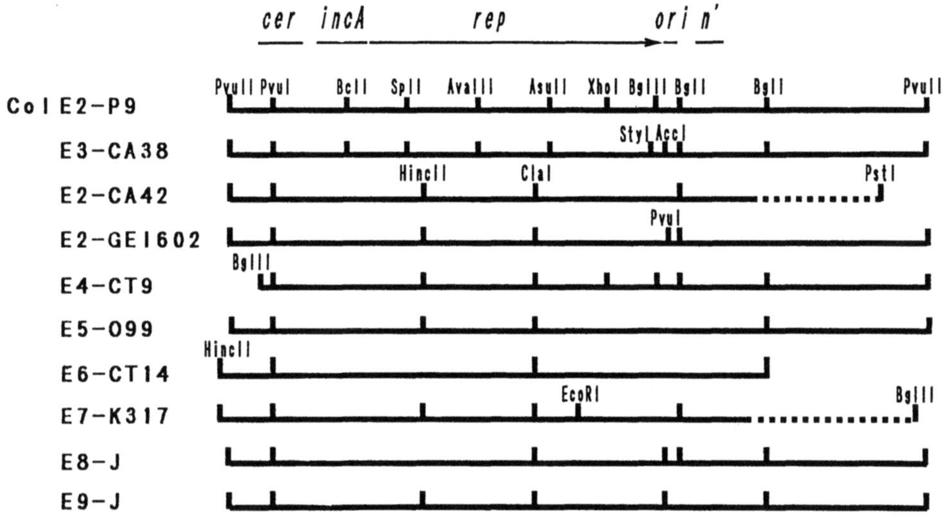

Figure 3. Restriction maps of the segments containing the replicons of the ten representative ColE2-related plasmids. Genes and functions identified for ColE2-P9 are shown above the maps. n', a site for priming of the lagging-strand DNA synthesis recognized by *E. coli* n' protein.

Assuming that all these plasmids share common mechanisms of initiation of replication and its control, we cloned the subsegments of these plasmids presumably carrying the *incA* or the *incB* genes and examined their effect on autonomously replicating ColE2-related plasmids (Table 1). We found four additional IncA specificity groups and two additional IncB specificity groups. In total there are five IncA specificity groups and four IncB specificity groups. We found nine combinations of the IncA and IncB specificity groups among twenty (that is 4 x 5) possible combinations, suggesting a possibility of their random combinations.

To further examine relatedness of these plasmids, we determined nucleotide sequences of the essential replication regions of the 10 representative plasmids (complete nucleotide sequences to be published elsewhere). The sequences are quite homologous all through the entire 1.3-kb regions. Comparison of the nucleotide sequences proves that all these plasmids share an identical organization of the replicon.

The Rep proteins of these ColE2-related plasmids are quite homologous (Figure 4). As to the amino-terminal regions of the Rep proteins they are divided into two groups. One consists of ColE2-P9 and ColE3-CA38 and the other of all the remaining plasmids. As to the carboxy terminal regions they are divided into four groups, which exactly correspond to the four groups of specificity of the IncB function. The sequences are almost identical in each group. A region with a deletion (ColE2-P9) or insertion (ColE3-CA38) of nine amino acid residues (Figure 4, indicated by a line above the amino acid sequence of the ColE2-P9 Rep protein) are involved in determination of specificity of interaction of the Rep proteins and origins. The amino acid sequences of the particular regions of the Rep proteins, however, are identical for ColE2-CA42, ColE3-CA38 and ColE5-099 groups.

```
ColE2-P9   MSAVLQRFREKLPHKPYCTNDFAYGVRILPKNIAILARFIQQNQPHALYWLPFDVDRTGASIDWSDRNCPAPNITVKNPRNGHAHLLYALALPVRTAPDASASA
4-CT9      ···A··Y·D·N···R··H·D·L·F·L··SG·GR·L··RY·······QF··V··V··E··A·····R-KLV····V···V··H·······NIA······S·VK·

2-CA42     ···A··Y·D·N···R··H·D·L·F·L··SG·GR·L··RY·······QF··V··V··E··A·····LNAPA····I···V··H·······NIA······A·VK·
2-GEI602   ···A··Y·E·N···R··H·D·L·F·L··SG·GR·L··RY·······QF··V··V··E··A·····RNAPA····V···V··H·······NIA······A·VK·

3-CA38     ···V··R·R·K···K··C·N·F·Y·V··LP·NI·I··RF·······LY··P··V··T··S·····RNCPA····V···R··H·······ALP······A·AS·
7-K317     ···A··Y·D·N···R··H·D·L·F·L··SG·GR·L··RY·······QF··V··V··E··A·····RNAPA····V···V··H·······NIA······S·VK·
8-J        ···A··Y·E·N···R··H·D·L·F·L··SG·GR·L··RY·······QF··V··I··E··A·····RNAPA····V···V··H·······NIA······S·VK·

5-O99      ···A··Y·E·N···R··H·D·L·F·L··SG·GR·L··GY·······QF··V··V··E··A·····RNAPA····V···V··H·······NIA······A·VK·
6-CT14     ···A··Y·E·N···R··H·D·L·F·L··SG·GR·L··RY·······QF··V··I··E··A·····RNAPA····V···V··H·······NIA······A·VR·
9-J        ···A··Y·E·N···R··H·D·L·F·L··SG·GR·L··RF·······QF··V··V··E··A·····RNAPA····V···V··Y·······NIA······A·VK·

ColE2-P9   LRYAAAIERALCEKLGADVNYSGLICKNPCHPEWQEVEWREEPYTLDELADYLDLSASARRSVDKNYGLGRNYHLFEKVRKWAYRAIRQGWPVFSQWLDAVIQR
4-CT9      ·K·····RS····C·············F·L··QVM····A··D········A·A···I··H··M···CH···MT······I·····V···········

2-CA42     ·K·····RA····G·············F·L··QVM····A··D········T·A···I··H··M···CH···MT······I·····A···········
2-GEI602   ·K·····RA····G·············F·L··LVM····A··D········A·A···I··H··M···CY···MT······I·····V···········

3-CA38     ·R·····RA····G·············C·P··QEV····P··D········A·A···V··N··L···CY···KG······I·····A···········
7-K317     ·K·····RS····C·············F·L··QVM····A··D········A·A···I··H··X···CH···MT·······M····E···········
8-J        ·K·····RS····C·············F·L··QVM····A··G········A·A···I··R··M···CH···MT······I·····A···········

5-O99      ·K·····RA····G·············F·L··LVM····A··D········A·A···I··H··M···CH···MT······I·····V···········
6-CT14     ·K·····CA····G·············F·L··LVM····A··D········A·E···I··H··M···CH···MT······I·····V···········
9-J        ·K·····RA····G·············F·L··QVM····A··D········A·A···I··H··M···CH···IT······I·····A···········

ColE2-P9   VEMYNASLPVPLSPAECRAIGKSIAKYTHRKFSPEGF---------SAVQAARGRKGG----TKSKRAAVPTSARSLKPWEALGISRATYYRKLKCDPDLAK
4-CT9      ·····A·······A·····K·······K·SP·G·---------SAV··A·········----TKSKRAAVPTSARSLKPWEALGISRATYYRKLKCDPDLAK

2-CA42     ·····A·······P·····K·······N·TA·T·AQYVADTHTPEI··K······IAKGEAYDDKRFMALCMLENGYSQKAIAAMLEVSTRTIRNWKSGK
2-GEI602   ·····A·······A·····K·······N·TP·T·AQYVATTHTPEI··A······IAKGAGYEDKRSTALEMLVNGCTQKAIAEMLGVSDRTIRNWKSGK

3-CA38     ·····A·······P·····K·······N·TP·T·AQYVADTHTPEI··T·········----SKSKRGTVATSARTLKPWEKLGISRAWYYQLKKRGLVE
7-K317     ·····A·······P·····R·······N·TP·T·AQYVADTHTPEI··A·········----SKSKRSTVATSARTLKPWEALGISRAWYYQLKKRGLVE
8-J        ·····A·······A·····K·······N·TP·T·AQYVADTHTPEI··T·········----SKSKRGTVATSARTLKPWEKLGISRAWYYQLKKRGLVE

5-O99      ·····V·······A·····K·······N·TP·T·AQYVADTHTPEI··A······KIGGAKSKRGAVATSARTLKPWETLGISRAWYYQLKKRGLVE
6-CT14     ·····V·······A·····K·······N·TP·T·AQYVADTHTPEI··A······KIGGAKSKRGAVATSARTLKPWETLGISRAWYYQLKKRGLVE
9-J        ·····A·······P·····K·······N·TP·T·AQYVADTHTPEI··A······KIGGAKSKRGAVATSARTLKPWETLGISRAWYYQLKKRGLVE
```

Figure 4. Comparison of Rep proteins. The predicted amino acid sequences of the nine Rep proteins were aligned with that of the ColE2-P9 Rep protein. Dots indicate positions where amino acids in the nine Rep proteins are identical to that in the ColE2-P9 Rep protein. The non-homology in the regions with about forty five amino acids at the carboxy termini is mainly due to amino acid substitutions in the Rep proteins of ColE2-CA42 and ColE2-GEI602. Lines above the ColE2-P9 sequence indicate the regions responsible for discrimination of ColE2-P9 and ColE3-CA38 in plasmid-specific recognition of the origins by the Rep proteins. A line below the ColE9-J sequence indicates the region of non-homology between the ColE3 and ColE5 groups.

The only significant difference between the ColE3-CA38 and ColE5-099 groups is a deletion (ColE3-CA38) or insertion (ColE5-099) of four amino acid residues in the adjacent region. The region must be involved in discrimination of the two groups. For ColE2-CA42 group the region nearer to the carboxy-terminus might be involved in discrimination. Anyway it seems that the carboxy terminal regions of these proteins, containing the regions for discrimination of the different IncB specificity groups, must contain the domain for recognition and binding of the Rep proteins to the homologous origin sequences. The amino terminal and middle regions of the Rep proteins might be involved in primer RNA synthesis and possible interaction with DNA polymerase I.

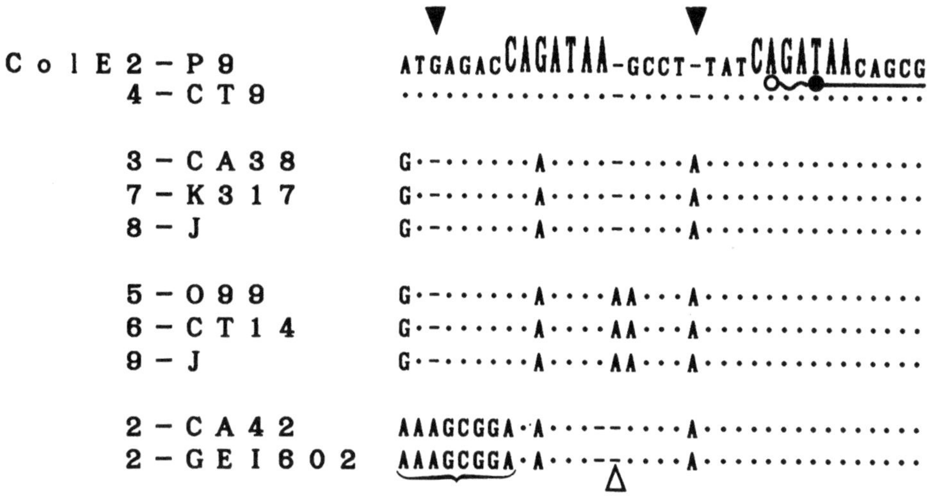

Figure 5. Comparison of replication origins. Sequences of the regions corresponding to the minimal origin region of ColE2-P9 with one additional nucleotide on each end are shown. The regions with non-homology are marked by arrowheads and a bracket. Primer RNA (wavy line) and newly synthesized leading-strand DNA (thin line) of ColE2-P9 are shown.

The origin regions of the sequenced ColE2-related plasmids are divided into four groups which exactly correspond to the four groups of specificity of the IncB function of these plasmids (Figure 5). The sequences are identical for the right halves of the origin regions where the synthesis of the primer RNA and the first leading-strand DNA and the termination of the synthesis of the lagging-strand DNA of ColE2-P9 take place. The nucleotide sequences of the left halves are identical in plasmids of the same groups but different in plasmids of the different groups. The two positions indicated by two black arrowheads (Figure 5) determine the specificity of interaction of the Rep proteins and origins between ColE2-P9 and ColE3-

CA38. The position indicated by a black arrowhead in the middle is essential for the weak but specific interaction (Yasueda *et al.*, 1989) and the proper combination of nucleotides at the two positions is essential for the strong and specific interaction (Itoh T, unpublished). The only difference in the nucleotide sequence of the origin regions of the ColE3-CA38 and the ColE5-099 groups exists at the position indicated by a white arrowhead. The specificity between the two groups must be determined by this position. This position must be also involved in determination of specificity between the ColE2-P9 and ColE5-099 groups and either or both of the positions with black arrowheads might be also involved. Finally the specificity between the ColE2-CA42 group and other groups must be determined either or both of the positions indicated by a white arrowhead and a bracket. The position indicated by a black arrowhead in the middle might be also involved. The origin sequence of the ColE2-CA42 group is considerably different from those of other groups, so are the amino acid sequences of the Rep proteins of this group.

The nucleotide sequences of the *incA* regions of these plasmids are divided into four groups which exactly correspond to the four groups of specificity of the IncAfunction. The sequences are identical in plasmids of the same groups. The antisense RNA (RNA I) molecule of plasmids of the ColE2-P9 group forms two stable stem-loop structures (Sugiyama T & Itoh T, unpublished), whereas those of other groups may form only one stable stem-loop structure with or without very short and unstable one (Figure 6). The larger stem-loop structure of ColE2-P9 is important for the regulatory activity of RNA I and the nucleotide sequence at the loop region determines specificity and efficiency of interaction of RNA I and Rep mRNA (Yasueda H, Takechi S & Itoh T, submitted; Sugiyama T & Itoh T, submitted). Surprisingly, however, the 12-nucleotide sequence (CAAUCUUGGCGG) around the loop regions of the stem-loop structures of RNA I are identical among plasmids of the ColE2-P9, ColE5-099 and ColE9-J groups. For plasmids of the ColE5-099 and ColE9-J groups the homologous regions around the loop regions are even longer. Therefore the specificity of interaction of RNA I and Rep mRNA among plasmids of these groups must be determined by the sequence differences in other regions, namely, those around the junction of the stem and loop or even in the stem.

The antisense RNA (inc RNA) molecule of plasmid Rl of IncFII family forms two stem-loop structures (Wagner & Nordstrom, 1986). The loop sequence of inc RNA of IncFII plasmids is important for the interaction of the incRNA and RepA mRNA and determines the specificity and efficiency of interaction (Rosen *et al.*, 1981; Persson *et al.*, 1988). The incRNA molecules of other plasmids in IncF and IncI families may also form either one or two stem-loop structures and the sequences at the loop regions are identical (CCGCCAA; see Couturier *et al.*, 1988) And surprisingly the sequence is complementary to the sequence at the loop region of RNA I of ColE2-P9. This might suggest some evolutionary significance of the particular complementary nucleotide sequences. They might affect the stability of the secondary or tertiary structures of RNA molecules or the efficiency of the interaction between two RNA molecules.

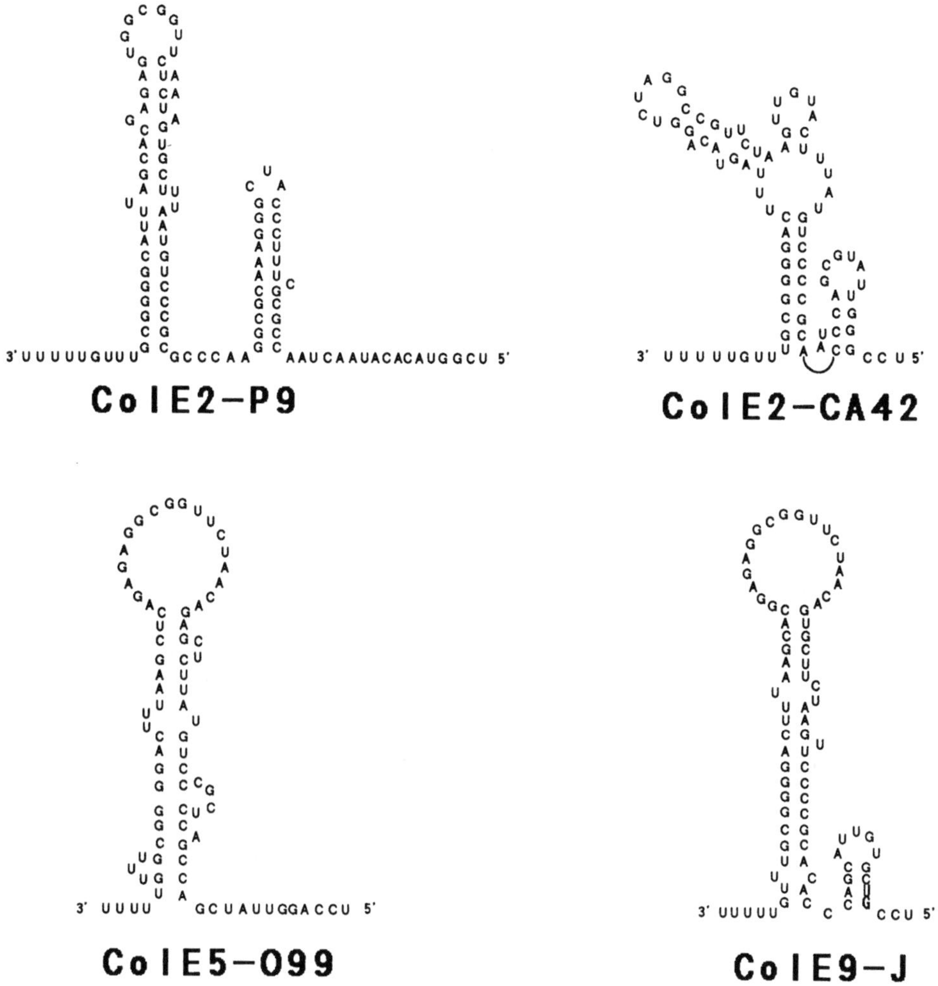

Figure 6. Comparison of antisense RNAs. Predicted secondary structures of the putative antisense RNAs of the three new IncA groups are shown with the secondary structure of ColE2-P9 RNA

The nucleotide sequences of the regions involved in the site-specific recombination for resolution of plasmid polymers to monomers in ColE2-P9 and ColE3-CA38 (Kantake N & Itoh T, unpublished) and those of the corresponding regions in other ColE2-related plasmids are quite similar. The regions in ColE2-P9 and ColE3-CA38 function for stable maintenance of these plasmids by host bacteria. It was confirmed that the corresponding regions in other ColE2-related plasmids are also involved in site-specific recombination and stabilization of plasmid maintenance. The nucleotide sequences of the regions containing the actual crossover sites for ColE2-P9 and ColE3-CA38 and that for ColEl are compared with the corresponding regions of other ColE2-related plasmids (Figure 7). The two bracketed segments indicate the conserved regions for all these plasmids. Note that the conserved regions are included within the crossover regions. The conserved regions extend further if the nucleotide sequences are compared only among the ColE2-related plasmids. Because of the sequence homology and involvement of the *xerA* gene product in site-specific recombination the same *E. coli* site-specific recombination system must be operating for all these plasmids. We know that the site-specific recombination takes place between the heterologous sites of ColE2-P9 and ColE3. We do not know, however, whether the site-specific recombination can takes place between the sites of any pair of ColE2-related plasmids.

```
ColE2-P9     AAGGGGGGCGTACAA--CGGGAGTTATGGTAAATGGATCGGGTTTTCGGGAAGG
    3-CA38   AAGGGGTGCGTACAA--CGGGAGTTATGGTAAATGGATCGGTTTTTCGGGAAGG
    2-CA42   AAGGGGTACGTACAA--CGGGAGTTATGGTAAATGGATCGGGTTTTCGGGAAGG
    2-GEI602 AAGGGGGGCGTACAA--CGGGAGTTATGGTAAATGGATCGGGTTTTCGGGAAGG
    4-CT9    ATGGGGTGCGTACAA--CGGGAATTATGGTAAATGGATCGGGTTTTCGGGAAGG
    5-099    AAGGGGTACGTACAA--CGGGAGTTATGGTAAATGGATCGGGTTTTCGGGAAGG
    6-CT14   AAGGGGTGCGTACAA--CGG-AGTTATGGTAAATGGATCGGATTTTGGGGAATA
    7-K317   ACGGGGTGCGTACAA--CGGGAGTTATGGTAAATGGATCGGGTTTTCGGGAAGG
    8-J      AAGGGGTACGTACAA--CGGGAATTATGGTAAATGGATCGGGTTTTGGGGAATA
    9-J      AAGGGGTACGTACAA--CAGGAGTTATGGTAAATGGATCGGGTTTGCGGGAAGA

ColE1        -AGCGGTGCGTACAATTAAGGGATTATGGTAAATCCACTTACTGTCTGCCTCGT
```

Figure 7. Comparison of *cer* regions. Portions of sequences corresponding to the *cer* region of ColEl are shown. Conserved segments are indicated by brackets. Wavy lines indicate the regions where actual cut-and-rejoin (crossover) events occur between ColE2-P9 and ColE3-CA38 (above the ColE2-P9 sequence) and between ColEl and ColK (below the ColEl sequence).

As described above all these ColE2-related plasmids are closely related and must have been derived from a common ancestor. Accumulation of base substitutions, deletions and insertions in the *incA* region might have created new IncA specificity groups. Because of complementarity between antisense RNA and its target mRNA even a single base substitution could have given rise to a new specificity group. The carboxy terminal coding region of the *rep* gene and the origin must have coevolved to maintain specific interaction between them for initiation of DNA replication. Accumulation of neutral and near neutral mutations might have changed the Rep proteins and the origins gradually and eventually created new IncB specificity groups.

Comparison of the nucleotide sequences of the replicon regions of the ten ColE2-related plasmids (Figure 8) reveals striking features in the sequence relationships among the subsegments of these replicons. The patterns of group-specific sequence homology in the *incA* regions are not necessarily linked to those in the *incB* regions. The patterns of group-specific sequence homology in the replicon regions of these plasmids are correlated neither to the species of colicins (colicins E2 to E9) nor to the types of colicins (DNase-type or RNase-type). Therefore we see typical mosaic structures not only in the replicon regions (Figure 8), but also in the almost entire genomes of these plasmids (see Table 1). It seems difficult to consider that all those structures have sequentially diverged from one ancestral structure as a result of simple accumulation of mutations, substitutions, deletions and insertions. Rather those structures must have arisen by exchanging functional parts of the replicons through mechanisms, such as homologous recombination and/or site-specific recombination. Somewhat similar structural organization of the replicon regions with exchangeable segments has been observed in plasmids of IncF and IncI families (Saadi *et al.*, 1987; Couturier *et al.*, 1988).

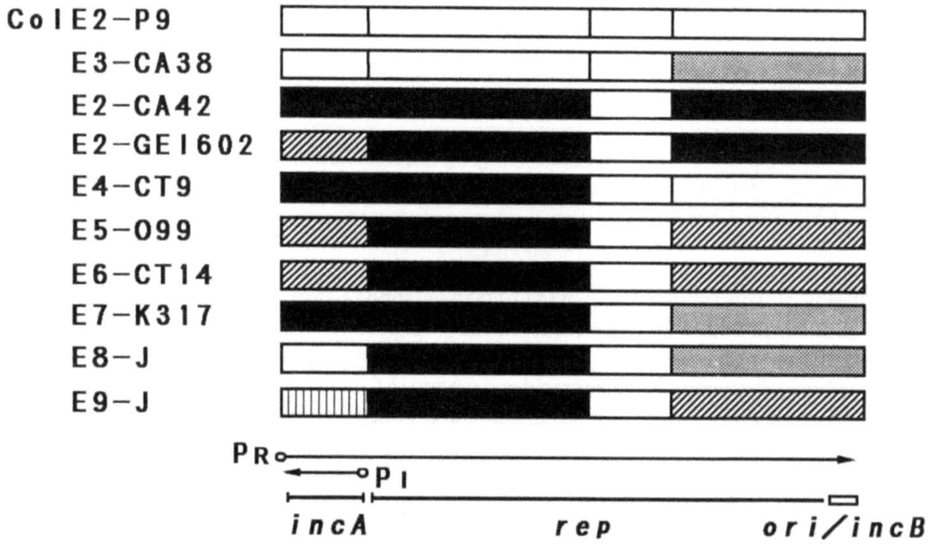

Figure 8. Mosaic structure of ColE2-related replicons. Genes and functions identified in the minimal ColE2-P9 replicon are shown at the bottom. The minimal replicon regions are divided into four segments according to the differences in distribution of sequence homology among these plasmids. For each segment portions marked in the same symbol (white, black, diagonal or longitudinal stripes and dots) are almost identical in sequence and those marked in different symbols significantly differ from each other.

The entire replicon regions of these ColE2-related plasmids are quite homologous and seem to be suitable substrates for homologous recombination events. In addition the site for the site-specific recombination which resolves plasmid dimer (polymer) molecules into monomer molecules is located just next to the replicon region of each plasmid. Suppose that two different ColE2-related plasmids happen to coexist in one bacterial cell which is proficient in homologous recombination. An homologous recombination event between certain homologous regions in the replicon regions might give rise to a composite plasmid, a heterodimer, carrying recombinant structures. If the site-specific recombination between the heterologous sites of the parental plasmids could occur properly, the heterodimer molecules would be resolved into two progeny monomer molecules, both of which are recombinant plasmids with new combinations of certain functional parts of the replicon. The probability of coexistence of the two different ColE2-related plasmids may be very low and the first homologous recombination event between the two plasmids under the usual rec^+ condition may be very rare. Once such conditions are met, however, the site-specific recombination will readily produce the progeny recombinant plasmids. We are currently trying to create, experimentally, plasmids with new combinations of the IncA and IncB specificity groups through the mechanism described above.

Acknowledgement

We thank Drs H. Ogawa, T. Ogawa and T. Horii for their interest and encouragement.

References

Chak K-F, James R (1986) Characterization of the ColE9-J plasmid and analysis of its genetic organization. J Gen Microbiol 132:61-71
Chak K-F, Kuo W-S, Lu F-M, James R (1991) Cloning and characterization of the ColE7 plasmid. J Gen Microbiol 137:91-100
Chan PT, Ohmori H, Tomizawa J, Lebowitz J (1985) Nucleotide sequence and gene organization of ColEl DNA. J Biol Chem 260:8925-8935
Clewell DB, Helinski DR (1972) Effect of growth conditions on the formation of the relaxation complex of supercoiled ColEl deoxyribonucleic acid and protein in *Escherichia coli. J* Bacteriol 110:1135-1146
Cooper PC, Hawkins FKL, James R (1986) Incompatibility between E colicin plasmids. J Gen Microbiol 132:1859-1862
Cooper PC, James R (1984) Two new E colicins, E8 and E9, produced by a strain of *Escherichia coli. J* Gen Microbiol 130:209-215
Couturier M, Bex F, Bergquist PL, Maas WK (1988) Identification and classification of bacterial plasmids. Microbiol Rev 52:375-395
Hakkaart MJJ, van den Elzen PJM, Veltkamp E, Nijkamp HJJ (1984) Maintenance of multicopy plasmid CloDF13 in *E. coli* cells: Evidence for site-specific recombination at par B. Cell 36:203-209
Horii T, Itoh T (1988) Replication of ColE2 and ColE3: The regions sufficient for autonomous replication. Mol Gen Genet 212:225-231
Inselburg J (1974) Incompatibility exhibited by colicin plasmids El, E2, and E3 in *Escherichia coli.* J Bacteriol 119:478-483

Itoh T, Horii T (1989) Replication of ColE2 and ColE3 plasmids: *In vitro* replication dependent on plasmid-coded proteins. Mol Gen Genet 219:249-255

Kido M, Yasueda H, Itoh T (1991) Identification of a plasmid-coded protein required for initiation of ColE2 DNA replication. Nucleic Acids Res 19:2875-2880

Kingsbury DT, Helinski DR (1970) DNA polymerase as a requirement for maintenance of the bacterial plasmid colicinogenic Factor El. Biochem Biophys Res Commun 41:1538-1544

Lawrence GMP, James R (1984) Characterization of the ColE8 plasmid, a new member of the group E colicin plasmids. Gene 29:145-155

Males BM, Stocker BAD (1982) Colicins E4, E5, E6 and A and properties of *btuB*$^+$ colicinogenic transconjugants. J Gen Microbiol 128:95-106

Mock M, Pugsley AP (1982) The *btuB* group Col plasmids and homology between the colicins they encode. J Bacteriol 150:1069-1076

Persson C, Wagner EGH, Nordstrom K (1988) Control of replication of plasmid Rl: kinetics of *in vitro* interaction between the antisense RNA, CopA, and its target, CopT. EMBO J 7:3279-3288

Rosen J, Ryder T, Ohtsubo E (1981) Role of RNA transcripts in replication incompatibility and copy number control in antibiotic resistance plasmid derivatives. Nature 290:794-799

Saadi S, Maas WK, Hill DF, Bergquist PL (1987) Nucleotide sequence analysis of RepFIC, a basic replicon present in IncFI plasmids P307 and F, and its relation to the RepA replicon of IncFII plasmids. J Bacteriol 169:1836-1846

Stirling CJ, Stewart G, Sherratt DJ (1988) Multicopy plasmid stability in *Escherichia coli* requires host-encoded functions that lead to plasmid site-specific recombination. Mol Gen Genet 214:80-84

Summers DK, Sherratt DJ (1984) Multimerization of high copy number plasmids causes instability: ColEl encodes a determinant essential for plasmid monomerization and stability. Cell 36:1097-1103

Summers D, Yaish S, Archer J, Sherratt D (1985) Multimer resolution systems of ColEl and ColK: Localization of the crossover site. Mol Gen Genet 201:334-338

Tacon W, Sherratt DJ (1976) ColE plasmid replication in DNA polymerase I-deficient strains of *Escherichia coli*. Mol Gen Genet 147:331-335

Tajima Y, Horii T, Itoh, T (1988) Replication of ColE2 and ColE3: Two ColE2 incompatibility functions. Mol Gen Genet 214:451-455

Toba M, Masaki H, Ohta T (1988) Colicin E8, a DNase which indicates an evolutionary relationship between colicins E2 and E3. J Bacteriol 170:3237-3242

Tomizawa J (1987) Regulation of ColEl DNA replication by antisense RNA. In: Inouye M, Dudock BS (eds) Molecular Biology of RNA: New perspectives. Academic Press, New York, pp 249-259

Wagner GEH, Nordstrom K (1986) Structural analysis of an RNA molecule involved in replication control of plasmid Rl. Nucleic Acids Res 14:2523-2538

Watson R, Rowsome W, Tsao J, Visentin LP (1981) Identification and characterization of Col plasmids from classical colicin E-producing strains. J Bacteriol 147:569-577

Watson RJ, Vernet T, Visentin LP (1985a) Relationships of the Col plasmids E2, E3, E4, E5, E6 and E7: restriction mapping and colicin gene fusions. Plasmid 13:205-210

Watson RJ, Vernet T, Visentin LP (1985b) Relationships of the Col plasmids E2, E3, E4, E5, E6 and E7: restriction mapping and colicin gene fusions. Erratum. Plasmid 14:97

Yasueda H, Horii T, Itoh T (1989) Structural and functional organization of ColE2 and ColE3 replicons. Mol Gen Genet 215:209-216

GENETIC DETERMINANTS FOR MICROCIN H47, AN *ESCHERICHIA COLI* CHROMOSOME-ENCODED ANTIBIOTIC

M. Laviña and C. Gaggero
División de Biología Molecular
Instituto de Investigaciones Biológicas C. Estable
Av. Italia 3318
11.600 Montevideo
Uruguay

A search for antibiotic activities produced by Gram-negative bacteria from Uruguay was performed. Strains isolated from different natural sources were screened and 23 out of 170 were able to inhibit the growth of an *E. coli* K12 indicator strain. Antibiotic activities were typed following the criteria of Baquero and Moreno (1984). They were classed as colicins when their production was induced with mitomycin, and as microcins when they were not mitomycin-inducible and were able to pass through a cellophane membrane which allows the passage of molecules smaller than 7.000 Da. Thus, fifteen strains producing colicins and eight strains producing microcins were isolated.

The ability to produce antibiotic was transferred to *E. coli* K12 cells by conjugation from eight wild type strains. Cross-immunity tests indicated that the transferred activities were colicin N (from 5 strains), colicin E7 (from two strains), and microcin B17 (from one strain). No antibiotic-producing *E. coli* K12 exconjugants, or transformants, were obtained when any of the remaining strains were used as a donor. One of them, H47, an *E. coli* producing a microcin-type antibiotic activity, was chosen for further studies. Its antibiotic production was called microcin H47 (MccH47). It is active on Gram-negative bacteria and exerts a bactericidal action on *E. coli* K12 cells.

To identify the genetic determinants encoding MccH47 a DNA library from strain H47 was prepared *in vivo* using the transposon Mu d5005 (Groisman & Casadaban, 1986). An *E. coli* K12 strain was transduced with the phage lysate and antibiotic producing clones were obtained. They all overproduced MccH47 and were immune to each other and to the antibiotic production of strain H47. In this genetic background it was confirmed, by cross-immunity assays, that MccH47 is indeed a novel antibiotic. Recombinant plasmids from these clones all contained a minimal common 13 Kb DNA stretch which was assumed to contain the genetic determinants for MccH47 production and immunity (Laviña *et al.*, 1990). To determine the origin of the DNA directing the MccH47 antibiotic functions Southern hybridization experiments were performed using two Mu d5005 derivative plasmids containing the antibiotic system as probes. They were shown to hybridize with restricted chromosomal DNA from the H47 strain, recovering the restriction pattern of the cloned DNA in the probes. There was no hybridization with plasmid DNA

NATO ASI Series, Vol. H 65
Bacteriocins, Microcins and Lantibiotics
Edited by R. James, C. Lazdunski and F. Pattus

from H47 or with genomic DNA from an *E. coli* K12 strain (Laviña *et al.*, 1990). These results strongly support the conclusion that MccH47 is encoded by the chromosome of the H47 strain. This would therefore be the first description of a chromosomal genetic system encoding an antibiotic in the family *Enterobacteriaceae.*

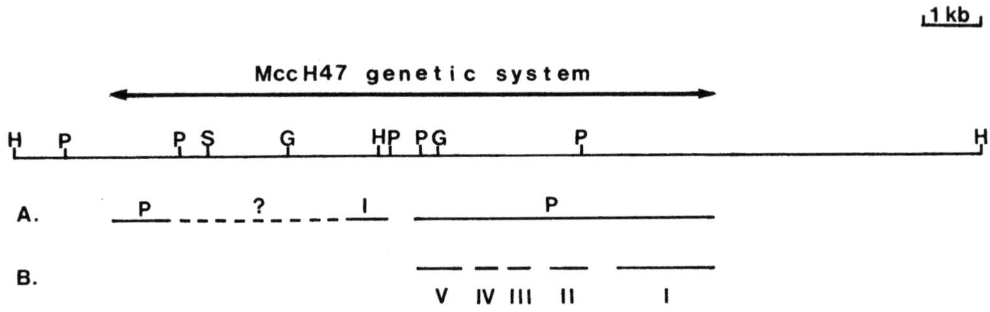

Figure 1. Cloned DNA containing the MccH47 genetic determinants. A physical map is represented. Restriction sites: H, *Hind*III; P, *Pst*I; S, *Sal*I; G, *Bgl*II. **A** - Functional regions; P, MccH47 production; I, immunity to MccH47. **B** - Complementation groups.

The MccH47 genetic determinants were located in a DNA segment of 16.6 Kb (Figure 1), and were sub-cloned into compatible high copy-number plasmid vectors. Insertion mutagenesis was then performed on the hybrid derivatives. The insertion sites were physically located and correlated with the mutant phenotypes. The results indicated that a 10 Kb long DNA region is required to express the antibiotic functions. It was shown that two sub-regions, of 1.1 and 5.5 Kb, are involved in MccH47 production, and that a DNA segment of 0.8 Kb determines MccH47 immunity. A 3 Kb region inside the system could not be clearly related to any antibiotic function.

Complementation experiments between plasmid mutants defective for MccH47 production led to the identification of five complementation groups within the 5.5 Kb DNA subregion. Preliminary results suggest that complementation groups I and II correspond to cistrons involved in MccH47 export to the medium, since mutants belonging to these groups do not produce MccH47 but contain the antibiotic intracellularly. The results are summarized in Figure 1 (manuscript in preparation). It appears that the genetic system encoding for MccH47 production, secretion, and immunity exhibits a higher degree of complexity than those encoding the previously described microcins B17 and C7 (Garrido *et al.*, 1988; Genilloud *et al.*, 1989; Novoa *et al.*, 1986).

Acknowledgement

The work is financed by International Foundation for Science grant F/1495-1 and 2 and by PEDECIBA.

References

Baquero F, Moreno F (1984) The microcins. FEMS Microbiol Lett 23:117-124

Garrido MC, Herrero M, Kolter R, Moreno F (1988) The export of the DNA replicator inhibitor microcin B17 provides immunity for the host cell. EMBO J 7:1853-1862

Genilloud O, Kolter R, Moreno F (1989) DNA sequence, products, and transcriptional patterns of the genes involved in production of the DNA replication inhibitor microcin B17. J Bacteriol 171:1126-1135

Groisman EA, Casadaban MJ (1986) Mini-Mu bacteriophage with plasmid replicons for *in vivo* cloning and *lac* gene fusing. J Bacteriol 168:357-364

Laviña M, Gaggero C, Moreno F (1990) Microcin H47, a chromosome-encoded microcin antibiotic of *Escherichia coli*. J Bacteriol 172:6585-6588

Novoa MA, Díaz-Guerra L, San Millán JL, Moreno F (1986) Cloning and mapping of the genetic determinants for microcin C7 production and immunity. J Bacteriol 168:1384-1391

BLIS PRODUCTION IN THE GENUS STREPTOCOCCUS

J.R. Tagg
Department of Microbiology
University of Otago
Dunedin
New Zealand

Streptococci have been found to comprise a substantial proportion of the indigenous microbiota and of the bacterial pathogens of humans and other animals. For 21 years now, the principal research focus of my laboratory has been the investigation of the production by streptococci of bacteriocin-like inhibitory substances (BLIS) (Tagg, Dajani & Wannamaker, 1976).

As these studies progressed, it soon became clear that the streptococci would prove to be a very rich source of BLIS and that there was a need to develop some means of comparing the BLIS activities of different strains. For this purpose we devised a streptococcal BLIS 'fingerprinting scheme' (Tagg & Bannister, 1979). This has as its basis the detection of the production by the test strains of particular patterns of inhibitory activity against a set of nine indicator bacteria (this was called P-typing) and also the testing of the sensitivity of the test strains to nine known BLIS-producing streptococci (called S-typing).

Although originally devised specifically for application to beta-hemolytic streptococci of Lancefield groups A, B, C, D and G, the P-typing procedure has subsequently also been found to detect BLIS production in a wide variety of other streptococcal species and also in other Gram-positive bacteria including lactobacilli, staphylococci, enterococci, lactococci, actinomyces, micrococci and stomatococci.

Presented here is a summary of the results of our testing of strains of various streptococcal species for BLIS production using the standardized P-typing procedure.

MATERIALS AND METHODS

The BLIS P-typing scheme currently in use in our laboratory is slightly modified from the original version (Tagg & Bannister, 1979). Briefly, the test involves first growing the test strain as a l-cm wide diametric streak culture on the surface of the typing medium contained in a petri dish. We now use Columbia agar base (GIBCO Laboratories) supplemented with 0.1% (w/v) calcium carbonate (to minimize acidic pH-mediated inhibition) and 5% (v/v) human blood (since production of some BLIS appears to be human blood-dependent). Incubation of these cultures is for 18h at 37 °C in an anaerobic atmosphere (to eliminate inhibition due to hydrogen peroxide). After removing the macroscopic bacterial growth, the surface of the test medium is exposed to chloroform vapours for 30 min and overnight liquid

NATO ASI Series, Vol. H 65
Bacteriocins, Microcins and Lantibiotics
Edited by R. James, C. Lazdunski and F. Pattus

cultures of the nine standard indicator strains are inoculated across the line of the original test strain culture. Following reincubation of the plate at 37 °C for 18 h, any inhibition of growth of the indicator strains is recorded. The pattern of inhibition of the nine indicators is then expressed according to a triplet code designation as the BLIS P-type of the test strain.

The streptococcal strains tested for BLIS production have been obtained from a wide variety of sources; some wild-type isolates from our own epidemiological surveys and many strains generously contributed by other investigators in the course of collaborative studies. These sources have been identified in the original publications listed in the accompanying table.

RESULTS AND DISCUSSION

Applications of the BLIS P-typing scheme under standardized conditions, excluding inhibition due to acidic metabolites and hydrogen peroxide, has shown that BLIS production is extremely common in some streptococcal species (Table).

The current scheme has been particularly effective when applied to strains of *Streptococcus salivarius*, *S. uberis*, mutans streptococci, Lancefield group E streptococci and enterococci (the latter being included as 'honorary' streptococci). For each of these groups of strains the proportion typable was at least 45 % and (except for the group E strains) 10 or more different P-types were detected.

A wide variety of P-types (5 to 9) was also identified within the BLIS-positive strains of Lancefield groups A, B and C streptococci and the milleri streptococci. However, here the incidence of BLIS production was somewhat lower, ranging from 29% in the group A streptococci to only 5% of the group B streptococcus strains.

Only a single P-type pattern has been detected to date in the tested group G streptococci and also amongst the *S. sanguis* isolates. None of the relatively small number of nutritionally variant streptococci examined have yet been found to be BLIS-positive.

One particularly interesting observation resulting from our application of the standardized BLIS P-typing scheme to a wide variety of streptococcal species is that individual patterns of BLIS activity (P-types) tend to be specific for strains of each particular species. Hence, as a conservative estimate, it appears that there may be more than 50 different BLIS produced by the 3075 strains of streptococci evaluated in the current study. It is conceded that some of the observed inhibition patterns may represent the combined activities of two or more different BLIS. Nevertheless, this impressive variety and proportion of BLIS-producers has resulted from the application of only nine indicator strains and one set of incubation conditions. With additional indicators and some "tinkering" with the incubation conditions even more impressive figures could be anticipated.

TABLE: BLIS PRODUCTION BY STREPTOCOCCI

Strain identity	Number of strains	BLIS-positive strains (%)	Number of BLIS types	Selected references
LANCEFIELD GROUP				
A	655	29	9	Ragland and Tagg (1990)
B	176	5	5	Schofield and Tagg (1983)
C	98	21	9	Barnham *et al.* (1987)
E	13	92	4	Tagg (1985)
G	105	15	1	Tagg and Wong (1983)
S. salivarius	1450	45	12	Tagg *et al.* (1983)
S. sanguis	58	21	1	unpublished
S. uberis	164	81	10	Tagg and Vugler (1986)
mutans streptococci	180	47	12	Crooks *et al.* (1987)
milleri streptococci	89	19	7	unpublished
nutritional variants	19	0	0	Tagg and van de Rijn (1991)
Enterococcus sp.	68	89	11	James and Tagg (1986)

Clearly the streptococci represent a rich source of previously undescribed BLIS. For example, streptococcin A-FF22, the first group A streptococcal BLIS to be described (Tagg & Wannamaker, 1978), has recently been established to be a lantibiotic, similar in many respects to nisin (Jack & Tagg, 1991). Our current research has become focussed on attempting to determine the possible ecological and pathogenic significance to the streptococci of their apparently ubiquitous capacity to produce BLIS.

ACKNOWLEDGEMENTS

This work has been supported by grants from the Health Research Council of New Zealand, the National Health Foundation of New Zealand and the New Zealand Dental Research Foundation.

REFERENCES

Barnham M, Cole G, Efstratiou A, Tagg JR, Skjold SA (1987) Characterization of *Streptococcus zooepidemicus* (Lancefield group C) from human and selected animal infections. Epidem Inf 98:171-182

Crooks M, James SM, Tagg JR (1987) Relationship of bacteriocin-like inhibitor production to the pigmentation and hemolytic activity of mutans streptococci. Zbl Bakt Hyg I Abt Orig A 263:541-547

Jack RW, Tagg JR (1991) Isolation and partial structure of streptococcin A-FF22. In: Jung G, Sahl H-G (eds) Lantibiotics. (in press)

James SM, Tagg JR (1986) The typing of enterococcus strains according to their production of bacteriocin-like inhibitory activity. Proc Univ Otago Med Sch 64:65-66

Ragland NL, Tagg JR (1990) Applications of bacteriocin-like inhibitory substance (BLIS) typing in a longitudinal study of oral carriage of β-haemolytic streptococci by a group of Dunedin schoolchildren. Zbl Bakt Hyg I Abt Orig A 274:100-108

Schofield CR, Tagg JR (1983) Bacteriocin-like activity of group B and group C streptococci of human and animal origin. J Hyg 90:7-18

Tagg JR (1985) An inhibitor 'fingerprinting' scheme applicable to Lancefield group E streptococci. Canad J Microbiol 31:1056-1057

Tagg JR, Bannister LV (1979) "Fingerprinting" β-haemolytic streptococci by their production of and sensitivity to bacteriocine-like inhibitors. J Med Microbiol 12:397-411

Tagg JR, Van de Rijn I (1991) An inverse correlation in nutritionally variant streptococci between the production of bacteriolytic activity and sensitivity to a *Streptococcus pyogenes* BLIS. J Clin Microbiol 29: 848-849

Tagg JR, Vugler LG (1986) An inhibitor typing scheme for *Streptococcus uberis*. J Dairy Res 53: 451-456

Tagg JR, Wannamaker LW (1978) Streptococcin A-FF22: nisin-like antibiotic substance produced by a group A streptococcus. Antimicrob Agents Chemother 14:31-39

Tagg JR, Wong HK (1983) Inhibitor production by group-G streptococci of human and of animal origin. J Med Microbiol 16:409-415

Tagg JR, Dajani AS, Wannamaker LW (1976) Bacteriocins of gram-positive bacteria. Bacteriol Rev 40: 722-756

Tagg JR, Pybus V, Phillips LV, Fiddes TM (1983) Application of inhibitor typing in a study of the transmission and retention in the human mouth of the bacterium *Streptococcus salivarius*. Arch Oral Biol 28:911-915

A NEW LEUCONOSTOC BACTERIOCIN, MESENTERICIN Y105, BACTERICIDAL TO *LISTERIA MONOCYTOGENES*

Y. Héchard*, C. Jayat[2], F. Letellier[3], M.H. Ratinaud[2], R. Julien[2] and Y. Cenatiempo
Lab de Biologie Moleculaire
CNRS UA 1172
Poitiers 1
FRANCE.

Summary

Screening of various samples of goat's milk lead to the identification of several lactic acid bacterial strains which inhibit *Listeria monocytogenes* growth (Hechard, 1990). Among them is a *Leuconostoc mesenteroides* strain producing a new bacteriocin, mesentericin Y105. The bacteriocin was characterized and purified. Pure mesentericin Y105 was obtained after 3 steps: affinity chromatography, ultrafiltration and reverse phase HPLC. The mode of action of mesentericin Y105 was studied by microbiological methods and by an original and efficient tool, i.e. flow cytometry (FCM). Using this technique, the membrane potential of *L. monocytogenes* could be estimated, via the amount of a fluorochrome, rhodamine 123 (Rh 123), bound to the cells. Thus mesentericin Y105 seems to act by a direct transmembrane depolarization of the sensitive cells. Finally, FCM was used to follow on - line the behavior of mixed cultures. The percentage of each population was estimated solely by light scattering. A decrease in the amount of *L. monocytogenes* cells pointed out that the bacteriocin displays a bactericidal effect.

Introduction

Lactic acid bacteria (LAB) are traditionally used as starters for food fermentation. Since they display a capacity to inhibit both spoilage and pathogenic bacteria, they are important in food preservation and intestinal prophylaxis. Such bacterial antagonism could arise from lactic acid production, hydrogen peroxide formation and secretion of specific proteinaceous components with a narrow inhibitory spectrum: bacteriocins (Tagg, 1976; Klaenhammer, 1988).

Several food-contaminating bacteria were identified as pathogens responsible for toxin-infections; amongst them are *Listeria monocytogenes* (Farber, 1991). The fact that they are closely related to *Lactobacillaccae* led to the identification of anti-Listeria lactic acid bacteria. Various LAB displaying

---

[2] Lab de Biotechnologie, Limoges; [3] Lab de Microbiologie Appliquee, La Rochelle.
* to whom correspondence should be addressed to Lab de Biologie Moleculaire, UFR Sciences, 40 av. du Recteur Pineau, 86022 Poitiers Cedex, FRANCE.

NATO ASI Series, Vol. H 65
Bacteriocins, Microcins and Lantibiotics
Edited by R. James, C. Lazdunski and F. Pattus

activity against *L. monocytogenes* have been isolated from goat's milk (Hechard, 1990). Among them a *Leuconostoc mesenteroides* strain synthetizes a bacteriocin named mesentericin Y105. In our study, we attempted to characterize and to purify the mesentericin Y105, in order to analyze its amino acid sequence and to later analyze its gene. Then, to study bacteriocin action on sensitive strains, we investigated changes in membrane potential by fluorocytometry. Analysis was performed by using rhodamine 123, a fluorescent probe, previously employed to measure mitochondrial membrane potential (Leprat, 1990) and now, available for bacteria (Matsuyama, 1984; Resnick, 1985). Moreover, mixed-cultures of bacteriocin-sensitive and producing strains were followed on-line by FCM, allowing an accurate and fast estimate of the % of each population in the mixture (Hechard, 1991). Microbiological methods were used to corroborate flow cytometry experiments.

Characterization of the mesentericin.

We found that mesentericin Y105 is a low molecular weight proteinaceous component (2.5 kDa by SDS-PAGE), excreted during the exponential phase of *Leuconostoc* growth (data not shown). Bacteriocin activity appears extremely heat stable, since it should be autoclaved for 15 min at 121°C, or heated for 120 min at 100°C, without a significant decrease in its activity. Moreover its titer remains unchanged over a pH 4.0 to 8.5 range. Finally, it could be stored for long periods (several months) at low temperature (-20°C) and low pH (about 4.5) without any noticeable loss of activity.

Purification of the mesentericin.

The bacteriocin was purified from supernatants of broth cultures prepared in a defined medium (methionine assay medium, DIFCO, supplemented with methionine) to minimize the presence of contaminating proteins and peptides, otherwise present in MRS. Purification was achieved in 3 steps: affinity chromatography, ultrafiltration and then reverse phase HPLC. Crude extract, obtained at the beginning of the stationary phase, was loaded onto a dye ligand (Mimetic Blue 2 similar to Cibacron Blue 3GA, ACL) chromatography column. After a wash, the proteins were eluted by increasing the ionic strength (Acetate buffer, NaCl 1 M).

The fractions containing bacteriocin active against *Listeria monocytogenes* (well diffusion method) were pooled and concentrated by ultrafiltration onto a 5 kDa cut-off membrane (Millipore). The sample was further purified by reverse-phase HPLC on a nucleosil C4 columm using a 6 to 60% acetonitrile linear gradient. In order to localize bacteriocin containing fractions, they were tested by the well-diffusion method. Mesentericin Y105 was detected at 220nm in a major peak (Figure 1), which was evaporated in a concentrator and sent to determine the N-terminal amino-acid sequence of the protein. No amino acid sequence could be determined, presumably because the N-terminus is modified. Therefore, we have to envisage enzymatic or chemical cleavage of the mesentericin prior to any sequencing experiment.

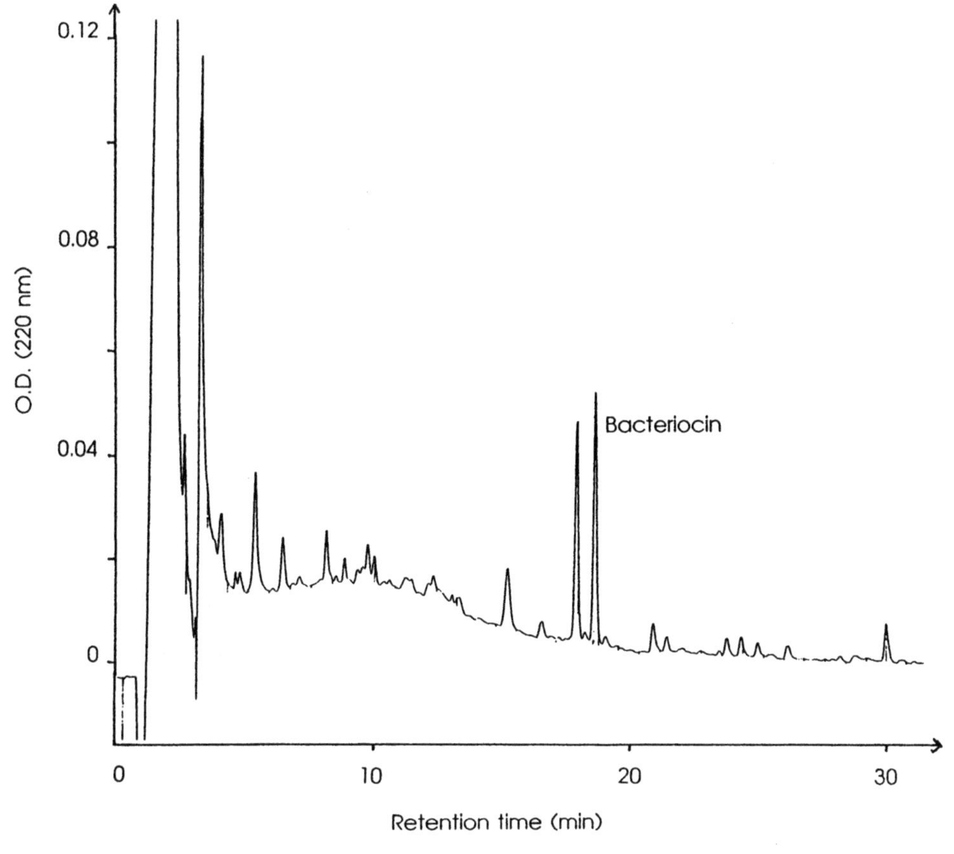

Figure 1. Elution of the bacteriocin from a reverse phase HPLC columm.
The sample loaded on a C4 column was obtained by submitting a crude extract (culture supernatant) to affinity chromatography and an ultrafiltration step. Bacteriocin containing fractions were detected by a well-diffusion method (Hechard, 1990) and had a 19 min retention time using a 0-60% acetonitrile gradient.

Mode of action

First, the membrane potential of *L. monocytogenes* was estimated by incorporation of a the dye, Rh 123. Bacteria collected from a lag phase culture were treated with various drugs: mesentericin Y105, valinomycin, a potassium ionophore, CCCP leading to bacterial membrane depolarization and nigericin, an H^+/K^+ exchanger leading to an hyperpolarization. Concomitantly to drug action, the Listeria were stained by Rh 123. After 5 min, bacteria were collected and washed before analysis by FCM (Table 1). On the one hand, mesentericin seems to act very quickly and more efficiently than CCCP or valinomycin (5 x 10^{-6}M). On the other hand, nigericin induces a 30% increase in dye binding to the *L. monocytogenes* membrane, correlated to its hyperpolarization. Therefore, bacteriocin is responsible for the decline of fluorochrome uptake, owing to bacterial membrane depolarization like some other well-known bacteriocins (Konisky, 1982; Pattus, 1990).

Second, mixed cultures of either bacteriocin - producing strain (Bac^+) or a non producing mutant (Bac^-), together with a large excess of bacteriocin-sensitive *Listeria* were studied (Hechard, 1991). The relative quantity of each bacterial species was estimated on-line only by light scattering at differents angles during 6 h (Figure 2). The relative concentration of the wild type *Leuconostoc mesenteroides* rapidly increases, whereas *L. monocytogenes* is dramatically affected. At the end of the experiment, the *Listeria* population represents only 38% of the overall bacteria, whereas its started at more than 92%. This effect is not observed in the presence of the mutant strain (Bac-), underlaying the bactericidal action of the bacteriocin.

Table 1: Drug action onto Rh 123 uptake by *Listeria monocytogenes* membranes.

Drugs	Relative fluorescence
None	100
Bacteriocin	57 ± 0.5
CCCP	74.5 ± 3
Valinomycin	71.1 ± 1
Nigericin	131 ± 2.5

Bacteriocin (80 AU/ml) and all other drugs (5x10^{-6}M) act during 5 min in the presence of Rh 123. The membrane potential was then estimated by the relative amount of fluorescent dye bound to each cell (flow cytometer ACR1500, Bruker, France).

Finally, a classical microbiological method, was used to validate the previous obsertion. Since only living bacteria were detected on selective media, a smaller number of viable *Listeria* were counted in the presence of the mesentericin-producing strain (data not shown). This indicates that, dead *Listeria* cells, still keeping their physical parameters, are disrupted only after a lag period, following mesentericin action, explaining the higher relative amount seen by flow cytometry.

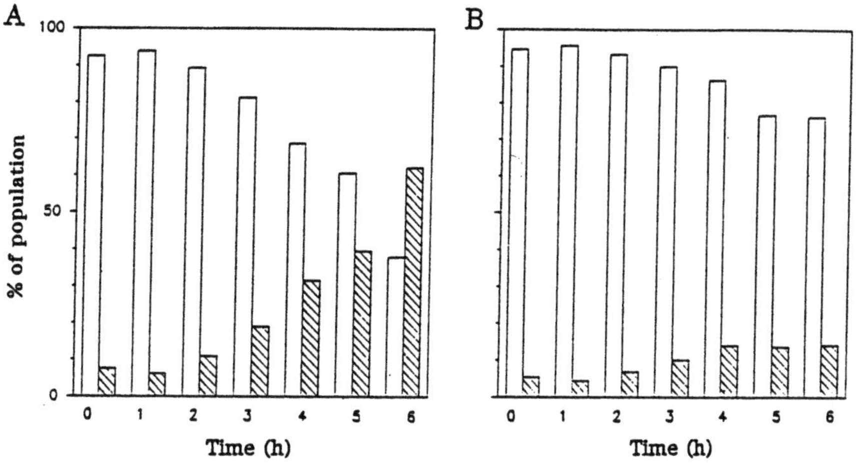

Figure 2: On-line behavior of mixed cultures.
Relative quantity of both the *Leuconostoc* bacteriocin producing strain (A) or mutant (B) and *Listeria* (dashed bars) were followed by light scattering at different angles during 6 hours. Foward angle scatter (FAS) and right angle scatter (RAS) respectively related to bacterial size and granulometry, allow visualization of individual bacterial species in the mixture.

Conclusions

Previous screening of goat's milk led us to a new *Leuconostoc mesenteroides* strain which synthesizes an anti-*Listeria* bacteriocin, the mesentericin Y105. Purification and characterization of the bacteriocin are in progress. This 2.5 kDa peptide is produced in the log phaseof growth and is extremely stable. We are now attempting to clone the bacteriocin gene.

The mesentericin Y105 acts via a bacterial membrane depolarization. It could be a direct action because it occurs rapidly, within 5 minutes. To our knowledge, this is the first time that flow cytometry has been used to follow, on - line, the behaviour of mixed cultures of antagonistic strains.

Oxymetry and molecular genetic studies are in progress to complete the flow cytometry investigations.

References

Farber JM, Peterkin P (1991) Microb Rev 55:476-511
Hechard Y, Dherbomez M, Letellier F, Cenatiempo Y (1990) Lett Appl Microbiol 11:185-188

Hechard Y, Jayat C, Letellier F, Julien R, Ratinaud MH, Cenatiempo Y (1991) Cytometry sup.5, 68, abstract 332-A.
Hechard Y, Jayat C, Letellier F, Julien R, Cenatiempo Y, Ratinaud MH (1991) Submitted to Appl Env Microbiol
Klaenhammer TR (1988) Biochimie 70:337-349
Konisky JA (1982) Ann Rev Microbiol 36:125-144
Leprat P, RatinaudMH, Maftah A, Petit JM, Julien R (1990) Exp Cell Res 186:130-137
Matsuyama T (1984) FEMS Microbiol Lett 21:153 -157
Pattus F, Massote D, Wilmsen HU, Lakey J, Tsernoglou D, Tucker A, Parker MW (1990) Experientia 46: 180-192
Resnick M, Schuldiner S, Bercovier H (1985) Curr Microbiol 12:183-186
Tagg JR, Dajani AS, Wannamaker LW (1976) Bocteriol Rev 40:722-756

CLONING AND CHARACTERISATION OF A LYSIN GENE FROM A *LISTERIA* BACTERIOPHAGE

John Payne and Michael Gasson
AFRC Institute of Food Research, Norwich laboratory,
Norwich Research Park
Colney Lane
Norwich, NR4 7UA
UK

Introduction

The concept of using bacteriophages as antimicrobial agents is well established but has generally been unsuccessful in practice due in part to the limited host range of most bacteriophage. The release of natural bacteriophage particles from host bacteria is effected by digestion of the bacterial cell wall by a lysozyme-like enzyme encoded for by the bacteriophage genome. These bacteriophage lysins may have potential as antimicrobial agents.

In recent years the problem of *Listeria monocytogenes* as a cause of foodborne listeriosis has been widely recognised (Lund, 1990). We have used gene cloning technology to isolate and characterise the gene for a bacteriophage lysin that attacks *Listeria*. Both the gene and the gene product have considerable potential as a novel means to control *Listeria*.

Cloning the bacteriophage ∅LM4 lysin gene.

A bacteriophage named ∅LM4 was isolated in this laboratory from a culture of *L. monocytogenes* serotype 4b and purified by single plaque isolation procedures using *L. monocytogenes* F6868 as host. *Hin*dIII fragments of purified bacteriophage ∅LM4 DNA were cloned into *Escherichia coli* TBl using pUC18 as vector. Recombinants containing an insert were screened for their ability to produce a bacteriophage lysin active against *Listeria* using a patch overlay technique. After incubation at 30°C for 18 h clear zones of lysis were apparent around patches of clones expressing the *Listeria* bacteriophage ∅LM4 lysin. Positive clones were recovered and the pUC18 derivative plasmid isolated and characterised by digestion with *Hin*dIII. One lysin expressing pUC18 clone that contained a 3.6Kb insert of ∅LM4 DNA was chosen for further analysis. This plasmid was designated pFI322.

Characterisation of the ∅LM4 lysin gene

Characterisation of pFI322 was undertaken by constructing a restriction and deletion map of this

NATO ASI Series, Vol. H 65
Bacteriocins, Microcins and Lantibiotics
Edited by R. James, C. Lazdunski and F. Pattus

insert with a variety of restriction enzymes. These results demonstrated that the structural gene for bacteriophage ∅LM4 lysin was contained within the left hand 1.2Kb of the DNA cloned in pFI322 and defined by a *Hind*III site at co-ordinate 0 and an *Eco*RI site at co-ordinate 1.2 of the map illustrated in Figure 1.

The orientation of the *Listeria* bacteriophage ∅LM4 DNA with respect to the *E. coli* lac α promoter that is present on vectors pUC18 and pUC19 is also indicated. It is apparent that a positive reaction in the lysin assay is only found when one orientation is maintained. This suggests that expression of the lysin gene depends on the use of the *E. coli* lac α promoter and that no *Listeria* bacteriophage ∅LM4 promoter is present and active in *E. coli*.

Identification of the lysin protein

A 2Kb fragment from plasmid pFI328 between the *Hind*III site at co-ordinate 0 and a unique *Bam*HI site present on the polylinker of pUC19 was isolated and cloned between the *Hind*III and *Bam*HI sites of the T7 expression vector pSP73. The constructed plasmid named pFI331 was transformed into the *E. coli* host strain JM109DE3.

The *E.coli* T7 promoter in this vector is expressed by the bacteriophage specific T7 RNA polymerase which is induced by addition of isopropyl-β-D-thiogalactopyranoside (IPTG) in the appropriate host strain *E. coli* JM109DE3. Cultures of this strain carrying pSP73 as a control or pFI331 were grown for 3 h and induced by the addition of IPTG. Incubation was continued for a further 3 h before the cultures were harvested and used to prepare cell extracts (Studier *et al.*, 1990). Proteins present in cell extracts were analysed using SDS-polyacrylamide gel electrophoresis (Laemmli, 1970). The results clearly demonstrate that the 2Kb fragment of pFI331 expresses a single protein with a molecular size of about 31 kilodaltons which represents the lysin enzyme.

DNA sequence of the *Listeria* bacteriophage ∅LM4 lysin gene

The region of DNA between co-ordinate 0 and 1.2 in Figure 1 was subjected to oligonucleotide sequence analysis using the dideoxy chain-termination method (Sanger *et al.*, 1980). Computer analysis of the nucleotide sequence using the GCG package (Devereux *et al.*, 1984) revealed an open reading frame (ORF) 864 nucleotides long and corresponding with the location of the lysin gene as determined by deletion analysis and bio-assay. This ORF would specify a protein of 287 amino acids with a molecular size of 32.9 kilodaltons which agrees well with the calculated 31 kilodalton size of the protein expressed by the T7 vector pSP73.

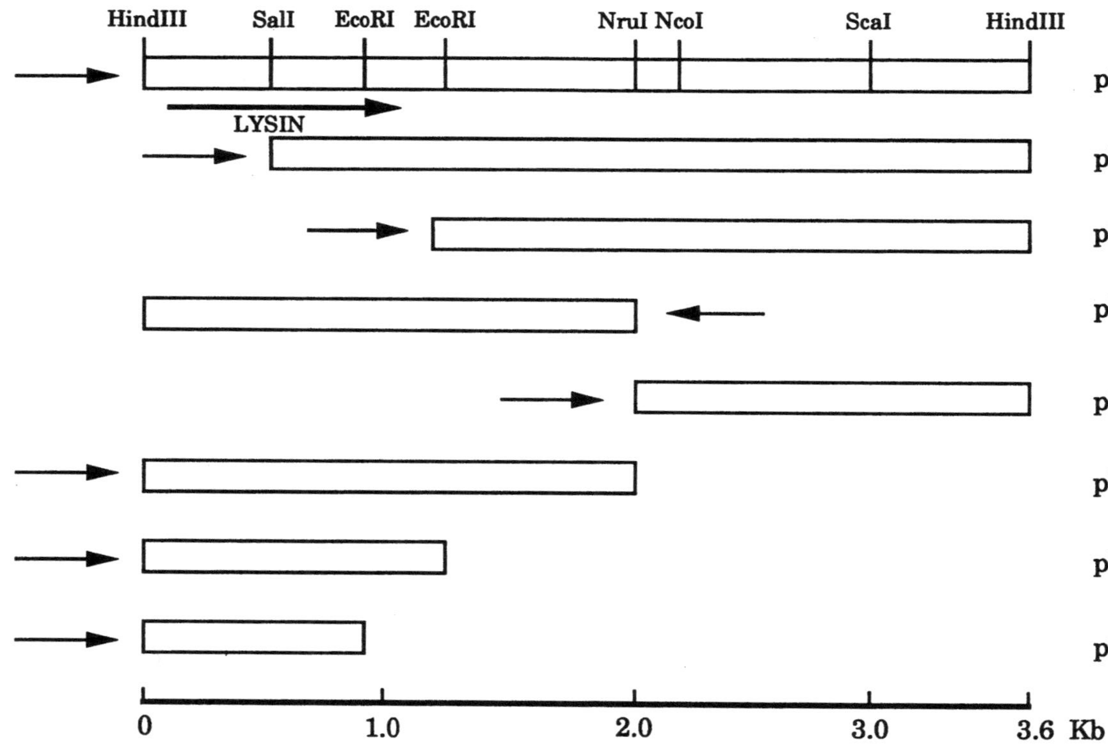

Figure 1. Restriction and deletion map of the ∅LM4 lysin expressing clone pFI322. The precise location of the lysin gene is deduced from the sequence analysis. In the deletion analysis regions of DNA retained in sub-clones are indicated by the bars. Arrows indicate the orientation of the lysin gene with respect to the lac α promoter of the pUC vector which is transcribed from left to right in this figure. The presence or absence of lysin activity is indicated by + or -.

Activity and specificity of the *Listeria* bacteriophage ∅LM4 lysin

Crude cell-free extracts of *E. coli* sub-clones were spectrophotometrically assayed for lysin activity as described above. Figure 2 shows the lytic activity of such extracts of *E. coli* TBl carrying the plasmids pFI322, pFI328, pFI329 and pUC19. This activity was related to units of commercially available mutanolysin (Shearman *et al.*, 1989). The crude cell extracts of lysin expressing clones typically contained 5000 mutanolysin equivalent units/mg protein.

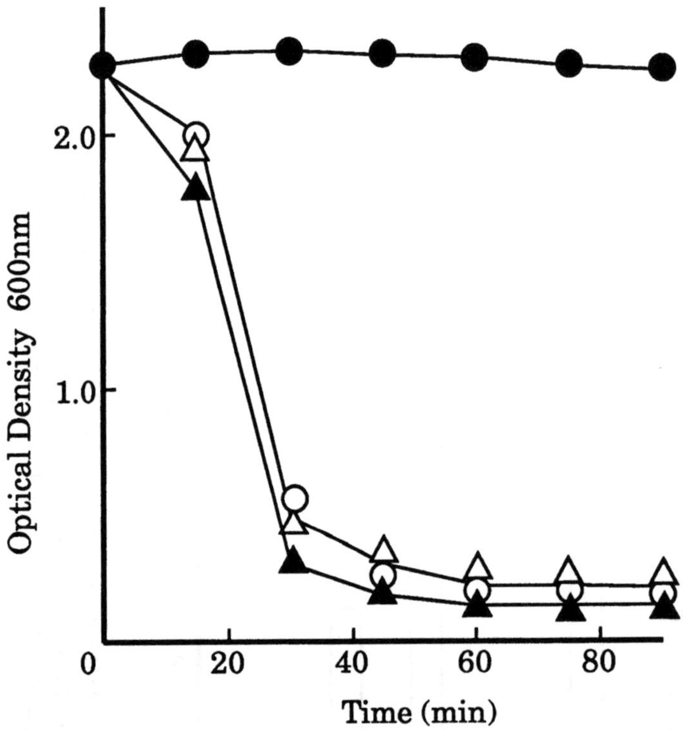

Figure 2. Spectrophotometric assay of cloned lysin activity in crude cell-free extracts of *E. coli.* A cell suspension of *L. monocytogenes* F6868 in 0.lM Tris buffer pH 7.5 was incubated at 37°C with crude cell-free extract (lmg protein) of *E. coli* strains carrying plasmids pUC19 (●); pFI329 (O); pFI322 (▲) and pFI328 (△). Optical density at 600nm was followed with time.

Strains of *L. monocytogenes* representing 16 serotypes together with the other species of the genus *Listeria* and a variety of other Gram positive and Gram negative organisms were tested in this assay system. All serotypes of *L. monocytogenes* tested were susceptible to the cloned lysin activity as were strains of other *Listeria* species and *Kurthia zopfii*. All other organisms tested were insensitive.

Conclusion

We have shown that undesirable micro-organisms in food, in this case isolates of *L. monocytogenes* from confirmed foodborne outbreaks of listeriosis, may be combated by cloning suitable bacteriophage lysin genes and exploiting the expressed enzyme activity. Future work will concentrate both on the practical use of this lytic enzyme in food and agriculture and in the expression and secretion of the lysin in genetically engineered lactic acid bacteria.

References

Devereux J, Haeberli P, Smithies O (1984) A comprehensive set of sequence analysis programs for the VAX. Nucleic Acids Res 12:387-395

Laemmli UK (1970) Cleavage of structural proteins during the assembly of the head of bacteriophage T4. Nature 227:680-685

Lund BM (1990) The prevention of foodborne listeriosis. British Food J 92(4):13-22

Sanger F, Coulson AR, Barrell BG, Smith AJH, Roe BA (1980) Cloning in single stranded bacteriophage as an aid to rapid DNA sequencing. J Mol Biol 143:161-178

Shearman C, Underwood HM, Jury K, Gasson MJ (1989) Cloning and DNA sequence analysis of a *Lactococcus* bacteriophage lysin gene. Mol Gen Genet 218:214-221

Studier FW, Rosenberg AH, Dunn JJ, Dubendorff JW (1990) Use of T7 RNA polymerase to direct expression of cloned genes. Methods Enzymol 185:60-89

TRANSFORMATION OF *Enterococcus faecalis* OGIX WITH THE PLASMID pMB2 ENCODING FOR THE PEPTIDE ANTIBIOTIC AS-48, BY PROTOPLAST FUSION AND REGENERATION ON CALCIUM ALGINATE

M. Martínez-Bueno, I.Guerra, M.Maqueda, A.Gálvez and E.Valdivia
Departamento de Microbiología
Facultad de Ciencias
Universidad de Granada
18071-Granada
Spain

The ability to introduce individual molecules of plasmid DNA into bacterial cells by transformation has been of central importance to the recent and rapid advance of plasmid biology and to the development of DNA-cloning methods. Protoplast fusion in the presence of polyethylenglycol (PEG) has proven an efficient method for transformation of yeasts (Svoboda & Ourednicek, 1990) and bacteria such as *Bacillus* (Chang & Cohen, 1979), *Lactococcus* (Rondo & McRay, 1982) and *Enterococcus* (Smith, 1985).

A crucial step for high yield transformation is protoplast regeneration. Efficient protocols have been developed for reversion of lacotococcal and enterococcal protoplasts by inclusion in solid matrixes, although these techniques have seldom been applied to bacteriocinogenic plasmids. In some cases, transformation with plasmids encoding for membrane-damaging bacteriocins and antibiotics may be troublesome since a functional cell wall may be required for immunity, making protoplast regeneration mandatory before application of the bacteriocin as selective pressure. Such is the case of the plasmid pMB2, which encodes the broad-spectrum peptide antibiotic AS-48 (Martínez-Bueno *et al.*, 1990). This peptide acts on the cell membrane of sensitive bacteria (Gálvez *et al.*, 1991). Although the producer strain *Enterococcus faecalis* S-48 is resistant to high concentrations of AS-48, its protoplasts are very sensitive. Therefore, the cell wall of any transformant recipient should be allowed to regenerate before application of selective pressure. To solve this problem, we have used an ionotropic matrix for protoplast regeneration and recovery of reverted cells before selection. By this procedure, transformation of *E. faecalis* OGIX with the plasmid pMB2 has been obtained with high efficiency.

Protoplast transformation, regeneration on alginate and selection of transformants

Formation of protoplasts from *E. faecalis* OGIX and transformation in the presence of PEG were carried out according to Wirth *et al.* (1986). Briefly, 10^9 cfu of the protoplast suspension were transformed with 100 ng of plasmid pMB2. The fused protoplasts were finally resuspended in 1 ml of Todd-Hewitt broth (THB) plus 0.5M sucrose, gently mixed with 20 ml of 0.5M sucrose- 2% alginate and poured on

NATO ASI Series, Vol. H 65
Bacteriocins, Microcins and Lantibiotics
Edited by R. James, C. Lazdunski and F. Pattus

plates containing 0.5M sucrose-50 mM $CaCl_2$-1.5 % agar. After 24 h incubation at 37°C, the alginate layers were dissolved with sodium citrate (2 % wt/vol). Reverted cells were separated by centrifugation, resuspended in THB and plated on selective medium containing AS-48 (100 AU/ml). Alternatively, protoplasts transformed with pMB2 were plated on THB agar-0.5M sucrose, or THB agar-0.5M succinate, with and without AS-48 (100 AU/ml).

When protoplasts were plated on media containing AS-48, no regeneration was obtained, even of putative transformants. On the contrary, when the fused protoplasts were plated on media lacking AS-48, more than 90% of the cells were able to revert, regardless of the media used (THB-agar succinate, THB-agar sucrose, or THB-alginate sucrose). When cells regenerated in alginate were recovered after solubilization with sodium citrate and plated on selective medium THB-agar + AS-48 (100 AU/ml), a high number of transformants (10^3-10^4 per μg DNA) was obtained. These transformants harboured the plasmid pMB2 and produced AS-48.

Accordingly, the use of ionotropic matrixes may be an adequate alternative to traditional methods when protoplast regeneration is mandatory before application of selective pressure, such as in the case of cloning the genes encoding for resistance to AS-48 or other lytic agents with the aid of protoplast fusion.

REFERENCES

Chang S, Cohen SN (1979) High frequency transformation of *Bacillus subtilis* protoplasts by plasmid DNA. Mol Gen Genet 168:111-115

Gálvez A, Maqueda M, Martínez-Bueno M, Valdivia E (1991) Permeation of bacterial cells, permeation of cytoplasmic and artificial membrane vesicles, and channel formation on lipid bilayers by peptide antibiotic AS-48. J Bacteriol 173: 886-892

Rondo JR, McRay LM (1982). Transformation of *Streptococcus lactis* protoplasts by plasmid DNA. Appl Env Microbiol 43:1213-1215

Martínez-Bueno M, Gálvez A, Valdivia E, Maqueda M (1990) A transferable plasmid associated with AS-48 production in *Enterococcus faecalis*. J Bacteriol 172:2817-2818

Smith MD (1985) Transformation and fusion of *Streptococcus faecalis* protoplasts. J Bacteriol 162:92-97

Svoboda A, Ourednicek, P (1990) Yeast protoplasts immobilized in alginate: cell wall regeneration and reversion to cells. Current Microbiol 20:335-338

Wirth R, An FY, Clewell DB (1986) Highly efficient protoplast transformation system for *Streptococcus faecalis* and *Escherichia coli-S. faecalis* shuttle vector. J Bacteriol 165:831-836

NMR STUDIES OF LANTIBIOTICS: THE THREE-DIMENSIONAL STRUCTURE OF NISIN IN AQUEOUS SOLUTION

Frank J.M. van de Ven, Henno W. van den Hooven, Cornelis W. Hilbers and Ruud N.H. Konings
Nijmegen SON Research Center
Laboratory of Biophysical Chemistry, University of Nijmegen
Toernooiveld
6525 ED Nijmegen
The Netherlands

SUMMARY

Nisin, a bacteriocin produced by *Lactococcus lactis* ssp., is a post-translationally modified pentacyclic polypeptide of 34 amino acids. It is a member of the class of bacteriocins, known as lantibiotics, that contain the unusual amino acid lanthionine. Its structure in aqueous solution has been determined on the basis of data obtained from Nuclear Magnetic Resonance Spectroscopy (NMR) studies. Translation of the interproton distance constraints, derived from Nuclear Overhauser Enhancement Spectroscopy (NOESY)1, and torsion angle constraints, derived from Double Quantum Filtered Correlated Spectroscopy (DQF-COSY), into a 3D structure was carried out with the distance geometry program DISMAN, followed by restrained energy minimization using CHARMm. Due to the internal mobility of the polypeptide chain a determination of the precise overal folding of the molecule was prohibited, but parts of the structure could be obtained albeit with sometimes low resolution. The structure of nisin can best be described as follows: the outermost N- and C- terminal regions appear quite flexible, the remainder of the molecule consists of an amphiphilic N-terminal fragment (residues 3-19), joined by a flexible 'hinge' region to a rigid double-ring fragment formed by residues 23-28. The latter fragment has the appearance of a somewhat overwound α-helix. It is postulated that i) the coupling between residues 23 and 26 as well as between 25 and 28 by thioether bridges and ii) the inversion of the C^{α} chiralities at positions 23 and 25 occurs via an intermediate α-helical structure of the prenisin molecule.

INTRODUCTION

Nisin, a bacteriocin produced by certain strains of *Lactococcus lactis* ssp., belongs to a class of antimicrobial polypeptides known as lantibiotics (Schnell *et al.*,1988). These bacteriocins are character-

---

[1]ABBREVIATIONS: NOESY, Nuclear Overhauser Enhancement Spectroscopy; COSY, Correlated Spectroscopy; DQF, Double Quantum Filter; Dhb(O), Dehydrobutyrine (β-methyldehydroalanine); Dha (U), dehydroalanine; Ala_s, alanine moiety of lanthionine; D-Abu, β-methylalanine moiety of β-mehyllanthionine

NATO ASI Series, Vol. H 65
Bacteriocins, Microcins and Lantibiotics
Edited by R. James, C. Lazdunski and F. Pattus

ized by their cationic properties and high content of unusual amino acids, such as lanthionine (DAla$_s$-S-Ala$_s$) β–methyllanthionine (D-Abu-S-Ala$_s$)[1] dehydroalanine (Dha) and dehydrobutyrine (Dhb) and are formed via posttranslational processing of a ribosomally synthesized precursor polypeptide (Buchman *et al.*, 1988, Schnell *et al.*, 1988, Banerjee & Hansen, 1988). Most probably prior to cleavage off of the N-terminal amino acid sequence of 23 amino acids of the primary translation product, or prepeptide, the amino acids serine, threonine and cysteine undergo post-translational modification. The modifications involve dehydration of the serines and threonines to produce α,β-unsaturated amino acids, followed by the addition to the double bonds of the sulfhydryl groups of the cysteines in a stereospecific way to create the lanthionines (Kusters *et al.*, 1984, Weil *et al.*, 1990). All lantibiotics isolated thus far, show bactericidal activity against Gram-positive bacteria. Antimicrobial activities against other cells and/or organisms have been reported as well (Hurst, 1981, Ruhr & Sahl, 1985).

Nisin, the most prominent lantibiotic, acts on a variety of Gram-positive bacteria such as lactococci, lactobacilli, *Listeria monocytogenes*, and several species of *Clostridia*. The outgrowth of *Clostridium* and *Bacillus* spores is inhibited by nisin (Klaenhammer, 1988), a property which has been applied in milk fermentations to prevent the 'blowing' of cheese caused by *Clostridia*. It furthermore has great potential for several additional applications in food fermentation and preservation in the canning and dairy industry as it has been approved by the FDA to be used for such purposes. Moreover, its use as a growth inhibitor of 'spoilage' bacteria in the brewing industry is anticipated.

Nisin is 34 amino acids long (molecular mass 3400 Da) and contains five rings, each closed by thioether bridged lanthionines (Figure 1).

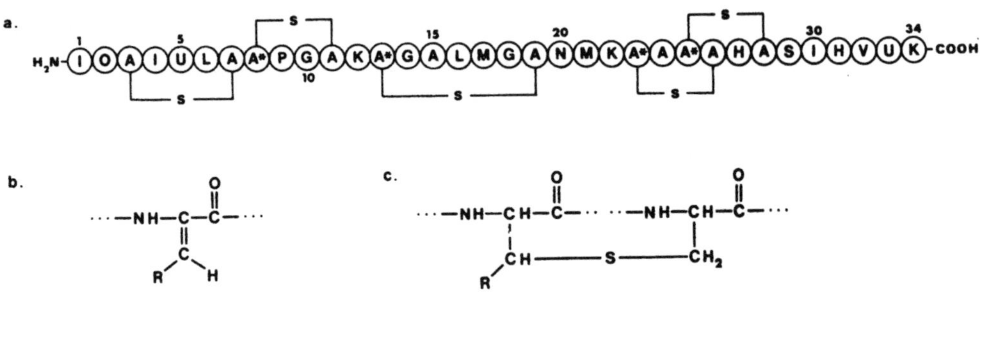

Figure 1. The primary structure of nisin. The unusual amino acids are indicated by the following one letter code: U, dehydroalanine; O, dehydrobutyrine; A-S-A, lanthionine and A*SA, β-methyllanthionine.

The α-carbon atoms of the N-terminal units of the five lanthionines are always in the D-configuration (Gross & Morell, 1971), and the β-carbon atoms of the β-methyllanthionines are in the S-configuration (Kusters *et al.*, 1984; Fukase *et al.*, 1988). In addition, nisin contains three α,β unsaturated amino acids: i.e. dehydroalanine (residues 5 and 33) and dehydrobutyrine (residue 2) (Gross& Morell, 1971). In aqueous solution nisin is most soluble and stable at pH 2. At higher pH values both the solubility and stability decrease, alkali inactivates the peptide (Liu & Hansen, 1990).

Studies on the mode of action of nisin have suggested that it increases the permeability of the cytoplasmic membrane (Ruhr & Sahl, 1985; Henning *et al.*, 1986; Kordel & Sahl, 1986), thereby dissipating the membrane potential, inhibiting amino-acid uptake and causing efflux of substrates from sensitive cells and vesicles derived from Gram-positive bacteria. The polypeptide requires a trans-negative threshold potential for membrane interaction (Sahl *et al.*, 1987). Effects of nisin on the cell wall synthesis has also been reported (Ruhr & Sahl, 1985; Henning *et al.*, 1986; Reisinger *et al.*, 1980), which might act synergistic to the disturbance of the membrane function.

Because nisin has great potential for several applications in food fermentation and preservation a fundamental insight into its structure/function relationship is of importance for the rational design of new lantibiotics with 'improved' physico-chemical and biological properties, such as solubility, stability or bactericidal activity. Therefore we embarked on a systematic NMR study of nisin and related bacteriocins, aiming to unravel their spatial structures and molecular mechanisms related to their biological function. The complete assignment of the ^{1}H NMR spectrum of nisin in aqueous solution has already been reported (Slijper *et al.*, 1989). Here we summarize our studies (Van de Ven *et al.*, 1991a,b) aimed to analyse the NMR data in terms of a three dimensional model.

MATERIALS AND METHODS

Nisin was purchased from Koch-Light Ltd. or NBS Biologicals. To make sure that all crosspeaks in the NMR spectra, that were used for the structural analysis, were indeed derived from intact molecules, it was further purified by HPLC chromatography. Samples were prepared by dissolving the purified material in $H_2O/^2H_20$ (90:10) up to a concentration of 3-5 mM, and adjusting the pH to 3.5 (pH meter reading).

Two-dimensional NMR spectra were recorded on Bruker AM spectrometers, interfaced to Aspect 3000 computers. NOESY spectra were recorded at 7°C and 600 MHz, employing mixing times of 0.05, 0.2 and 0.4 s. A DQF-COSY spectrum was recorded at 7°C and 400 MHz. Time domain matrices were 512 x 2048 and 800 x 8192 (t1 x t2) for NOESY and DQF-COSY, respectively. The solvent resonance was suppressed by continuous irradiation during the relaxation delays (1.5 to 2.0 s) and in the NOESY experiments also during the mixing time. The carrier frequency was placed at the solvent resonance, in the center of the spectrum; phase sensitive spectra were obtained by making use of the Time Proportional Phase Incrementation method (Redfield & Kunz, 1975; Marion & Wuthrich, 1983). Fourier transformation, plotting and crosspeak integration were performed using the Bruker 'uxnmr' program. Data manipulation consisted of phase shifted sine-bell window multiplication and zero filling prior to Fourier transformation. The final digital resolution, in the f2 dimension, was 3 Hz/pt for NOESY and 0.5 Hz/pt for DQF-COSY.

The assignment of all the NOEs was performed with the aid of the computer program 'ASSIGNOE' which for each NOESY crosspeak consults a table containing the chemical shifts of all protons to generate all possible assignments for the f1 and the f2 position of the crosspeak. The conversion of NOESY crosspeak volumes to interproton distances was done using our program N02DI (Van de Ven *et al.*, 1991c). In the case at hand it was found that the different NOESY mixing times yielded interproton distances that typically did not differ more than 10 %. Upper bounds were generated by adding 15 % to the interproton distances obtained. In some selected cases lower bounds were calculated by subtracting 15 % from the distances. For most residues no stereospecific assignments for CH_2 protons were available. For these, and for CH_3 groups, an extra 0.5 Å was added to the upper bounds which were then referenced pseudo atoms, situated at the geometric means of the methyl(ene) protons.

The vicinal J-coupling constants of the NH-H^a proton pairs were estimated from the splittings, in the f2 dimension, of the DQFCOSY crosspeaks. Based on these coupling constants, constraints for the normal L amino acids were generated for the Φ torsion angle as described by Kline *et al.* (1988). Coupling constants larger than 8.0 Hz were found for all four β-methyllanthionines; these allowed Φ to be restrained to the region [80°, 160°]. For three of these residues, at positions 13, 23 and 25, the $J_{\alpha\beta}$ could be measured as well. In combination with NOESY this allowed us to constrain X^1 in the region [-110°,-10°] for these residues. The J-coupling constants and NOEs of all residues with $C^{\beta}H_2$ moieties, with the exception of those at positions 13, 26 and 28, clearly indicated conformational averaging, caused by rotation around the C^{α}-C^{β} bond. Structures, satisfying the experimental interproton constraints and the backbone torsion angle constraints, were generated using the program DISMAN (Braun & Go, 1985), which was kindly provided by Dr. W. Braun. We used version 2.0, running on Convex C120 computer. The original DISMAN program was adapted be able to handle the special residues present in nisin. Constraints for thioether bridge formation were added as follows: lanthionine was constructed via an upper bound of 0.0 Å between the S atoms of D-Ala_s and L-cysteine, and a range of 2.6-2.8 Å between the C^{β} atoms of the pairing residues. β-Methyllanthionine was constructed in the same way, using D-Abu and L-Cys. As the atoms of the two contributing residues are considered non-bonding by the program, the sulphur atoms were defined as dummy atoms in order to prevent steric clashes. The C^{β}-C^{β} distance of 2.7 Å constrains the C-S-C angle around 100°, which was taken from the standard C-S-C geometry of methionine.

The final DISMAN structures were further refined by restrained molecular mechanics calculations. To this end the program package QUANTA/CHARMm, from Polygen, was used, which ran on Silicon Graphics workstations 4D/70 and 4D/25. The force field contained only those terms which are necessary for maintaining the covalent structure and the Van der Waals non bonded term; electrostatic interactions and hydrogen bonds were excluded. The torsion energy expression for Φ and Ψ of Dha and Dhb had minima at 0° and 180°, allowing for an all-planar arrangement of these residues and their flanking amide bonds (Ajo *et al.*, 1971).

RESULTS and DISCUSSION

Due to the presence of no less than five thioether rings, two of which are even intertwined, the conformational freedom of the nisin molecule is already fairly limited. As a consequence, one does not expect to find regular secondary structure elements such as α-helices or β-sheets. In NOESY spectra recorded with short mixing times, up to 100 ms, we cannot observe any crosspeaks that would give a clue regarding the tertiary structure of the molecule. The only observable 'long range' NOEs are between residues that are adjacent in the thioether rings, such as D-Ala_s3 and Ala_s7 etc., thus confirming the covalent linkages shown in Figure 1. Only in NOESY spectra recorded with longer mixing times, e.g. 400 ms or more, and at low temperature, a few 'long range' NOEs become visible. Thus, the relative orientation of the first two thioether bridged rings follows from NOEs between residues Ile4 and Pro9, and between Ala_s7 and Pro9. The third ring is constrained relative to the second one by NOEs between Pro9 and Ala 15, and by very weak ones, between Ala_s11 and Gly14. These NOEs only allow for a rather crude description of the spatial structure of the molecule as a whole. The scant data regarding the tertiary fold of nisin only allow for a rather crude description of the spatial structure of the molecule as a whole. Moreover, we cannot exclude the possibility that these long range NOEs represent only momentary contacts between the residues. Thus, the conformational freedom may even be larger than proposed in the present study since the distance geometry approach necessarily requires all the constraints to be simultaneously satisfied in a single molecule.

The DISMAN procedure yielded thirteen structures that exhibited sub-Ångstrom distance violations and torsion angle violations smaller than 5°. The total computation took about three cpu hours. After further refinement by means of restrained molecular mechanics the bound violations in the structures were further reduced below 0.2 Å.

In Figure 2 the overall shapes of the various structures are shown. At first sight the structures all have very different shapes, and it makes no sense to try to superimpose them. However, when we concentrate on the region encompassing residues 3 to 19, i.e. the first three thioether bridged rings, a global folding pattern emerges (Figure 2a). Hence, our NMR data allow us to propose a low resolution structure for this region. The average root mean square deviation (RMSD) of the backbone of this region is 2.7 Å; pairwise RMSDs vary between 0.76 Å and 4.38 Å. A much better superposition, and consequently higher resolution, is obtained for the region 23 to 28, which comprises the fourth and fifth thioether rings (Figure 2b). This fragment forms one structural element, residues 25 and 26 being part of both rings. Here the average RMSD is only 0.23 Å. The N- and C-terminal regions are poorly defined, as is the section 20-22, which forms a flexible 'hinge' region between the sections 3-19 and 23-28 (c.f. Figure 2). It is difficult to ascertain whether there is in fact enhanced flexibility in the 'hinge' region 20-22, or if there is scarcity of NMR data for another reason. The vicinal $J_{NH\alpha}$ coupling constants of Asn20 and Met21 are both 7.0 Hz. This may correspond to a Φ torsion angle of about -75° or -155° (Pardi *et al.*, 1984), but is also the value expected for conformational averaging. Typical of a regular rigid peptide backbone is the observation of either a strong sequential $d_{\alpha N}$ NOE or a strong d_{NN} connectivity, but not both. We observe fairly weak connectivities for both $d_{\alpha N}$ and d_{NN} for steps 19-20 and 20-21. Although, again, this is not conclusive evidence, we take these features as indications of motional averaging. The distribution of polar and apolar residues over the molecular surface of nisin may be of relevance with respect to its biological activity. E.g. for other lantibiotics, gallidermin (Freund *et al.*, 1991) and Ro 09-0198 (= cinnamycin) (Kessler *et al.*, 1988), structures have been presented that are clearly amphiphilic. As can be gleaned from Figure 1, nisin already has amphiphilic character as far as the amino acid sequence is concerned, with a cluster of bulky hydrophobic residues at the N- and hydrophilic ones at the C-terminal end. We find that, despite their mutual differences, in all the structures obtained the first three rings are oriented relative to one another in such a way that a hydrophobic face is formed by the residues Ile1, Ile4, Leu6, Pro9 and Leu16 at one side, while Lys12 sticks out on the opposite side. In Figure 3 some representative structures are shown. The C-terminal region contains the positively charged side chains of lysines 22 and 34, and histidines 27 and 31. The relative orientations of the charged groups and hydrophobic groups, such as Ile30 and Val32, of the C-terminal end vary strongly from one structure to another. As can be seen in Figures 3a and 3b there is a possibility that the amphiphilic character of the N-terminus is extended all the way throughout the whole molecule.

The overall fold of nisin in aqueous solution can only be defined with rather low precision, as described above. However, some details of the structure have become quite clear in the present study. The obvious structural entities to be considered in more detail are the five thioether bridged rings. The structures described for the first and the third ring should be interpreted as 'snap shots' of fairly flexible entities; the

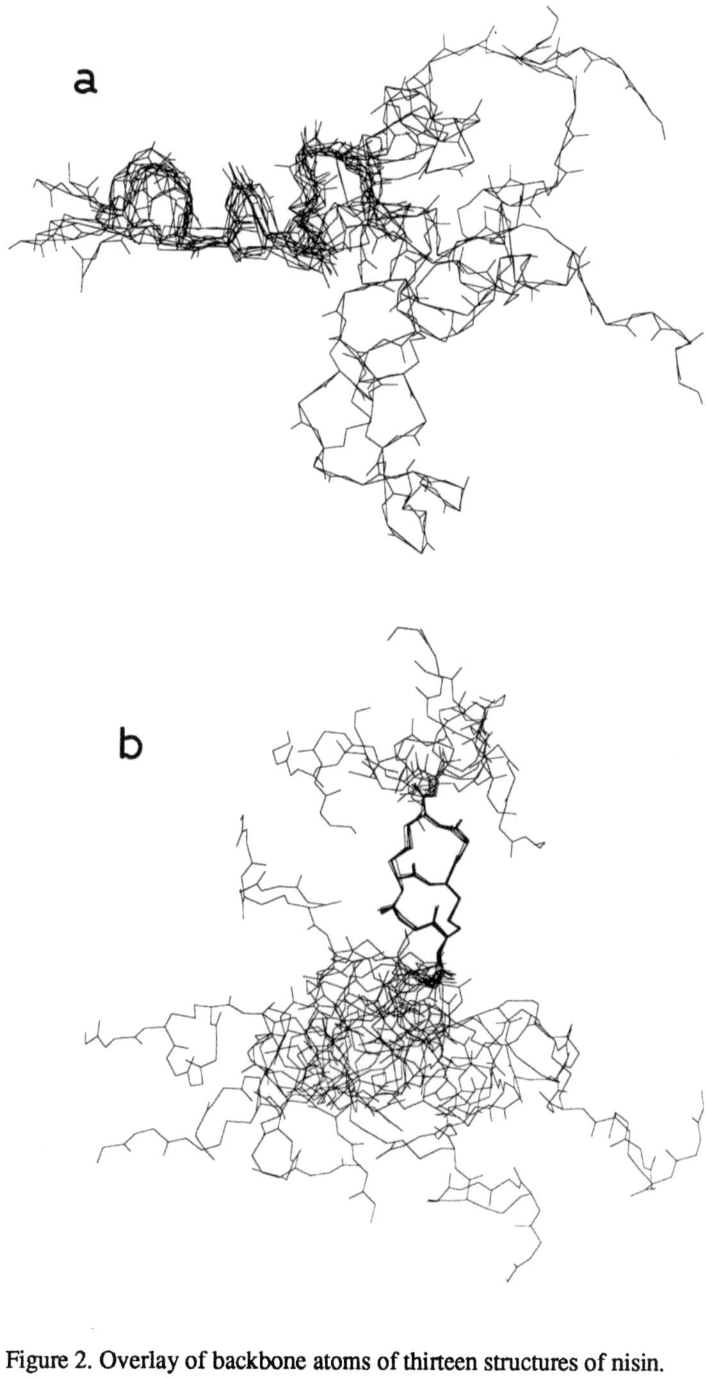

Figure 2. Overlay of backbone atoms of thirteen structures of nisin.
a) Optimal superposition of fragment 3-19, b) of fragment 23-28.

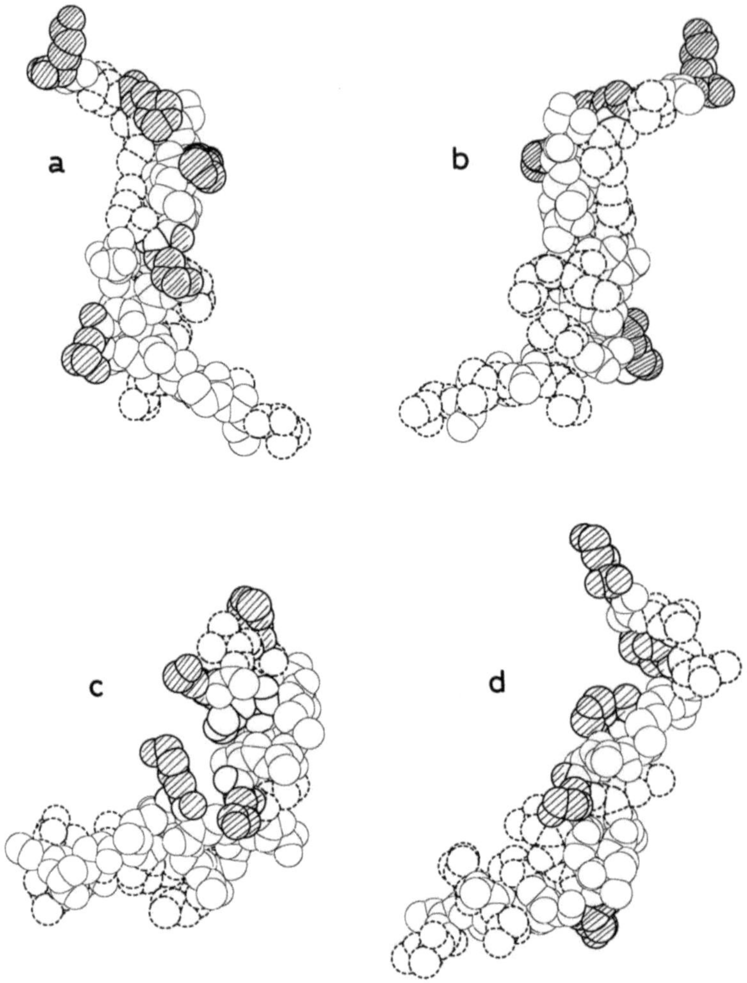

Figure 3. Space filling models of three structures. Bulky hydrophobic side chains are indicated by dashed contours, positively charged residues by shading. a,b) Two sides of same structures; c,d) two other ones.

thirteen structures obtained in the present study exhibited significant differences in these regions, with RMSDs of 0.85 Å and 1.47 Å for the backbone atoms of the first and third ring respectively. The second, fourth and fifth ring are all well defined with RMSDs of 0.25 Å for the backbone atoms of region 9-13 and 0.23 Å for the region 23-28. The amide proton ofAla_s11 points inwards into the second ring, as does the carbonyl oxygen of D-Abu8, thus providing a suitable orientation for hydrogen bond formation. The structure strongly resembles a type II turn. A remarkable feature of the double ring structure, which has been determined with high precision, is the roughly parallel orientation of the three peptide bonds joining residues 23-24, 25-26 and 27-28, which generates the overall appearance of two consecutive β-turns or of a somewhat overwound α-helix (c.f. Figure 4e). In Figure 4 we show an attempt to rationalize this structure by assuming an α-helical structure for residues 23 till 28 in prenisin. In Figure 4a it is shown how in a regular α-helix the side chains of Thr23 and Cys26 as well as of Thr25 and Cys28 are positioned relative to each other to promote thioether bridge formation. In Figure 4b it is seen how inversion of the chiralities of the Cα carbons of Thr23 and Thr25 orients the Cβ carbons of these residues towards the sulphur atoms of Cys26 and Cys28, so that Cβ-S bonds can be formed. Figure 4c shows the result of such bond formation. This structure was obtained by computer-chemistry: removing two H_20 moieties from Figure 4b, joining the atoms Cβ and S, to establish the 23-26 and 25-28 links, followed by a few steps of energy minimization to obtain correct bond lengths and -angles. The actual mechanism rather involves the conversion of threonines 23 and 25 to Dhb, as shown in Figure 4d. The structure obtained in the present study, Figure 4e, is remarkably similar to Figure 4c. Thus, by assuming an α-helical structure for prenisin 23-28 we can rationalize a) the pairing 23-26 and 25-28 (i, i+3) of the thioethers, b) the fact that inversion of the Cα chiralities of threonines 23 and 25 is required for optimal thioether formation, and c) the resulting structure of the two coupled rings.

To our knowledge this is the first proposal for a complete structure of nisin in aqueous solution. Studies on isolated lanthionine ring fragments 3-7 and 8-11 have been reported by Palmer *et al.* (1989). Those studies were based on a rather limited set of NOEs, obtained in DMSO solution, that was used in restrained molecular dynamics calculations. The results obtained for the first ring appear somewhat different (c.f. Figure 5 in Palmer *et al.*, 1989), which may reflect the difference in experimental conditions, the computational methodology, or both. The structure reported for the second ring closely resembles the one shown here. Roberts and coworkers (Lian *et al.*, 1991) have also conducted a combined NMR, restrained molecular dynamics study of nisin in aqueous solution. The major difference between their study and the one reported here is the absence of long range NOEs in their data set. For the individual thioether rings essentially the same results were obtained, including the effect of conformational averaging caused by rotation around the Cα-Cβ bonds of residues Ala_s7 and 19.

We find nisin to consist of two domains: the first one ranges from residue 3 till 19 and comprises the first three lanthionine rings, and the second one consists of the coupled ring system formed by residues 23 till 28. The two domains are connected by a flexible 'hinge' region around methionine 21. The first domain is defined with rather low precision, most likely due to the fact that the nisin molecule is far from being rigid. It features an amphiphilic N-terminus, with Ile4, Leu6, Pro9, Ile16 and Met17 protruding from the hydrophobic side, and Lys12 sticking out on the opposite side. The amphiphilic character of the C-terminus is less clear. However, we have noted that in a mixed solvent of water and trifluoroethanol the α-helical character of the fragment 23-28 is extended all the way to the C-terminus (unpublished observation). Such a helix is clearly amphiphilic with His27, His31 and Lys34 all on one side.

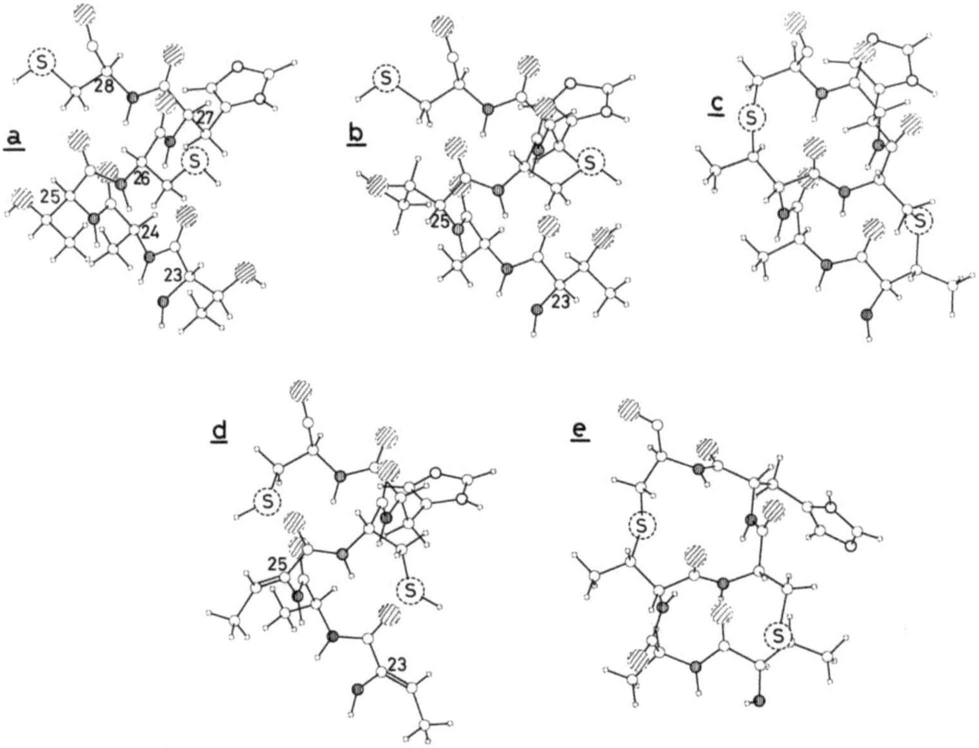

Figure 4. Rationalization of the structure of the fragment DAbu23 to Ala_s28, based on α-helix structure for Thr23 to Cys28 in prenisin. a) Helix 23-28 in prenisin; b) the same helix with the inversion of the chiralities of the Cα carbons of Thr23 and Thr25; c) a structure for nisin fragment 23-28 obtained from Figure 4b by removing two H_20 moieties, followed by linking the Cβ atoms to the sulphur atoms and a few steps of energy minimization to obtain correct bond length and angles, this structure is to be compared with the experimentally determined one (Figure 4e); d) the intermediate likely to be involved in thioether bridge formation, with Dhb at positions 23 and 25; e) NMR structure of fragment D-Abu23 to Ala_s28.

The most likely target for the biological activity of nisin is the cytoplasmic membrane (Ruhr & Sahl, 1985; Henning *et al.*, 1986; Kordel & Sahl, 1986; Sahl *et al.*, 1987). Taking into account its amphiphilic character, which has also been reported for other lantibiotics (Freund *et al.*, 1991; Kessler *et al.*, 1988), the mode of action of nisin may well be analogous to that of other amphiphilic peptides such as melittin. For the molecular mechanism of this peptide both an 'insertion model', in which membrane inserted molecules form clusters around a central pore (Hanke *et al.*, 1983; Tosteson & Tosteson, 1984), and a 'wedge model' in which an amphiphilic molecule adheres to the outside of a membrane causing destabilisation of the bilayer structure and thus promoting pore formation (Terwilliger *et al.*, 1982; Brown *et al.*, 1982) have been proposed. It may well be that either one of both of these models applies to nisin as well. Of course, the conformation of nisin at the membrane may differ from the one in aqueous solution described here. Therefore we are currently in the process of determining the structure of nisin in a more lipophilic environment, i.e. in other solvents, and complexed with detergent micelles.

ACKNOWLEDGEMENTS

Dr. Werner Braun, ETH Zurich, is gratefully acknowledged for making DISMAN available to us, and for giving advice on making the program suited for the unusual amino acids in nisin. The HPLC purified nisin was a generous gift from Dr. Harry Rollema, Dutch Institute for Dairy Research. Dr. Sybren Wijmenga and Jos Joordens, of the Dutch national SON-NMR facility are thanked for their help and advice in running 2D NMR experiments as optimally as possible. This research was sponsored by the Netherlands Foundation for Scientific Research (NWO), *in casu* the Foundation for Chemical Research (SON) and the Foundation for Applied Science (STW).

REFERENCES

Ajo D, Granozzi G, Tondello E, Del Pra A, Zanotti G (1979) Crystal structure and conformational flexibility of 2(acetylamino)prop-2-enoic acid (N-acetyldehydroalanine). J Chem Soc Perkin 2:927-929

Banerjee SS, Hansen NJ (1988) Structure and expression of a gene encoding the precursor of subtilin, a small protein antibiotic. J Biol Chem. 263:9508-9514

Braun W, Go N (1985) Calculation of protein conformations by proton-proton distance constraints. A new efficient algorithm. J Mol Biol 186:611-626

Brown LR, Braun W, Kumar A, Wuthrich K (1982) High resolution nuclear magnetic resonance studies of the conformation and orientation of melittin bound to a lipid-water interface. Biophys J 37:319-328

Buchman GW, Banerjee S, Hansen NJ (1988) Structure, expression and evolution of a gene encoding the precursor of nisin, a small protein antibiotic. J Biol Chem. 263:16260-16266

Freund S, Jung G, Gutbrod O, Folkers G, Gibbons WA, Allgaier H, Werner R (1991) The solution structure of the lantibiotic gallidermin. Biopolymers 31:803-811

Fukase, T, Kitazawa M, Sano A, Shimbo K, Fujita H, Horimoto S, Wakamiya T, Shiba T (1988) Total synthesis of peptide antibiotic nisin. Tet Letters 29:795-798

Gross E, Morell JL (1971) The structure of nisin. J Am Chem Soc 93:4634-4635

Hanke W, Methfessel C, Wilmsen H-U, Katz E, Jung G, Boheim G (1983) Melittin and a chemically modified trichotoxin from alamethicin-type multi-state pores. Biochim Biophys Acta 727:108-144

Henning S, Metz R, Hammes WP (1986) Studies on the mode of action of nisin. Intern J Food Microbiol 3:121-134

Hurst A (1983) Nisin and other inhibitory substances from lactic acid bacteria. In: Series of Food Science 10, Marcel Dekker, New York, p327-351

Hurst A (1981) Nisin. Adv Appl Microbiol 27:85-123

Kessler H, Steuernagel S, Will M, Jung G, Kellner R, Gillesen D, Kamiyama T (1988) The structure of the polycyclic nonadecapeptide Ro 09-0198. Helv Chim Acta 71:1924-1929

Klaenhammer TR (1988) Bacteriocins of lactic acid bacteria. Biochimie 70:337-349

Kline AD, Braun W, Wuthrich K (1988) Determination of the complete three-dimensional structure of the α-amylase inhibitor tendamistat in aqueous solution by nuclear magnetic resonance and distance geometry. J Mol Biol 204:675-724

Kordel M, Sahl HG (1986) Susceptibility of bacterial, eukaryotic and artificial membranes to the disruptive action of the cationic peptides Pep5 and nisin. FEMS Microbiol Lett 34:139-144

Kusters E, Allgaier H, Jung G, Bayer E (1984) Resolution of sulphur-containing amino acids by chiral phase gas chromatography. Chromatographia 18:287-293

Lian L-Y, Chan WC, Morley SD, Roberts GCK, Bycroft BW, Jackson D (1992) NMR studies of the solution structure of Nisin A and related peptides. In: Jung G & Sahl HG (eds) Nisin and novel lantibiotics. ESCOM Science Publishers B.V., Leiden, p 43

Liu W, Hansen JN (1991) Some chemical and physical properties of nisin, a small protein antibiotic produced by *Lactococcus lactis*. Appl Environ Microbiol. 56: 2551-2558

Marion D & Wuthrich K (1983) Application of phase sensitive two-dimensional correlated spectroscopy (COSY) for measurements of ^{1}H-^{1}H spin-spin coupling constants in proteins. Biochem Biophys Res Commun 113:967-974

Palmer DE, Mierke DF, Pattaroni C, Goodman M, Wakamiya T, Fukase K, Kitazawa M, Fujita H, Shiba T (1989) Interactive NMR and computer simulation studies of lanthionine-ring structures. Biopolymers 28:397-408

Pardi A, Billeter M, Wuthrich K (1984) Calibration of the angular dependence of the amide proton-C^α proton coupling constants, $^3J_{HN\alpha}$ in a globular protein. J Mol Biol 180:741-751

Redfield AG, Kunz SD (1975) Quadrature Fourier NMR Detection: simple muliplex for dual detection and discussion. J Magn Reson 19:250-254

Reisinger P, Seidel H, Tschesche H, Hammes WP (1980) The effect of nisin on murein synthesis. Arch Microbiol 127:187-193

Ruhr E, Sahl HG (1985) Mode of action of the peptide antibiotic nisin and influence on the membrane potential of whole cells and on cytoplasmic and artificial membrane vesicles. Antimicrob Agents and Chemother 27:841-845

Sahl HG, Kordel M, Benz R (1987) Voltage-dependent depolarization of bacterial membranes and artificial lipid bilayers by the peptide antibiotic nisin. Arch Microbiol 149:120-124

Schnell N, Entian KD, Schneider F, Gotz F, Zahner R, Kellner R, Jung G (1988) Prepeptide sequence of epidermin, a ribosomally synthesized antibiotic with four sulphur rings. Nature (London) 333:276-278

Slijper M, Hilbers CW, Konings RNH, Van de Ven FJM (1989) NMR studies of lantibiotics. Assignment of the ^{1}H-NMR spectrum of nisin and identification of inter residual contacts. FEBS Lett 252:22-28

Terwilliger TC, Weissman L, Eisenberg D (1982) The structure of melittin in the form I crystals and its implication for melittin's lytic and surface activities. Biophys J 37:353-361

Tosteson MT, Tosteson DC (1984) Activation and inactivation of melittin channels. Biophys J 45:112-114

Van de Ven FJM, van den Hooven HW, Konings RNH, Hilbers CW (1991a) NMR studies of lantibiotics. The structure of nisin in aqueous solution. Eur J Biochem (in press)

Van de Ven FJM, van den Hooven HW, Konings RNH, Hilbers CW (1991b) The spatial structure of nisin in aqueous solution. In: Jung G & Sahl HG (eds) Nisin and Novel Lantibiotics. ESCOM Science Publishers B.V., Leiden, p 35

Van de Ven FJM, Blommers MJJ, Schouten RE, Hilbers CW (1991c) Calculation of interproton distances from NOE intensities. A relaxation matrix approach without requirement of a molecular model. J Magn Reson 94:140-151

Weil HP, Beck-Sickinger A, Metzger J, Stevanovic S, Jung G, Joster M, Sahl HG (1990) Biosynthesis of the lantibiotic Pep5. Isolation and characterization of a prepeptide containing dehydro amino acids. Eur J Biochem 194:217-223

LOCALIZATION AND PHENOTYPIC EXPRESSION OF GENES INVOLVED IN THE BIOSYNTHESIS OF THE *LACTOCOCCUS LACTIS* subsp. *LACTIS* LANTIBIOTIC NISIN

Luc De Vuyst and Erick J. Vandamme
Laboratory of General and Industrial Microbiology
Faculty of Agricultural Sciences
University of Gent
Coupure links 653 B-9000 Gent Belgium

INTRODUCTION

Nisin is a pentacyclic lantibiotic of 34 amino acids having a molecular mass of about 3500 and displaying bacteriocin-properties (Jung & Sahl, 1991). The bioactive peptide is synthesized by certain strains of *Lactococcus lactis* subsp. *lactis* (De Vuyst, 1990). As with other lantibiotic-containing peptide antibiotics, nisin is synthesized as a prepeptide which undergoes post-translational modification to generate the mature antibiotic (Buchman *et al.*, 1988; Kaletta & Entian, 1989). Nisin is commercially produced only by microbial fermentation (Vandamme, 1984) and it is exclusively used as a food preservative in processed cheese and canned foods (Hurst, 1981; Rayman & Hurst, 1984). Furthermore, because of its activity against various food spoiling and pathogenic bacteria, it has a great potential to be widely used as a biological food preservative (Daeschel, 1989; Delves-Broughton, 1990). In view of this, it would be worthwhile to improve nisin yields in cultures not only by metabolic engineering and process technological control (De Vuyst, 1990; De Vuyst *et al.*, 1988, 1990; De Vuyst & Vandamme, 1991, 1992), but also by genetic means. A fundamental knowledge of the genetic location, structure, expression and regulation of genes involved in the biosynthesis of the lantibiotic nisin is a prerequisite to achieve this.

Several research groups have studied the genetics of both nisin production and nisin resistance, but up to now, there still exist ambiguities about the cellular location of genes associated with these mentioned properties. By using classical genetic approaches such as curing and conjugation experiments, several laboratories have reported that nisin production and nisin resistance are associated with plasmids (Kozak *et al.*, 1974; Fuchs *et al.*, 1975; Leblanc *et al.*, 1980; Gasson, 1984; Gonzalez & Kunka, 1985; Steele& McKay, 1986; Tsai & Sandine, 1987). However, the physical isolation of a single plasmid was never successful, except in the study of Tsai and Sandine (1987), which indicated that the nisin phenotype of *L. lactis* subsp. *lactis* 7962 is mediated by a plasmid of 17.5 MDa. Moreover, other investigators have obtained conflicting results which suggested that nisin production is mediated by either chromosomal DNA or plasmid DNA. Davey and Pearce (1982) and Harris *et al.* (1990) even reported on apparently plasmid-free variants which still produced nisin. Finally, Buchman *et al.* (1988), Kaletta & Entian (1989),

NATO ASI Series, Vol. H 65
Bacteriocins, Microcins and Lantibiotics
Edited by R. James, C. Lazdunski and F. Pattus

Dodd *et al.* (1990) and Rauch *et al.* (1990) have independently cloned and sequenced the structural nisin precursor gene, *nisA*. The *nisA* gene codes for a 57 amino acid prepeptide, with a 23-residue leader region and a 34-residue structural region. However, Buchman *et al.* (1988) and Dodd *et al.* (1990) localized the nisin genes on the chromosomal DNA of *L. lactis* subsp. *lactis* ATCC 11454 and *L. lactis* subsp. *lactis* FI 5876, respectively, while Kaletta and Entian (1989) localized the nisin gene on a plasmid of *L. lactis* subsp. *lactis* 6F3. The reasons for these inconsistent results may be strain divergency, different indicator strain responses, phenotype-testing methods, procedures for cell lysis, and use of different plasmid DNA isolation and purification protocols. By applying a common DNA isolation method on different nisin-producing strains of *L. lactis* subsp. *lactis* and by using a hybridization probe for the prenisin structural gene (*nisA*), we could proof that the genetic determinant of the nisin/sucrose genes is situated on a chromosomal location.

MATERIALS AND METHODS

Strains and media

Lactococcal strains used in this study are shown in Table 1. Cultures were maintained as frozen stocks held at -80°C in liquid medium plus 25% glycerol. Lactococcal cultures were propagated at 30°C in M17 medium (Terzaghi & Sandine, 1975) supplemented with 0.5 % (w/v) glucose (GM17) instead of lactose for nisin-non-producing strains, or sucrose (SM17) for nisin-producing strains. When necessary, the appropriate antibiotic(s) were added to sterile medium to maintain selection for resistance markers. Selective antibiotic concentrations were as follows (in micrograms per ml): streptomycin, 100; rifampicin, 50. Agar media were prepared with 1.5 % granulated agar; overlay agar was prepared with 0.75 % agar.

Phenotypic characterization of lactococcal strains

Carbohydrate fermentation (glucose, sucrose or lactose) was determined by using Elliker agar medium (Elliker *et al.*, 1956) containing the desired filter-sterilized carbohydrate at a final concentration of 0.5 % and bromocresol purple at a final concentration of 0.004 %. Strains capable of fermenting the applied carbohydrate as a sole carbon source produced yellow colonies, while non-fermenters produced blue colonies.

Nisin production was detected by a plate diffusion assay (Tramer & Fowler, 1964), using as an indicator the nisin-sensitive *L. lactis* subsp. *cremoris* IP5 strain.

Nisin immunity was determined by controlling the growth of the tested strain in M17 broth supplemented with gradient concentrations of nisin.

Isolation and restriction of L. lactis subsp. lactis chromosomal DNA

Overnight *L. lactis* subsp. *lactis* cultures were 1/20 diluted in 45 ml of GM17 medium and incubated at 30°C until an optical density (at 600 nm) of 0.6 was reached. Cells were collected by centrifugation (10 min, 4000 rpm), resuspended in 15 ml THMS-buffer (30 mM Tris, 3 mM $MgC1_2$, 25% sucrose, pH 8.0) and subsequently digested with 2 mg/ml lysozyme for 1 h at 37°C. The centrifuged protoplasts were resuspended in 0.5 ml THMS buffer and then 2 ml TES-buffer (10 mM Tris, 100 mM NaCl, 1 mM EDTA, pH 8.0; 0.1% SDS) was added. Subsequently, the DNA was extracted with 1 volume of phenol, 1 volume of 50:50 phenol:chloroform and finally with 1 volume of chloroform. The DNA was ethanol-precipitated in the presence of 3 M sodium acetate. The DNA was then washed with 80% ethanol and finally resuspended in 1 ml TE-buffer (10 mM Tris, 1 mM EDTA, pH 8.0). For restriction digestion, conditions recommended by the commercial supplier (Bethesda Research Laboratories) were followed. Resultant DNAfragments were separated on 0.7 % agarose gels.

DNA-probes and hybridization procedures

Natural DNA probes were employed to search for the nisin precursor and sucrose-6-phosphate hydrolase genes in *L. lactis* subsp. *lactis* DNA. The nisin and sucrose probe were a chromosomal 0.83 kb *Hinc*II/*Sau*3A restriction fragment containing the *nisA* gene and a 1.5 kb *Eco*RV/*Hind*III fragment containing the *sacA* gene, respectively. Both genes had previously been cloned from chromosomal DNA of *L. lactis* subsp. *lactis* NIZO R5 (Peter Rauch, NIZO, The Netherlands). The probes were labeled either radioactively via nick translation using alpha-^{32}P-ATP or random primed using digoxigenin-labeled dUTP.

DNA-transfer was performed as described by Sambrook *et al.* (1989) with lOxSSC (lxSSC is 0.15 M NaCl plus 0.015 M sodium citrate) as transfer buffer and a Gene Screen Plus nylon membrane (Dupont, NEN Research Products, Boston, Mass., USA). Radioactive hybridizations were carried out as described by Sambrook *et al.* (1989). The nonradioactive DNA Labeling and Detection Kit (Boehringer Mannheim) based on random-primed DNA synthesis and digoxigenin-labeled dUTP was used according to the instructions of the manufacturer.

Solid-surface conjugation and characterization of transconjugants

To prepare cells for conjugation, 1 % inoculations into fresh M17 broth from an overnight culture were made. Donors were grown in SM17 broth and recipients were grown in GM17 broth containing the appropriate antibiotic(s). Donor and recipient strains were incubated for about 4 h at 30°C (to reach an OD at 600 nm of about 0.3-0.6), collected by centrifugation, washed with sterile M17 broth to remove residual nisin, and finally resuspended in the original volume of fresh M17 medium (10 x diluted). Donors and recipients were then mixed 1:2 to a final volume of 0.2 ml. Alpha-chymotrypsin (stock solution is 16 mg/ml in 0.1 N HCl) was occasionally added to the conjugal mixture to degrade residual nisin.

This mixture was spread on the surface of glucose/milk agar plates (5% non-fat dry milk, 1 % glucose, 1.5 % agar) and incubated at 30°C for 4 h. This short-term plate mating was necessary to overcome killing of the recipient strains by nisin produced from the donor strain. The conjugation mixture was subsequently harvested with 1 ml of 0.85 % saline and then 0.1 ml volumes were plated onto selective media (SM17-agar for donor cells, GM17-agar with 0.1 mg/ml streptomycin and/or 0.05 mg/ml rifampicin for recipient cells and SM17-agar with 0.05 mg/ml streptomycin and/or 0.05 mg/ml rifampicin for transconjugants). Donor and recipient controls received the respective cell ratio blended with saline. Plates were finally examined for large yellow transconjugant colonies. Transfer frequencies were expressed as the number of transconjugants per donor CFU, and the values reported were the average of at least four separate experiments.

Transconjugants, selected by chromosomal resistance and sucrose fermentation, were further characterized phenotypically as to their bacteriophage sensitivity, their capacity to ferment sucrose and lactose and to produce nisin. Cleared lysates for plasmid DNA examination were obtained by the lysis procedure of Anderson and McKay (1983) and were examined by 1.2 % agarose gel electrophoresis. Nisin production by transconjugants was verified by an agar overlay technique.

RESULTS AND DISCUSSION

Phenotypic linkage between sucrose fermentation capacity, nisin production ability and nisin immunity

Twenty different *L. lactis* subsp. *lactis* strains were compared as to their carbohydrate fermentation capacity, nisin production ability and nisin immunity. The results are presented in Table 1. All strains tested fermented glucose. The ability to ferment sucrose and lactose varied with the strain tested. Indeed, whereas glucose is a stable property of lactic acid bacteria, lactose fermentation capacity is encoded by a plasmid (McKay, 1983). The *L. lactis* subsp. *lactis* MG 1363, MG 1390, MG 1614 and NZDRI 584 strains are plasmid-free and consequently lac$^-$. For all strains listed in table 1, except for *L. lactis* subsp. *lactis* NIZO P2A, the capacity to ferment sucrose was always linked with the ability to produce nisin. No consistent linkage could be detected between lactose fermentation capacity and nisin production ability. Nisin producing (and sucrose fermenting) strains were of course also immune to nisin (see table 1), otherwise

Table 1. Phenotypic and genotypic characterization of some nisin-producing and non-producing *Lactococcus lactis* subsp. *lactis* strains

Lactococcus lactis subsp. *lactis*	Phenotype Glu[1]	Suc[1]	Lac[1]	Nip[2]	Nim[3]	Genotype *nisA*
ATCC 2054	+	+	+	+++	++++	1
ATCC 11454	+	+	+	+	++++	1
6F3	+	+	-	+	++++	1
MG 1363	+	-	-	-	-	0
MG 1390	+	-	-	-	-	0
MG 1614	+	-	-	-	-	0
NCFB 606	+	-	+	-	+	0
NCFB 615	+	-	-	-	+	0
NCFB 894	+	+	+	+++	++++	1
NCFB 912	+	+	-	+	++++	1
NCFB 917	+	+	+	+	+++	1
NCFB 1404	+	+	+	++	++++	1
NCFB 2754	+	+	-	+	++	1
NIZO 22186	+	+	+	+++	++++	1
NIZO R5	+	+	+	+	++++	1
NIZO R5L0	+	+	-	+	++++	1
NIZO R512	+	+	-	+	+++	1
NIZO P2A	+	+	+	-	+	1
NIZO TS165.6	+	+	+	+	+++	2
NZDRI 584	+	-	-	-	-	0

1) fermentation (+) of glucose (Glu), sucrose (Suc), lactose (Lac)
2) nisin production (Nip): > 1500 IU/ml (+++), 500-1500 IU/ml (++), < 500 IU/ml (+), 0 IU/ml (-)
3) nisin immunity (Nim): > 2000 IU/ml (++++), 1000-2000 IU/ml (+++), 100-1000 IU/ml (++), 10-100 IU/ml (+), < 10 IU/ml (-).

they should not survive. The linkage of nisin production (and nisin immunity) with sucrose metabolism was observed as early as 1951, when it was reported that 12 nisin-producing strains were also able to ferment sucrose (Hirsch & Grinsted, 1951). Later on, many authors have suggested a genetic linkage between those three metabolic properties. With the exception of the Suc$^+$, Nip$^-$, Nim$^-$ *L. lactis* subsp. *lactis* NIZO P2A mutant strain, other natural Suc$^-$, Nip$^+$, Nim$^+$ or Suc$^+$, Nip$^-$, Nim$^-$ *L. lactis* subsp. *lactis* strains were never found. However, nisin resistance can occur separately. For example, the plasmid pNP40 (60 kb) of *L. lactis* subsp. *lactis* var. *diacetylactis* DRC3, codes for nisin resistance and a temperature sensitive phage defense mechanism, but not for nisin production or sucrose fermentation (McKay & Baldwin, 1984). Plasmid pTR1040 (64 kb) from *L. lactis* subsp. *lactis* ME2 links lactose fermentation capacity with nisin resistance (Klaenhammer & Sanozky, 1985) and the plasmid pSFOl (46 kb) from *L. lactis* subsp. *lactis* 10.084 codes for nisin resistance, lactose fermentation and proteinase activity (Von Wright *et al.*, 1990). These results indicate that the three metabolic properties sucrose fermentation, nisin production and nisin immunity are genetically linked within the nisin-producing *L. lactis* subsp. *lactis* strains.

Genetic location of the *suc, nip, nim* gene block

Localization of the *nisA* gene.

Chromosomal DNA and plasmid DNA were isolated from different *L. lactis* subsp. *lactis* strains: the nisin-producing *L. lactis* subsp. *lactis* NIZO 22186, NIZO R5, NIZO R512 and TS165.6 strains, and the nisin-non-producers *L. lactis* subsp. *lactis* NIZO P2A and MG1614. Chromosomal DNA was then digested with the restriction endonuclease *Eco*RI. The resulting fragments were separated by agarose gel electrophoresis, blotted and probed with a natural prenisin gene, i.e. a radioactively labelled 0.83 kb *Hinc*II/ *Sau*3A-fragment of the *L. lactis* subsp. *lactis* NIZO R5 chromosomal DNA containing the *nisA*-gene. Hybridization was carried out at 65°C. The nisin blot was washed once in 6xSSC at room temperature for 30 min and a second time in 2xSSC at 40°C for 20 min. The autoradiogram is shown in Figure 1.

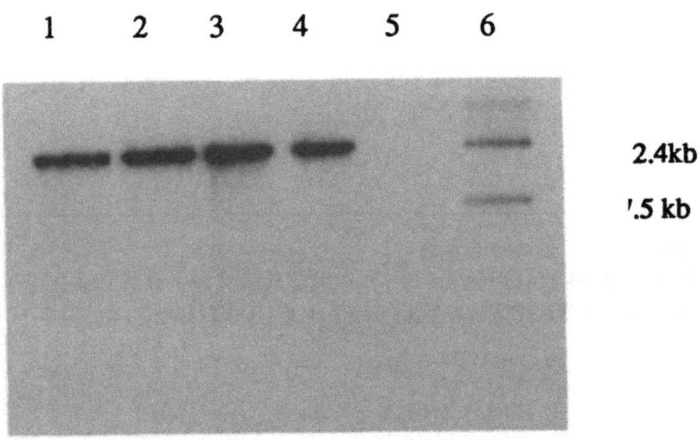

Figure 1. Hybridization pattern with the radioactively labelled prenisin probe (*nisA*) of *Eco*RI-digested chromosomal DNA of *Lactococcus lactis* subsp. *lactis* NIZO 22186 (lane 1), NIZO R512 (lane 2), NIZO P2A (lane 3), NIZO R5 (lane 4), MG 1614 (lane 5) and NIZO TS 165.6 (lane 6).

The natural prenisin gene probe hybridized with a 12.4 kb *Eco*RI restriction fragment of the chromosomal DNA of all nisin producers tested. The plasmid DNA did not give any hybridization signal (data not shown). Non-producing strains as MG 1614 did not reveal any hybridizing band, indicating that those strains lack the precursor nisin gene and consequently, are not able to synthesize the nisin precursor. The chromosomal DNA of the *L. lactis* subsp. *lactis* NIZO P2A mutant strain, a sucrose fermenter but a nisin-non-producer, yet gave a hybridization signal with the prenisin probe. Consequently, this strain did contain the *nisA* gene in its chromosomal DNA; its Suc$^+$, Nip$^-$, Nim$^+$ phenotype thus reflects a *suc*$^+$, *nip*$^+$, *nim*$^+$ genotype. Finally, in case of the transconjugant *L. lactis* subsp. *lactis* TS 165.6, a strain obtained via conjugation of the Suc$^+$, Nip+, Nim$^+$ phenotype from the donor *L. lactis* subsp. *lactis* NIZO R5 to the *L. lactis* subsp. *lactis* MG 1614 recipient strain, a second but smaller hybridizing chromosomal DNA-fragment of 7.5 kb could be detected besides the 12.0 kb *Eco*RI-fragment. Plasmid DNA could not be detected at all. These results together clearly indicate that the *nisA* gene is chromosomally encoded. This is in contrast to earlier reports, which indicated that genes for nisin production were located on plasmids (Tsai & Sandine, 1987; Kaletta & Entian, 1989). Only very recently, Steen *et al.* (1991) and Broadbent & Kondo (1991) have also reported that the *nisA* gene is associated with chromosomal DNA. Steen *et al.* (1991) even reported that the nisin gene is part of a polycistronic operon with a size greater than 8.5 kb.

Localization of the *sacA*-gene

The autoradiogram of *Eco*RI-digested chromosomal DNA of the different nisin-producing and non-producing *L. lactis* subsp. *lactis* strains mentioned above, probed with the natural *sacA* gene, a radioactively labelled 1.5 kb *Eco*RV/*Hin*dIII fragment of the *L. lactis* subsp. *lactis* NIZO R5 chromosomal DNA containing the *sacA* gene, is presented in Figure 2.

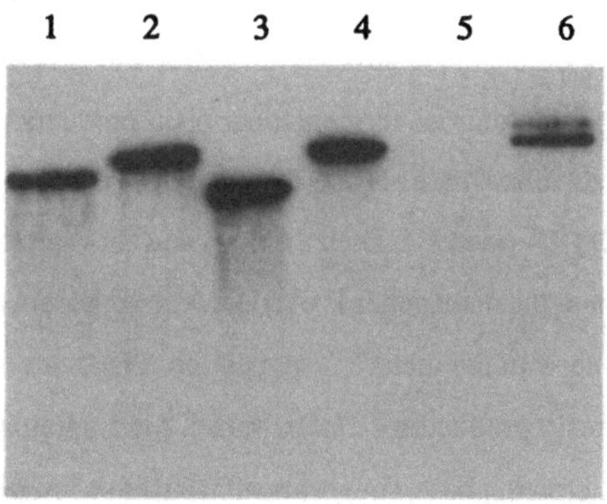

Figure 2. Hybridization pattern with the radioactively labelled sucrose-6-phosphate hydrolase probe (*sacA*) of *Eco*RI-digested chromosomal DNA of *Lactococcus lactis* subsp. *lactis* NIZO 22186 (lane 1), NIZO R512 (lane 2), NIZO P2A (lane 3), NIZO R5 (lane 4), MG 1614 (lane 5) and NIZO TS 165.6 (lane 6)

Hybridization was performed at 65°C. The sucrose blot was washed in 6xSSC at 65°C for 20 min and in 2xSSC at 50°C for 50 min. The sucrose probe hybridized to a DNA-fragment of 17.5 kb in case of *L. lactis* subsp. *lactis* R5, *L. lactis* subsp. *lactis* R512 and *L. lactis* subsp. *lactis* TS 165.6 and of 9.4 kb in case of *L. lactis* subsp. *lactis* NIZO 22186 and *L. lactis* subsp. *lactis* NIZO P2A. *L. lactis* subsp. *lactis* MG1614 (Suc$^-$, Nip$^-$) did not give any hybridization signal with the sucrose probe. No hybridization signal could be detected with plasmid-DNA. Consequently, the sucrose-6-phosphate hydrolase gene is localized on the chromosomal DNA as well. The fact that both genes, the *nisA* and *sacA* genes, hybridize to chromosomal DNA-fragments of a comparable size, underlines their genetic linkage. These experiments clearly confirm the results of Rauch *et al.* (1990) and Rauch & de Vos (1990).

Genotypic linkage between sucrose fermentation, nisin production and nisin immunity

To confirm the genetic linkage between sucrose fermentation capacity, nisin production ability and nisin immunity, chromosomal DNA of the 20 different nisin-producing and non-producing *L. lactis* subsp. *lactis* strains mentioned above was isolated and subsequently digested with the restriction endonuclease *Eco*RI. The restriction fragments were separated on a 0.7% agarose gel, blotted and hybridized to a digoxigenin-labelled prenisine probe, followed by immunodetection. The results are presented in Table 1. Hybridizations with the *sacA* gene were not repeated anymore, because a strain fermenting sucrose evidently has the sucrose-6-phosphate hydrolase gene, since *L. lactis* subsp. *lactis* strains have to hydrolyse sucrose-6-phosphate into glucose-6-phosphate and fructose by means of a sucrose-6-phosphate hydrolase to be able to ferment sucrose. Table 1 clearly shows that all sucrose fermenting *L. lactis* subsp. *lactis* strains - which contain the sucrose-6-phosphate-hydrolase gene (*sacA*-gene) in their genome - possess the chromosomally localized prenisine gene (*nisA* gene) as well. Even the chromosomal DNA of *L. lactis* subsp. *lactis* 6F3 revealed a hybridization signal with the prenisin probe. However, Kaletta and Entian (1989) linked nisin production of this strain to a plasmid. From our results, it is clear that the prenisin structural gene is located on chromosomal DNA. Our experiments also strongly underline the genetic linkage between the industrially important metabolic properties sucrose fermentation, nisin production and nisin immunity, not necessarily phenotypically (*L. lactis* subsp. *lactis* NIZO P2A is indeed Suc$^+$ but Nip$^-$) but always genotypically (*L. lactis* subsp. *lactis* NIZO P2A possesses both the *sacA* and *nisA*-genes in its chromosomal DNA). The ultimate result of expression of the nisin genes is of course the appearance of the antibiotic activity in the fermentation broth associated with the mature nisin peptide. The latter is produced mainly in the late logarithmic growth phase of nisin producing *L. lactis* subsp. *lactis* strains, although the nisin gene is expressed early in the exponential growth phase, suggesting that it takes a longer time before sufficient quantities of the required processing enzymes have been synthesized. However, production of nisin follows primary metabolite kinetics (De Vuyst & Vandamme, 1991, 1992).

Conjugative transposition of the *suc, nip, nim* gene block

The results of conjugal matings of the Suc$^+$, Nip$^+$, Nim$^+$ *L. lactis* subsp. *lactis* NIZO R5 and NZDRI

1402 donor strains with respectively *L. lactis* subsp. *lactis* MG 1614 and NZDRI 584 as recipients (Suc$^-$, Nip$^-$, Nim$^-$), are summarized in Table 2. Transfer of sucrose fermenting ability was selected on SM17-agar, supplemented with bromocresol purple as pH indicator, containing streptomycin-rifampicin and streptomycin as counterselective antibiotics, respectively. The transfer frequency of the Suc$^+$, Nip$^+$ phenotype for both recipients was low, approximately 10^{-8} transconjugants per donor cell. Levels of resistant mutants were however substantially lower than resistant colonies from conjugal mating. This low transfer frequency can be explained by the fact that Suc$^+$, Nip$^+$ transfer is dependent on the type of donor and/or acceptor strains used (De Vuyst, 1990). Broadbent andKondo (1991) indicated furthermore that the applied method does not only influence the frequency of transfer, but also the transfer itself. When assayed, all Suc$^+$ transconjugants were all found to be Nip$^+$. Their authenticity was confirmed by their sensitivity to their appropriate bacteriophages and by their plasmid profiles. The transconjugants generated possessed the recipient phenotype for phage sensitivity. Plasmid analysis of transconjugants showed that all Suc$^+$, Nip$^+$ isolates screened contained no additional plasmid DNA. Because the resulting transconjugants were not only Suc$^+$, Nip$^+$ but also Nim$^+$, the transferred DNA-element must also encode nisin immunity. This is in agreement with the hypothesis that those three metabolic properties are controlled by one common gene block.

Table 2. Conjugative transfer of the sucrose and nisin genes between *Lactococcus lactis* subsp. *lactis* strains

Lactococcus lactis subsp. *lactis* donor (phenotype)	*Lactococcus lactis* subsp. *lactis* acceptor (phenotype)	selected phenotype	conjugation frequency (per donor)	selected transcon-jugants
NIZO R5 (Suc$^+$ Nip$^+$ Nim$^+$ Str$^-$ Rif$^-$)	MG 1614 (Suc$^-$ Nip$^-$ Nim$^-$ Str$^+$ Rif$^+$)	Suc$^+$ Str$^+$ Rif$^+$	0.5×10^{-8}	TC 165.1
NIZO R5 (Suc$^+$ Nip$^+$ Nim$^+$ Str$^-$ Rif$^-$)	MG 1614 (Suc$^-$ Nip$^-$ Nim$^-$ Str$^+$ Rif$^+$)	Suc$^+$ Str$^+$ Rif$^+$ (+ 0.005 ml alfa-chymot.)	0.6×10^{-8}	TC 165.2 TC 165.3 TC 165.4
NZDRI 1402 (Suc$^+$ Nip$^+$ Nim$^+$ Str$^+$)	NZDRI 584 (Suc$^-$ Nip$^-$ Nim$^-$ Str$^-$)	Suc$^+$ Str$^+$	5.0×10^{-8}	TC 5802.1

Subsequently, DNA/DNA hybridizations were performed to clearly demonstrate the transfer of the nisin precursor gene (*nisA*) during conjugation. Restriction digests with the enzyme *Eco*RI were performed on chromosomal DNA isolated from the donors, the recipients and the Suc$^+$, Nip$^+$ transconjugants. The DNA fragments were separated in an agarose gel and then transferred to a hybridization membrane and probed with the digoxigenin-labeled prenisin probe to detect the *nisA*-gene. The results are presented in Table 3.

Table 3. Phenotypic and genotypic characterization of *Lactococcus lactis* subsp. *lactis* donor, acceptor and transconjugant strains

Lactococcus lactis subsp. *lactis* strain	nisin production (IU nisin/ ml medium)	nisin immunity (IU nisin/ ml medium)*	number of *nisA* genes	hybridizing DNA-fragment (kb)
NIZO R5	277	2000	1	12.0
NZDRI 1402	nd	nd	1	12.0
MG 1614	0	0	0	-
NZDRI 584	0	0	0	-
TC 165.1	175	2000	1	12.0
TC 165.2	222	2000	2	12.0, 7.5
TC 165.3	288	2000	2	12.0, 7.5
TC 165.4	328	2000	2	12.0, 7.5
TS 165.6	134	2000	2	12.0, 7.5
TC 5802.1	nd	nd	1	12.0

* 2000 IU/ml was the highest tested concentration because of solubility. nd = not detected

Autoradiography (figure 3) showed that the probe hybridized to chromosomal DNA bands of the Suc$^+$, Nip$^+$ donor and transconjugants, but did not hybridize to Suc$^-$, Nip$^-$ recipient DNA. Because those genes were indeed absent in the original recipient strain, but were present in the donor strain (Figure 1), the transconjugants thus gained these sequences as a result of the conjugation event in which the nisin/sucrose gene block was transferred. Hybridization to plasmid DNA was not detected at all. Although, some investigators still locate the sucrose and nisin genes on plasmid DNA, we conclude that plasmid DNA is even not involved in the transfer of the *suc,nip,nim* gene block. Moreover, since these genes could be transferred to plasmid free recipient strains (such as *L. lactis* subsp. *lactis* MG 1614 and NZDRI 584) and were always found in the chromosomal DNA after transfer, it is clear that stable chromosome integration took place after transfer. Our hybridization data indicated furthermore that at least 20 kb of DNA was transferred with the sucrose and nisin genes. Very recently, it has indeed been shown that the sucrose/nisin gene block is part of a conjugative transposon of about 70 kb (Dodd *et al.*, 1990; Rauch *et al.*, 1990; Rauch

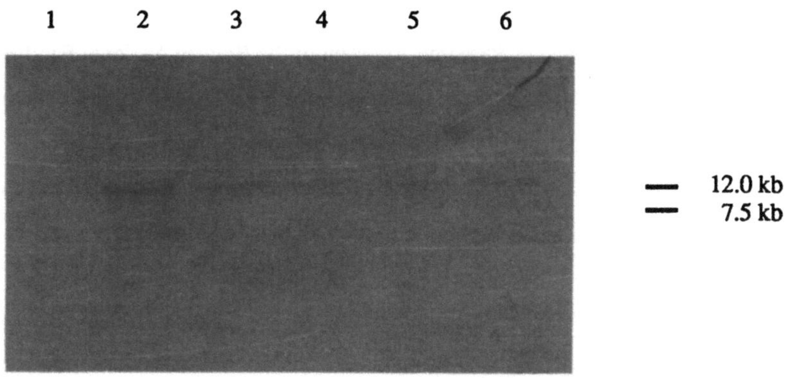

Figure 3. Hybridization pattern with the digoxigenin-labelled prenisin probe (*nisA*) of *Eco*RI-digested chromosomal DNA of *Lactococcus lactis* subsp. *lactis* MG 1614 (lane 1), NIZO R5 (lane 2), TC 165.1 (lane 3), TC 165.2 (lane 4), TC 165.3 (lane 5) and TC 165.4 (lane 6).

& De Vos, 1990). The *nisA* gene is located on one end of the conjugative element. Analysis of DNA regions located upstream of the nisin determinant revealed nucleotide sequences characteristic of an IS element (IS904), carrying a gene coding for a putative transposase. This element may play a role in mediating its transfer between strains (Dodd *et al.*, 1990). The probe always hybridized to a chromosomal DNA-fragment of 12.0 kb and sometimes to a second fragment of 7.5 kb in *Eco*RI digests. Rauch and De Vos (1990) suggested that there exists a hot-spot integration site on the recipient chromosomal DNA. Secondary integration of the conjugative transposon is responsable for the second copy in the lactococcal chromosomal DNA. Finally, the capacity of parental and transconjugant strains to produce nisin was evaluated. None of the *L. lactis* subsp. *lactis* recipients produced nisin. However, the transconjugants produced nisin. Production levels obtained after a 12 h period of fermentation ranged from 134 IU/ml by *L. lactis* subsp. *lactis* TS 165.6 to 328 IU/ml by *L. lactis* subsp. *lactis* TC 165.4 (table 3). These levels are approximately the same and even less than the nisin production level of the donor strain (277 IU/ml), even when two copies of the transposon are present in the recipient chromosome. One obvious reason for the variable and often less nisin production by the transconjugants may be the limiting maturation enzymology in these strains. It is possible that the gene block is unable to encode all genes for enzymes involved in modification of precursor nisin. There is no evidence to suggest that these genes are linked to the nisin structural gene and the possibility exists that this maturation function is genetically separate, perhaps fulfilling other functions in the cell.

ACKNOWLEDGEMENTS

Dr. ir. L. De Vuyst is a Senior Research Assistant of the Belgian National Fund For Scientific Research (NFWO). This research was partially financed by the NFWO and by the Research Fund of the University of Gent (UG). He also thanks Prof. Dr. Willem de Vos, Dr. John Mulders and ir. Peter Rauch (NIZO, The Netherlands), as well as Dr Graham Davey (DRI, New Zealand) for providing mutant strains and DNA probes, for the experimental support with the radioactive hybridizations and for the helpful discussions

REFERENCES

Anderson DG, McKay LL (1983) Simple and rapid method for isolating large plasmid DNA from lactic streptococci. Appl Environ Microbiol 46:549-552

Broadbent JR, Kondo JK (1991) Genetic construction of nisin-producing *Lactococcus lactis* subsp. *cremoris* and analysis of a rapid method for conjugation. Appl Environ Microbiol 57:517-524.

Buchman GW, Banerjee S, Hansen JN (1988) Structure, expression, and evolution of a gene encoding the precursor of nisin, a small protein antibiotic. J Biol Chem 263:16260-16266

Daeschel MA (1989) Antimicrobial substances from lactic acid bacteria for use as food preservatives. Food Technol 164-167

Davey GP, Pearce LE (1982) Production of diplococcin by *Streptococcus cremoris* and its transfer to nonproducing group N streptococci. In: Schlessinger D (ed) Microbiology-1982. American Society for Microbiology, Washington DC, 221-224

Delves-Broughton J (1990) Nisin and its application as a food preservative. J Soc Dairy Technol 43:73-76

De Vuyst L (1990). Biosynthesis, Fermentation and Genetics of the *Lactococcus lactis* ssp. *lactis* Lantibiotic Nisin. Ph.D-thesis, University of Gent

De Vuyst L, De Poorter G, Vandamme EJ (1990). Metabolic control of nisin biosynthesis in *Lactococcus lactis* subsp. *lactis*. In: Pak-Lam Yu (ed) Fermentation Technologies, Industrial Applications. Elsevier Applied Science, London New York, 166-172

De Vuyst L, Joris K, Beel C, Vandamme EJ (1988). Physiological Characterization of the Nisin Fermentation Process. In: Breteler H, Van Lelyveld PH, Luyben KChAM (eds) Proceedings 2nd Netherlands Biotechnology Congress. Netherlands Biotechnology Society, The Netherlands, 436-442

De Vuyst L, Vandamme EJ (1991) Microbial manipulation of nisin biosynthesis and fermentation. In: Jung G, Sahl HG (eds) Nisin and novel lantibiotics. ESCOM Science Publishers, Leiden. p397-409

De Vuyst L, Vandamme EJ (1992) Influence of the carbon source on nisin production in *Lactococcus lactis* subsp. *lactis* batch fermentations. J Gen Microbiol 138: in press

Dodd HM, Horn N, Gasson MJ (1990) Analysis of the genetic determinant for production of the peptide antibiotic nisin. J Gen Microbiol 136:555-566

Elliker PR, Anderson A, Hannessen GH (1956) An agar culture medium for lactic acid streptococci and lactobacilli. J Dairy Sc 39:1611-1612

Fuchs PG, Zajdel J, Dobrzanski WT (1975) Possible plasmid nature of the determinant for production of the antibiotic nisin in some strains of *Streptococcus lactis*. J Gen Microbiol 88:189-192
Gasson MJ (1984) Transfer of sucrose fermenting ability, nisin resistance and nisin production into *Streptococcus lactis* 712. FEMS Microbiol Lett 21:7-10
Gonzalez CF, Kunka BS (1985) Transfer of sucrose-fermenting ability and nisin production phenotype among lactic streptococci. Appl Environ Microbiol 49:627-633
Harris LJ, Fleming HP, Klaenhammer TR (1990). Characterization of two nisin-producing *Lactococcus lactis* strains isolated from a commercial vegetable fermentation. Abstract Book 3rd International Conference on Streptococcal Genetics, 23, A/47
Hirsch A, Grinsted E (1951) The differentiation of the lactic streptococci and their antibiotics. J Dairy Res 18:198-204
Hurst A (1981) Nisin. Adv Appl Microbiol 27:85-123
Jung G, Sahl HG (eds) (1991) Nisin and novel lantibiotics. ESCOM Science Publishers, Leiden
Kaletta C, Entian KD (1989) Nisin, a peptide antibiotic: cloning and sequencing of the *nisA* gene and post-translational processing of its peptide product. J Bacteriol 171:1597-1601
Klaenhammer TR, Sanozky RB (1985) Conjugal transfer from *Streptococcus lactis* ME2 of plasmids encoding phage resistance, nisin resistance and lactose-fermenting ability: evidence for a high-frequency conjugative plasmid responsible for abortive infection of virulent bacteriophage. J Gen Microbiol 131:1531-1541
Kozak W, Rajchert-Trzpil M, Dobrzanski WT (1974). The effect of proflavin, ethidium bromide and an elevated temperature on the appearance of nisin-negative clones in nisin-producing strains of *Streptococcus lactis*. J Gen Microbiol 83:295-302
Leblanc DJ, Crow VL, Lee LN (1980) Plasmid mediated carbohydrate catabolic enzymes among strains of *Streptococcus lactis*. In: Stuttard C, Rozee KR (eds) Plasmids and transposons: environmental effects and maintenance mechanisms, Academic Press, New York, p 31-41
McKay LL (1983) Functional properties of plasmids in lactic streptococci. Antonie van Leeuwenhoek 49:259-274
McKay LL, Baldwin KA (1984) Conjugative 40-megadalton plasmid in *Streptococcus lactis* subsp. *diacetylactis* DRC3 is associated with resistance to nisin and bacteriophage C2. Appl Environ Microbiol 47:68-74
Rauch PJG, Beerthuyzen MM, de Vos WM (1990) Nucleotide sequence of IS1904 from *Lactococcus lactis* subsp. *lactis* strain NIZO R5. Nuc Acids Res 18:4253-4254
Rauch PJG, de Vos WM (1990) Molecular analysis of the *Lactococcus lactis* nisin-sucrose conjugative transposon. Abstract book 3rd International Conference on Streptococcal Genetics, 23, A/46
Rayman K, Hurst A (1984) Nisin: properties, biosynthesis and fermentation. In: Vandamme EJ (ed) Biotechnology of industrial antibiotics, Marcel Dekker Inc, New York, p 607-628
Sambrook J, Fritsch EF, Maniatis T (eds) (1989) Molecular cloning: a laboratory manual. Cold Spring Harbor Laboratory Press, New York
Steele JL, McKay LL (1986) Partial characterization of the genetic basis for sucrose metabolism and nisin production in *Streptococcus lactis*. Appl Environ Microbiol 51:57-64
Steen MT, Chung YJ, Hansen JN (1991) Characterization of the nisin gene as part of a polycistronic operon in the chromosome of *Lactococcus lactis* ATCC 11454. Appl Environ Microbiol 57:1181-1188
Terzaghi BE, Sandine WE (1975) Improved medium for lactic streptococci and their bacteriophages. Appl Microbiol 29:807-813
Tramer J, Fowler GG (1964) Estimation of nisin in foods. J Sc Food Agr 15:522-528
Tsai HJ, Sandine WE (1987) Conjugal transfer of nisin plasmid genes from *Streptococcus lactis* 7962 to *Leuconostoc dextranicum* 181. Appl Environ Microbiol 53:352-357
Vandamme EJ (ed) (1984) Biotechnology of industrial antibiotics. Marcel Dekker Inc, New York.
Von Wright A, Wessels S, Tynkkynen S, Saarela M (1990) Isolation of a replication region of a large lactococcal plasmid and use in cloning of a nisin resistance determinant. Appl Environ Microbiol 56:2029-2035

EXPRESSION OF NISIN IN *BACILLUS SUBTILIS*

Helena Rintala, Lars Paulin, Tytti Graeffe, Tiina Immonen and Per E.J. Saris
The Institute of Biotechnology
Valiomotie 7D
00380 Helsinki
Finland

SUMMARY

Bacillus subtilis ATCC6633 expresses subtilin, a lantibiotic very homologous to nisin, a lactococcal lantibiotic used as a food preservative. We inserted the nisin gene, with a subtilin-nisin leader fusion, into the subtilin operon of a subtilin producer, which we had previously made nisin resistant. The nisin gene was transcribed from the subtilin promoter. The nisin precursor was modified and secreted by the subtilin maturation machinery and could be detected in the growth medium of bacterial cultures of the integrant.

INTRODUCTION

Of the lantiobiotics described, the only one used as a food additive is nisin, expressed by *Lactococcus lactis* subsp. *lactis* (Hurst, 1981). The preservative effect of nisin is due to its action on gram-positive bacteria, a group of bacteria including *Clostridium botulinum* and *Listeria monocytogenes*. In its pure form nisin is very expensive and this inhibits the increase of the usage of nisin, a "natural" preservative which could be used to safely decrease the amount of nitrate added to food. As *lactococcus lactis* is normally fermented in anaerobic conditions the achieved cell densities are low, resulting in low levels of nisin.

We were interested in evaluating the possibility of producing nisin in a bacteria, for which high cell density growth conditions are available. For this purpose we chose *Bacillus subtilis* ATCC6633, which expresses subtilin (Banerjee *et al.*, 1988), a lantibiotic very homologous to nisin. Both peptide antibiotics contain unusual amino acids, such as lanthionine, β-methyllanthionine, dehydroalanine and dehydrobutyrine, which are formed post-translationally. Because of this structural similarity the nisin precursor might be modified by the subtilin maturation machinery. This would enable production of nisin in *B.subtilis* with high cell densities, which could result in high amounts of nisin being produced. We inserted the nisin gene into the subtilin operon in order to test if the subtilin maturation components could function with the nisin precursor as substrate.

NATO ASI Series, Vol. H 65
Bacteriocins, Microcins and Lantibiotics
Edited by R. James, C. Lazdunski and F. Pattus

MATERIALS AND METHODS

Bacteria, plasmids and media

Bacterial strains and plasmids are listed in table 1. *L. lactis* was cultured on M17-medium (Gasson, 1984) supplemented with 0.5 % (w/v) glucose instead of lactose. *Escherichia coli*, *B. subtilis* and *Micrococcus luteus* were grown on Luria broth. For the HPLC-analysis the bacteria were grown for 40 h on buffered M17 medium supplemented with 10% sucrose and 0.03% cysteine.

Table 1. The bacteria and plasmids

Bacteria/Plasmids	Description	Reference
E.coli DH5a	cloning host	Hanahan, 1983
L.lactis ATCC 11454	nisin producer	
M.luteus	indicator strain	Valio
B.subtilis ATCC6633	subtilin producer	
B.subtilis BRB 1	control strain	Palva, 1982
B.subtilis BRB 779	nisin resistant, subtilin producer	this study
B.subtilis BAC 4	nisin resistant, nisin producer	this study
Bluescript +KS	cloning vector for *E.coli*	Melton, 1984
pHP13	*E.coli*, *B.subtilis* shuttle vector	Haima, 1987
pKTH 1948	*cat*	Palva, I.
pKTH 1908	pE194 ts-replicon	Palva, I.
pKTH 2038	Bluescript, *nisA*, *cat*, *spaS* flanking seq.	this study
pLEB 1	pHP13, *nisA*, *cat*, *spaS* flanking seq	this study
pLEB 4	pKTH 1908, *nisA*, *cat*, *spaS* upstream seq.	this study

Valio = Finnish Co-operative Dairies' Association

DNA-techniques

Transformation of *E. coli* was according to Hanahan (1983). Electroporation of the *B. subtilis* BRB 779 strain was performed as described by Vehmaanperä (1989) with a Gene Pulser apparatus from Bio-Rad. Polymerase chain reaction from colonies was done according to Saris *et al.* (1990). The oligonucleotides used are listed in figure 1 with their hybridization area shown.

0396 GATTTTAACTTGGATTTGGTATCTGTTTCGAAGAAAGATTCAGGTGCATCACCACGCATTACACGTA TTTCGCTATTACACCCGGTTGTAAAACAGGAG

SalI

0397 ATGCAAGTCGACTTATTTGCTTACGTGAATACTACAATGACAAGTTGCTGTTTTCATGTTACAACCC ATCAGAGCTCCTGTTTTACAACCGGGTGT

ClaI

0402 TCTTAGATCGATGATTTCAACTTGGATTTGGTATCTGT

EcoRI

0406 TCTGGAATTCCTAGAGTGATCAATCTCAT

EcoRI

0407 ATCTGAGAATTCCGGACAGGAGTATTTTAAGGA

XbaI

0408 TGTGTATCTAGAGCATTGCAAACTTG

T3 ATTAACCCTCACTAAAG

T7 AATACGACTCACTATAG

XbaI

SK TCTAGAACTAGTGGATC

EcoRI

NIS-1 TTTTCTAGAGAATTCATGGAAGGGGACGAAGC

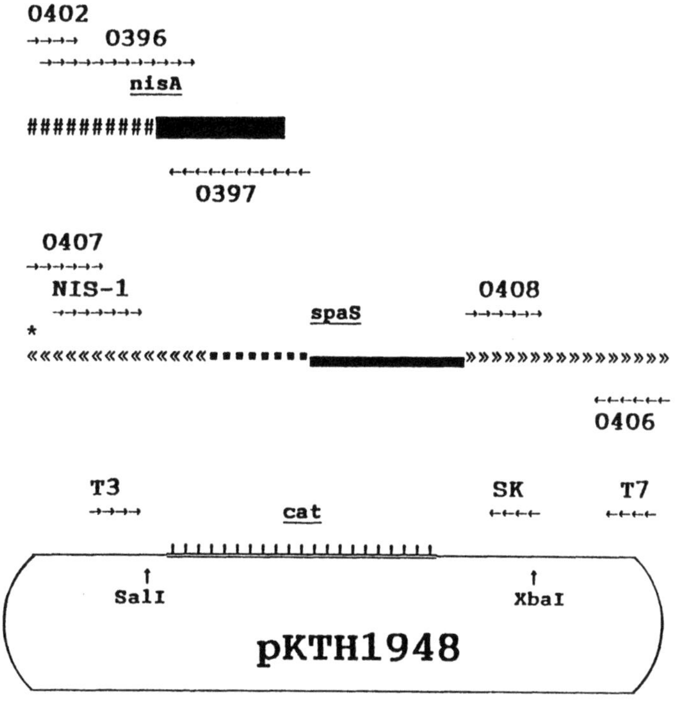

Fig. 1. The oligonucleotide primers and their hybridization areas. The underlined sequences represent the hybridizing part in the PCR, except for 0396 and 0397 were the underlined part represent tne complementary sequence of these two primers.

Microbial tests

Strains were tested for nisin or subtilin production by streaking them out or dropping 5 µl of culture supernatant on a Luria agar plate which had been spread with a suspension of *M. luteus*. The HPLC-fractions (1 ml) were dried and resuspended in 30 µl water (pH 2.5). Thereof 3 µl aliquots were spotted on plates as described above.

HPLC

For HPLC, 200 µl of cell-free growth culture was loaded on a K73 column. The column was washed for 10 min with 15 % acetonitrile followed by elution with an acetonitrile gradient raising from 15% to 60% in 30 min. 1 ml fractions were collected.

RESULTS AND DISCUSSION

Construction of a replacement vector.

We intended to change the subtilin production of *B. subtilis* ATCC6633 to nisin production by a gene replacement. For this purpose we constructed a replacement vector (figure 2). By PCR techniques with primers 0406, 0407 and 0408 we cloned the flanking sequences of *spaS*, the subtilin gene. These sequences were to provide homology for the double crossing-over in the replacement. The nisin gene was synthesized and the subtilin-nisin leader fusion was joined by PCR with primers 0402 and 0397. The subtilin-nisin leader fusion coded for seven amino acids from the beginning of subtilin leader peptide and the nisin leader peptide missing the first seven amino acids. This part was joined to the *spaS* upstream region by *Taq*I and *Cla*I restriction enzyme digestions (*Taq*I and *Cla*I form compatible ends), followed by ligation and amplification by PCR with primers 0407 and 0397. The *cat* gene from pKTH1948 was amplified by PCR, with primers T7 and T3, and joined to the *spaS* downstream sequence by *Xba*I digestion followed by ligation and amplification by PCR with primers T7 and 0406. The two fragments, one including the *spaS* upstream sequence with *nisA* and the other consisting of the selection marker (*cat*) and the *spaS* downstream sequence, were joined by their *Sal*I site and cloned as an *Eco*RI fragment in the *Eco*RI site of the Bluescript +KS vector, yielding pKTH2038.

A nisin resistant strain of *B.subtilis* ATCC6633 was isolated by natural selection and named BRB779. This strain was not competent for physiological transformation or protoplast transformation. By electroporation we could transform BRB779, but at low frequency (10^2 transformants/ µg plasmid). The

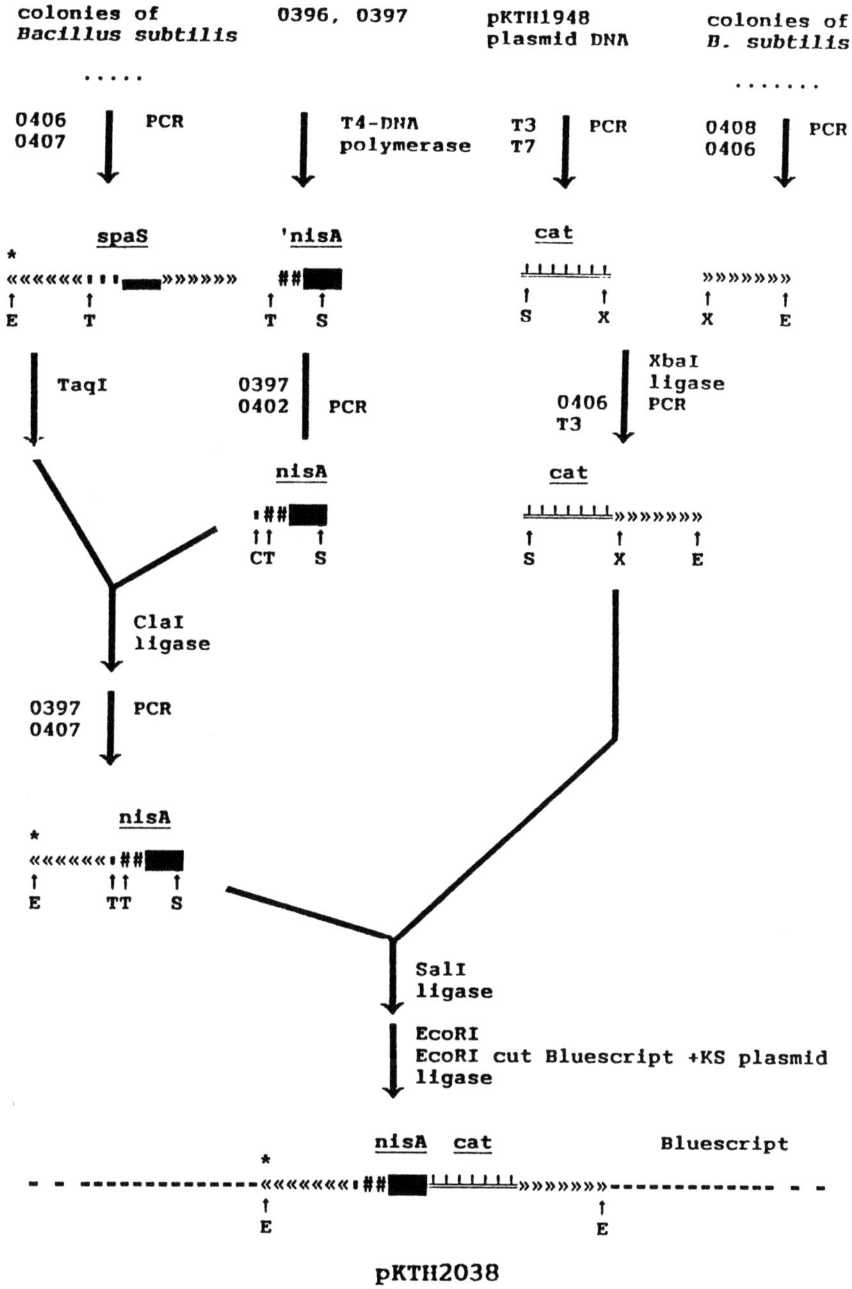

Fig. 2. Construction of the replacement vector. - -, the subtilin leader sequence; ▬, the structural gene of subtilin; ##, the nisin leader sequence; █, the structural gene of nisin; ⊥⊥, the *cat* gene; *, the putative *spaS* promoter; E, EcoRI; T, TaqI; S, SalI; X, XbaI; C, ClaI.

transformation frequency was too low to select directly for rare double crossing-over events, for which we would have been selecting with the replacement vector.

To test if the insert of pKTH2038 could be harbored in BRB779, we cloned the insert into the *E. coli*, *B. subtilis* shuttle vector pHP13 (figure 3), resulting in pLEBl. BRB779 could not be transformed by this plasmid, suggesting that expression of *nisA* from the plasmid could be lethal in BRB779.

Construction of an integration plasmid

Because of the described problems, we set up a new strategy. To avoid potential lethal expression of *nisA*, we cloned the insert from pLEBl, without the putative *spaS* promoter (Banerjee *et al.*, 1988) and the *spaS* downstream sequence, in the pKTH1908 plasmid (figure 3), which contains the pE194 termosensitive replicon. This was done by an PCR amplification of a specific part of the insert defined by the primers SK and NIS-1, followed by an *Eco*RI, *Xba*I digestion and ligation with the *Eco*RI, *Xba*I cut pKTH1908 plasmid, resulting in pLEB4.

Integration of pLEB4

Because of the thermosensitive replicon pLEB4 could be transformed into BRB779 at the permissive temperature. The transformants were then grown on plates at the restrictive temperature, which resulted in the integration of the plasmid. As pLEB4 did not contain the *spaS* downstream sequence, pLEB4 integrated into the chromosome of BRB779 by a single crossing-over using the *spaS* upstream homology. This strain was named BAC4.

Analysis of BAC4

The correct integration of pLEB4 was confirmed by PCR, with one oligonucleotide (SK) hybridizing in the plasmid sequence and the other (0407) hybridizing in the putative *spaS* promoter, using a colony of BAC4 as template. By correct integration, a 1580 kb fragment should be amplified as shown in figure 4.

The ability of BAC4 to produce any inhibitory substance was tested by streaking it on a Luria agar plate plated with *M. luteus*. BAC4 produced a substance that inhibited the growth of *M. luteus* (figure 5). Because the integration of pLEB4 was a single crossing over event, the *spaS* gene lost its promoter to the *nisA* gene, but remained in the chromosome. Therefore it was possible that the *spaS* gene could be transcribed in BAC4 by some readthrough from one of the plasmid promoters.

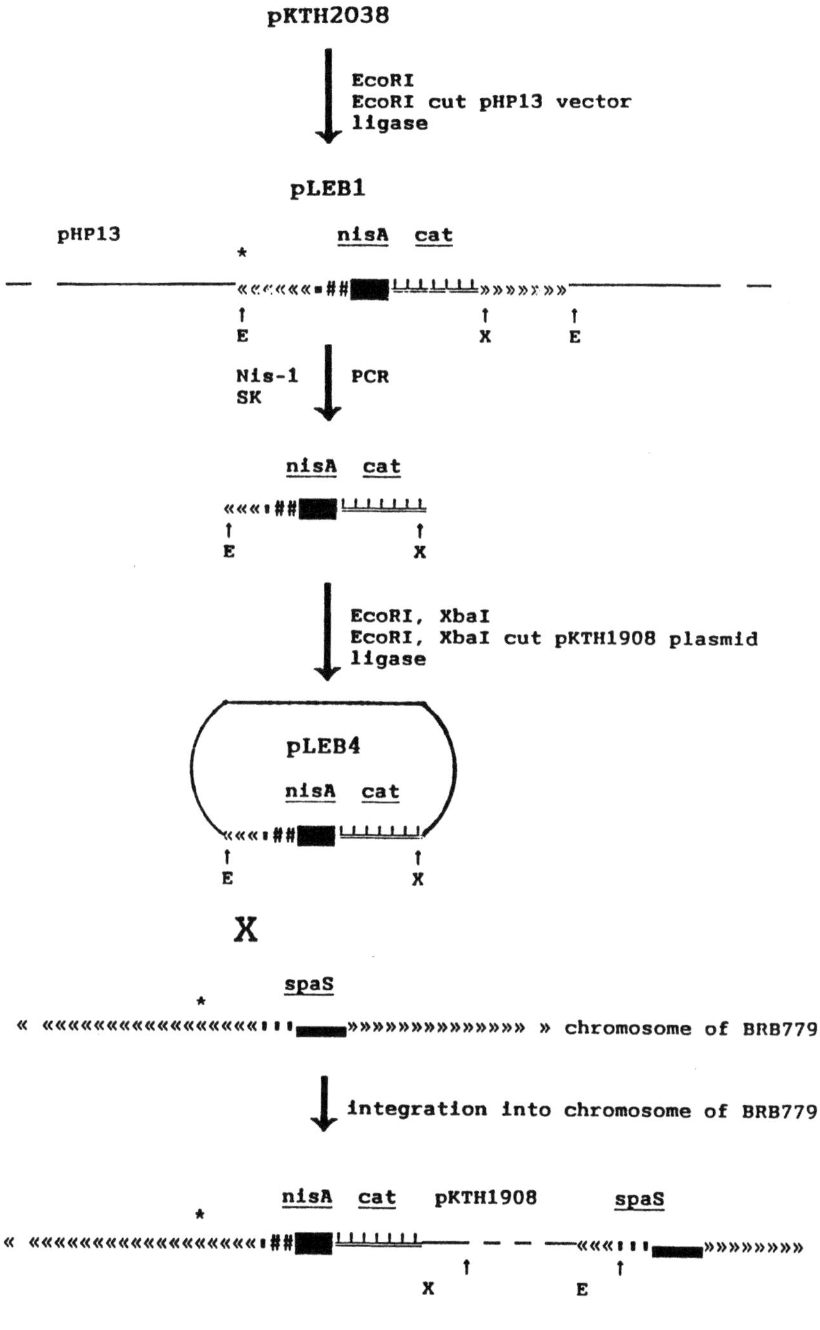

Fig. 3. **Construction of the termosensitive integrative plasmid and the integration event.** ▪##; the subtilin-nisin fusion leader sequence; █, the structural coding part of *nisA*; ⊥⊥, the *cat* gene; ▪▪, the subtilin leader sequence; ▬, the structural coding part of *spaS*, *, the putative *spaS* promoter; X, indication of the crossing over; --, plasmid sequence; »», chromosomal DNA of BRB779; E, EcoRI; X; XbaI.

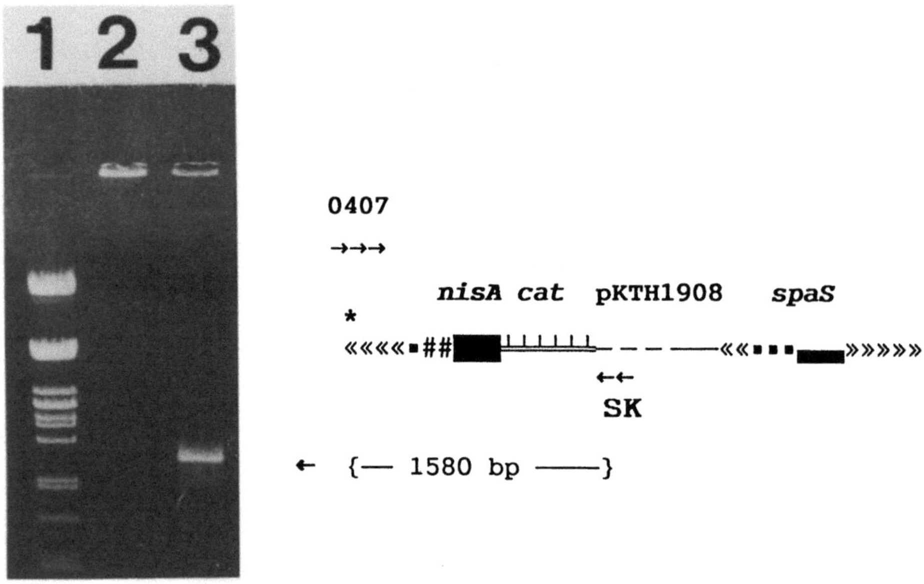

Figure 4. The location of *nisA* in BAC4. The PCR amplification products with primers 0407 and SK using colonies of BRB779 (lane 2) or BAC4 (lane 3) as templates. Lane l; lambda DNA digested with *Pst*I. The fragment (1580 bp), which should be amplified in a PCR, using primers 0407 and SK with BAC4 as template, if the integration of pLEB4 had occurred correctly, is indicated with an arrow.

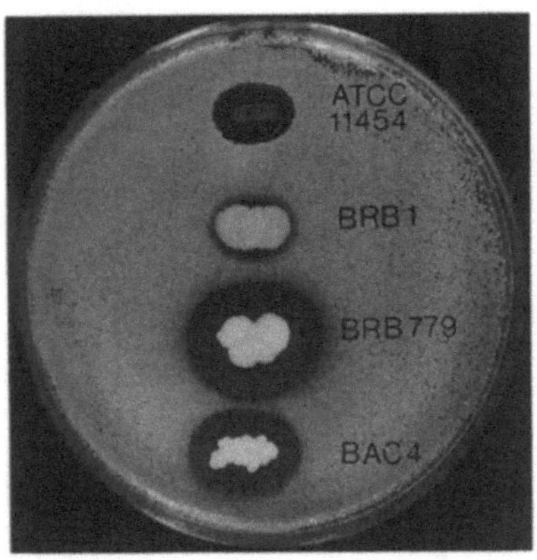

Figure 5. Inhibition zones with *M.luteus* as indicator.

To determine whether the inhibition zone of BAC4 was due to nisin or subtilin, an HPLC analysis from the culture supernatant was performed. The results are shown in table 2. BAC4 produced an inhibitory substance that eluted like nisin in the HPLC. The subtilin producer BRB779 did not produce any inhibitory substance eluting like nisin fraction. Some of the BAC4 inhibitory activity was also present in the fractions where the inhibitory substance of BRB779 eluted (table 2). This could represent subtilin, originating from the readthrough of a plasmid promoter. However, this was not supported by the result of an HPLC analysis of a replacement construct, with the *spaS* gene eliminated (results not shown). In this case an inhibitory substance was still present in the last fractions. Maybe this inhibitory substance represents incorrectly modified nisin.

Table 2. Inhibition of *Micrococcus luteus* by HPLC fractions

Growth medium of strain	HPLC fractions 10	11	12	13	14	15	16	17	18	19	20	21	22	23	24	25
L. lactis ATCC11454	-	-	-	-	+	++	+	-	-	-	-	-	-	-	-	-
B. subtilis BRB779	-	-	-	-	-	-	-	-	-	-	-	-	-	-	+	+
B. subtilis BAC4	-	-	-	-	+	++	+	-	-	-	-	-	-	-	+	+

+ = inhibition of *M.luteus*; - = no inhibition of *M.luteus*

We are now in the process of purifying the inhibitory substances produced by BAC4 and analyzing the modifications of nisin in order to verify if they are correct.

REFERENCES

Banerjee S, Hansen JN (1988) Structure and expression of a gene encoding the precursor of subtilin, a small protein antibiotic. J Biol Chem 263:9508-9514

Gasson M (1984) Transfer of sucrose fermenting ability, nisin resistance and nisin production into *Streptococcus lactis* 712. Fems Microbiol Lett 21:7-10

Haima P, Bron S, Venema G (1987) The effect of restriction on shotgun cloning and plasmid stability in *Bacillus subtilis* Marburg. Mol Gen Genet 209:335-342

Hanahan D, Meselson M (1983) Studies on transformation of *Escherichia coli* with plasmids. J Mol Biol 166: 557-580

Hurst A (1981) Nisin. Adv Appl Microbiol 27:85-123

Melton DA, Krieg PA, Rebagliati MR, Maniatis T, Zinn K, Green MR (1984) Efficient *in vitro* synthesis of biologically active RNA and RNA hybridization probes from plasmids containing a bacteriophage SP6 promoter. Nucleic Acids Res 12:7035-7056

Palva I (1982) Molecular cloning of α-amylase gene from *Bacillus amyloliquefaciens* and its expression in *B. subtilis*. Gene 19:81-87

Saris PEJ, Paulin LG, Uhlen M (1990) Direct amplification of DNA from colonies of *Bacillus subtilis* and *Escherichia coli* by the polymerase chain reaction. J Microbiol Methods 11:121-126

Vehmaanperä J (1989) Transformation of *Bacillus amyloliquefaciens* by electroporation. Fems Microbiol Lett 61:165-170

Tn*5301*, A LACTOCOCCAL TRANSPOSON ENCODING GENES FOR NISIN BIOSYNTHESIS

H.M. Dodd, N. Horn. S. Swindell and M.J. Gasson.
AFRC Institute of Food Research, Norwich Laboratory,
Colney Lane
Norwich, NR4 7UA
UK.

SUMMARY

The genes for nisin biosynthesis are encoded by the conjugative transposon Tn*5301*. The element is 70kb in size and is located in the chromosome of the nisin-producing strain FI5876. Sequence analysis demonstrated that the left and right termini of Tn*5301* were not defined by inverted repeats and that the integrated element was flanked by a directly repeated hexanucleotide target sequence. Integration of the element in the recipient chromosome is site and orientation specific. Two preferred sites have been identified which show some sequence homology. Tn*5301* shares features common to other gram-positive elements, eg. Tn*916* and Tn*554*. However, a number of unique properties suggest it may represent the prototype of a novel class of conjugative element.

INTRODUCTION

The peptide antibiotic nisin is produced by certain strains of *Lactococcus lactis*. Due to its broad spectrum activity it has an important role as a biopreservative in the food industry (Fowler & Gasson, 1991). The ability of nisin production determinants to be transferred in conjugations between strains has long been known and in the past the possible involvement of plasmids has been suggested. Recent characterisation of the nisin biosynthetic region in a nisin producing transconjugant FI5876 has established that the gene for prenisin (*nisA*) is located in the chromosome (Dodd *et al.*, 1990). This gene, together with other nisin determinants, a copy of IS*904* and the sucrose metabolising genes, were cotransferred to the plasmid free recipient MG1614 to produce a series of independent nisin-producing transconjugants (Gasson *et al.*, 1984). A range of molecular techniques were employed to establish that, in conjugations, a discrete segment of DNA was transferred which had properties characteristic of a transposable element (Horn *et al.*, 1991; Dodd *et al.*, 1991). Some of the properties of this element, designated Tn*5301*, are described here.

NATO ASI Series, Vol. H 65
Bacteriocins, Microcins and Lantibiotics
Edited by R. James, C. Lazdunski and F. Pattus

RESULTS AND DISCUSSION

Chromosomal location of Tn*5301*

To confirm the location of the genes involved in nisin biosynthesis it was demonstrated that a specific region of the chromosome in the non nisin-producing strain MG1614 gained a large segment of DNA as a result of acquiring the nisin determinants in conjugations (Dodd *et al.*, 1990; Horn *et al.*, 1991). Southern transfer hybridizations were carried out on restriction enzyme digested total DNA from both nisin (FI5876) and non nisin (MG1614) producing strains. A series of probes were generated (figure l) from sequences specific for *nisA* (probe 3), IS*904* (probe 2) and the recipient target site (probes 1 and 4). These experiments identified the left and right junction fragments between the newly acquired region encoding the nisin determinants (i.e. Tn*5301*) and the original recipient chromosomal DNA sequences. In addition an IS*904* probe (figure 1, probe 2) revealed the presence of 8 copies of this element in the MG1614 genome and an extra copy in the nisin producing transconjugant FI5876, located close to the left end of Tn*5301* (Horn *et al.*, 1991).

A combination of pulsed field gel electrophoresis (PFGE) and Southern transfer hybridization provided further confirmation of the chromosomal location of Tn*5301*. A 180kb MG1614 fragment was shown to increase in size to approximately 250kb. This indicated the region of integration and demonstrated that the element is about 70kb in size. PFGE analysis of a series of independently isolated transconjugants revealed that, whilst most Tn*5301* insertions were in the same *Sma*I fragment (termed a type 1 insertion), a second 330kb fragment could also receive the incoming element increasing in size to 400kb (type 2 insertion). A third type of transconjugant appeared to have resulted from a combination of the above two integration events in which both *Sma*I fragments involved in type 1 and type 2 insertions were absent. However, a further, as yet uncharacterised, rearrangement had occurred in these strains, possibly to stabilise the genome which would otherwise contain two large regions of homologous DNA (see Horn *et al.*, 1991)

Tn*5301* junctions sequences

A comparison of nucleotide sequences of DNA fragments specified by probes 1 and 4 from strains FI5876 and MG1614 (figure 1) established the precise termini of Tn*5301* (figure 2; Horn *et al.*, 1991). The sequence data revealed that the IS*904* copy was located 251 bases in from the left end indicating that this IS element does not play a role in Tn*5301* transposition. Unlike many transposons the left and right termini were not inverted repeats and a computer search did not find any sequences displaying homology with

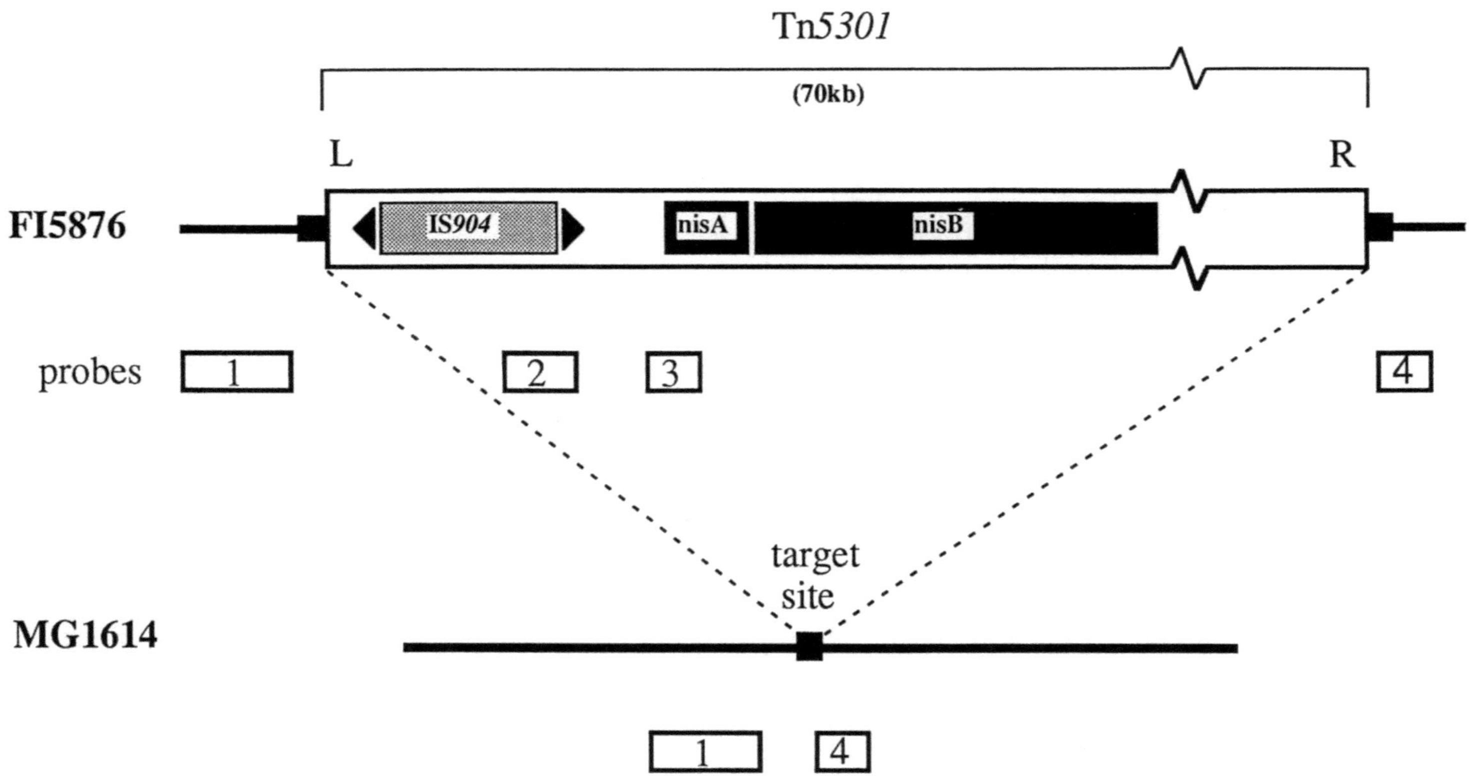

Figure 1. Integration of Tn*5301* in the chromosome of *L. lactis* MG1614 to generate *L. lactis* FI5876. Probes are indicated by open boxes under the maps. Probes 1 and 4, which are specific for the left and right of the integration site respectively, were used to identify the target region in the recipient strain MG1614 and the left and right junctions of Tn*5301* in the transconjugant strain FI5876. Probe 2 is specific for IS*904* sequences and Probe 3 is specific for the *nisA* gene.

```
CAATTCCGCCACCACCGCAAAACAGTACTTTAATATAGTACCATTTCTAGATTAATTGTGCAAGTAT

                                     5'-GAGGCAATTTCGGCAGC-3' ->
GAACTAGGGCTGATTTTTTATTGGAAATAAGAGGCAATTTCGGCAGCATGTTTTCCAGCGACAAAAC

CTGTGACAAAGGCACAAGTAATATTATAACCACCAGTATATCCATTAATGTCTAGCAGTTCCCCAGT

TCATAAATAAGCCTTGTGTTTTTTTGATACACATATAAAGTGCGTTTTAACCTAGTTTTAAAAGATT

TACTGTTAATAAAAATAAAATGAACCAAAGAACTAACCAAAGCAAAACTATACAGCATTTACTTTAT

CCAAAACCTACTGTAAAATTTATAGTAGGTTTTGTAATTTAAAATATTAAAGTAAATTTACAAATAA

CTCTTAAGTGAGATAGTCTAAATAAATGAATTGATGTGCACCCCAAAAGTTAGACTTTTTATCCAGG
                                                      <- 3'-AAATAGGTCC
TATTTATTGGAAAGGTTATAATAAACTAGAC....................................
ATAAATA-5'
...................................................................
                                                         5'-CATGAGGAAGA
..............................................AAAAGAATCATGAGGAAGA
TTAGGGA-3' ->
TTAGGGAATATCAATTTTTGAAATTCTTCTAAAAACTATTGTTAGAGACTTTTGAGGAAGAATAAGG

AAGTATAAATAAAAATATAATGGAAATTTTTGCTTTCCATACTTTTAGGATTCACTTCTTTCAAAGA

AACACCGCCACCAGTAACAAATGACTTCTCAATTGGGAGAGTTTTATTTACCATGATTTGCCAGTTT

TGAACAGTTGAACGAGTTTTCGATTTGTTGTTCAGATAAGATTGGCTGGTAAGGCAGGATTAATGCC

AGCTTTCAGATAAGAGAAGCATACGTCTTGGGTTAGTCCAGAGAAGGCATTTTTAATTGATTTATCA

CTTGAAGCTAGCTTGGTCAGTTCATTTTTTAAAAACCCAGCCGATTTATCTGGGAATTGGTCTAAGC

TGACCATGACAGGAGTTTCACCTTTTTCAAGAAGTTGGTTAATGAAACTTGAACAGCGTAAAGCAGC
                                                      <- 3'-GCATTTCGTCG

CGGACCAGACAAACCAAATGAGTGAAG
GCCTGG-5'
```

Figure 2. Nucleotide sequence of the left and right junctions of Tn*5301* with chromosomal DNA in *L. lactis* FI5876. Tn*5301* is represented by the shaded boxed sequences and the dotted line indicates the internal region of the element. The directly repeated hexanucleotide is indicated by arrows below the sequence at both termini. The left end of the copy of IS*904*, within the left end of Tn*5301*, is underlined. Synthetic oligonucleotides employed as primers for sequence determination and PCR analysis are indicated above and below the sequence.

either end. As is common with many transposons a duplication of the target site, in this case the sequence 5'TTTTTG-3', was found flanking the insertion as a direct repeat (figure 2). A model for chromosomal insertion and excision of Tn*916* has been proposed involving recombination between homologous sequences at the target site and at one or other end of the element. A similar mechanism for Tn*5301* transposition would generate the observed directly repeated hexanucleotide flanking the inserted element. Analysis of further independent insertion events is needed to establish whether Tn*5301* transposition is analogous to the mechanisms of other conjugative elements.

The Tn*5301* insertion in FI5876 represents a type 1 insertion. Using sequence data at the left and right junctions (figure 2) pairs of primers were designed which would amplify a 386bp and a 486bp fragment respectively in polymerase chain reactions (PCR). These same primers were used to analyse a number of independently isolated transposon derivatives (including type 1, 2 and 3 Tn*5301* insertions). In all type 1 and type 3 insertions both fragments were strongly amplified whereas only weak amplification of the left junction was detected in type 2 insertions. This result indicated that not only were the same terminal sequences involved in Tn*5301* transposition, but also that integration occurred at a specific target site. Furthermore, the weak amplification of type 2 insertions suggested that while a distinct site is involved in these transconjugants there may be some sequence homology with the type 1 target site (Horn *et al.*, 1991). The copy of Tn*5301* in FI5876 was able to retranspose to a new site in a non-nisin producing recipient strain. This indicated that all the information required for transposition was contained within the element. Furthermore, PCR analysis, using the same primers as above, indicated that not only were the same termini used in sequential transposition, but also that integration of Tn*5301* occurred at the same type 1 target site (Dodd *et al.*, 1991) .

The nisin determinants are encoded by a lactococcal transposon which shares similar properties to the class of element typified by Tn*916*. However, a number of novel features became apparent as a result of analysis of Tn*5301*. Further molecular analysis of transposon/chromosome junctions generated by independent insertions will lead to a greater understanding of the mechanism of this new type of conjugative transposon.

REFERENCES

Dodd HM, Horn N, Gasson MJ (1990) Analysis of the genetic determinant for production of the peptide antibiotic nisin. J Gen Microbiol 136: 555-566

Dodd HM, Horn N, Swindell S, Gasson MJ (1991) Physical and genetic analysis of the chromosomally located transposon Tn*5301*, responsible for nisin biosynthesis. In "Lantibiotics" Procedings of the First International Workshop.

Gasson MJ (1984). Transfer of sucrose fermenting ability, nisin resistance and nisin production into *Streptococcus lactis* 712. FEMS Microbiol Lett 21:7-10

Fowler GG, Gasson MJ (1991) Antibiotics - nisin. In "Food Preservatives" (eds. N.J. Russell & G.W. Gould) pp 135-152. Blackie, London, UK.

Horn N, Swindell S, Dodd HM, Gasson MJ (1991) Nisin biosynthesis genes are encoded by a novel conjugative transposon. Molec Gen Genet 228:129-135

DEVELOPMENT OF YEAST INHIBITORY COMPOUNDS FOR INCORPORATION INTO SILAGE INOCU LANTS

S.A.Goodman, C.Orr, P.J.Warner
Biotechnology Centre
Cranfield Institute of Technology
Bedford, MK43 OAL
England.

INTRODUCTION

Microorganisms cause considerable losses of ensiled animal fodder. Anaerobic deterioration is mediated principally by bacteria but this can be minimised by careful practice during silage making. Aerobic deterioration is however difficult to control, especially when ambient temperatures are high and feeding rates are low as in a mild winter for example. The principal cause of this type of spoilage is the growth of yeast and, to a lesser extent, moulds. The development of strategies to control this would thus be advantageous. Previously we have screened several strains of lactic acid bacteria for production of antifungal compounds without success. In this study, strains of *Bacillus* were screened for the production of compounds inhibiting growth of yeast. Many species of *Bacillus* produce antibiotics of which Gramicidin S and Tyrocidine are the best characterised. Antifungal peptides are also known to be produced. The major family of these are lipopeptides which comprise iturins (Delcambe & Devignat, 1957), surfactins (Kluge *et al.*, 1988), bacillomycins (Peypoux *et al.*, 1981) and mycosubtilins (Besson & Michel, 1990). Most of these compounds are cyclic peptides containing a sequence of D and L amino acids closed by a beta-amino acid carrying a long aliphatic chain. They all appear to be active against fungi and yeasts. Although there are numerous lipopeptidic antibiotics produced by *B.subtilis* strains only few hydrophilic and antifungal metabolites have been reported in the literature, including the dipeptides bacilysin (Walker & Abraham, 1970) and chlorotetain (Rapp *et al.*, 1988). Another similar substance, previously named Rhizoctonia factor (Michener & Snell, 1949), has recently been identified as a hydrophilic phosphono-oligopeptide usually existing as a dipeptide and named rhizocticin (Rapp *et al.*, 1988). Some isolates of *B.licheniformis* have been found to produce bacilysin (Kugler *et al.*, 1990), but there are no other reports on the production of di- and tri-peptides in this organism.

Here we describe the identification of an inhibitory compound towards *Saccharomyces* spp. which is produced by *B.licheniformis* CH200.

NATO ASI Series, Vol. H 65
Bacteriocins, Microcins and Lantibiotics
Edited by R. James, C. Lazdunski and F. Pattus

MATERIALS AND METHODS

Culture conditions

B.licheniformis CH200 was grown in malt extract medium (malt extract 30g/1, mycological peptone 3g/1, pH6.2, autoclaved, 115°C, 15mins) or in defined medium for the isolation of the inhibitory compound (glucose 20g/1, L-glutamic acid 5g/1, $MgS0_4.7H_20$ 1.02g/1, K_2HP0_4 1g/1, KCl 0.5g/1, trace salts (1ml, $MnS0_4.H_20$ 0.5g, $CuS0_4.5H_20$ 0.16g, $FeS0_4.7H_20$ 0.015g/1 in 100ml water) pH6.2. Cultures were grown with aeration at 37°C. Growth curves were performed in malt extract medium. pH and thermal stability experiments were performed in defined medium. *Saccharomyces* spp. were maintained on malt extract agar (malt extract 30g/1, mycological peptone 5g/1, agar 15g/1, pH5.4). Cultures for use in pour plates were grown in malt extract broth, pH5.4, for 6 hours at 37°C.

Plasmid isolation

Plasmid isolations were performed according to the method of Birnboim and Doly (1979).

Determination of inhibition

Pour plates of *Saccharomyces* were prepared using fresh culture (50μ1) and malt extract agar (70μl). Wells were cut using a sterile cork borer (4mm diameter) and the supernatant to be tested was inoculated into the wells. Zones of inhibition were measured from the perimeter of the well to the edge of the area of the growth of the test organism.

Correlation of growth phase and production of the inhibitory substance

An overnight culture (1ml) of *B.licheniformis* was inoculated into fresh pre-warmed and aerated malt extract broth, pH6.2 (100ml). The culture was incubated (37°C) with shaking and at hourly intervals the A_{600} was measured. Samples were taken simultaneously to test for inhibitory activity against *Saccharomyces exiguus*. Supernatants (1 ml), from which bacteria were removed by centrifugation, were treated with chloroform (10μ1) to kill remaining viable bacteria.

Thermal stability

The culture supernatant was adjusted to the required pH using 11.6M HCl or 5M NaOH. A sample was autoclaved (121°C,15min) and aliquots (10μ1) were tested for activity as described above.

Enzyme digestion

Enzyme solutions (Sigma Chemical Co.) were prepared (source in parentheses) in the appropriate buffer at a concentration of 1mg/ml; Pronase (*Streptomyces* spp.), Proteinase K (*Tritirachium album*), alpha chymotrypsin (bovine pancreas), lipase (wheat germ), alpha amylase (*Bacillus* spp.), pepsin (porcine stomach mucosa), ficin (fig tree latex), papain (papaya latex), protease II (*Aspergillus oryzae*), and catalase (bovine liver). Equal volumes of culture supernatant and enzyme solution were mixed and incubated (37°C, 2h). Inhibitory activity was assayed as described above.

Purification of the inhibitory substance

Culture supernatant was passed through an ultrafiltration membrane (YM3, molecular weight cut off 3,000, Diaflo, Amicon), lyophilised, resuspended and then further purified on a Biogel P2 column (Biorad), fractionation range 100-1800. The fractions were assayed for activity (20μl from 5ml) as previously described, the active fractions pooled and lyophilised and then passed through the column again. The active fractions were pooled and stored at -20°C for future use.

Gel exclusion chromatography

A Biogel P2 column (578x25mm) was prepared with deionised water as eluant. The column was calibrated using peptides of molecular weights between 1,355 and 307. A sample (1ml, starting volume 80ml) of the crude filtered substance was chromatographed and fractions collected. The fractions were bioassayed as previously described to estimate the molecular weight of the active substance.

RESULTS AND DISCUSSION

The production of the yeast antagonist by *B.licheniformis*, grown in complex medium, occurs in the late log/early stationary phase as seen in Figure 1.

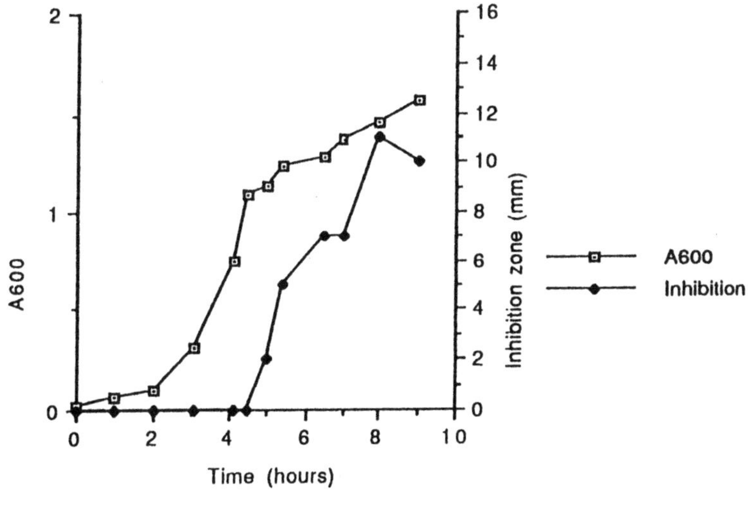

Figure 1. Correlation of growth phase with production of antagonist in *B.licheniformis* CH200

This is consistent with the general observation that secondary metabolites are produced in response to nutrient deprivation, a suggestion perhaps confirmed by the production of the inhibitory substance at an early stage of growth in defined medium (results not shown).

Examination of *B.licheniformis* CH200 revealed no plasmid DNA. This suggests that the production of the yeast antagonist may be chromosomally encoded, although alternatively, the technique may have failed to detect any plasmids present.

Resistance to elevated temperatures, acid and alkali conditions are common features of small peptides (Hurst, 1981). The substance described here appears stable in culture supernatant over a pH range of 2-10. When autoclaved it loses approximately 50% activity below pH5 and above pH 9 (Figure 2).

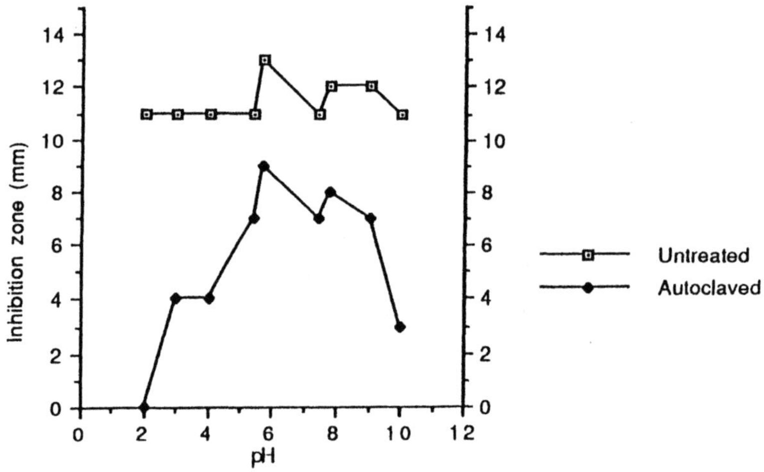

Figure 2. Thermal and pH stability of antagonist from *B.licheniformis* CH200.

Resistance to digestion with various proteases suggest the absence of particular amino acids. Only pronase a non-specific protease resulted in the loss of activity of the inhibitory substance. Treatment with lipase and amylase had no effect suggesting that neither carbohydrate nor lipid moieties were involved in activity.

Initially gel exclusion chromatography indicated that the active substance had an approximate molecular weight of 1200. However subsequent gel filtration of these samples only recovered active substance with a molecular weight close to 600. In later purifications, two peaks of activity were eluted from one sample, Peak I and Peak II, with approximate molecular weights of 1200 and 600 respectively

(Figure 3a), Active fractions spanning each peak were pooled and lyophilised. Peak I and Peak II were then chromatographed separately. Peak I has shown a shift in molecular weight to approximately 600 (Figure 3b) whilst Peak II still has an approximate molecular weight of 600.

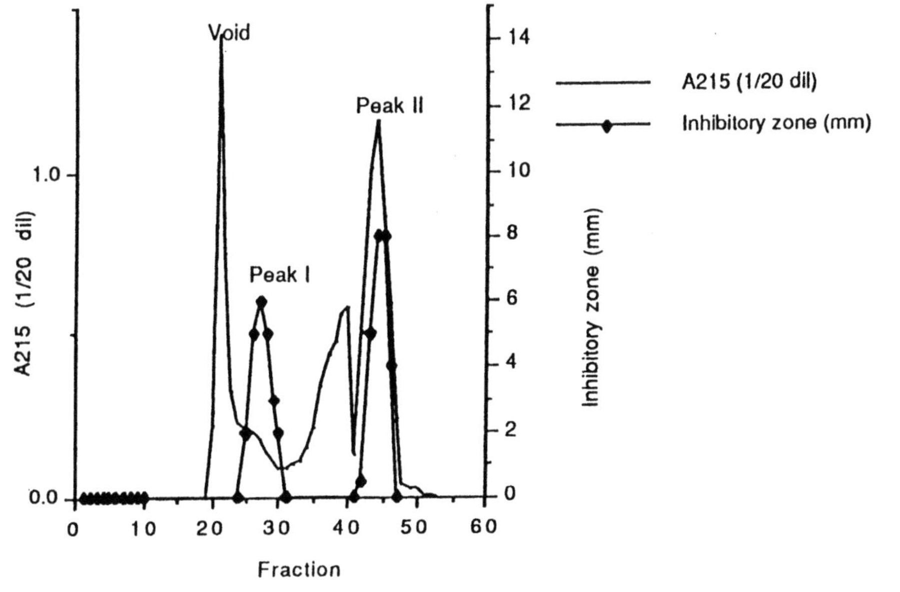

Figure 3a. Biogel P2 elution profile of *B.licheniformis* CH200 culture supernatant. Elution was carried out with deionized water at a flow rate of 55ml/hr. Bed volume was 283ml.

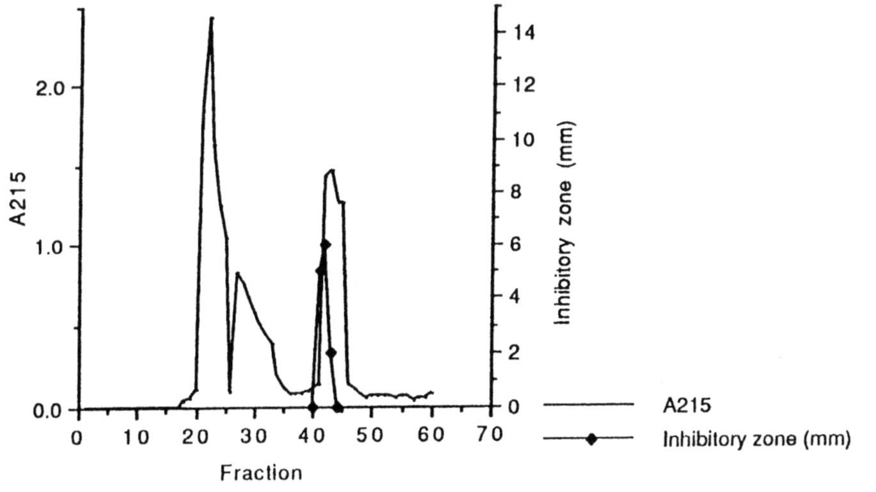

Figure 3b. Biogel P2 elution profile of Peak I. Elution was carried out as for Figure 3a.

The change in molecular weight of Peak I could be due to instability of a dimer, where the total peptide concentration in the solution is the critical factor. Alternatively repeated lyophilisation of the peptide may destabilise the structure. However, it is not yet proven that Peak I and Peak II, although showing similar activity and an approximate molecular weight of 600, are the same substance.

HPLC is currently being employed in the further purification of this molecule. Elucidation of its constituent amino acids, mode of action, range of inhibitory activity, possible molecular cloning and characterisation of the genes encoding its production are part of future studies. Cloning into *Lactobacillus* spp. to form a novel but natural biocontrol agent for silage fermentations is the long term aim. Alternatively it may be possible to chemically synthesise the peptide to provide a spray for the open face of the silo to control yeast growth.

SUMMARY

A small peptide produced by *Bacillus licheniformis* CH200 capable of inhibiting the growth of *Saccharomyces* spp. has been identified. It is produced in late log/early stationary phase in complex medium. It is acid and alkali resistant and, in addition, retains 50% of its activity when autoclaved at pH less than 5 and more than 9. It appears to be stable to a range of proteases including pepsin, papain, proteinase K, alpha chymotrypsin, ficin. It shows two peaks of activity; Peak I with a molecular weight of approximately 1200 which shows a shift to approximately 600, and Peak II with a molecular weight of approximately 600 as determined by gel exclusion chromatography.

REFERENCES

Delcambe L, Devignat R (1957) L'iturine, nouvel antibiotique d'origine congolaise. Acad Roy Sci Coloniales 6:1 -77

Kluge B, Vater J, Salnikow J, Eckart K (1988) Studies on the biosynthesis of surfactin, a lipopeptide antibiotic from *Bacillus subtilis* ATCC 21332. FEBS Lett 231:107-110

Peypoux F, Besson F, Michel G, Delcambe L (1981) Structure of bacillomycin a new antibiotic of the iturin group. Eur J Biochem 118:323-327

Besson F, Michel G (1990) Mycosubtilins B and C: minor antibiotics from mycosubtilin-producing*Bacillus subtilis*. Microbios 62:93-99

Walker JE, Abraham EP (1970) The structure of bacilysin and other products of *Bacillus subtilis*. Biochem. J 118:563-570

Rapp C, Jung G, Katzer W, Loeffler W (1988) Chlorotetain from *Bacillus subtilis*, an antifungal dipeptide with an unusual chlorine-containing amino acid. Angew Chem Internat Ed 27:1733-1734

Michener HD, Snell N (1949) Two antifungal substances from *Bacillus subtilis*. Arch Biochem 22:208-214

Rapp C, Jung G, Kugler M, Loeffler W (1988) Rhizocticins - new phosphono-oligopeptides with antifungal activity. Liebigs Ann Chem 655-661

Kugler M, Loeffler W, Rapp C, Kern A, Jung G (1990) Rhizocticin A, an antifungal phosphono-oligopeptide of *Bacillus subtilis* ATCC6633: biological properties. Arch Microbiol 153:276-281
Birnboim HC, Doly J (1979) A rapid alkaline extraction procedure for screening recombinant plasmid DNA. Nucl Acids Res 7:1513-1523
Hurst A (1981) Nisin. Adv Appl Microbiol 27:85-123

THE *excC* AND *excD* GENES OF *ESCHERICHIA COLI* K-12 ENCODE THE PEPTIDOGLYCAN-ASSOCIATED LIPOPROTEIN (PAL) AND THE TolQ PROTEIN, RESPECTIVELY.

J.C. Lazzaroni, A. Vianney, C. Amouroux and R. Portalier
Laboratoire de Microbiologie et Génétique Moléculaire
UMR 106 CNRS, University of Lyon I
Bat 405
69622 Villeurbanne cedex
France

The *tol* genes of *Escherichia coli* K-12 are involved in the import of macromolecules such as group A colicins or filamentous phage DNA, and in the maintenance of cell envelope integrity (for a review, see Webster, 1991). Four genes, *tolQRAB*, located at min 17 in the genetic map have been identified and sequenced (Sun & Webster, 1987; Levengood & Webster, 1989). Mutants altered in the *tolQRA* genes are tolerant to groupA colicins and resistant to filamentous phages while mutants in the *tolB* gene are only tolerant to colicin A, E2 and E3. In addition, *tol* mutants are hypersensitive to drugs and release most of the periplasmic proteins into the extracellular medium.

We are interested in identifying the genes responsible for the maintenance of the cell envelope integrity and have isolated a series of periplasmic-leaky mutants (Lazzaroni & Portalier, 1981 ; Fognini-Lefebvre & Portalier, 1984). Most of the corresponding *lky* mutations are located at min 17 in the *E. coli* genetic map. Complementation analysis showed that some *lky* mutations map either in the *tolA* and *tolB* genes, or in two other complementation groups named *excC* and *excD* (Lazzaroni *et al.*, 1989). Here we report the fine genetic mapping of *excC*, the identification of its product and the characterization of 4 *excD* mutations.

Characterization of the *excC* gene.

A 1884 bp fragment carrying the *excC* gene was isolated and sequenced. It contained the 3' end of the *tolB* gene and an open reading frame encoding the 18,748 Da ExcC protein (Lazzaroni & Portalier, 1991). This protein was composed of a hydrophobic region of 22 residues and displayed an overall hydrophilic configuration. The isolation of enzymatically active ExcC-PhoA fusions showed that ExcC was an exported protein. The potential signal peptide cleavage site was highly homologous to the consensus sequence shared by bacterial lipoproteins (Regue & Wu, 1988). The lipoprotein nature of ExcC was confirmed by labeling the protein with palmitic acid or glycerol. Comparison of the DNA sequence of the *excC* gene with sequences present in the Data banks revealed that ExcC was indeed the PAL (peptidoglycan-associated lipoprotein) protein first described by Mizuno (1979), then sequenced by Chen

NATO ASI Series, Vol. H 65
Bacteriocins, Microcins and Lantibiotics
Edited by R. James, C. Lazdunski and F. Pattus

and Henning (1987). The *pal* gene had not yet been characterized on the *E. coli* linkage map since no obvious phenotype could be identified until now.

A topologic analysis of the PAL protein was carried out using PAL-PhoA translational fusions. The PAL protein was associated with the outer membrane by its N-terminal part. The enzymatically active fusions were randomly distributed along all the protein, indicating that the PAL protein was integrated to the outer membrane only via the lipid moiety linked to the first amino acid (cysteine) of the mature protein which contained 151 amino acids. The C-terminal part of the protein in particular region 101-116 was necessary for a correct interaction of the PAL protein with the peptidoglycan layer.

Characterization of *excD* mutations

excD mutants are hypersensitive to drugs and release periplasmic proteins into the extracellular medium, some of them are tolerant to colicin A but all are sensitive to colicin El. *excD* mutations are only partially complemented by plasmid pTPS308 carrying the *tolQ* and *tolR* genes (Sun & Webster, 1986). We have cloned the *excD* gene on plasmid pNJ301 (Lazzaroni *et al.*, 1989). This plasmid was sequenced and shown to be identical to pTPS308. Plasmid pTPS308 is derived from pUC8 whereas pNJC301 is derived from pBR328. Difference in the plasmid copy number could explain this discrepancy.

We constructed a series of genomic banks by cloning fragments of chromosomal DNA from *excD* mutants into pUC18. The 1800 bp *Eco*RI-*Hin*dIII chromosomal fragment of pNJ301 was used to make a probe and screen for the plasmids carrying the *excD* mutations. Sequencing of the corresponding plasmids was carried out.

The *excD856* and *excD875* mutations corresponded to the substitution of glycine 107 to arginine in the sequence of the TolQ protein which contained 230 amino acids. Expression of tolQ was normal in the presence of these mutations as shown in minicell experiments. In this case, the change in amino acid probably directly altered TolQ structure or activity.

The *excD890* mutation corresponded both to a change of amino acid 131 from asparagine to lysine within the ORFl which is located upstream *tolQ* (see Figure l) and an alteration in the ribosome binding site just upstream of the initiation codon of *tolQ* from GGAG to GAAG. Expression of the TolQ protein was greatly reduced. Moreover, a plasmid carrying the ORFl gene was unable to complement the *excD890* mutation, whereas a multicopy plasmid carrying the *excD890* allele fully complemented a chromosomal *excD890* mutation. Thus, the *excD890* defect was due to a decrease in *tolQ* expression which was the consequence of the alteration of the *tolQ* ribosome binding site.

The *excD925* mutation led to the substitution of the alanine in position 177 to a valine. These amino acids have quite similar properties although the valine has a slightly larger R-chain. Surprisingly, the TolQ and TolR proteins could not be visualized in minicells in the presence of *excD925* mutation This could be the result of an increased sensitivity of TolQ to proteolytic degradation and a subsequent TolR hydrolysis which could be explained only if TolQ was necessary for TolR stability. Another alternative could be that the *excD925* mutation led to an alteration in the *tolQ* DNA or mRNA secondary structure which resulted in a decrease of transcription or translation of both *tolQ* and *tolR* genes. This could indicate that these genes belong to the same operon. According to our results the *excD* mutations have been renamed *tolQ*.

Conclusion

This study led us to a better understanding of the very complicated cluster *tolQRABpal* which encodes several membrane proteins involved in filamentous phage DNA entrance, group A colicin uptake as well as in envelope integrity (Figure 1).

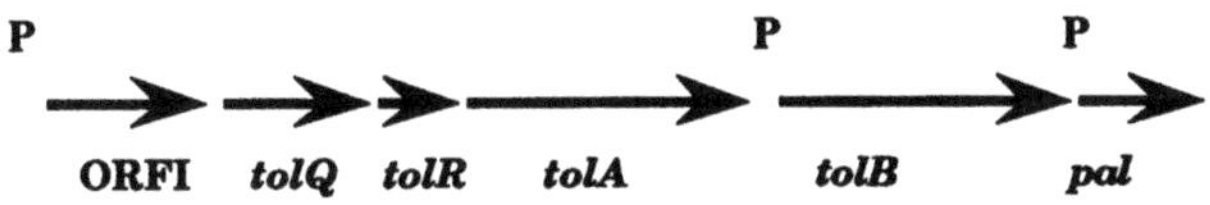

Figure 1. Organisation of the *tolQRAB pal* cluster located at min 17 on the *E. coli* linkage map. The position of the putative promoter boxes are indicated by P; the arrows indicate the direction of transcription.

Five genes have been characterized: they are transcribed in the same direction and are all necessary for the maintenance of cell envelope integrity. The *tolQRA* genes are involved in the uptake of group A colicins as well as in the entry of filamentous phage DNA, whereas the *tolB* gene is indispensable for the uptake of some of the group A colicins. The genetic organisation of this cluster is unknown. Three potential promoter boxes can be identified from the DNA sequence, upstream of ORFI, *tolB* and *pal*, but an operon organization of even the *tolQRA* genes remains hypothetical. Some of the *tol* genes show sequence or functional homologies with proteins involved in the uptake of group B colicins. The TolQR and ExbBD proteins show extensive amino acid homologies (Eick-Helmerich & Braun, 1989), while the TonB and TolA proteins share the same unusual overall structure including an uncleaved N-terminal sequence anchoring these proteins to the inner membrane, a long stable alpha-helix spanning the periplasm and a C-terminal part supposed to interact with outer membrane components (Brewer *et al.*, 1990; Levengood *et al.*, 1991). Both systems show some kind of cooperativity with respect to colicin tolerance (Braun, 1990).

The *tolQ* mutations we have characterized showed a different pattern of colicin tolerance when compared with other *tolQ* mutations (Sun & Webster, 1986). All our *tolQ* mutants are sensitive to colicin El, moreover, in the presence of *tolQ890* and *tolQ925* mutations which result in a decrease in the content of TolQ, cells are sensitive to colicin A. As these mutants are periplasmic-leaky, our result indicate that a lower TolQ content is still sufficient for the entry of some colicins while more protein are required for the maintenance of cell envelope integrity. The difference in sensitivity towards colicins could also be interpreted as a partial compensation of the loss of TolQ activity by the homologous ExbB protein.

The Tol/PAL proteins are cell envelope proteins (Webster, 1991; Lazzaroni & Portalier, 1991). TolQRB proteins are located into the cytoplasmic membrane; TolA, although anchored to the cytoplasmic membrane, spans the periplasm and might interact with the outer membrane; PAL is anchored to the outer membrane and spans the periplasm interacting with the peptidoglycan layer.

The main difference between the Tol and TonB systems is that the primary role of the TonB system is the import of iron siderophores across the cell envelope while the exact role of the *tol* genes is unknown. A very attractive hypothesis could be that all these proteins form a complex able to attach the outer membrane to the cytoplasmic membrane, thus ensuring the physical stability of the outer membrane. A second alternative could be that these proteins play a role in the translocation of some outer membrane components. The colicins and filamentous phages may then use this structural complex to enter the cell. These hypotheses are currently being studied in our group.

References

Braun V (1989) The structurally related *exbB* and *tolQ* genes are interchangeable in conferring *tonB*-dependent colicin, bacteriophage and albomycin sensitivity. J Bacteriol 171:2458-2465

Brewer S, Tolley M, Trayer IP, Barr GC, Dorman CJ, Hannavy K, Higgins CF, Evans JS, Levine BA, Wormald MR (1990) Structure and function of X-Pro dipeptide repeats in the TonB proteins of *Salmonella typhimurium* and *Escherichia coli*. J Mol Biol 216:883-895

Chen R, Henning U (1987) Nucleotide sequence of the gene for the peptidoglycan-associated lipoprotein of *Escherichia coli* K-12. Eur J Biochem 163:73-77

Eick-Helmerich K, Braun V (1989) Import of biopolymers into *Escherichia coli*: nucleotide sequence of the *exbB* and *exbD* genes are homologous to the *tolQ* and *tolR* genes respectively. J Bacteriol 171:5117-5126

Fognini-Lefebvre N, Portalier R (1984) Isolation and preliminary characterization of beta-lactamase excretory mutants of *Escherichia coli* K-12. FEMS Microbiol Lett 21:323-328

Lazzaroni JC, Portalier R (1981) Genetic and biochemical characterization of periplasmic-leaky mutants of *Escherichia coli* K-12. J Bacteriol 145: 1351-1358

Lazzaroni JC, Fognini-Lefebvre N, Portalier R (1989) Cloning of *excC* and *excD* genes involved in the release of periplasmic proteins by *Escherichia coli* K-12. Mol Gen Genet 218:460-464

Lazzaroni JC, Portalier R (1992) The *excC* gene of *Escherichia coli* K-12 re quired for the cell envelope integrity encodes the peptidoglycan-associated lipoprotein (PAL). Mol Microbiol, in press

Levengood SK, Webster RE (1989) Nucleotide sequence of the *tolA* and *tolB* genes and localization of their products, components of a multistep translocation system in *Escherichia coli* K-12. J Bacteriol 171:6600-6609

Levengood SK, Beyer WF, Webster RE (1991) TolA: a membrane protein involved in colicin uptake contains an extended helical region. Proc Natl Acad Sci 88:5939-5943

Mizuno T (1979) A novel peptidoglycan-associated lipoprotein found in the cell envelope of *Pseudomonas aeruginosa* and *Escherichia coli*. J Biochem86:991-1000

Regue M, Wu HC (1988) Synthesis and export of lipoproteins in bacteria. In: Das RC, Robbins PW, (eds) Protein transfer and organelle biogenesis Academic Press Inc, San Diego, pp 587-606

Sun TP, Webster RE (1986) *fii* a bacterial locus required for filamentous phage assembly and its relation to colicin-tolerant *tolA* and *tolB*. J Bacteriol 165:107-115

Sun TP, Webster RE (1987) Nucleotide sequence of a gene cluster involved in the entry of E colicins and single stranded DNA of non filamentous bacteriophages into *Escherichia coli*. J Bacteriol 169:2667-2674

Webster RE (1991) The *tol* gene products and the import of macromolecules into *Escherichia coli*. Mol Microbiol 5: 1005-1011

RESISTANCE AND TOLERANCE OF BACTERIA TO E COLICINS

J. Smarda
Deparment of Biology, Faculty of Medicine
Masaryk University
Jostova 10
CS-66244 Brno
Czechoslovakia

It is easy to select colicin non-sensitive mutants from a population of a colicin sensitive strain. In most strains, they occur in relatively high frequencies without induction. It is also easy to exclude colicin immunity as a possible cause of non-sensitivity. However, it is not easy to distinguish between colicin resistance and colicin tolerance. In this study, experimental evidence concerning this problem is presented; we used the indicator strain *Escherichia coli* K12-Row and its spontaneous mutants selected for decreased sensitivity to colicins El-E9 as a model system.

Resistance to colicins El-E9, which is generally considered to be a qualitative marker of the strain (clone) in question, is due to the absence of a biologically efficient BtuB protein in the outer membrane of the Gram-negative bacterial cell wall (DiMasi *et al.*, 1973). This protein functions as a common receptor also for vitamin B_{12}, cyanocobalamin (White *et al.*, 1973), and for the bacteriophage BF23 (Bradbeer *et al.*, 1976). It is a globular protein of Mr 66,412.

Tolerance (formerly refractivity) to colicins El-E9 is due to mutations in the poorly understood colicin translocation (uptake) system which transports the colicin molecule from its receptor to its target in the inner, plasma membrane. Belonging to the group A, E colicins require for their translocation the inner membrane and periplasmic proteins TolA, TolB, TolQ and TolR. Colicin El does not require the function of TolB but requires that of TolC in the outer membrane. (Only the uptake mechanisms of colicins El, E2 and E3 have been analysed in some detail.) In most cases, colicin tolerance is just one of the phenotypic markers of a pleiotropic mutation involving several functions of the plasma membrane. For all these reasons, it is doubtful whether colicin tolerance is a qualitative or a quantitative marker. Anyway, standard E colicin uptake is a function of several structural genes.

Theoretically, colicin tolerant clones should dispose of an unaltered colicin binding ability because of their functioning receptors, but should be defended from colicin lethal action. With a few exceptions, colicin tolerance is less specific of the colicin type than the colicin-receptor interaction.

Strains and their mutants

From clear inhibition zones formed by colicins El-E9 in the lawn of indicator bacteria *Escherichia*

NATO ASI Series, Vol. H 65
Bacteriocins, Microcins and Lantibiotics
Edited by R. James, C. Lazdunski and F. Pattus

coli K12-Row, we isolated small colonies of surviving bacteria. Grown-up pure mutant clones were checked for basic markers of the *E.coli* K12 strain and their possible colicinogeny was excluded. In this way, we isolated a series of about 50 clones of *E. coli* K12-Row characterized by a strikingly decreased sensitivity to E colicins in the agar test. For further research, 14 of them were chosen at random. In our set of spontaneous E-colicin non-sensitive mutants, 55 % were tolerant, 25 % were resistant and 20 % were tolerant to some and resistant to other E colicins. (Only concordant results of repeated experiments are stated here. However, some mutants were shown to be unstable.)

For comparison we also had at our disposal the strain *Escherichia coli* KLR320 and its 7 *btuB* deletion mutants, kindly donated by Dr. R. E. Glass (Department of Biochemistry, University of Nottingham, England). All of these mutants bear nonsense mutations in the NH_2-terminus of the BtuB protein and are therefore resistant to colicins El-E9. Finally, we used 5 *tolC* mutants (with their parent strains), kindly supplied by Dr. P. Reeves (Department of Microbiology, University of Adelaide, South Australia) and by Dr. A. Pugsley (Institut Pasteur, Paris, France).

Inhibition zones

We checked the capacity of the sensitive indicator Row and of its 14 mutants to form inhibition zones in agar cultures around the macrocolonies of the strains producing colicins El-E9 (Table 1). The high sensitivity of the strain Row to all E colicins was confirmed. The mutants of decreased colicin E sensitivity fell into 3 distinct categories. Three mutants formed no inhibition zones at all (by analogy to the E resistant mutants from R.E. Glass); they were E resistant. Two mutants showed distinctly decreased widths of inhibition zones formed by colicin El, and no zones formed by colicins E2-E9; they were El tolerant and E2-E9 resistant. (In one mutant the inhibition zones were not followed.) Eight mutants, just as all *tolC* mutants, formed inhibition zones narrower than the parent strain Row, with all the E colicins. They were E tolerant. The decreased diameter was specific for most colicin/mutant combinations.

Survival

We followed the survival rates of the same mutants exposed for 30 min to colicins in suspensions of $2x10^9$ cells/ml (Table 2). Only $1x10^{-5}$ from this amount of sensitive Row bacteria survived. In resistant mutants, the survival was 1000 times to 10000 times higher and reached $1.5x10^{-2}$ - $1 x 10^{-1}$. In no resistant mutant did the survival exceed 10 %. (The same holds for the Glass KLR320 mutants.) This finding is quite unexpected and suggests that at least 90 % of bacteria in a colicin E receptor mutant population are E sensitive and hence killed by colicin E in some process different from that mediated by the receptor.

Table 1.
E tolerant and E resistant mutants Row: inhibition zones produced by colicins E1 to E9

Strain	Strain Status	Average widths (mm) of inhibition zone rings Colicin								
		E1	E2	E3	E4	E5	E6	E7	E8	E9
Row	E1-E9 sensitive	5.0	8.3	6.6	6.5	6.8	6.7	7.0	10.1	10.1
Row/E6a	E1-E9 tolerant	4.6	7.7	5.7	5.9	5.6	5.5	6.6	9.7	9.7
Row/E5c	E1 tol., E2-E9 resistant	4.3	0	0	0	0	0	0	0	0
Row/E5a	E1-E9 resistant	0	0	0	0	0	0	0	0	0
Row/E7b	E1-E9 resistant	0	0	0	0	0	0	0	0	0

Table 2.

E resistant and E tolerant mutants Row: sensitivity to colicins E1-E9 (proportion of colony-formers in about $2x10^9$ cells/ml)

Strain	Mean survival with colicin								
	E1	E2	E3	E4	E5	E6	E7	E8	E9
Row (sensitive)	$1x10^{-5}$	$1x10^{-5}$	$1x10^{-5}$	$1x10^{-5}$	$1x10^{-5}$	$1x10^{-5}$	$1x10^{-5}$	$1x10^{-5}$	$1x10^{-5}$
Row/E5a (res.)	$7.9x10^{-2}$	$8.3x10^{-2}$	$5.4x10^{-2}$	$7.8x10^{-2}$	$6.7x10^{-2}$	$6.1x10^{-2}$	$7.2x10^{-2}$	$7.3x10^{-2}$	$5.6x10^{-2}$
Row/E5c (E1 tolerant, E2-E9 resistant)	$1.0x10^{-3}$	$7.0x10^{-2}$	$4.9x10^{-2}$	$8.0x10^{-2}$	$4.7x10^{-2}$	$4.1x10^{-2}$	$5.7x10^{-2}$	$2.6x10^{-2}$	$2.3x10^{-2}$
Row/E6a (tol.)	$1.1x10^{-3}$	$9.2x10^{-4}$	$6.2x10^{-4}$	$7.3x10^{-4}$	$7.6x10^{-4}$	$8.9x10^{-4}$	$6.7x10^{-4}$	$6.2x10^{-4}$	$5.1x10^{-4}$

Different survival patterns of resistant mutants treated with colicins El-E9 indicate their binding on more or less different epitopes in the BtuB receptor. They offer a basis of numerical input data for computer construction of a map of binding epitopes on the BtuB protein for these colicins. They show clearly that colicin resistance - though due to a mutation in a single gene - is a quantitative marker.

On the other hand, in tolerant mutants the survival was only 10 times to 100 times enhanced and reached $9x10^{-5}$ - $1.4x10^{-3}$. The same results were obtained in populations of *tolC* mutants. The survival patterns varied considerably for various mutants and colicins, indicating that tolerance, too, is a quantitative marker. All tolerant mutants displayed the highest survival with colicin El and the lowest one with colicin E3. The two El tolerant and E2-E9 resistant mutants showed a slightly lower survival to most of them than the "true" resistant mutants. - The survival counts of tolerant mutants exposed to all E colicins can be used as data for computer construction of a theoretical diagram of amalgamated distances of uptake mechanisms for colicins El-E9. Preliminarily, we have constructed a dendrogram of this kind on the basis of decreased inhibition zone diameters. The closest amalgamated distance of uptake mechanisms was shown for colicins E2 and E3 and for E6 and E9, all belonging to one cluster.

Table 3. E resistant and E tolerant *E.coli* K12-Row mutants: phage BF23 efficiency of plating

Strain	Status	Efficiency of BF23 plating (related to e.o.p. on Row)
Row	E sensitive	100%
Row/Ela	E resistant	< 0.3%
Row/E4a	"	< 0.3%
Row/E5a	"	< 0.4%
Row/E7b	"	< 0.4%
Row/E5c	El tolerant, E2-E9 resistant	14.9 %
Row/E6b	"	14.1 %
Row/Elc	E tolerant	78.0 %
Row/E2a	"	79.3 %
Row/E5b	"	55.8 %
Row/E5e	"	65.7 %
Row/E6a	"	64.1 %
Row/E8a	"	61.2 %
Row/E9c	"	64.5 %
Row/E9f	"	67.9 %

Phage BF23 plating efficiency

If we consider the e.o.p. in the parent strain Row (i.e. $2.52x10^{11}$/ml) equal to 100 %, then the e.o.p. in all E resistant mutants, including the KLR320 mutants, corresponded to less than 0.4 % (practically no plaques could be seen). On the other hand, the e.o.p. in all E tolerant mutants corresponded to 56 % - 79% (Table 3). The BF23 e.o.p. in our two El tolerant, E2-E9 resistant mutants was 14 % - 15 %, i.e. intermediate between those in "true" resistant and in "true" tolerant ones.

Colicin binding

We followed the binding capacity of the Row mutants for colicins El-E9 by directly titrating the proportion of free, unbound colicin from a standard solution of 100 lethal units per cell, after contact with a standard suspension of $2x10^9$ cells/ml (Table 4). The parent strain Row was able to bind about 57 % of any E colicin offered, which points at a binding capacity of about 57 lethal units per cell. [It implies that one lethal unit of any colicin E - in the system and under the experimental conditions given - equals to 4-5 molecules (Smarda & Damborsky, 1991).]

In general, the binding capacity of resistant mutants was decreased significantly; from the 57 % of the parent strain Row to as little as 11 %, displaying a specific pattern for each colicin and for each mutant. However, the binding decrease was not proportional to the increase in survival. In these experiments, an experimental error of the colicin assay used of ± 5 % must be taken into account and a constant proportion of up to 10 % of colicin used in the system given can get adsorbed unspecifically. Thus a majority of colicin gets bound unspecifically or on non-efficient receptors in mutants of low binding ability (Smarda 1975).

The same experience was achieved in the experiments on BtuB NH_2-terminal truncated mutants of the strain KLR320. And the same results, in general, were achieved by checking the binding capacity of E tolerant mutants. Grossly, tolerant mutants did not differ from resistant ones as to their colicin-binding capacity.

Thus, most resistant mutants keep and dispose of residual - though, in some of them, heavily impaired - binding capacity, a high proportion of which falls to the account of unspecific binding. Clearly, the extent of impairment is given by the location and extent of the deletion or missense reading of the NH_2-terminus of the BtuB protein chain. And most tolerant mutants show a more or less profound impairment of their binding capacity, too. It may be assumed that, in most cases, the pleiotropy of the mutations causing tolerance bears - in addition to impairment in the translocation mechanism - also impairment of the binding function of the BtuB receptor and hence overlaps in its consequences with a receptor mutation. Vice versa, the BtuB protein is required not only for binding, but also for translocation of colicins E (Bénédetti *et al.* 1989).

Table 4.
E resistant ($BtuB^-$) and E tolerant ($BtuB^+$) strains: adsorption capacity of $2x10^9$ cells for colicins E1 to E9

Strain	Relative amounts of colicins bound at 37° C									
	E1	E2	E3	E4	E5	E6	E7	E8	E9	Remark
E susceptible $BtuB^+$										
Row	57 %	51 %	50 %	55 %	55 %	59 %	70 %	57 %	58 %	K12 58-161 ~57 %
E resistant $BtuB^-$										
Row/E5a	52 %	46 %	43 %	37 %	55 %	47 %	55 %	43 %	43 %	parent: Row
Row/E7b	45 %	30 %	37 %	38 %	34 %	24 %	38 %	36 %	24 %	"
Row/E5c	39 % x)	32 %	13 %	28 %	34 %	42 %	49 %	46 %	22 %	"
E tolerant $BtuB^+$										
Row/E6a	28 %	29 %	33 %	39 %	25 %	38 %	53 %	38 %	12 %	parent: Row

x) E1 tolerant   100 % = 10^3AU   1 AU = $2x10^8$LU

Table 5.

E tolerant mutant Row/E6a and its E tolerant double, triple and four-selection step mutants: proportion of cells (%) surviving the action of colicins E1 to E9

	Colicin								
Strain	E1	E2	E3	E4	E5	E6	E7	E8	E9
Row	0.001 %	0.001 %	0.001 %	0.001 %	0.001 %	0.001 %	0.001 %	0.001 %	0.001 %
Row/E6a	0.1 %	0.1 %	0.1 %	0.1 %	0.1 %	0.1 %	0.1 %	0.1 %	0.1 %
Row/E6a/E3b	0.4 %	0.4 %	0.4 %	0.4 %	0.3 %	0.4 %	0.4 %	0.4 %	0.3 %
Row E6a/E3b/ E2c	0.4 %	0.4 %	0.4 %	0.4 %	0.4 %	0.4 %	0.4 %	0.4 %	0.4 %
Row E6a/E3b/ E2c/E7g	0.4 %						0.4 %		

It follows from the results stated that E colicin resistance cannot be decisively and clearly distinguished from E colicin tolerance in many mutants. A mutant can be "truly" tolerant to some E colicins and "truly" resistant to others; and the tolerance to any E colicin usually impairs the receptor binding function for all E colicins. And both E colicin resistance, as well as E colicin tolerance, are quantitative markers.

Selection of double-step mutations

We confirmed that the survival of colicin E resistant mutants cannot be increased by repeated selection of their clones exposed to high doses of any colicin E. However, the low survival of most (not all!) tolerant mutants of $1x10^{-3}$ was enhanced 4 times in the second selection step (Table 5); this increase could be confirmed by narrower inhibition zones in agar cultures. In further successive selections there was no more increase of tolerance. These results confirm that at least two genes are engaged in the uptake of any E colicin.

Tolerance presents a much less efficient mechanism of survival of bacterial cells from colicin inhibition than resistance. Tolerant mutants are also more frequent in natural populations of sensitive bacteria than the resistant ones. Tolerance makes use of less specific, less elaborate, more easily fragile and hence probably more ancient molecular mechanisms than the mutated receptor-driven resistance.

Methodological advice

In laboratory practice, the efficiency of plating of the bacteriophage BF23 can serve as the simplest and most reliable criterion for differentiating E colicin resistant and E colicin tolerant mutants. In "true" E resistant mutant, the BF23 e.o.p. will not exceed 0.5 % - as related to that in the parent strain. The BF23 e.o.p. of tolerant mutants will fluctuate between 40 % - 80 %. And mutants displaying a BF23 e.o.p. of 1% - 20 % deserve closer attention; most probably, they are tolerant to some E colicins and resistant to others. This must be distinguished in the next step - by counting survivals. (Survival counts should be routinely used for distinguishing resistance from tolerance to colicins other than E as well.)

References

Bénédetti H, Frenette M, Baty D, Lloubès R, Geli V, Lazdunski C (1989) Comparison of the uptake systems for the entry of various BtuB group colicins into *Escherichia coli*. J Gen Microbiol 135:3413-3420

Bradbeer C, Woodrow M L, Khalifah L I (1976) Transport of vitamin B_{12} in *Escherichia coli*: common receptor system for vitamin B_{12} and bacteriophage BF23 on the outer membrane of the cell envelope. J. Bacteriol 125:1032-1039

DiMasi D R, White J C, Schnaitman C A, Bradbeer C (1973) Transport of vitamin B_{12} in *Escherichia coli*: common receptor sites for vitamin B_{12} and the E colicins on the outer membrane of the cell envelope. J Bacteriol 115:506-513

Smarda J (1975) Novel approaches to the mode of action of colicins. Folia Microbiol 20:264-271

Smarda J, Damborsky J (1991) A quantitative assay of group E colicins. Scripta Medica (Brno) 64:111-118

White J C, DiGirolamo P M, Fu M L, Preston Y A, Bradbeer C (1973) Transport of vitamin B_{12} in *Escherichia coli*: location and properties of the initial B_{12}-binding site. J Biol Chem 248:3978-3986

CONSTRUCTION AND CHARACTERIZATION OF CHIMERIC PROTEINS BETWEEN PYOCINS AND COLICIN E3.

M. Kageyama , M. Kobayashi , Y. Sano , T. Uozumi** and H. Masaki**
Mitsubishi Kasei Institute of Life Sciences
11, Minamiooya
Machida-shi
Tokyo 194
JAPAN

Pyocins Sl, S2 and AP41 are the most frequently found protease-sensitive bacteriocins in *Pseudomonas aeruginosa*. They are classified by the different receptor specificities. A peculiar feature of these bacteriocins is that their genetic determinants are all located at definite sites on the the chromosome. This is in sharp contrast to the case of other bacteriocins, such as colicins, which are of plasmid origin. We cloned these pyocin genes on appropriate plasmids and determined their DNA base sequences, and purified their proteins.

Three pyocin genes are organized in a similar manner, encoding two proteins transcribed in the same direction: Sl, 64.6 and 10 kilodaltons (kDa); S2, 74 and l0kDa; AP41, 83.6 and 10 kDa. The larger components carry the killing activity and the smaller components confer immunity. Striking homology in the amino acid sequence was found among these pyocins and colicin E2.

The larger (killing) components seemed to be composed of three domains; (from N-terminus to C-terminus) receptor binding, translocation and nuclease (and immunity binding). The amino acid sequences of the receptor-binding domains are different from each other. High homology was found in the translocation and the nuclease domains of these pyocins. Particularly, only two amino acid differences were found between pyocins Sl and S2 in these domains. Furthermore, the nuclease domains of these pyocins showed considerable homology to that of colicin E2; more than 50% of the C-terminal 130 amino acids were homologous among them. The immunity proteins were also very similar among these pyocins and colicin E2. Only one amino acid change was found between the immunity proteins of pyocin Sl and S2. In fact, pyocin Sl and S2 shared complete immunity. Slight homology was found in the translocation domains of these pyocins and colicin E2, E3 and cloacin DF13, although the arrangement of domains are different between pyocins and colicins.

Pyocin proteins were purified to study their properties and killing mechanisms. All pyocins were obtained as a 1:1 complex of the activity and the immunity components. In the cells treated with any pyocin, breakdown of chromosomal DNA was observed. The larger, killing components (with the immunity

---

** Faculty of Agriculture, University of Tokyo, Yayoi, Bunkyo-ku, Tokyo 113.

NATO ASI Series, Vol. H 65
Bacteriocins, Microcins and Lantibiotics
Edited by R. James, C. Lazdunski and F. Pattus

protein removed), but not the pyocin complex, revealed an *in vitro* DNase activity. Furthermore, pyocins Sl and S2 showed peculiar features in their action.

1) The susceptibility of the indicator strain to either pyocin was affected by the iron concentration in the growth medium; a higher susceptibility was seen under iron deficiency.

2) Either pyocin, but not AP41, inhibited the synthesis of phospholipid in the sensitive strain.

To establish the domain structure of pyocins (assignment of the functions to specific domains and their alignment), construction of chimeric proteins was attempted between pyocins and colicins. In this paper, properties of chimeras between pyocin Sl/S2 and colicin E3 are reported. We succeeded in exchanging the nuclease domain (along with the immunity gene) between colicin E3 and pyocin Sl or S2. Such chimeric proteins exhibited killing activity. Chimeras consisted of the translocation and the receptor domains of E3 and the nuclease domain of pyocin Sl or S2 killed *E. coli* but not *P. aeruginosa*. Their immunity specificities were of the pyocin type. On the other hand, chimeras composed of the receptor and the translocation domains of pyocin Sl or S2 and the nuclease domain of colicin E3 killed *Pseudomonas* strains with Sl or S2 specificity, but did not kill *E. coli*. Their immunity specificities were of the colicin E3 type. Pyocin Sl or S2 causes DNA breakdown of the sensitive cells, whereas Sl-E3 or S2-E3 did not possess this activity. However such chimeric proteins exhibited an *in vitro* RNase activity when their immunity proteins were removed. Thus new kinds of bacteriocin were constructed *in vitro*. These results indicate that the nuclease domain of colicin E3 and pyocin Sl or S2 can function independently of the other domains of the molecule.

A chimera comprising the complete Sl gene under the translocation and the receptor domains of colicin E3 showed a strong colicin activity but practically no pyocin activity. Another chimera, consisting of the receptor-binding and the translocation domains of pyocin S2 followed by the entire colicin E3 molecule, except the 12 N-terminal amino acids, killed both *E. coli* and *P. aeruginosa*, the efficiency being better with *E. coli*. When deletions were introduced into each domain of these chimeras, the resulting proteins revealed pyocin or colicin activity depending upon the position of deletion. These results indicate that although the nuclease domains can function in either species, the translocation domains seem to be more species-specific, and that the steric structure of the molecule modulate the specific activity. Properties of the killing process of these chimeras are under investigation.

COLICINS AS ANTI-TUMOUR DRUGS

J. Smarda
Department of Biology, Faculty of Medicine
Masaryk University
Jostova 10
CS-66244 Brno
Czechoslovakia

Colicins have not only bactericidal but also cytocidal effects: they inhibit and kill animal tissue cells *in vitro* (Smarda, 1983); the sensitivity of tumour cells is a most remarkable feature. The colicins used by us were produced as purified and lyophilized substances by the Institute of Sera and Vaccines, Prague. Their purity was checked as described earlier (Smarda *et al.*, 1975). In addition, several controls were run in each experiment to check the specificity of colicin effects.

In vitro treatment of tumour cells

Colicin E3 (at an activity at least 10^4 bacterial lethal units per cell) exerted a dramatic cytocidal effect towards the stable line HeLa of human carcinoma colli uteri cells (Smarda & Obdrzálek, 1977), killing 100% of HeLa cells in the exposed culture. The growth phase of the culture was of no importance. Be it HeLa cells in a freshly inoculated culture (i.e. the inoculum itself), or cells already attached to glass and growing (in a 1-day or 2-day culture), the culture would be destroyed completely. LD_{50} was as low as $3x10^3$ bacterial lethal units per cell (Smarda *et al.*, 1978). Preliminary results of our recent experiments suggest that the biosynthesis which is affected primarily by colicin E3 is not protein synthesis, as it is in bacteria, but DNA synthesis.

A little less dramatic, but also considerable, was the cytocidal effect of colicin A on HeLa cells, while analogous effects of colicins El, E2 or D were not detected. Colicin E3 also distinctly decreased the viability of cells of a non-stabilized line of human breast carcinoma 23lMB.

In most cases noted so far, tumour cells were one- to several log-orders more sensitive to colicins than the standard ones, although their lines, too, showed considerable differences in colicin sensitivity. Similarly to the action on prokaryotic cells, colicins act on eukaryotic cells in a specific way mediated by the cell membrane.

Rat glioma cells of the stable line C6 treated with 10^4 bacterial lethal units of colicins E7 or E8 showed a distinctly reduced proliferation rate. The two colicins, E7 as well as E8, decreased DNA synthesis in them by about 47 % and 26 %, respectively, under the experimental conditions given. On the other hand,

NATO ASI Series, Vol. H 65
Bacteriocins, Microcins and Lantibiotics
Edited by R. James, C. Lazdunski and F. Pattus

neither colicin D nor, surprisingly enough, colicins E2 or E3 (under the same experimental conditions) influenced the rat glioma cells.

Experiments on mouse T-lymphocytes brought results testifying that the primary intervention of colicin E3, and probably also of other colicins, is directed towards the plasma membrane of eukaryotic cells. Colicin E3 inhibited almost completely the mitogenic activation of T-lymphocytes induced by the lectin concanavalin A; this activation was inhibited to the same extent as late as 30 hours after coupling the mitogen on the cell membrane. Thus, the inhibition cannot be explained by a receptor competition between the mitogen and the colicin (Viklick *et al.*, 1979). Furthermore, colicin E3 reduced significantly the homing of T-lymphocytes into lymphatic nodes of syngenic recipient mice. In appropriate low doses it caused a "respiration burst", a substantial enhancement of the oxidoreductive activity of mouse peritoneal exudate white blood cells, in direct proportion to its concentration (Lokaj *et al.*, 1982).

Colicin E3 reduced the proliferation rate of mouse myeloic leukemia cells of the stable line P388 (Fuska *et al.*, 1979). Within wide limits, the inhibition degree was directly proportional to the colicin dose and to the duration of colicin treatment. In these model cells again, DNA synthesis was more profoundly and perhaps primarily affected, compared to RNA synthesis. Extensive experimental knowledge on the inhibitive effects of a colicin on blastic cells, especially on those of human lymphatic leukemia, has been accumulated by Farkas-Himsley and her co-workers (Farkas-Himsley & Kuzniak, 1978). However, she suspects that in her colicin preparations these effects may be due to another undefined protein closely associated with the colicin itself (Farkas-Himsley, 1988).

In our *in vitro* studies we paid attention to the viability of malignant white blood cells of mice exposed to colicins. The tumour blood cells were of four lines: lymphosarcoma Gardner 6C3 HED, lymphoma Németh-Kellner, lymphoma Skalski, and plasmocytoma LP-2. All four of them were treated with colicins A, D, E2 and E3 alone, as well as with a mixture of these. The viability of cells was checked, at different intervals, using their ability to take up hexavalent chromium Cr^{51} as a criterion; this chromium isotope is taken up only by living cells.

The best results were obtained with colicin A (Table 1). It decreased incorporation of the radioactive chromium by 50 % to 60 %, which means that it killled 50 % to 60 % of cells in experimental cultures of all the lines tested. By the same degree, colicin E2 decreased the viability of cells of both lymphomas, and to a lesser degree (40 % to 45 %) it exerted a cytocidal effect on the two other tumour cell lines. Colicins E3 and D showed their anti-tumour activities as well: colicin E3 within the limits of a 23 % - 51 % loss of cell viability (the most sensitive was the plasmocytoma LP-2) and colicin D by 27% to 46 % (here the most sensitive was the Skalski lymphoma). The cytocidal effects of colicins A and D were proportional to the doses applied within broad limits. A general conclusion which could be drawn from all these and other *in vitro* studies was that tumour cells were considerably sensitive to colicins. This conclusion encouraged us to perform *in vivo* experiments in laboratory animals.

Table 1. Mean decrease of Cr^{51} incorporation by tumour cells treated with bacteriocins for 100 minutes

	Cells of			
Bacteriocin	Lymphosarcoma Gardner	Lymphoma Németh-Kellner	Lymphoma Skalski	Plasmocytoma LP-2
Colicin A	53 %	58%	54%	54%
Colicin D	27%	31%	46%	31%
Colicin E2	43%	55%	53%	40%
Colicin E3	27%	23%	30%	51%
Staphylococcin A	58%	20%	50%	56%
Pyocin VI	58%	61%	44%	42%

Experimental therapy of malignant tumours in mice

We undertook very extensive studies on the treatment of adenocarcinoma HK in mice. This solid tumour was transplanted subcutaneously into the dorsal region of mice. Starting on the 3rd to 8th days after transplantation, when the tumours reached the size of a cherry stone, solutions of colicin E3 were injected transcutaneously into them at regular intervals. A very broad range of colicin concentrations and doses was tested.

Colicin E3 exerted a profound oncolytic activity towards the adenocarcinoma HK. Tumours were reduced in size in relation to the daily doses applied. The histological findings showed de-differentiation of tumour cells and extensive central necroses of the tumours, compact or blotch-like. Small clusters of preserved tumour cells remained only in the superficial tumour layers.

According to the international convention put forward by the National Cancer Institute in Bethesda, Maryland, U.S.A., the decisive criterion of attractivity of any anti-tumour substance for closer research is its effect on the survival of animals with malignant tumours. Using suitable low doses (0.095 μg each) of colicin E3, we were able to reach a reduction of the average HK adenocarcinoma tumour mass by 49 % during 20 days (Table 2). At the same time, we reached a prolongation of the mean survival rate of tumour mice from the usual 16 days in controls to 28 days, which means by 75 %. Using higher doses we were able to reduce the solid tumour nodes of HK carcinoma - in mass and size - by up to 75 % (at 8.75 μg daily), but at such doses colicin E3, and probably also toxic products of massive disintegration of killed tumour cells, interfered with prolongation of survival.

Table 2. Oncolytic action of colicin E3 (in low doses) on mouse adenocarcinoma HK

Experimental series	Colicin E3 Daily dose (μg)	Treatment interval (days)	Mean Survival (days)	Mean Tumour wt (g)	Weight decrease (%)
Control	0	0	16	10.6	0
1	0.0095	20	25	8.3	22
2	0.095	20	28	5.4	49
3	0.19	17	18	3.3	69

In the last phase of this research, we undertook attempts at experimental therapy of those malignant tumours of mice lymphoid tissue whose cells were successfully treated with colicins *in vitro*, as mentioned above. Lymphosarcoma 6C3 HED, plasmocytoma LP-2 and lymphomas Németh-Kellner and Skalski were transplanted subcutaneously to mice of appropriate inbred lines. Four days after transplantation the application of colicins to the mice was started. Solutions of colicins A, D, E2 or E3 were applied intraperitoneally at regular low doses (2 to 6 doses; 0.5 - 1 mg of colicin each).

The results were unequivocally positive again, though their quantitative variation was broad. Highly significant effects were repeatedly obtained, especially for colicins A and E2 (Table 3). Thus the average weight of lymphoma Németh-Kellner, amounting to 276 mg at the beginning of colicin A treatment, grew up to 492 mg in control untreated mice during the following 4 days, while it decreased to 221 mg in animals treated with two doses of colicin A (0.5 mg each) within the same time limit. Similarly, in mice treated with colicin E2, under the same experimental conditions, the average mass of the same tumour reached 500 mg in comparison to the tumour mass of 734 mg in control animals. In most experiments, a rise in specific tumoricidal antibodies was noted following application of colicins A or E2.

The survival experiments confirmed, at last, the positive results of colicin therapy of these mouse tumours. These experiments were performed on female mice of the strain BALB/C suffering from transplanted plasmocytoma LP-2. Starting on the 6th day following transplantation, the mice received

Table 3. Weight of Tumours in Mice Treated with Colicins

Colicin	Mice Line	Mice Number	Tumour	Colicin Dosis i.p.	Time Interval	Tumour Weight (mg)	Significance Level (P)
No	B 10	28	Lymphoma		4th day	276 ± 14.9	
		30	Németh-		8th day	492 ± 46.3	
A		30	Kellner	2 x 0.5mg	8th day	221 ± 30.2	0.001
No	CBA	28	Lympho-		4th day	400 ± 33.9	
		27	sarcoma		8th day	799 ± 47.9	
A		29	Gardner	3 x 1.0mg	8th day	425 ± 42.0	0.001
No	CBA	10	Lympho-		4th day	254 ± 50.7	
		26	sarcoma		9th day	917 ± 79.8	
A		20	Gardner	2 x 0.5mg	9th day	590 ± 47.5	0.001
No	C 57 BL	15	Lymphoma		4th day	307 ± 36.2	
		15	Németh-		8th day	670 ± 54.8	
E2		18	Kellner	4 x 0.5mg	8th day	409 ± 39.7	0.01

colicin A intraperitoneally at doses of 0.25 mg each in 2-day intervals, with the following results: on the average, the untreated mice survived 44 days after tumour transplantation while the treated ones lived 63 days, i.e. 43 % longer; there was some variation, but the difference was statistically highly significant.

It thus may be concluded that colicins present a new, unexplored group of natural substances displaying clear cytocidal, tumoricidal and oncolytic effects; seemingly, in this respect they stand side by side with certain bacterial toxins. In addition, this conclusion is supported by the unequivocally positive results of treating blastic lymphoma, lymphosarcoma and plasmocytoma cells with further bacteriocins: with a pyocin and a staphylococcin or with a mixture of these and colicins, both *in vitro* and *in vivo*. For instance, the mass of Németh-Kellner lymphoma *in vivo* was limited to 498 mg on the 6th day following transplantation by six successive intraperitoneal doses of staphylococcin A (0.5 mg each), while it reached 1408 mg during the same time interval in control untreated animals. Application of pyocin VI also raised the titre of tumoricidal antibodies in blood of the animals tested.

In close agreement with the experience of Farkas-Himsley *et al.* (1988) our results supply an argument for continuing and further developing this topic in the area of applied research.

Acknowledgements

The author wishes to express his most sincere thanks to his co-workers: A. Drozdová, C. Oravec, J. Keprtová, V. Obdrzálek, V. Viklicky, J. Lokaj and J. Fuska.

References

Farkas-Himsley H (1988) Sensitivity of various malignant cell lines to partially purified bacteriocins: inherent and proliferative dependence. Microbios Lett 39:57-66

Farkas-Himsley H, Kuzniak S (1978) Bacteriocins as inhibitors of neoplasia. Curr Chemother: 1188-1191

Fuska J, Fusková A, Smarda J, Mach J (1979) Effect of colicin E3 on leukemia cells P388 *in vitro*. Experientia 35:406-407

Lokaj J, Smarda J, Mach J (1982) Colicin E3 enhances the oxidoreductive activity of guinea-pig leucocytes. Experientia 38:1352-1353

Smarda J (1983) The action of colicins on eukaryotic cells. Toxin Rev 2:1-76

Smarda J, Ebringer L, Mach J (1975) The effect of colicin E2 on the flagellate *Euglena gracilis*. J Gen Microbiol 86:363-366

Smarda J, Obdrzálek V (1977) The lethal effect of colicin E3 on HeLa cells in tissue cultures. IRCS J Med Sci 5:524

Smarda J, Obdrzálek V, Táborsky I, Mach J (1978) The cytotoxic and cytocidal effects of colicin E3 on mammalian tissue cells. Folia microbiol (Praha) 23:272-277

Viklicky V, Smarda J, Dráber Petr, Pokorná Z, Mach J, Dráber Pavel (1979) The cytoplasmic membrane as a site of the primary effect of colicins on eukaryotic cells. Folia microbiol (Praha) 25:116-125

List of Participants

Dr R. JAMES	Norwich	UK	(Director)
Prof C. LAZDUNSKI	Marseille	France	(Co-Director)
Dr F. PATTUS	Heidelberg	Germany	(Co-Director)
Dr T. ABEE	Groningen	The Netherlands	
Dr D. BATY	Marseille	France	
Ms A. BAYER	Tubingen	Germany	
Dr G. BIERBAUM	Bonn	Germany	
Prof V. BRAUN	Tubingen	Germany	
Dr D. CAVARD	Marseille	France	
Dr M. CHIKINDAS	Groningen	The Netherlands	
Prof W. CRAMER	Purdue	USA	
Dr M. CURTIS	Norwich	UK	
Dr L. DE VUYST	Gent	Belgium	
Dr H. DODD	Norwich	UK	
Mr D. DUCHE	Marseille	France	
Prof K-D. ENTIAN	Frankfurt	Germany	
Prof A. FINKELSTEIN	New York	USA	
Dr C. FREMAUX	Raleigh	USA	
Dr S. FREUND	Tubingen	Germany	
Dr A. GALVEZ del POSTIGO	Granada	Spain	
Dr V. GELI	Marseille	France	
Dr R. GLASS	Nottingham	UK	
Dr J. GONZALES MANAS	Heidelberg	Germany	
Mrs S. GOODMAN	Cranfield	UK	
Dr A. GUDMUNDSDOTTIR	Reykjavik	Iceland	
Dr RE. HARKNESS	Ontario	Canada	
Mr Y. HECHARD	Poitiers	France	
Dr C. HERNANDEZ-CHICO	Madrid	Spain	
Dr C. HILL	Fermoy	Ireland	
Mr H. HOLO	As-NLH	Norway	
Prof SP.HOWARD	Newfoundland	Canada	
Dr J. HUGENHOLTZ	Ede	The Netherlands	
Dr T. ITOH	Osaka	Japan	
Dr K. JAKES	New York	USA	
Dr RJ. KADNER	Charlottesville	USA	
Prof M. KAGEYAMA	Tokyo	Japan	
Prof TR. KLAENHAMMER	Raleigh	USA	
Dr R. KOLTER	Boston	USA	
Prof RNH. KONINGS	Nijmegen	The Netherlands	
Dr OP. KUIPERS	Ede	The Netherlands	
Dr J. LAKEY	Heidelberg	Germany	
Dr PCK. LAU	Montreal	Canada	
Dr M. LAVINA	Montevideo	Uruguay	
Dr JC. LAZZARONI	Villeurbanne	France	
Dr RA. LEDEBOER	Vlaardingen	The Netherlands	
Dr KS. LEE	Suwon	Korea	
Prof L. LETELLIER	Orsay	France	
Dr G. LIMSOWTIN	Palmerston North	New Zealand	
Dr R. LLOUBES	Marseille	France	
Dr J. LUIRINK	Heidelberg	Germany	
Dr M. MARTINEZ-BUENO	Granada	Spain	
Dr JD. MARUGG	Vlaardingen	The Netherlands	
Dr H. MASAKI	Tokyo	Japan	
Prof F. MORENO	Madrid	Spain	
Ms CI. MORTVEDT	Osloveien	Norway	
Prof A. NAKAZAWA	Yamaguchi	Japan	

Dr IF. NES	As-NLH	Norway
Dr B. OUDEGA	Amsterdam	The Netherlands
Mr JC. PIARD	Jouy-en-Josas	France
Dr M. REGUE	Barcelona	Spain
Prof HG. SAHL	Bonn	Germany
Dr PE. SARIS	Helsinki	Finland
Dr M. SHOHAM	Cleveland, Ohio	USA
Dr MSM. SIBAKOV	Helsinki	Finland
Dr S. SLATIN	New York	USA
Prof J. SMARDA	Brno	Czechoslovakia
Miss LH. STANBRIDGE	Bath	UK
Prof ME. STILES	Edmonton	Canada
Dr JR. TAGG	Dunedin	New Zealand
Dr M. VAN BELKUM	Groningen	The Netherlands
Dr GF. VAN DER GOOT	Heidelberg	Germany
Dr K. VENEMA	Groningen	The Netherlands
Mr R. WALLIS	Norwich	UK
Mr RJ. WARD	Nottingham	UK
Prof RE. WEBSTER	Durham, NC	USA
Mr S. YAJIMA	Tokyo	Japan
Dr P. YORGEY	Boston, MA	USA

SUBJECT INDEX

NATO ASI Series H

Vol. 1: **Biology and Molecular Biology of Plant-Pathogen Interactions.** Edited by J.A. Bailey. 415 pages. 1986.

Vol. 2: **Glial-Neuronal Communication in Development and Regeneration.** Edited by H.H. Althaus and W. Seifert. 865 pages. 1987.

Vol. 3: **Nicotinic Acetylcholine Receptor: Structure and Function.** Edited by A. Maelicke. 489 pages. 1986.

Vol. 4: **Recognition in Microbe-Plant Symbiotic and Pathogenic Interactions.** Edited by B. Lugtenberg. 449 pages. 1986.

Vol. 5: **Mesenchymal-Epithelial Interactions in Neural Development.** Edited by J. R. Wolff, J. Sievers, and M. Berry. 428 pages. 1987.

Vol. 6: **Molecular Mechanisms of Desensitization to Signal Molecules.** Edited by T M. Konijn, P J. M. Van Haastert, H. Van der Starre, H. Van der Wel, and M.D. Houslay. 336 pages. 1987.

Vol. 7: **Gangliosides and Modulation of Neuronal Functions.** Edited by H. Rahmann. 647 pages. 1987.

Vol. 8: **Molecular and Cellular Aspects of Erythropoietin and Erythropoiesis.** Edited by I.N. Rich. 460 pages. 1987.

Vol. 9: **Modification of Cell to Cell Signals During Normal and Pathological Aging.** Edited by S. Govoni and F Battaini. 297 pages. 1987.

Vol. 10: **Plant Hormone Receptors.** Edited by D. Klämbt. 319 pages. 1987.

Vol. 11: **Host-Parasite Cellular and Molecular Interactions in Protozoal Infections.** Edited by K.-P. Chang and D. Snary. 425 pages. 1987.

Vol. 12: **The Cell Surface in Signal Transduction.** Edited by E. Wagner, H. Greppin, and B. Millet. 243 pages. 1987.

Vol. 13: **Toxicology of Pesticides: Experimental, Clinical and Regulatory Perspectives.** Edited by L.G. Costa, C.L. Galli, and S.D. Murphy. 320 pages. 1987.

Vol. 14: **Genetics of Translation. New Approaches.** Edited by M.F. Tuite, M. Picard, and M. Bolotin-Fukuhara. 524 pages. 1988.

Vol. 15: **Photosensitisation. Molecular, Cellular and Medical Aspects.** Edited by G. Moreno, R. H. Pottier, and T. G. Truscott. 521 pages. 1988.

Vol. 16: **Membrane Biogenesis.** Edited byJ.A.F Op den Kamp. 477 pages. 1988.

Vol. 17: **Cell to Cell Signals in Plant, Animal and Microbial Symbiosis.** Edited by S. Scannerini, D. Smith, P. Bonfante-Fasolo, and V. Gianinazzi-Pearson. 414 pages. 1988.

Vol. 18: **Plant Cell Biotechnology.** Edited by M.S.S. Pais, F. Mavituna, and J. M. Novais. 500 pages. 1988.

Vol. 19: **Modulation of Synaptic Transmission and Plasticity in Nervous Systems.** Edited by G. Hertting and H.-C. Spatz. 457 pages. 1988.

Vol. 20: **Amino Acid Availability and Brain Function in Health and Disease.** Edited by G. Huether. 487 pages. 1988.

Vol. 21: **Cellular and Molecular Basis of Synaptic Transmission.** Edited by H. Zimmermann. 547 pages. 1988.

NATO ASI Series H

Vol. 22: **Neural Development and Regeneration. Cellular and Molecular Aspects.** Edited by A. Gorio, J. R. Perez-Polo, J. de Vellis, and B. Haber. 711 pages. 1988.

Vol. 23: **The Semiotics of Cellular Communication in the Immune System.** Edited by E.E. Sercarz, F. Celada, N.A. Mitchison, and T. Tada. 326 pages. 1988.

Vol. 24: **Bacteria, Complement and the Phagocytic Cell.** Edited by F. C. Cabello und C. Pruzzo. 372 pages. 1988.

Vol. 25: **Nicotinic Acetylcholine Receptors in the Nervous System.** Edited by F. Clementi, C. Gotti, and E. Sher. 424 pages. 1988.

Vol. 26: **Cell to Cell Signals in Mammalian Development.** Edited by S.W de Laat, J.G. Bluemink, and C.L. Mummery. 322 pages. 1989.

Vol. 27: **Phytotoxins and Plant Pathogenesis.** Edited by A. Graniti, R. D. Durbin, and A. Ballio. 508 pages. 1989.

Vol. 28: **Vascular Wilt Diseases of Plants. Basic Studies and Control.** Edited by E. C. Tjamos and C. H. Beckman. 590 pages. 1989.

Vol. 29: **Receptors, Membrane Transport and Signal Transduction.** Edited by A. E. Evangelopoulos, J. P. Changeux, L. Packer, T. G. Sotiroudis, and K.W.A. Wirtz. 387 pages. 1989.

Vol. 30: **Effects of Mineral Dusts on Cells.** Edited by B.T. Mossman and R.O. Begin. 470 pages. 1989.

Vol. 31: **Neurobiology of the Inner Retina.** Edited by R. Weiler and N.N. Osborne. 529 pages. 1989.

Vol. 32: **Molecular Biology of Neuroreceptors and Ion Channels.** Edited by A. Maelicke. 675 pages. 1989.

Vol. 33: **Regulatory Mechanisms of Neuron to Vessel Communication in Brain.** Edited by F. Battaini, S. Govoni, M.S. Magnoni, and M. Trabucchi. 416 pages. 1989.

Vol. 34: **VectorsasTools for the Study of Normal and Abnormal Growth and Differentiation.** Edited by H. Lother, R. Dernick, and W. Ostertag. 477 pages. 1989.

Vol. 35: **Cell Separation in Plants: Physiology, Biochemistry and Molecular Biology.** Edited by D. J. Osborne and M. B. Jackson. 449 pages. 1989.

Vol. 36: **Signal Molecules in Plants and Plant-Microbe Interactions.** Edited by B.J.J. Lugtenberg. 425 pages. 1989.

Vol. 37: **Tin-Based Antitumour Drugs.** Edited by M. Gielen. 226 pages. 1990.

Vol. 38: **The Molecular Biology of Autoimmune Disease.** Edited by A.G. Demaine, J-P. Banga, and A.M. McGregor. 404 pages. 1990.

Vol. 39: **Chemosensory Information Processing.** Edited by D. Schild. 403 pages. 1990.

Vol. 40: **Dynamics and Biogenesis of Membranes.** Edited by J. A. F. Op den Kamp. 367 pages. 1990.

Vol. 41: **Recognition and Response in Plant-Virus Interactions.** Edited by R. S. S. Fraser. 467 pages. 1990.

NATO ASI Series H

Vol. 42: **Biomechanics of Active Movement and Deformation of Cells.**
Edited by N. Akkas. 524 pages. 1990.

Vol. 43: **Cellularand MolecularBiologyof Myelination.**
Edited by G. Jeserich, H. H. Althaus, and T. V. Waehneldt. 565 pages. 1990.

Vol. 44: **Activation and Desensitization of Transducing Pathways.**
Edited by T. M. Konijn, M. D. Houslay, and P. J. M. Van Haastert. 336 pages. 1990.

Vol. 45: **Mechanism of Fertilization: Plants to Humans.**
Edited by B. Dale. 710 pages. 1990.

Vol .46: **Parallels in Cell to Cell Junctions in Plants and Animals.**
Edited by A. W Robards, W. J . Lucas, J . D. Pitts, H . J . Jongsma, and D. C. Spray. 296 pages. 1990.

Vol. 47: **Signal Perception andTransduction in Higher Plants.**
Edited by R. Ranjeva and A. M. Boudet. 357 pages. 1990.

Vol. 48: **Calcium Transport and Intracellular Calcium Homeostasis.**
Edited by D. Pansu and F. Bronner. 456 pages. 1990.

Vol. 49: **Post-Transcriptional Control of Gene Expression.**
Edited by J. E. G. McCarthy and M. F. Tuite. 671 pages. 1990.

Vol. 50: **Phytochrome Properties and Biological Action.**
Edited by B. Thomas and C. B. Johnson. 337 pages. 1991.

Vol. 51: **Cell to Cell Signals in Plants and Animals.**
Edited by V. Neuhoff and J. Friend. 404 pages. 1991.

Vol. 52: **Biological Signal Transduction.**
Edited by E. M . Ross and K . W. A. Wirtz. 560 pages. 1991.

Vol. 53: **Fungal Cell Wall and Immune Response.**
Edited by J. P. Latge and D. Boucias. 472 pages. 1991.

Vol. 54: **The Early Effects of Radiation on DNA.**
Edited by E. M. Fielden and P. O'Neill. 448 pages. 1991.

Vol. 55: **The Translational Apparatus of Photosynthetic Organelles.**
Edited by R. Mache, E. Stutz, and A. R. Subramanian. 260 pages. 1991.

Vol. 56: **Cellular Regulation by Prote in Phosphorylation.**
Edited by L. M. G. Heilmeyer, Jr. 520 pages. 1991.

Vol. 57: **MolecularTechniques in Taxonomy.**
Edited by G . M . Hewitt, A. W. B. Johnston, and J. P. W. Young . 420 pages. 1991.

Vol. 58: **Neurocytochemical Methods.**
Edited by A. Calas and D. Eugene. 352 pages. 1991.

Vol. 59: **Molecular Evolution of the Major Histocompatibility Complex.**
Edited by J. Klein and D. Klein. 522 pages. 1991.

Vol. 60: **Intracellular Regulation of Ion Channels.**
Edited by M. Morad and Z. Agus. 261 pages. 1992.

Vol. 61: **Prader-Willi Syndrome and Other Chromosome 15q Deletion Disorders.**
Edited by S. B. Cassidy. 277 pages. 1992.

NATO ASI Series H

Vol. 62: **Endocytosis. From Cell Biology to Health, Disease and Therapie.**
Edited by P. J. Courtoy. 547 pages. 1992.

Vol. 63: **Dynamics of Membrane Assembly.**
Edited by J. A. F. Op den Kamp. 402 pages. 1992.

Vol. 64: **Mechanics of Swelling. From Clays to Living Cells and Tissues.**
Edited by T. K. Karalis. 802 pages. 1992.

Vol. 65: **Bacteriocins, Microcins and Lantibiotics.**
Edited by R. James, C. Lazdunski, and F. Pattus. 530 pages. 1992.

Vol. 66: **Theoretical and Experimental Insights into Immunology.**
Edited by A. S. Perelson and G. Weisbuch. 497 pages. 1992.